Landscape ecology in the Dutch context: nature, town and infrastructure

Editors T.M. de Jong, J.N.M. Dekker, R. Posthoorn

Edited by T.M. de Jong, J.N.M. Dekker, R. Posthoorn.
Lay-out and cover by T.M. de Jong and C. Looijen.
Printing by Ponsen en Looijen, Wageningen, The Netherlands.

This publication was made possible by contributions of:
University of Technology Delft, Ministry of Transport, Public Works and
Water Management, V&W (Rijksinstituut voor Integraal Zoetwaterbeheer
en Afvalwaterbehandeling RIZA and Dienst Weg- en Waterbouw DWW),
Milieu- en NatuurPlanbureau MNP and Vereniging Natuurmonumenten.

A scientific committee of peers judged the content as scientifically relevant:

Concerning the section NATURE:
    Dr. Olaf Bastian, Sächische Akademie der Wissenschaften Leipzig.

Concerning the section TOWN:
    Univ. Prof. Dr. H. Jürgen Breuste, Universität Salzburg;
    Jack Ahern, University of Massachusetts, Amherst, USA.

Concerning the section INFRASTRUCTURE:
    Bella Davies, Oxford Brookes University;
    Rinus Jaarsma, Wageningen University;
    E. Ashley Steel, NW Fisheries Science Center, Seattle.

© KNNV Publishing, Zeist, The Netherlands, *2007*

KNNV Publishing is a foundation of the Royal Dutch
Society for Natural History
ISBN 978 90 5011 257 4
NUR: 940
www.knnvpublishing.nl

AUTHORS AND INSTITUTIONS

Bakker[a], A. de s.s.t.t.; Chapter 26
adebakker@anwb.nl

Barendregt[n], Dr. A.; Chapter 3, 6, 29 a.barendregt@geo.uu.nl

Berg[t], Dr. A. van den; Chapter 15
Agnes.vandenberg@wur.nl

Bohemen[l], Dr. ing. H.D. van; Chapter 23, 24 H.D.vanbohemen@tudelft.nl

Bruin[g(retired)], Ir. D. de; Chapter 21
dick.debruin@wanadoo.nl

Dekker[o], Dr. J.N.M.; Chapter 3, 13; editor section Nature;
j.n.m.dekker@uu.nl

Denters[b], Drs. T.; Chapter 14
tondenters@hetnet.nl

Eerden[h], Dr. M.R. van; Chapter 11
Mennobart.van.eerden@rws.nl

Glastra[e], Ir. M.J.; Chapter 4
glastra@utrechtslandschap.nl

Hinsberg[i], Dr. A. van; Chapter 5
arjen.hinsberg@mnp.nl

Hooimeijer[k], Drs. F.L.; Chapter 19
F.L.Hooimeijer@bk.tudelft.nl

Jong[k], Prof.dr.ir. T.M. de; Chapter 12, 20, 21, 22; editor;
T.M.deJong@tudelft.nl
http://team.bk.tudelft.nl

Jongman[t], Dr. R.H.G.; Chapter 9, 29 rob.jongman@wur.nl

Jonkhof[t], Ir. J.F; Chapter 25
Jos.Jonkhof@wur.nl

Lammers[i,now j], Drs. G.W.; Chapter 5
w.lammers@staatsbosbeheer.nl

Loonen[i], Ir. W.; Chapter 5
willem.loonen@mnp.nl

Pelk[f], Ir. M.L.H.; Chapter 8
M.L.H.Pelk@minlnv.nl

Reijnen[i], Dr. M.J.S.M.; Chapter 5
rien.reijnen@wur.nl

Renes[p, s], Dr. J.; Chapter 7
j.renes@geo.uu.nl

Sanders[i], Dr. ir. M.E.; Chapter 5
marlies.sanders@wur.nl

Schouten[n], Drs. M.A.; Chapter 6
M.Schouten@geo.uu.nl

Schrijnen[k], Prof.ir. J.A.M.; Chapter 28 J.M.Schrijnen@tudelft.nl

Smidt[c], Dr. J.T. de; Chapter 1, 29
jt.de.smidt@hetnet.nl

Tjallingii[k], Dr. S.P.; Chapter 17, 25
s.p.tjallingii@bk.tudelft.nl

Toorn[k], Ir. M.W.M. van der; Chapter 27 M.W.M.vandenToorn@bk.tudelft.nl

Veen[q], Drs. Ing. P.H.; Chapter 9
Peter.Veen@VeenEcology.nl

Veen[r], Ir. P.J.; Chapter 18
vista@vista.nl; www.vista.nl

Verweij[o], Dr. P.A.; Chapter 6
P.A.Verweij@uu.nl

Wardenaar[r], Drs. K.J.; Chapter 18
vista@vista.nl; www.vista.nl

Wassen[n], Prof.dr. M.J.; Chapter 10, 29 m.wassen@geo.uu.nl

Zande[t, f], Prof.dr.ir. A. van der; Chapter 2 Andre.vanderZande@wur.nl

Zoest[d, m], Drs. J. van; Chapter 16
zst@dro.amsterdam.nl

[a] ANWB Cycle and Automobile association; Wassenaarseweg 220 2596 EC Den Haag
[b] Bureau Ecostad; Waterpoortweg 401b, 1051 PX Amsterdam

Authors and institutions

[c] Chairman Werkgemeenschap Landschapsecologisch Onderzoek (WLO); PO Box 80123, 3508 TC Utrecht

[d] Dienst Ruimtelijke Ordening, Amsterdam; PO Box 2758 1000 CT Amsterdam

[e] Het Utrechts Landschap; PO Box 121, 3730 AC De Bilt

[f] Ministry of Agriculture, Nature and Food Quality (LNV); Bezuidenhoutseweg 73, 2594 AC The Hague

[g] Ministry of Transport, Public Works and Water Management (V&W); PO Box 20901 2500 EX Den Haag

[h] Ministry of Transport, Public Works and Water Management (V&W), Rijkswaterstaat RIZA; PO Box 19, 8200 AA Lelystad

[i] Netherlands Environmental Assessment Agency (MNP); PO Box 303, 3720 AH Bilthoven

[j] Staatsbosbeheer; PO Box 303, 3720 AH Bilthoven

[k] University of Technology Delft, Faculty of Architechture, Department of Urbanism; Berlageweg 1, 2628 CR Delft

[l] University of Technology Delft, Faculty of Civil Engineering; Stevinweg 1, 2628 CN Delft

[m] University of Technology Eindhoven, Faculty of Architecture; PO Box 513, 5600 MB Eindhoven

[n] Utrecht University, Department of Environmental sciences, Copernicus Institute for Sustainable Development and Innovation; PO Box 80115, 3584 CS Utrecht

[o] Utrecht University, Department of Science, Technology and Society, Copernicus Institute for Sustainable Development and Innovation; Heidelberglaan 2, 3584 CS Utrecht

[p] Utrecht University, Faculty of Geosciences; PO Box 80115, 3508 TC Utrecht

[q] VeenEcology; Gruppelderweg 17, 8071 WK Nunspeet

[r] Vista Landscape and Urban Design; Prinsengracht 253, 1016 GV Amsterdam

[s] Vrije Universiteit Amsterdam; De Boelelaan 1105, 1081 HV Amsterdam

[t] Wageningen University and Research centre, Alterra; PO Box 47, 6700 AA Wageningen

4

Contents

Contents

1 INTRODUCTION

Jacques de Smidt[a]

1.1 Origin of the book

1.1.1 WLO conferences

This book is the result of three symposia of the Dutch Society for Landscape Ecology (in Dutch: WLO, Werkgemeenschap voor Landschapsecologisch Onderzoek). The first symposium in 2005 was about the National Ecological Network in the Netherlands (NEN or in Dutch: EHS, Ecologische Hoofdstructuur). The reason was that the implementation of the NEN, decided upon in 1990, was halfway. The second symposium, in 2006, was about urban ecology (stadsecologie) and the third one, to be held in 2007, will be about civil infrastructure. This book does not cover the conferences completely and new contributions are added.

Three themes in the Dutch context
The three themes are important contexts in which landscape ecologists do their research and apply their knowledge and skills. Of course, there are many more subjects to hold conferences about, for example climate change, urbanisation, agriculture, landscape ecology itself etc.
The focus of the conferences is on the Netherlands. Although many WLO members do their work abroad or in an international context, these conferences offer a window on what happens in the Dutch context. The experiences may be of value for other contexts and that is why we present the results in English.
The selected themes and the focus on the Dutch context are serious demarcations of what landscape ecology in the Netherlands is all about. The book does not represent all research and applications of landscape ecology.

Different disciplines
On the other hand the scope of the contributions is very broad. The conferences showed presentations from different disciplines a landscape ecologist meets in practice: scientists, engineers and policy makers being researchers, designers and decision makers. They are representatives of scientific probability, technical options and public desirability.
These disciplines use different languages with different concepts, logics, methodologies and tools, and different ways of arguing, telling stories or publishing. As a consequence, the contributions are very different in style. Moreover, some contributions show a historical overview or state of the art,

---

[a] Chairman WLO

others show scientific or technical analyses, and, finally, some argue for a specific approach.

Because of this the contributions are very different in their approach of the theme, their view on the subject or their presentation. The editors showed respect for those differences, because they are characteristic for the whole field of (applied) landscape ecology in the Netherlands. That does not mean that everything can be said, the contributions should meet general and discipline specific standards.

Sections of the book

It starts with the core business of actual landscape ecology research, a section on 'nature', be it in the context of the most sucessful part of Dutch nature conservation policy until now, the NEN. Then it offers a section on 'town', the largest human artifact on Earth, accommodating another kind of nature, object of urban ecology. In that section human motives, activities and artifacts play a role. It cannot avoid the question if urban ecology is a part of landscape ecology. Then it offers a section about 'infrastructure' connecting and separating pieces of land motivated by humans. By its fragmentation and disturbance of the landscape it is usually seen as an enemy of 'nature', but to what extent it could be an ally?

Every main section is introduced by the editors giving a summary of the contributions concerned and their relations. The last chapter of the book gives a reflection on the task of landscape ecology gathered on a meeting of landscape ecologists discussing the contents of this book. To make these contents as accessible as possible an extended list of key words referring to page is added. This list includes names of places and names of cited authors to find back pages where they are cited and where their results are discussed.

1.1.2   Why this book

Landscape ecology appeared to be a very successful new branch of science in the seventies of last century. This success can be attributed to two factors. One is the great need in society for new concepts and methods to understand the causes of environmental problems and to design solutions. The other is the very open character of the term landscape ecology. This openness made it flexible to respond adequately on many kinds of environmental problems. There was great readiness in society to make money free for this type of research. The output was strong, because the new concepts and methods appeared also to be valid in areas with no environmental problem. The landscape ecological approach inspired other ecological branches and environmental studies. This success created also a repercussion. It has become unclear what is the explicit field of landscape ecology and where are the boundaries to other disciplines. This reflection tries to understand the mechanisms and consequences of this process and to find a way out.

Reflection on task and identity of landscape ecology
The question: "Where am I busy with", knocks on every ones door from time to time. The same question will turn back after a while, because of the changing times. Although question and answer belong to the repetitive burden of life, they may also help us to enjoy being a conscious part of it, also when working in the field of landscape ecology. The Dutch Society for Landscape Ecology (WLO) is since its foundation in 1972 active in this branch of science. In high speed the discipline deepened and broadened its concept, knowledge and application. Landschap, the professional journal published by the WLO, reflects in its files this complex development. It is interesting stuff for ecologists and for researchers in history of science. The WLO is happy and grateful to have found a redaction team that took the job to prepare a book on today's state of this field of science and its applications.

To communicate with colleagues in other countries, the English language is chosen in. A forum for reactions could be the Bulletin of the International Association for Landscape Ecology (IALE). A tradition appears to exist.
A first book in English appeared as the proceedings of the founding congress of IALE in the Netherlands 1981 (Tjallingii and Veer, 1982). The second was to celebrate the 25 years existence of WLO in 1997 (Klijn and Vos, 2000). This, the third book, appears at the occasion of IALE's 25th anniversary, taking place in the Netherlands.

The WLO journal Landschap will continue to be published in Dutch. The articles are peer reviewed. The choice for the regional language is to maximise the accessibility to planners, decision makers, concerned citizens and other sectors of society. An interesting support to this choice comes from the empirical fact that the geographic area where Dutch is understood comprises the lowland around the river deltas of Rhine, Maas and Schelde. This region is surprisingly uniform in its landscape ecological pattern and processes and different from the surrounding parts of Europe.

The identity
"What is landscape ecology and what are the urgencies?" are logic questions, 35 years after the start of WLO and 25 years after the start of IALE. These questions are part of the ongoing debate in landscape ecology. This book presents new reflections on the state of the art and views on the future. The first section reports on the concept of ecological networks and its successful implementation in policy and planning. The second section reflects on the question: "What is urban ecology?" and the third section on the role of civil infrastructure.

In accordance with today's world characterised by the confusing mix of openness and lack of certainties, the book provides intellectual food and then leaves the conclusions to the reader.

This lack of a clear answer on the questions may be connected to the rapidly growing complexity of the total body of knowledge. The main reason, however, seems to be the term landscape ecology. The term can be understood as an open invitation to introduce new aspects. This openness is caused by the fact that landscape ecology is neither an explicit aspect science nor an explicit object science.

Ecology as such is the aspect science of a concrete object: plants (plant ecology), the sea (marine ecology), et cetera. From this point of view it is logic to see landscape as the object of landscape ecology. Landscape, however, is by far not as clearly defined as object as for instance plant or sea. For that reason Ies Zonneveld (2000) preferred the term land ecology. The definition generally accepted by landscape ecologists is: A landscape is an area of the earth's surface with a characteristic arrangement of land units (Klijn and Vos, 2000, p.150). This definition contains a large amount of openness. That is the effect of the following three components of the definition. The area, because it depends on scale. Land units, because they depend on scale and parameters. Characteristic arrangement, because it depends on parameters and criteria.

Projects
Of course, in each individual research project can be made explicit which scale, parameters and criteria have been chosen. Then the validity of the outcome of that research can be tested. So the landscape ecological results will belong to the objective knowledge produced by science. This explains the value and importance of this science to understand - and to a certain extent to manage - the earth.

What is not offered, however, is a defined difference between landscape ecology and other branches of ecology. This is a disadvantage in the communication between scientists. On the other hand this appears to be an advantage in the communication with other sectors of society when urgent scientific help is needed to solve environmental problems of an unknown complexity. Landscape ecologists are not afraid of complex systems, are flexible in finding new methodology and are able to approach the system on different scales and on different levels of integration.

In chapter 2, page 13 of this book André van der Zande explains that because of these properties landscape ecology had its boosting period during the wave of environmental awareness in the sixties and the seventies of last century. This could mean that landscape ecology is a response science rather than a wake up science. Examples of wake up sciences are at the moment for instance sciences working on the monitoring biodiversity, on climate changes and on sources of energy. These sciences have an explicit object: species, atmosphere, the earth's crust. This clearness in object is the outcome of reductionism, by subdividing our complex planet into well defined components.

The holistic approach
Thinking into the other direction it might become obvious that landscape ecology is the outcome of the holistic approach. If we accept this, the logic object of landscape ecology is: a holos that we call landscape. The consequence is that we accept that landscape includes everything. Although this sounds strange, we might find support in the fact that landscape ecology studies do not explicitly exclude any living or non living component as (part of the) study object. This open concept could explain the ability of landscape ecology to respond rapidly on unexpected or unpredicted events, by developing relevant research and if necessary with new methods and new concepts.

On the other hand, in times of less prominent public concern with nature and environment, the urgency for landscape ecological research might fade away. As continuity is vital for each scientific discipline, this is an important risk factor. This connection between new waves in landscape ecology and periods of public concern, could tell us that landscape ecology belongs to the category of problem solving sciences together with pharmacy, medicine and environmental studies. To fit logical in this category, the object of landscape ecology should be: landscape problems. This is interesting, because methodologically there is no clear difference between a problem in the landscape and a problem in the environment. Pollution of air and water, noise, lowering water of tables, construction of new roads and towns will all have a negative impact on the landscape.

1.2   Landscape as the environment of the human species

The consequence of this coincidence leads to the question: are landscape and environment the same thing? Affirmation is found in the etymology of the Dutch word *landschap* (Vries, 1971). The German word *Landschaft* has the same roots. The English word landscape, however, was introduced from the Dutch into the English language in the time that Dutch paintings called *landschap* became popular in England (see also Renes, chapter 7, page 106). This explains why word variations like townscape and seascape can be used in English and not in Dutch.

Landschap has two separate roots: *land* and *schap*. *Land* means originally the place where people live with the group they belong to. Later this meaning is expressed by the word *vaderland* (fatherland, home country). *Schap* is connected with the verb *scheppen* (to create). *Schap* was the name of a chisel. In mythology the *schap* was used by the gods to *schep* (cut) humans from tree stems. By men it was used to cut branches from the trees. The branches were used to make a wall to protect the home place against dangers from the surrounding wilderness. Later the word *schap* was used to indicate the wall itself and after that to indicate the area protected by the wall. Around the year 1200 the two words melted together into *landschap*, with the meaning of home settlement area. In modern words: the home country that is designed and managed to provide safety, food, housing and community life to the people that live there and are the owners.

An animal species
This description looks amazingly alike the ecological description of the needs of an animal species living in social groups. From the point of view that the human species is an animal species, we come to the interesting conclusion that *landschap* is the term for human habitat in the ecological sense. A strange consequence of this conclusion is that uninhabited parts of the earth have no *landschap*. Another difficult point is: do nomads have a *landschap*?
In spite of these difficulties in solving the problem with the term landscape ecology, there is a strong argument to use this word, being the product of a long cultural history.
This would lead to the recommendation to use landscape in the meaning: the habitat of a human population.

It must be admitted that this creates a very rare type of term, because it is restricted to only one species: Homo sapiens. For no other species we combine both home range and population in one term. The same was true for town, until that concept appeared to make sense not only for humans but also for a number of insect species. So we can start with the landscape concept only for humans, and may be that it can also be applied for other species.

Link to the humanities
Linking landscape to the human species makes clear that villages, towns, industrial areas are also landscapes. In this concept social activities like politics, government, sports, art, and religion are processes with effect on design and management of the landscape. It gives support of the ongoing action to develop methods that include social sciences in landscape studies.

Landscape science and spatial ecology
A consequence of the landscape concept as human habitat is, that the word ecology as part of the term landscape ecology loses its logic, because the concept of habitat includes ecology already. This double in the term landscape ecology is avoided in the term: landscape science.

Our relief to have found the term landscape science to avoid confusion by the term landscape ecology is however disturbed by the difficult question how to indicate the vast and successful field of science what we call landscape ecology and is not directly linked to the study of human habitat. Spatial ecology is a term that is already used. This could be the answer, because it is a clear and explicit term.

2   HERITAGE INTERESTS AS A CONTEXT OF LANDSCAPE ECOLOGY

is the WLO really developing into a community of Landscape Scientists?

Andre N. van der Zande[a]

Summary: *Landscape ecology in the Netherlands from the very start in 1972 has been positioned on the nature-culture continuum, being an exponent of the dominant Van Leeuwen-Westhoff paradigm. Gradually the wilderness paradigm and the (eco)system approach became more dominant, the latter being represented by the ecohydrologists and the metapopulation ecologists. Landscape (heritage) has lost its integrative impact on the Netherlands landscape ecologist community in spite of the existence of the Working group on Historical Ecology since 1991. This development within the landscape ecology community marginalises this community from the dominant trends and needs in society and hampers a further scientific development towards a real interdisciplinary landscape science.*

---

[a] The author thanks Roel During, his assistant professor, for his valuable remarks and comments on the draft of this article.

2.1   Introduction

Landscape ecology: nature conservation and nature management transformed
The founding of the WLO (Werkgemeenschap Landschapsecologisch Onderzoek) in 1972 as a scientific community in the Netherlands marked a moment in history where nature conservation ecology and nature management ecology transformed into landscape ecology. This transformation covered three fundamental dimensions that reflect the ambitions and expectations of the newly formed community of scientists.

A shift of scale and multidisciplinary context
The first dimension is about scaling: from the site or object to the scale of the landscape. It is also the scale of the spatial planning of regions and provinces. It is also when a lot of ecologists (including myself) find jobs with provinces and regional authorities, in most cases within the spatial planning units (see also Cramer, Kuiper and Vos, 1984).

A shift from separate causal relations into system
The second is the shift from causal relation studies to ecosystem or landscape system studies. The causal relation studies covered auto-ecological or population ecological or abiotic physical studies. The interaction between ecological processes and ecological patterns   becomes integrated into the systems approach where processes and patterns are interrelated. This also integrates the plant research community (dominance of patterns research) with the animal research community (dominance of processes research) and the abiotic research community (both processes and patterns research).

A shift including human impact
The third and for this article most important transition is the inclusion of the influence of man on landscape, both in its historical impacts and in its present or threatening impacts. It was also the period in time when Environmental Impacts Assessments as a policy tool was established. So this change is not only dealing with the inclusion of the (historical) impacts of man on the landscape, but also marks the development of ecology from a descriptive science into a forecasting science.

Developments in the Dutch Community of landscape ecologists (WLO)
In this article evidence will be provided that from the launch in 1972 and thereafter, there have been major shifts in the dominant insights and concepts within the landscape ecology community, shifts that addressed the three fundamental dimensions of the scientific challenges. These shifts will be analysed using several bodies of information. First is the Magazine/Journal of the WLO, a magazine that has evolved from a grassroots hand collected stencil towards a fully fledged scientific journal in 1984:

Landschap. The second body of information is the rise and fall of the (active) Working Groups of the WLO.

The shifts will be described and analysed within the scientific community of WLO itself. I will clarify the state of ideas and concepts at the beginning of the WLO community and will go somewhat deeper into a few important dilemmas, these being indicators of change. Some indicators of shifts in general society will be compared with the shifts within the WLO and from this comparison conclusions will be drawn that reflect on the ambitions and expectations in the founding phase.

First of all it is important to go back to the early seventies, when WLO started.

2.2 Who were the founding fathers and what was their dream?

Four types of professionals

It is helpful that the early written magazine of the WLO (then called: "Mededelingen van de Werkgemeenschap Landschapsecologisch Onderzoek" later called "WLO-Mededelingen") contained a list of members, together with their professional addresses. What we see is roughly a mix of four types of professionals: those working in research institutes, those working in nature conservation management, those working as a consultant and those working in spatial planning institutes. The "research" institutes in question were the national government institutes for nature conservation and management, land use, aerial cartography and mapping, landscape, soil mapping and spatial planning. Of course a large number of members worked at universities, covering a whole array of academic disciplines such as biology, geography, agronomy, environmental sciences and planning sciences.

Communities of conservationists, planners and developers brought together

To simplify the matter I consider the start of the WLO as a ménage à trois of the ecological community ("*conservationists*") with the spatial planning community ("*planners*") and the land use community ("*developers*"). All three communities had a different background and state of development, resulting from the issues they were involved in, their ambitions and their ideals. See also the three citation communities that are distinguished by Cramer and Van der Wulp (1989) on the basis of an international quantitative analysis of citations between journals.

Conservationists

The *conservationists* were dominated by what I call the Westhoff-Van Leeuwen doctrine (see also: Van der Windt, 1995). They acknowledged the fact that nature in the Netherlands is somewhere halfway on the wilderness-culture continuum and that active human management is necessary for the survival of these specific succession stages of ecosystems. They struggled with the phenomenon of dynamics, both the human induced dynamics, most of the time coming in the form of threats from outside the reserve areas, and with the dynamics of the abiotic processes, especially the hydrological ones. Their ambition was to include the challenging promise of the new scientific systems approach (see Werkgroep Theorie, 1986 and Zonneveld, 1989) in conservation ecology and this challenge made it necessary to go beyond the borders of the reserve areas. Their step towards landscape ecology was part of fulfilling this promise: ecosystem processes were often on a landscape scale and not on a site scale. They recognised or expected that the planning community was in charge of these landscape scale processes and therefore the planners were an interesting partner.

Planners

The *planners* had reached the stage of recognising that the post-war rebuilding ideology was becoming outdated. This ideology was, basically, a short-sighted war against the housing shortage, and the post-war reconstruction (known in Dutch as'de oorlog tegen de woningnood en de Wederopbouw'). This "war" had had no consideration for historical landscapes and cities, for ecological values and, most of the time, not even for what people really wanted (see also Van der Zande, 2006). The planners had developed a non-political technical and rational planning approach, which had put them in an influential position and given them leverage (see also Hidding, 2006). The more clever thinkers in this field of planners recognised that it was necessary to include ecology (the new kid on the block) in their methods and of course they urged the ecologists to adopt their technical and rational appraisals and valuations of the new spatial deficits. Now, ecological deficits could also become part of the planning game. Most of all they liked the nature value maps and mapping processes, because they fitted seamlessly into their planning language repertoire (see also De Haas, 2006). The planners were also inspired by the thinking in systems and one of their holy grails in that period was the idea of developing an all-encompassing spatial model (see Van der Maarel on the so-called GEM, 1974, Dauvelier and Little, 1977; and Maarel and Dauvelier, 1978). Because social scientists were almost absent in those days in the WLO family, none of this, including ambition, was challenged by these disciplines (in 1981 Saris and Wilders estimate the number of social scientists at 8% of the WLO community). And of course in those days there still was a strong belief that society could somehow be 'made'.

Developers

The third community of the *developers* was the most colourful group. They had worked as scientists for the post-war development of agriculture in the Netherlands and for rural development in the Third World. From the start their focus of attention had been on the region or the project (most of the time also being large areas). They never had the luxury of dealing only with conservation or only building houses; instead they had to take full account of the scarcity of budgets for investment and other resources and the need to optimise the use of these. The developers had something of a dubious reputation because of their role in the drastic post-war agricultural reallotment projects.

They developed from disciplines like soil mapping, agronomy and physical and historical geography, but had to deal with real people in their projects. They therefore developed the landscape concept and land use science concepts. Their ambition was to learn from the historical man-nature interaction on a landscape scale and to develop interesting paths for future agro-eco-system developments. Their ideal was not so much the systems approach but the holistic approach (see Zonneveld, 1982). The ambition of holism lay more in a qualitative understanding of the world around us, whereas the systems approach had a more quantitative ambition, which later resulted in the tsunami of modelling and models in the physical environmental sciences.

Including social sciences

I think it was no coincidence that a prominent representative of this third group, Prof. I.S. Zonneveld, took the moral leadership and that of the WLO in its first decade. He chaired the first international conference of the landscape ecologists (IALE) in Veldhoven in 1981. Though the developers wanted to analyse and understand the land use processes of the past they also recognised the inevitability and even the desirability of land use changes. They advocated that these changes should be planned and executed sensibly and wisely, knowing what the impacts of any possible interventions were. This third community cherished the most sincere wish to merge with social sciences, because they had themselves felt in the field the shortcomings of their own purely physical and technical approach. The developers themselves mostly had a more physical technical background.

2.3 The forces working upon and within WLO

Two dilemmas

Now that they had found love, what was the ménage a trois going to do with it?

As in any marriage there were ups and downs and dilemmas about the road to follow.

The Board of WLO described in 1978 the position of WLO in a short article (Anonymous, 1978) and in the tenth anniversary year, 1982, Zonneveld was asked to give an overview (Zonneveld, 1982), which he did.

The WLO community seemed to face two important dilemmas:

1. One was the dilemma *of action and research*.
2. The second was the dilemma of broad or deep, *between landscape and ecology (the dilemma of specialising versus integrating, in fact)*.

2.3.1   Between action and research

Predicting and influencing or describing and analysing
This dilemma is explicitly referred to by the WLO Board in the last paragraph of the 1978 article. It should be realised that nature conservation in the Netherlands was at that time still in a fierce process of emancipation and most nature researchers were also very active in non-governmental organisations for nature (see Van der Zande, 2004). Positions within the WLO were fiercely divided between the purely positivistic neutral scientists on the one hand and the action researchers on the other, and of course there were some intermediate positions. The scientists just wanted to describe and analyse the patterns and processes in the landscape to fully comprehend them. The action researchers not only wanted to describe and analyse, but also to predict and to influence. Between these extremes, some strong bridges were built, of which I will name three.

Bridged by impact assessments and predictions
One is the development of methodology for impact assessments and predictions. A working group was formed (see also below) that focused upon what was then called interaction studies (in Dutch: relatie onderzoek). The traditional scientists felt comfortable because it was neutral empirical science (resulting in good publications and models) and the action researchers were content because these studies were very useful in Environmental Impact Statements and thus in influencing decision making. During the yearly WLO study day on Environmental Impact Assessment on November 24 1978 the whole WLO community was united, whereas on other occasions there were severe clefts within the membership (especially on the subject of ecological valuation).

Bridged by project studies
The second bridge is the practice of large "integrated" project studies (or region approach). Landscape ecology in its early years was stimulated by the joint research (and action) in some mega projects. Very well-known icons of this type of projects were Midden-Brabant (Lier, 1975; Van Oostrom, 1977), Kromme Rijn (Kromme Rijn Project, 1974), Duinvalleien (Zadelhoff, 1978) and Drentse Aa (Grootjans, 1979). This approach also had another effect, which I mentioned earlier. Landscape ecologists became involved in sociological processes, decision making and political processes and were encouraged to widen their scope of disciplines.

Bridged by challenging decision makers
The third bridge was more an intermediate political position formula of the WLO Board. They said it was the task of the WLO to challenge the decision makers to use and include state of the art landscape ecological insights in their decision making, but that it was not the task of the WLO to influence the outcome of the decision making itself.

These three bridges had enough impact as peacemakers and WLO grew and flourished in the first few decades of its existence.

2.3.2   Between Landscape and Ecology

Broad versus deep
This dilemma was also the subject of discussion and reflection in the early years. Even in the 1978 article the Board wonders whether it is to be expected that the important issues of discussion and debate will stabilise. The key word of the Board message is to professionalise by improving methodologies and databases. The landscape side of landscape ecology calls for including those sectors and disciplines that are relevant to the use of and to developments in landscape. The ecological side of landscape ecology calls for improving the ecological scientific basis of the landscape ecology. This dilemma of broad versus deep is still present today and I will present some semi-quantitative information on this dilemma below.

A trend of narrow and deep
In the addendum to the 1982 article, Zonneveld writes (my translation), "During discussions it appeared that a substantial number of WLO members objected to too broad a landscape ecology concept. Nature conservation values should be the focus, according to these members. I insist however that an integrated approach to the land-ecosystem complex integrating several disciplines, is the essence of landscape ecology whether this takes place in landscapes with or without nature conservation values and even with or without life...". Zonneveld is consistent and refers to his ideal of transdisciplinarity and the urgency of widening up the scope of landscape ecology to people (Barendregt and Klein, 2006). The present chairman of WLO, Jacques de Smidt, also supports this long-standing plea by Zonneveld that landscape ecology cannot only be a merger of earth sciences and life sciences, but should also include people and therefore the people sciences (Smidt, 2006). However, the trend has been in another direction.

In the following paragraphs I shall present some semi-quantitative evidence about the development of the WLO in the last decade.

2.4   What did the WLO community write about?

Articles in the WLO journal 'Landschap' and its predecessor
To gain a representative sample of the dominance of insights, concepts and paradigms within the WLO community, I have analysed the Magazine/Journal and classified the articles in the years 1975/76, 1985, 1995 and 2005.

Fig. 1 shows the number and the percentage of articles dealing with heritage and the historical influence of man on the landscape. It also shows the percentage dealing with ecology concepts (such as ecohydrolgy, wilderness, meta-population and city ecology), and of course there is an "other" category of articles, the content of which I have not specified. (see Fig. 1).

Year	1975/6	1985	Average 75-85	1995*	2005	Average 95-05
Landscape/ history	14	4	9	3	4	4
Ecology concepts	8	7	8	34*	10	22
Other	13	16	15	2	9	6
Number of articles	35	27	31	39	23	31

* including theme issues on ecological networks and on manure

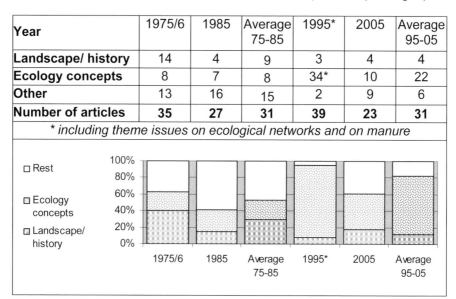

Fig. 1 Articles in the WLO Journal in 5-year intervals in three categories

The emergence of ecology as the dominant stream

These figures illustrate that ecology within landscape ecology has grown into the dominant stream within the WLO community. The history of landscape research has been marginalised into a small minority. This is not surprising if one studies the development of the WLO membership, which has not been analysed recently, but in 1981 Saris and Wilders already describe a clear dominance of biologists (35%) over geographers (24%), technical scientists (25%), and social scientists (a mere 8%).

This finding is consistent with those of the editorship of *Landschap* in 1988, when 110 articles in the first 5 years of the Journal (Redactie Landschap, 1988) were evaluated. Only 3 of these 110 articles covered the landscape system as a whole.

Browsing through 30 years of WLO journal issues, there are a few more relevant observations to be made.

Nature Conservation Values

One of these relates to the discussion on General Nature Conservation Values (ANK in Dutch) by Derckx (1995) and Buys (1995), among others.

In fact they do not discuss the values of landscapes and their relevance to present and future society, but just the methodological differences about measuring the ecological values of landscapes. Landscape as an integrative ambition is completely absent from this discussion and it is illustrative for the dominance of nature above landscape.

The Working Group on Historical Ecology also seems more focused on the history of nature management *sensu stricto* than on the history of man-made landscapes, and the desire to understand man-landscape history and possible sustainable paths of future development (Dirkx *et al.*, 1992). The Working Group tried for a period of time to find supporters within the WLO for making a historical atlas of the Netherlands on a landscape ecological basis. This initiative failed where other atlases and maps succeeded (pers. inform. R. During).

Integrating policies rather than landscapes
A very interesting contribution by Boersema and Kwakernaak in 1993 on integration does not use the landscape concept as a means for integration at all but focuses directly on the wish for integration between environmental policy, water policy and spatial planning policy as such. Here, the pull factor of the integrative landscape concept seems obsolete or is avoided.

Promising exception on identity
One of the very few promising mini debates in *Landschap* of the last years is the trio of articles about the identity of landscapes (Boerwinkel, 1994; Haartsen, 1995 and Van Bolhuis, 1995). Here we see the whole scope of landscape types involved, the dilemma of how to deal with the historical patterns and processes and the input of designers as part of the integration puzzle. A rare gem of growth towards landscape science, one that I believe is very welcome.

2.5   For which subjects did the WLO community create Working Groups?

Reports on Working Groups
The most interesting or important topics within the WLO community were the reasons for the formation of the so-called Working Groups. I have been able to follow the rise and fall of the Working Groups in the WLO Magazine, because in that first decade there were regular reports on their work.. Unfortunately these reports appeared less frequently, or a particular Working Group was mentioned only occasionally in the Chairman's Column. From 1986, *Landschap* was no longer a reliable source for the activities of Working Groups.

Fig. 2 below I give my (incomplete) reconstruction of the life of the Working Groups in this period.

Period	Start and finish of Active Working Groups	N
1972-74	Theory, Activation and Application, National Landscape Parks and Nature building are active (from the start of the WLO)	4
1972-77	Plus Nuclear Energy	5
1978	Plus Interaction Studies and Education	7
1979	Minus Nuclear Energy: they present a final report Minus Education: they did not become viable	5
1982	Minus National Landscape Parks* Minus Nature building* Minus Interaction Studies*	2
1983	Plus Vlaanderen (Flanders)	3
1986	Minus Vlaanderen (Flanders): seceded from WLO Minus Activation and Application	1
?	Minus Theory	0 ?
*1988***	*Plus Ecohydrology*	*1*
*1993***	*Plus Historical Ecology (formed outside WLO in 1991)*	*2*
*2000***	*Plus City Ecology*	*3*

* Termination of these Working Groups was decided as part of the 1982 WLO Policy Plan (1982-1987) see Anonymous (1982).
** The start of the last three Working Groups are based on information on the WLO website and not from *WLO- Mededelingen* or *Landschap* as a source of information.

Fig. 2 Activity of WLO Working Groups in chronological order

From diverging into narrowing down
It is remarkable that even the issue of nuclear energy entered into the scope of landscape ecologists in those early years of the WLO. Remarks by the Board of the WLO can also be found, regretting the inability to sustain a Working Group on Education and the lack of success of the Working Group on International Affairs . In the case of the latter, it should be remembered that many WLO members became active in IALE and found their international contacts and debates there.
In the first decade we see a diverging interest, but in the second half of the nineties the number and scope of the Working Groups narrows down. No Working Group has the whole landscape as a focus of discussion and there are also no policy or issue driven Working Groups any more, such as was the case with the Working Group on National Landscape Parks.

External Issues
External issues that prompted the setting up of active Working Groups were:

1.   Ecohydrology: related to the political issue of water management; an issue of growing importance and recently refuelled by the climate change debate.

2. Historical Ecology: related to the issue of the destruction of historical landscapes as a result of large-scale nature restoration and nature creation projects and to the discussion about historical reference situations for present-day nature management.
3. City Ecology: related to the political issue of our environmentally degraded cities, and as a response, the movement to make cities greener.

Responding to growing concerns about the deterioration of our historical landscapes

Apparently there is no counterpart within WLO to the growing concern about the deterioration of our historical landscapes, which have been "promoted" in our national spatial planning document (VROM et al., 2004) by incorporating 20 National Landscapes and our UNESCO World Heritage sites. We see Dutch historical geographers regrouping themselves in a recent initiative called "Netwerk Historische Cultuurlandschappen" (Historical Landscapes Network). Perhaps this group can become part of the WLO structure just as historical ecologists did after spending 2 years working independently.

It is interesting to see whether or not the WLO community is developing in a way and a direction that is comparable with the general trend in present-day Dutch society.

2.5.1 What did science journalists in the Netherlands write and talk about?

One of the largest newspapers

To gain a representative sample of the articles appearing in general interest publications aimed at the more discerning levels of Dutch society,, I analysed Saturday editions of *De Volkskrant* (one of the largest newspapers in the Netherlands, and originally a catholic left-wing newspaper) from approximately the same years as the editions of *Landschap* that I analysed. I selected the scientific sections in the Saturday papers and counted the number and percentage of articles on history, ecology and 'others' (see *Fig. 3*).

Year	1995	2002	2004	2006	Average
History	5	2	9	6	6
Ecology	5	5	4	2	4
Others	19	20	20	30	22
Total	29	27	33	38	32

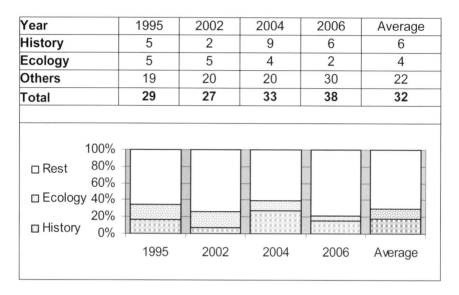

Fig. 3 Articles in the Science Section of De Volkskrant in three categories: History, Ecology and Others, in four years.

Diverging concern on history

It is unfortunate that *De Volkskrant* Science Sections from the early seventies are not easily accessible, but we can see that the history percentage in *De Volkskrant* (average of 17%) is higher than in *Landschap* (12.5% on average in the same period). In contrast to *Landschap,* this percentage does not show a decrease. In *Landschap,* there is a drop from 20% in the whole period and only 12.5% in the '95-'05 period, while ecology grows from 42.5% in the whole period to 65% in the '95-'05 period. In *De Volkskrant* the percentages for history and ecology are of the same order of magnitude (17% versus 13.5 %), whereas in the same period the percentage for ecology in *Landschap* is three times higher (65% versus 12.5 %).

We see that the WLO community publishing in *Landschap* is becoming more ecologically focused, whilst the scientific journalists in the Netherlands are not.
It seems strange that this change in the WLO has taken place. Is the situation in the Netherlands different to the international one? Can we find explanations for it?
Apparently ecology has beaten landscape and specialisation has beaten integration.
But it was also Zonneveld who as early as in his 1982 article feared this development within the WLO and recently in his interview in the Zonneveld/Biesbosch themed edition of *Landschap.*

2.5.2   Is the situation of Landscape Ecology in Europe different?

State of the art of European landscape ecology
Recently, there has been a review of the state of the art of European land-scape ecology in *Landscape Ecology*. Based on the results of the 6th IALE World Congress in July 2003 in Darwin, Australia, the Landscape Europe network published 7 state of the art articles, written by 27 authors. Pedroli, Pinto-Correira and Cornish (2006) give an analysis of the situation in a review article. This review shows that there are more interdisciplinary ambitions in Europe; 4 out of the 7 articles are classified as interdisciplinary, and this seems more than in the WLO/*Landschap* situation. On the other hand the authors observe that there is "…the classical issue of the interface between hard and soft sciences and the need to establish bridges…".

Covering design and decision studies or not
Their position on this issue is described as follows:
"But this does not mean that the definition of landscape as an "area perceived by people" needs to become dominant in landscape ecology. Studies on perception and motivations can establish the link between landscape classification and typology, and the management of landscapes."
Their position is comparable with the definition of De Groot and Udo de Haes (1984) of landscape ecology as a technology. I think this position is not satisfactory and does not do justice to the fact that several authors in the special issue come from EU countries where the (historical) landscape is changing rapidly and drastically. Pedroli *et al.* do recognise the feeling of "loss and grief" accompanying this process and also that there is "…a tension between the rhetoric of bottom-up participatory approaches and the reality of homogenous top down approaches to managing landscapes…". In their concluding remarks they emphasise that "…others have recently contributed to the continuing debate on the range of landscape ecology and its covering or not of design and decision studies…".

Driving forces in landscapes
I have the impression that the situation in the Netherlands for landscape ecologists is indeed different from that of their European counterparts. I think it has to do with the different orientation of landscape ecologists towards the processes of change in the landscape and the position of nature conservation (and restoration).
I feel affinity with Burgi *et al.* (2004), mentioned in Pedroli *et al.,* who depart from the driving forces of landscape change as the most important object of landscape ecology study (see also below) - driving forces which are strongly related to the European Union as a developing force, especially in agricultural landscapes.

2.5.3   Possible explanations for the WLO Landscape Ecology situation

Opening windows by public awareness meeting science and technology in the NEN

As a scientist and as a civil servant I am much impressed by the policy window concept of Kingdon (Kingdon, 1984). If people with awareness of a problem in society meet people with a good scientific and technical solution to that problem, they together have an influence on politics so that a policy window can open up. I experienced such a period of a policy window opening up in the late eighties and early nineties. The people with awareness of the acute loss of biodiversity and the people enthusiastically promoting the solution of creating ecological networks teamed up and found a response in politics. As a result a substantial investment flow came about to create the National Ecological Network in the Netherlands. It now amounts to up to 1 billion euros a year. In that light, it is very plausible that the ecological aspect of landscape ecology has become more dominant and that for example the meta-population concept was developed as the scientific tool to cope with scattered habitats (Opdam et al., 2002).

Utilising demand pull for integration

My own hypothesis is that integrative sciences only thrive through the existence of substantial investment programmes aimed at urgent issues in society. It is not the pure disciplinary science push but the issue-driven demand, the pull, from society that makes integration work. Because the baby boomers are aging and worried about their health there is a growing investment flow in better food and nutrition and thus in better health research focusing on prevention. There is also a growing awareness of the lack of water safety and of water management issues and thus a growing investment flow in water management measures and thus also in water management sciences. De Groot and Udo de Haes (1984) make an interesting distinction between what they call technologies (in my definition integrative sciences) and science, the first being oriented toward solutions (to problems in society) and the latter being oriented towards truth finding. Cramer and Van der Wulp (1989) describe a more dialectic tension between what they call the orientation towards society and an orientation towards theory of landscape ecologists.

The urgency of identity

The question is whether there is a strong enough sense of urgency when we are dealing with landscape deterioration and lack of historical identity of our present-day landscapes. And whether there are scientifically sound concepts for solutions to this problem, in order to fulfil the Kingdon policy window conditions?

There are some interesting signals. As mentioned before the National Landscapes are back on the national spatial planning stage. There is also a concept for a solution developed in the Belvedere Memorandum (Feddes ed., 1999). This concept is the bridge between pure conservation of land-scape, which is rejected as too drastic and too ineffective (what results are no living landscapes, but museums of landscapes) and reckless develop-ment on the other side. The concept is called "Conservation by develop-ment" and it seems to have roots in the way of thinking of the *developers* community of the WLO of the early years. Perhaps a new policy window can indeed open up here? The WLO as a whole is still stuck in the earlier policy window of the National Ecological Network, in my opinion.

2.6 Conclusions

A local trend from integration into specialisation
My conclusion is that the WLO has shifted from an integrative force based on the concept of landscape into a dissipative force where landscape ecol-ogy has become just a sub-discipline of ecology. The integrative position of the developers has been lost and the conservationists (ecologists) have taken over the dominant position. I see a growing dominance of hard sys-tem ecologists, especially ecohydrologists and an exclusion or marginalisa-tion of (heritage and historical) geographers and historical ecologists. In spite of the consistent pleas of the last 30 years by Zonneveld, the WLO has not developed further towards a landscape science, integrating for example the social and political sciences. I wonder whether the similar plea from the present chairman, De Smidt (2006), will have an impact.

A general trend in reverse?
My thesis is that in society at large the trend is the other way: more towards a growing importance of history and of the (historical) identity of landscape and less of the hard systems ecology of the landscape (see also Van der Zande, 2006). But perhaps this is only a recent development in society in general. And there are of course the new more physical urgencies such as climate change, which re-emphasise the ecological primacy.

An integrative science or a sub discipline of ecology
In conclusion, the WLO and the landscape ecologists need to take the next step on their path of development towards an integrative landscape science and should change their name into landscape science (according to Klijn and Vos, 2000) or marginalise into a sub-discipline of ecology and leave the integration to others.

Nature

NATURE

Jos Dekker

Editorial introduction

In 1990 the Dutch government decided to develop a national ecological network (NEN; in Dutch: Ecologische Hoofdstructuur, EHS). It should be realised in 2018. As for the time, in 2004 the implementation was halfway. Landscape ecological concepts and knowledge played a role in the design of the NEN and its elaboration. For that reason, the WLO organised a conference "Halfway the NEN (Halverwege de EHS)" in 2005. All but one lectures are included in the Nature part of this book. Four chapters are added (Schouten, Verweij and Barendregt; Van Eerden; Wassen; De Jong).

Barendregt and Dekker (chapter 3, page 35: *Landscape ecology and the National Ecological Network in the Netherland*) tell the story of the conception and the development of the NEN. The NEN was a strong strategy, based on relevant landscape ecological concepts and knowledge. However, important landscape ecological relations were not worked out well. Its implementation was difficult, because it had to be organised in the context of a densely populated and intensively used country. The result is a practically shaped NEN with uncertain ecological profits.

Glastra, a conservationist at the province level, expresses his enthusiasms for the concept of the NEN (chapter 4, page 65: *The National Ecological Network in practic*). The NEN provided an offensive agenda for conservation, fitting perfectly within the Dutch tradition of delta projects. However, in practice the implementation of the NEN was not always easy, one of the problems being the voluntary condition of acquiring new areas. Creative solutions were necessary for realizing spatial interconnection. At the current pace, the NEN will not be complete by 2018. The story of the NEN must be continually retold, with the original enthusiasm revived.

Reijnen, Van **Hinsberg**, **Lammers**, **Sanders** and **Loonen** (chapter 5, page 74: *Optimising the Dutch National Ecological Networor*) identify environmental and spatial problems, which require further optimisation of the NEN. Their analysis is based on the nature target types allocated to the NEN. Environmental problems relate to water tables and atmospheric nitrogen deposition. Spatial problems concern total area and spatial configuration. They assess the urgency of the problems, based on the principles of the EU Habitats Directive, and suggest a method to prioritise them. The spatial coherence and environmental conditions are as yet insufficient to meet the international commitments. Large continuous habitat areas offer the best opportunities for sustainable protection. Mosaics of smaller habitat areas may function as one large natural core area as well.

Schouten, Verweij and Barendregt (chapter 6, page 92: *Biodiversity and the Dutch National Ecological Networor*) analyse the distribution of hotspots of species richness and uniqueness for a number of species groups (Syrphidae, Odonata, Orthoptera, herpetofauna and Bryophytes). They compare the occurrence of these hotspots with the NEN. Patterns of both types of hotspots differ. Hotspots of species richness of Syrphidae and Bryophytes are poorly covered by the NEN. Hotspots of uniqueness are not very well represented within the NEN. They recommend a better and larger selection of target species, especially including less well-known groups (cryptobiota) and 'common' species.

Renes (chapter 7, page 104: *Landscape in the Dutch 'National Ecological Network*) focuses on the impact of the NEN on cultural landscapes. He states that landscape is often wrongfully neglected in discussions on the NEN, in favour of 'new nature'. The old, ecologically rich cultural landscapes, once the basis of the protection movement, are situated both inside and outside the NEN. However, over the last fifteen years the landscape part of the NEN has been very disappointing, partly because nature development projects (new nature) damaged historic landscape features, partly because priority of funding for new nature. He advocates a careful assessment of landscape values before starting new nature development. Agriculture can become a partner in landscape management.

Pelk (chapter 8, page 119: *The natura 2000 network in the Netherland*) explains the philosophy and approach of the Dutch government regarding the development of conservation targets for Natura 2000. Natura 2000 is another ecological network, resulting from the EU Birds and Habitats Directives. Most of the Natura 2000 sites on land (nearly 100 %) are covered by the NEN. About 40% of the NEN is also Natura 2000 area. After the selection and designation of the sites, allocation of conservation targets is needed to direct the implementation of management plans, which are obligatory for the Netherlands. Despite its small area, the Netherlands is of crucial relative importance for a number of species and habitat types in Europe, especially for non-breeding birds. However, most of the habitat types and selected species do not have a favorable conservation status. Both at the national and site level conservation targets have been determined.

The Dutch NEN is part of European ecological networks, Natura 2000 in the EU and Emerald in other European countries. Together they constitute the Pan-European Ecological Network (PEEN). **Jongman and Veen** (chapter 9, page 141: *Ecological networks across Europ*) explain the principles of the design of these networks, like connectivity, and hierarchy of spatial scale. Upscaling and downscaling of ecological networks is important. There are differences between and within countries in the planning of these networks. Public support is important, because of the impact of ecological networks in the wider countryside. Dutch landscape ecologists are involved in many projects for the development of Natura 2000 in the new member states. Jongman and Veen show a bottom up approach in the landscape ecological support for the development of national ecological networks by research on (semi-)natural grasslands in Bulgaria and virgin forests in Romania.

Wassen (chapter 10, page 169: *Ecohydrology; contribution to science and practic*) defines ecohydrology, a landscape ecological specialisation, and reviews its history and state of the art, both from a Dutch and international perspective. The Dutch focal point was understanding of spatial relations via water in regional landscapes and the impact of changes in cycling of water and nutrients on ecosystems. Environmental problems such as desiccation, acidification, and eutrophication were assessed. However, explicit attention for the fragmentation problem did not come from ecohydrologists. They were involved in the design of the NEN, but their advice was limited to the regional scale. At present, the Dutch contribution to ecohydrology in the international arena is changing. Ecologists retract themselves to (restoration) ecology. The central position of ecologists in ecohydrology is taken over by hydrologists inspired by global issues like global change.

Eerden (chapter 11, page 187: *Habitat scale and quality of foraging are*) focuses on the carrying capacity of Dutch wetlands for water birds. Wetlands like IJsselmeer, Lauwersmeer and the Delta are important components of the NEN. Habitat scale and foraging area are factors determining carrying capacity. He translates his findings into management directives at species and habitat level. Legislative protection is not sufficient, selective management measures are needed to maintain carrying capacity, such as: preservation of large scale open water, restoration of hydrological connectivity between large lakes and with marine systems, linkage of marshes with open water, and addressing relations with (agricultural) surroundings.

Connectivity was and still is a leading principle in the design of the NEN. Van Leeuwen, a Dutch ecologist, showed the importance of boundaries and environmental gradients for the diversity of species. De **Jong** (chapter 12, page 208: *Connecting is easy, separating is difficul*) explains some of his theoretical principles and stresses their importance for spatial design at different levels. Gradients suppose both connectedness and unconnectedness, combined in 'selection' as a steering principle. However, the ecological theory of Van Leeuwen hardly plays a role in pure landscape ecological research anymore. Stressing boundaries means less emphasis on adjacent areas to be characterised by target species asked for by contemporary policy. Van Leeuwen's philosophy still is an important principle in landscape architecture.

There is much more to say about the development of the NEN, the many research activities and results of Dutch landscape ecology and its support to the NEN. For example, the biggest part of the NEN is not on land, but on the Dutch part of the North Sea, the Wadden Sea and other big waters. But this is not yet an object of Dutch landscape ecology. Moreover, much research is focused on the management of reserves and their protection against negative environmental impacts; and on the development of new nature areas and their results. A growing body of research concerns the changes to be expected by climate change and strategies to encounter its impact. Important adaptations of the design of the NEN might be necessary.

Dutch landscape ecology is strongly application driven. That is why the NEN, as the most important conservation project in the Netherlands, is the focus of much landscape ecological research. However, its focus does not mean that this research is without theoretical significance, as the chapters of Renes, Van Eerden, Wassen and De Jong show.

The NEN is more than a habitat for species. It is a project of conservationists, (landscape) ecologists and politicians. It fulfils many functions, more than conservation and research. It provides services like leisure, drinking water production, agriculture, forestry, fishery, transport, a place to live, water management and CO_2 storage. However, these functions are not the main focus of landscape ecology in this part of the book. That focus is the (a)biotic layer of the landscape with its biodiversity. The other, more social, layers are the subject of the other two parts of the book.

3 LANDSCAPE ECOLOGY AND THE NATIONAL ECOLOGICAL NETWORK IN THE NETHERLANDS

Aat Barendregt[a] and Jos Dekker

---

[a] Special words of thanks are given for the cooperation with Sander van Opstal (Ede).

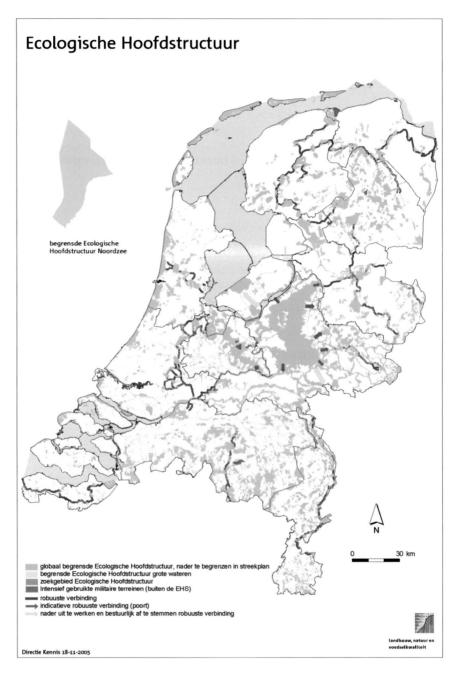

Ecologische Hoofdstructuur

begrensde Ecologische
Hoofdstructuur Noordzee

N

0   30 km

☐ globaal begrensde Ecologische Hoofdstructuur, nader te begrenzen in streekplan
☐ begrensde Ecologische Hoofdstructuur grote wateren
☐ zoekgebied Ecologische Hoofdstructuur
■ Intensief gebruikte militaire terreinen (buiten de EHS)
▬ robuuste verbinding
➡ indicatieve robuuste verbinding (poort)
▬ nader uit te werken en bestuurlijk af te stemmen robuuste verbinding

landbouw, natuur en
voedselkwaliteit

Directie Kennis 18-11-2005

*Fig. 4 Gross National Ecological Network (LNV, 2005) of the Netherlands;
on land, on large water expanses and in the North Sea; lines and arrows
indicate (indicative) robust corridors; update November 2005 (EC-LNV, 2005)*

3.1 Introduction

Landscape ecology

Landscape ecology aims to understand the patterns and processes in the landscape and this science incorporates a number of aspects such as soil, water, chemical substances, biodiversity, spatial arrangement and various aspects of human management. The research in landscape ecology focuses on the relations in the landscape. It describes and explains the spatial arrangement of visual elements in a region organised by human (cultural and historical) activities, but it can also analyse the nutrient availability organised by site conditions that are caused in a pattern of groundwater flows. Ecology is incorporated in the classical definition with the interrelation of a-biotic and biotic components and mostly (even in urban ecology) the distribution of plants and animals is a principal element in research. This means nature is a prominent object to landscape ecology and, at the same time, landscape ecology learns about the conditions how the variety in nature is organised.

National Ecological Network (NEN)

Landscape ecology might become a perfect knowledge base for the conservation of nature with its characteristic processes and spatial relations. When society wishes to maintain biodiversity and even wishes to restore populations of plants and animals that have declined, the stipulating factors for the diversity in species can be postulated from landscape ecology. This is also true when society wishes to perform at a national level the construction of a National Ecological Network (NEN; in Dutch: EHS, Ecologische HoofdStructuur, see *Fig. 5*). The central question in this paper will be whether this NEN is organised in such a way that key landscape ecological relations are incorporated and whether by these relations biodiversity is conserved in a sustainable way.

History and future

This paper starts with a description how nature conservation developed during the last century, to be continued with an analysis what landscape ecology in the Netherlands learns about the functioning of the ecosystems we wish to conserve or stimulate. In a later section, we review how from 1990 the concept of the Network developed during the last 15 years. Finally we will bring these two last aspects together in a discussion about what can be added from landscape ecology to the present implementation of the Network.

3.2   Development of nature conservation in the Netherlands

History

The conservation of nature in the Netherlands changed from safeguarding a few bird and mammal species a century ago to a policy for the maintenance of total biodiversity nowadays (for example Windt, 1995; Dekker, 2002; Koppen, 2002). This development was needed since the conditions in the country changed from the presence of many wastelands and free natural processes to a reclaimed country where most processes are human controlled, moreover with a lot of environmental problems (Eggink et al., 2002; Duuren et al., 2003). The more natural areas become scarce, the more society wishes to conserve its remaining nature. It is an economic law, valid for this non-value item. Since human management controls most processes in the Netherlands at present, for example by water management and spatial arrangements, we also can improve these processes so that they will be profitable for nature again.

The first Dutch nature reserve Naardermeer was established in 1905. This area was bought by persons who wished to maintain this wetland area for its fine nature and especially for the presence of famous nesting birds such as the spoonbill. It was the starting point for the establishment of the Society for the Preservation of Nature in the Netherlands (in Dutch: Natuurmonumenten). About 25 years later also other societies for the conservation of nature and landscape were established in each province of the country. They got or bought areas where special animals or plants were present and protected this nature by not allowing any disturbance from outside. At this starting period most acquired areas were woodlands and the management incorporated forestry to obtain benefits from the available natural resources. In wetlands, harvesting of reed and fishery was common for the same reason.

In the first decades, lack of finances was limiting the possibilities to buy new reserves; the number of members of the conservation societies was still restricted. At the same time the technical facilities to reclaim wastelands increased and intensification in land use was taking place almost everywhere in the country. In the period 1900-1960 some 500,000 ha of wastelands (mostly heathlands in the eastern half of the country) was converted to arable land, more than 10 % of the land area of the Netherlands. But not only heathlands were reclaimed; also bogs, salt marshes, fens, reed beds and other types of natural ecosystems were incorporated. Through these changes, many nature reserves became nature islands surrounded by intensively managed urban or agricultural areas. Moreover, the urban expansion increased so that at present 11.5 % of the country is occupied by roads and buildings.

In the second half of the 20th century the number of nature reserves in-creased for various reasons. The national government started its own con-servation of nature; a task performed by the State Forestry Service (in Dutch: Staatsbosbeheer). At the same time the government facilitated by loan or subsidy the available finances for the conservation societies to buy nature areas. In these new reserves the former agricultural or forestry management with restricted input of nutrients and resource use was mostly continued, since biodiversity was dependent on those conditions for dec-ades and nature could be maintained by that management. However, some decades ago, the societies for conservation observed that the number of species in the preserved nature islands was decreasing, although the inter-nal management appeared to be optimal. External factors from the sur-rounding environment appeared to be having a negative impact on the conditions within the reserves. The physical fence around the reserve was unable to stop acidification, eutrophication, pollution and desiccation in the area. Negative spatial and functional relations were detected by investiga-tions from landscape ecology. These negative processes caused the site conditions within the reserve to change in a way that many characteristic species declined or disappeared.

Biodiversity

This failure to conserve the diversity in species became well known in soci-ety; the number of members of the societies for nature conservation in-creased and, as a result, the political pressure to undertake a positive ac-tion. At the end of the eighties two new developments were incorporated at the national level of conservation policy. First, the Island theory (MacArthur and Wilson, 1967) was accepted as an explanation of the declining Dutch biodiversity: the remaining areas for nature became too small and too dis-persed to be sustainable for viable populations of plants and animals. The area of the nature reserves should expand and reserves should be con-nected in an ecological network (Opdam, 1978, 1987; Verboom et al., 2001).

The second event was introduced in 1973, when a heavy storm demolished a great number of forests, so much that it was technically impossible to clear-up all fallen trees from the forests. Ecologists' argued that removing the fallen trees was not necessary because real woodland species, such as woodpeckers, insects, and fungi, need dead trunks for their reproduction. This way of thinking was accepted by the society for nature areas. It led to a new kind of nature management: doing nothing. However, later the real event was that a large new area planned for industry but subsequently not used for that, changed over 15 years into a famous area for rare bird spe-cies. This area with the name 'Oostvaardersplassen' illustrated that nature could recover without human support, an unknown phenomenon in a coun-try where man created the whole landscape. The 'Oostvaardersplassen' became the paradigm of a new conservation strategy: nature development. The aim of this strategy was to give nature room to develop in its own way by stopping any management of nature reserves. Moreover agricultural lands should be converted into new nature areas. The strategy is incorporated in a plan to restore nature in the river forelands (Bruin et al., 1987) and for the first time applied in 1992 in a famous development project along the river Rhine with the name 'Blauwe Kamer'. This area is a fine illustration of how river nature can recover in a dynamic river system within a decade (Dam, 2002).

National Ecological Network (NEN)

Those new ideas were incorporated in a national plan to take care for nature: the Netherlands Nature Policy Plan, written in 1989 and accepted by the national parliament in 1990 (LNV, 1990). Essential in this policy plan is that nature is accepted as a real value, to be incorporated in the decisions of the government. The aims are to stimulate both the presence of species and the presence of natural processes. Essential elements from the plan are: core areas for nature should be maintained; new areas with nature should be developed on former agricultural lands; ecological corridors should connect the nature areas; special plans to protect species (groups) should be developed; and environmental problems should be solved. Most important was a map indicating the already existing nature areas in the Netherlands, the areas to develop new nature and the ecological connections to be developed: the National Ecological Network (NEN, see *Fig. 5*).

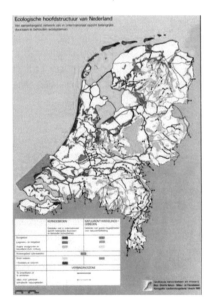
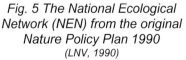
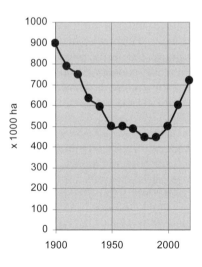

Fig. 5 The National Ecological Network (NEN) from the original Nature Policy Plan 1990
(LNV, 1990)

Fig. 6 Indication of total area with nature in the Netherlands from 1900 to 2000 and the aimed increase in area till 2020.
(data: MNP, 2005)

The map did not define everywhere the precise location of the NEN, but indicated a bigger area in which the government should realise the planned number of hectares of the NEN. A consequence of the plan including the map was that physical planning in the Netherlands should protect these nature areas (LNV, 1993).

The plan changed the way of thinking in nature conservation drastically. Instead of complaining that species and nature areas were disappearing, it facilitated an offensive strategy to develop or to restore nature. The original plan had to be finished in 2020, finally increasing the total area of terrestrial nature by 275,000 ha up to a total NEN area of 728,000 ha (*Fig. 6*) and thereby enhancing sustainable conditions for Dutch nature. In the Nature Policy Plan the Dutch government recognised the value of a more common nature outside the NEN.

Species protection
Another line in the development in nature conservation is the new legislation in species protection. EU nature policy induced a stricter protection of species. It really started with the implementation of the species annexes of the EU Birds and Habitats Directive at the end of the nineties. The species protection rules of theses directives were transposed into the national Flora and Fauna Act (in 2004), which prohibited the disturbance of these species inside and outside nature reserves. Most mammals, birds, reptiles and amphibians in the Netherlands are included in this law, as well as many plant, fish and invertebrate species. The law requires that a location with a species of this list cannot be disturbed in a direct way. The National Ecological Network is an important tool to preserve the species of the EU Birds and Habitats Directives, but some of them need protection outside the NEN as well (see chapter 8, page 119 by Pelk).

Implementation
The implementation of the National Ecological Network requires many activities, since new nature areas and corridors have to be developed. This evokes not only opposition from the present owners of the areas who wish to maintain their activities, but also from others indirectly involved, such as local communities restricted in their development by physical planning. In reality it means that mostly agricultural areas have to be bought to create nature: most new nature will be located on former pastures and arable land. Not only can nature conservation societies manage these areas, but another possibility is that private owners and farmers try to manage and to develop nature on their own land. The government supports this financially, so that rural areas managed by farmers become an element of the Network. In *Fig. 7* the areas of new nature to be realised by 2018 are reviewed.

Type of nature / management	Task 2018	Comments
New nature	112,200 ha	To be bought before 2016
Private nature management	42,800 ha	
Agricultural nature management	117,700 ha	Including 20,000 ha outside NEN
Wet nature	9,500 ha	6,500 ha realised in 2010
Robust corridors	13,500 ha	Another 13,500 ha later

Fig. 7 The areas of new nature to be realised by 2018 (after MNP, 2005)

To realise the National Ecological Network a period of 30 years (later reduced to 28 years) was scheduled. In 2005 we were half way through that period and it was therefore a perfect moment to evaluate different aspects of the Network. The Milieu- en NatuurPlanbureau, the Dutch environmental agency (MNP, 2005: 'Natuurbalans'), provides information every year about the state of art with nature, the development of the NEN, and the trends to be expected. Algemene Rekenkamer (2006), in cooperation with the MNP, recently evaluated the political and technical consequences. From the point of view of landscape ecology, the question can be asked whether the key relations that influence the conditions in the NEN are sufficiently incorporated. Or, explained the other way around, can landscape ecology contribute to the development of a sustainable Network?

3.3   Knowledge from landscape ecology

From landscape typology into dynamics
During the last decades landscape ecology in the Netherlands developed from a descriptive science of types of landscapes, to a descriptive science including ecological information (Stevers et al., 1987; Runhaar, 1989) and finally into an analysis of landscape dynamics resulting in the prediction of site conditions (for example Opdam, 1993). Specialisation with modelling of the relations in the landscape has been the main development during the last two decades, facilitating new branches such as historical landscape ecology, urban ecology, eco-hydrology and spatial evaluation with assistance of the metapopulation theory. This section will highlight some available knowledge. In the next four steps the most basic information from landscape ecology in the Netherlands will be illustrated: patterns, processes, cultural-historical aspects and spatial evaluation.

3.3.1   Patterns in the landscape

Soil patterns
The main soil types and the land elevation provide an illustration about the most basic division of the Dutch landscape pattern (*Fig. 8*).

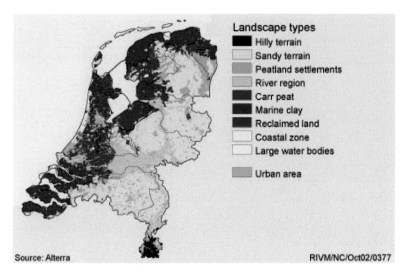

Source: Alterra   RIVM/NC/Oct02/0377

Fig. 8 Landscape types in the Netherlands
(MNP, 2006, Landscape typology)

The western half of the country is at (or even several metres below) sea level, with the exception of the sandy dunes along the coast, which have an elevation of at most some dozens of meters. In the most western parts the sea deposited sediments with clay soils; in the eastern parts of this region the rivers were the main source of the clay sediments. Long ago extensive bogs and fens were present in this region; erosion of peat layers by sea and human excavation changed this soil type and many areas were re-claimed as polder. The eastern half of the country consists of non-buffered sandy soils, with in the northern half ice-pushed hill-ridges up to 80 m above sea level. Locally, where rainwater stagnated, bogs developed. The extreme southeastern part of the country differs totally from the former types, by clay and loess soils rich in lime and with higher elevations (up to 300 m).

Physical-geographic diversity
Observing the landscape in more detail, soil characteristics indicate far more variation (*Fig. 9*). As a result of local differences in processes that occurred long ago, soil developed under wetter conditions, or with more erosion, or just without processes so that organic matter could accumulate. Differences in water tables caused by local variation in elevation, added another dimension to the local pattern. This differentiation within the region, for example with valleys and dryer parts, created the real diversity in condi-tions that is characteristic in the landscape. Human activities added ac-cents to this variation, for instance by creating pastures in the lower parts, arable land on the higher areas and leaving the highest dry parts as waste-land with heathlands and forest.

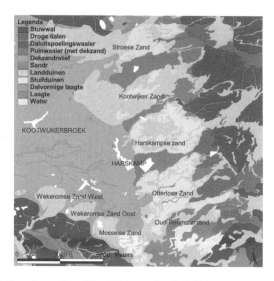

Fig. 9 Sandy soils of the Veluwe in detail (Koomen et al., 2004)

At the national level, the sea influences the climate. Parallel to the coast-line, the buffered climatic conditions from the sea disappear gradually. The eastern parts have colder winters, whereas the western parts experience more wind. However, there is also a gradient present from the warmer south to the colder north. The amount of precipitation is partly related to elevation, with the central hill ridges receiving most rain. When these global data from climate, soil and physical conditions are merged on a map at national level, the country can be subdivided in eight physical-geographic regions.

From landscape ecological research we are informed that the gradients between higher and lower parts in the landscape are important for the distribution of species (Leeuwen, 1966) (see chapter 12, page 208 by De Jong). The major gradients in the landscape are located at the transitions between the physical-geographic regions, river valleys and soil types. High diversity in species and landscape conditions are represented at these gradients, especially when nutrient poor soils are on top of the gradient (see chapter 5, page 74 by Reijnen et al.).

Species distributions

The distribution of species in the Netherlands (= biogeography) is related to the patterns in landscape conditions. The plant-districts in the Netherlands (Meijden, 1996) can directly be compared with the physical-geographic regions. The same regional subdivision has been valid for some fauna. Research is being performed to prove this (for example, Kwak and Berg, 2004); we illustrate this relation with an example from the combined distribution pattern of 325 hoverfly species (*Syrphidae*) over the country in grid cells of 5x5 km, with for each combination of species a different colour (*Fig. 10*).

Fig. 10 Combined distribution pattern of the 325 hoverfly species
(Syrphidae) (data: Reemer et al., in prep.)

Chapter 6, page 92 by Schouten will elucidate the distribution in biodiversity, in terms of hotspots of species.

Metapopulations
A different dimension in the landscape pattern is the spatial constellation of the available ecosystems, facilitating the populations in plants and animals to survive. The Island theory (McArthur and Wilson, 1967) explains that large areas contain populations with relatively many specimens, and that as a result these populations of species are more viable. From research with bird species (Opdam, 1987) we learn that populations cannot occupy relatively small habitats in an optimal way. In the fragmentised habitats in the Dutch cultural landscape the connection and/or spatial arrangement of the available patches of habitat is important since absence of connectivity will result in isolation of the subpopulations. The populations respond to this fragmented landscape pattern according to the metapopulation theory (Levins, 1969; Hanski, 1999; Vos et al., 2001; Opdam et al., 2002, 2003). This theory explains that patches can be temporarily occupied from sta-

45

tionary populations at core areas. The area of the isolated patch and the distance to the core area are variables that determine the effectiveness of the occupation of the patch by a species and by that the contribution to the total population. Recent research (for example, Opdam et al., 2003) concentrates on the connection between existing and new nature areas, to stimulate the relations in nature, and to the consequences of climate change (Opdam et al., 2004).

3.3.2   Processes in the landscape

Changing site conditions
The site conditions influence the presence or absence of species. The causal relation can be a direct physiological aspect (for example limiting pH-value), a semi-direct relation (for example competition with other species at an optimal pH-value) or an indirect relation (for example animal species depending on the presence of a special plant species or vegetation structure). Processes at the scale of the landscape influence most of the site conditions, such as sedimentation and erosion, settlement and succession of species and ecosystems, and human activities (mostly interrelated, for example, Belle et al., 2006). These processes are dependent on transport of (a)biotic material by air, water or animals. The transport by animals on the landscape scale is very limited; migrating mammals and birds might spread around manure and seeds, but in biomass this transport is restricted.

Transport by air
Transport by air is mostly diffuse and might be for long distances. The national production of ammonia is transported some kilometers by air, but might also be deposited up to Scandinavia. The prevailing wind direction determines the transport; many references from acid rain and air pollution are available (for example, RIVM, 2000), including the ecological consequences (for example, Bobbink et al., 1998). In other cases there is a short distance relation, for example the salt spray close to the sea determines the absence of trees in the outer dunes.

Transport by water
The major force for transporting sediments is water. Glaciers transport stones, rivers transport stones, sand and many diluted elements or clay particles and the seas transport sand and some clay. Sedimentation and erosion processes with water from rivers and the sea are the most important in the physical and chemical moulding of the landscape (Zonneveld, 1974; WNF, 1993). The absence of transport of sediments by groundwater seems to indicate an unimportant factor, however, the soluble minerals in seeping water are very important in the areas where this groundwater discharges.

Water conditions

So, water processes appear to be important for physical transportation, but induces the most important conditions in many major ecosystems too. The processes and spatial relations in river valleys and fens are investigated many times and they are the object of a special branch in landscape ecology: the ecohydrology (for example, Grootjans et al. 1985, 1988, 1996; Kemmers et al., 2003; Wirdum, 1991; Wassen et al., 1990, 1992, 1996; see chapter 10, page 169 by Wassen). In the higher elevated areas without buffer capacity nutrient-poor rainwater dominates the conditions in the ecosystem; this water infiltrates to deeper aquifers. On the lowest elevated areas surface water from the river dominates, and since the river is transporting water rich in calcium with some additional natural waste (nutrients and organic matter), these locations are buffered and rich in nutrients. Just in the middle of the gradient where groundwater discharges, the conditions are both buffered and nutrient-poor, caused by the calcium and iron ions from the discharging anaerobic groundwater, without nutrients. Since the groundwater flow is a process on the scale of the regional landscape, the patterns originate a process at that scale. Incorporation of these relations in conservation and policy will result in sustainable conditions for nature (for example, Nieuwenhuis et al., 1991, Barendregt et al., 1995). A next step in interdisciplinary research is to incorporate economic research and planning in society, to develop options for water management (for example, Barendregt et al., 1992; Turner et al., 2000; Bergh et al., 2001).

Small-scale variations

In valleys of brooklets, rivers and fen areas the pattern in hydrological processes determines the ecological development of the vegetation on a local level. Acidy and nutrient availability are the two most important controlling site factors. The water table itself controls for example, the accumulation and mineralisation of the organic material in the soil. The water tables control aeration and redox values and, together with the acidity, thereby define processes in many cycles in nitrogen, sulphur and phosphorus (Wassen and Verhoeven, 2003). The site factors controlling the conditions in the ecosystem appear to be complex and include local water table, acidity, organic matter and water and soil chemistry (Kemmers et al., 2002; Delft et al., 2002). Various processes in the landscape impact this complex directly.

Seasons

Superimposed on this spatial complexity in conditions is the variation in time by the seasons, which creates different conditions and also can induce a pattern in the landscape. For instance, research for the presence of special rare insect species on special spots in a large bog, learnt that heterogeneity in groundwater flow, nutrient poor conditions, depth of the water and dynamics from the water tables during the seasons influenced the distribution of species (Verberk et al., 2001, 2002; Duinen, 2002).

3.3.3   Cultural-historical aspects in the landscape

Land use and management

In previous sections we learnt that the landscape has characteristic patterns and that the processes at the landscape level incorporate many variables. Land use and management such as reclamation change those patterns and processes. Wassen et al. (1996), and Bootsma and Wassen (1996) compared the conditions in the reference of an undisturbed natural fen area in Poland and the conditions of a partly reclaimed wetland region in the Netherlands. The original processes and the species responding on these processes are exactly the same in both areas. However, in the Netherlands as an example of a country with many land use activities, the characteristic processes are interrupted and fragmented. Large-scale regional processes are subdivided in sub-regional processes and the locations where characteristic conditions are present are fragmented. Land use is changing the original conditions in nature and establishing a man-made landscape where other spatial conditions favour another set of species.

People influenced the natural conditions during many ages and as a result created by this a man-made landscape, with many different types of landscape, characterised individually by visual and spatial arrangement of elements (*Fig. 11*).

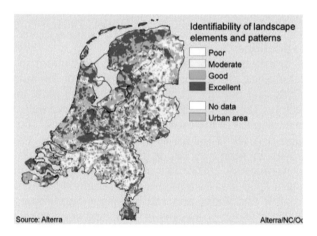

Fig. 11 Identifiability of landscape elements and patterns
(MNP, 2006, Identifiability of landscape elements and patterns)

This man-made landscape has a cultural value since we can observe how man created this landscape (see chapter 7, page 104 by Renes). Some of those landscapes are highly appreciated in an international context for their scenery or history (*Fig. 12*).

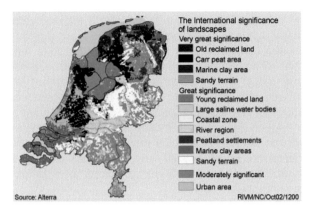

Fig. 12 The international significance of Dutch landscapes
(MNP, 2006, The international significance of landscapes)

By using the landscape people introduced, stimulated or removed species. Weeda et al. (1990) calculated that half of the 541 species on the Dutch Red List of plants lives in vegetation types actively in use by man, comparable with the conditions as they were a century ago. Conservation of these endangered species requires the continuation of this management. By preservation of these landscapes, including its management, a set of endangered species can be preserved. So, reserving ancient landscapes offer possibilities to maintain two different aspects at the same time: culture-history and ecology. National landscapes can offer challenges in this area.

An application of history in the landscape ecological relations and in explanation of conditions in the landscape is given by research on the impact of human alterations in the hydrology of the eastern sandy parts of the country. It appeared that farmers could increase the nutrient levels and decrease the acidity on their fields by flooding their lands with surface water from rivulets (Baaijens et al., 2000, 2004). The use of these natural processes resulted in better agricultural conditions and higher yields. Various by-passes in the rivulets made by human interventions caused an organised flooding. This system disappeared after the introduction of fertilisers a century ago. However, this practice may still have value for modern landscape management.

Historical priority
Old stationary ecosystems (old landscapes) have become rare since modern management practices interrupt the conditions in the landscape. Land re-allotment plans changed the landscape at most locations in the Netherlands and created younger ecosystems. In theory, old and not-disturbed ecosystems should be higher in species numbers than young and disturbed ones (Shear McCann, 2000).

So, the age of the landscape might be important for biodiversity. The impact of landscape age was tested with on the one hand a grid map with the time the landscape was installed (*Fig. 13*) and on the other hand a dataset with the number of hoverflies per grid cell (Cormont et al., 2004). The results show that really old agricultural landscapes, formed in the Middle Ages, are not very rich in hoverfly species; the landscapes created from wastelands 100-150 years ago appeared to be the ones with the highest mean hoverfly species number. Landscapes from the Middle Ages were probably never the highest in species number due to their conditions; the recent landscapes are used very efficiently and lack small elements such as hedgerows and ponds that facilitate conditions for a range of species; here the species number is limited.

Fig. 13 The age of the landscape (Cormont et al., 2004)

3.3.4   Spatial evaluation

Monitoring and modelling
During the last decades landscape ecology developed to an applied science of ecological relations, in a multifunctional landscape (Brandt and Vejre, 2004). The internal and external relations in a landscape are frequently modelled with the application of Geographic Information Systems, visualising the spatial consequences of land use change. Present day techniques allow for stratifications that can be applied over large areas (Jongman et al., 2006; Groot et al., 2007). Landscape ecology produces the practical translation of human activities to impacts on biodiversity, by way of modelling of groundwater flows (Schot and Molenaar, 1992) and, nutrient flows (Verhoeven et al., 1996) or on soil and chemical composition (Kemmers et al., 2002). Landscape ecological principles support the as-

sessment of the impacts of human interventions and for that reason they are widely applied by consultancy companies, governmental research institutes and universities.

3.4 The development of the National Ecological Network

When landscape ecology can describe the patterns and processes, it explains the ruling conditions for the presence of plants and animals. This is the reason why it is a good starting point for the development of the NEN. In this section we will discuss the question how principal ecological relations were incorporated in the basic structure of the NEN, how the first drafts of the maps were discussed, and how during the implementation of the NEN ideas in society changed.

3.4.1 The principal components

Island Biogeography
Till 1990 the nature conservation societies tried to increase the area by buying land on the free market. However, it was only possible on a voluntary base and in many cases the competition with agricultural and urban forces was prominent. Moreover, most new reserves were selected for the presence of special species and not for their strategic landscape ecological location, offering defined relations. There was no planned development of an ecological network but an increasing area of reserves on occasionally available wastelands.

During the second half of the eighties ecologists were convinced that the theory of Island Biogeography (MacArthur and Wilson, 1967) was very important to preserve the correct conditions for nature in our urbanised country. The first Dutch paper about this theory and consequences in application (fragmentation) is by Opdam (1978); the first broad distribution of these ideas is by Brussaard and Weijden (1980). The problem of fragmentation was incorporated in a national governmental report (CRM and VRO, 1981) and here, for the first time, an Ecological Infrastructure was defined to solve this problem (Klijn, 1983; Brenninkmeijer and Dekker, 2001). The WLO organised a symposium about the ecological infrastructure and a special issue with presentations (Klundert and Saris, 1983) became a famous summary of problems and solutions. This knowledge was incorporated in the NEN to stimulate the spatial relations in nature (personal comm. Van Opstal). The new aim was to expand the area of the reserves and to develop a real ecological network.

The incorporated key-factors
When the ministry responsible for nature developed a NEN in 1988, they defined the main conditions top-down. The philosophy to solve the question was derived from landscape ecological information (personal communica-

tion Van Opstal). The starting point was the definition of the three basic types of processes and the ecosystems that depend on them:

1. ecosystems depending on rain water
2. ecosystems depending on groundwater
3. ecosystems depending on surface water.

At this phase the knowledge of the main landscape ecological processes facilitated the description of the main variety in nature. The three groups of ecosystems were related to the main ecological districts in the Netherlands: the rain water dependent ecosystems in the sandy recharge areas in the eastern parts of the country, the groundwater dependent ecosystems at the slopes of these sandy areas where gradients with discharge of groundwater were prominent, and the surface water depended ecosystems in the lower parts in the western half of the country where river ecosystems and fens were present in conditions of clay sedimentation and peat accumulation.

Types of spatial relation
To enhance ecological connectivity between areas, three types of spatial relations were defined on a national scale (*Fig. 14*).

First, all coastal areas should be connected so that the dunes and the tidal marshes from southwest to northeast in the country would develop to one connected area with facilities for migrating species.

Second, most gradients with discharging groundwater were represented in a zone bordering the main sandy regions. In that way a connection should be developed from southwest to the north including the fens and seepage areas. At the same time, some main areas for the infiltration of rainwater were indicated as buffer zones to enhance the relation between these areas and the discharge zones.

Third, the main rivers and rivulets were indicated to obtain an ecological connection between the different systems with fresh surface water, to facilitate migration of species. In the western half of the country the main peat systems depending on surface water were indicated.

As well as these three relations, the areas with high concentration of nesting waders on pasturelands were indicated.

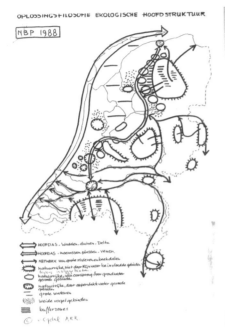

Fig. 14 Three types of spatial relations on a national scale (1988: main ax 1 – Wadden, dunes, Delta; main ax 2 - marshes, lakes, fens; small arrows - network of rivers and brooklet valleys
(comm. Van Opstal)

Fig. 15 The first (handmade) draft map of the NEN (in Dutch: Ekologische Hoofdstruktuur)
(comm. Van Opstal)

A map of nature areas and corridors

This philosophy of landscape ecology was worked out on a map with the nature areas present in 1988. The core areas of nature conservation were indicated with their relations. At locations where the core areas were too small, nature development areas were designated. Where core areas were isolated from other areas, corridors were designed. The result was the first handmade map with the NEN for the Netherlands (*Fig. 15*). This map includes buffer zones, which were removed later in the development. One assumption in 1988 was that the acid deposition in the Netherlands would be on an ecologically acceptable level in the year 2000, so that the NEN would be a sustainable investment for nature.

In conclusion, a first blue-print of the NEN was constructed on principles of landscape ecology, without intensive investigations on biodiversity and its distribution. Klijn and Harms (1990) wrote a critical review to which by Zadelhoff and Zande (1991) replied. At the same time, many experts from

53

nature conservation were consulted, so that the indication of important nature areas was also based on expert judgement. At the same time, a definition of key ecological relations with their areas on a national scale was presented for the first time. Philosophy was replaced by a blue-print.

3.4.2   The discussion about the first map and how to realise the NEN

Restrictions

This map with ecologically the best solution for the NEN was discussed with governments at the provincial level. A set of restrictions came from these authorities since the development of new nature areas and corridors would impact the vested interests of the inhabitants of those areas, and the future development given by the physical planning as well. One very practi- cal application of these consultancies was that the new nature areas and corridors were indicated on the map as search-areas in which the real loca- tions had to be found later. Since the number of hectares to be realised was indicated, the implementation could be reached later.

The Nature Policy Plan 1990

The NEN was incorporated in the first Nature Policy Plan (LNV, 1990). In this Plan the starting-points, the priorities and the implementation of the policy were elucidated. This Plan was accepted by the parliament, including the investment of a great amount of money for the next 30 years. More- over, the national physical planning should incorporate the NEN. The Struc- ture plan for the green (rural) areas in the Netherlands (LNV, 1993) en- compassed this incorporation.

It was the first time that a real structure was given to the protection and development of nature in the Netherlands. There was one crucial step to be taken, since the values of nature were not defined. The Plan incorporated the areas but did not inform about the ecological aims of each area on the map. The implementation needed a correct definition of the area goals. The protection of endangered species was well defined at a national level, so it was clear for which species action should be undertaken. The real problem was that not all species could be preserved in the same area; differentiation for each ecosystem was needed, including the supporting processes.

Target types

It took a long period to develop a valid strategy. The solution was offered by the definition of nature target types, in which the presence of target spe- cies, characteristic site conditions and management regime were merged (LNV, 1993). The description of 136 target types for nature in the Nether- lands (Bal et al., 1995) was based on the eight physical-geographic areas. The target type systems (improved in Bal et al., 2001, with 92 target types) appeared to be a way to communicate on nature management, both in policy and in practical management. The target type system became a standard, although it was at a high abstraction level, to define which nature had to be protected, restored or developed at each site of the NEN.

3.4.3   Changes in society over 17 years

Acceptance
During the first years the NEN appeared to be accepted in society and in physical planning; the first areas had been developed according to the policy. Nature conservation societies applied the concept in many regions, to remove the fragmentation of small nature areas, to stimulate landscape ecological processes and to enlarge the core areas. The gradual change in agriculture, where the less productive areas became available, offered new opportunities.
As well as the establishment of new nature reserves, another strategy to realise the NEN was to apply agri-environmental schemes on agricultural lands with natural values, so that nesting waders or special types of vegetation could be conserved. The financial support by the government ("Relatienota"), (L&V, 1975) to reach this goal was already legal, but for economic reasons many farmers were not interested in these contracts for six years. This attitude changed drastically in the mid-nineties when many farmers could no longer expand their farms because of new environmental legislation (for example from the EU). About 55,000 ha of agricultural land changed from 1990 to 1995 from intensive management to nature management on the farm (RIVM/CBS/DLO, 2003).
The 12 provinces were due to implement 224 national corridors and 43 international corridors between the nature areas. The designated number and length of the corridors appeared to be more than indicated by the NEN, however the efficiency and the possibility of target species to migrate appeared to be restricted (RIVM/DLO, 1998). From a test with a selection of areas only 25% of the corridors were classified as effective.

Nature Policy Plan 2000
After 8 years each policy plan was due to be updated and this was performed with the Nature Policy Plan 2 (LNV, 2000). The name of the plan indicated already a change in perception: "Nature for people, people for nature". The NEN was still valid but the implementation should get different highlights. First of all, as the name indicates, leisure (and other kinds of natural resource use) in nature areas became more popular so it was expected that there would be greater support from society to develop the NEN. Most nature reserves are nowadays opened for visitors. The argument is that society paid for nature and that the same society is consequently allowed to use and enjoy these areas.
Farmers who wish to maintain their lands, but are willing to adapt their management to produce nature and to take care of the landscape were able to give another support. As a consequence the financial support in nature management changed. Not only could the nature conservation so-

cieties be supported financially but also farmers and private owners of nature areas, when they produce the desired target types. Through this change the government encouraged farmers and private owners to realise parts of the NEN, and conversely, farmers with a nature oriented farming style or limited possibilities to expand their economical activities were supplied with finances.

Five emphases
In the 2000-plan five emphases (perspectives) are given. Realisation of the NEN, and stimulation of wetland nature, agricultural nature, urban nature and internationally important nature are mentioned. Since the development of the corridors appeared to be difficult in practice, the government concluded that 12 major (robust) corridors should be developed with priority; extra finances became available for these activities. They aimed for broad connections between core areas, so that for example mammals like the red deer could migrate between these areas. As a consequence, local governments no longer supported the establishment of small corridors.

National Parks
For many decades only two National Parks were present in the Netherlands. In the 2000-plan the task is formulated to realise 20 new National Parks all over the country. A National Park is defined as a region with an area of thousands of ha where nature, leisure, landscape and education are the major functions in land use. New activities in the Park should support these functions. In practice those National Parks are regions with a landscape that is appreciated by many persons, for its scenery, its nature and its history; a number of possibilities are stimulated for leisure (footpaths, cycling paths, boating, etc.). The National Parks are a real possibility to maintain a landscape including its values for history and ecology and at the same time a trial to incorporate the inhabitants of those areas in its developments.

Decentralisation
Another political development was that the national government delegated the responsibility to realise the NEN in 2018 towards the 12 provinces, including the finances needed for the purpose. This was the result of replacing the responsibilities for the physical planning to a lower level ("Nota Ruimte", VROM, 2005). Most activities are no longer steered by the national government, but by the provincial ones. It can be explained in a positive way since people closer to the realisation of the new areas and corridors are involved; on the other hand, local economic considerations might result in other priorities such as the development of built-up areas or agricultural intensification.

Water management
Desiccation is a serious threat to many NEN areas. To tackle this problem each province has to define the aimed ground water level taking into account the designated nature target types of the NEN areas. Besides, the

EU Water Framework Directive needs to be implemented, so that on a general level water quality should be correct, but in special (nature) areas it should be perfect. At many locations an improvement can be expected.

Species protection
For a long time the Dutch nature protection focused on safeguarding areas. Different developments made it necessary to pay more attention to species protection. On the one hand a strong movement for animal protection and on the other hand the europeanisation of nature protection. The EU Birds and Habitats Directives forced member states to protect all wild birds and a selection of important other species. The Dutch government combined the new attitude towards animals, a number of existing species oriented laws and the species oriented rules of the EU directives into the new Flora and Fauna Act (2004). The management of the NEN areas should become more species oriented and outside the NEN some of the species should be protected as well.

Natura 2000
The other task given by the Habitats Directive is the installation of the Natura 2000 Network. The NEN appeared to be the answer; the Natura 2000 areas are mostly (nearly 100%) included in the NEN and they cover 40% of the NEN. It can be discussed whether this decision is favourable for the species and ecosystems of the Directive. The same development holds true for the Pan-European Ecological Network of the Council of Europe (see chapter 9, page 141 by Jongman and Veen).

Living Countryside
The most recent development is the policy to combine agricultural recon-struction, nature development, leisure and National Landscapes into one integrated policy for the rural areas, spatially differentiated per region (LNV, 2006). In this policy plan a map is incorporated (see *Fig. 20*) with the de-fined aims for 34 types of nature (ecosystems), stating the regional aimed differentiation in nature development.

3.4.4   The state of the art

Recently an independent audit institution of the government reported in cooperation with the Dutch environmental agency (in Dutch: MNP) with an audit about the realisation of the NEN (Algemene Rekenkamer, 2006). Their first conclusion is that the NEN is a good instrument to protect biodi-versity and at the same time to fulfil the tasks given by the Rio-Conference, the EU Birds and Habitats Directives and the Pan-European Network. However, their next conclusion is that the aims of the NEN are too general, so that it is not clear what should be developed and as a result of this, it is impossible to test whether the NEN is effective. Moreover, there was not a baseline (zero) description present of the conditions in 1990 so that it will never become possible to test the positive effects.

Actual state
Other conclusions concern the 275,000 ha of new nature areas to be real-ised in 2018. Halfway (2005) only 38% is effective nature. Especially new core areas and the new areas by private owners are less developed in numbers. Remarkable is that a quarter of the new nature areas are outside the borders of the officially designated NEN. The environmental conditions improved in the last 15 years, but not enough to realise most aims; espe-cially nitrogen deposition, surface water quality and desiccation are still problems. Another task for the future is a better physical planning, since this is not organised in a proper way.

The report of Algemene Rekenkamer ends up in a set of suggestions for improvement. It is discussed that national coordination should be improved and the goals of the Nature Policy Plan need to be defined in detail. The voluntary base of the realisation is discussed too; especially the contribu-tion from private owners is too restricted. When they contribute, they select a management option on a low ecological level, since this is easy to reach. Another problem from their analysis is that for ecological reasons the conti-nuity in agricultural management for nature needs a longer period than the six years-contracts.

3.5 Discussion

Back to science after a success story
The NEN appears to be a perfect medium to communicate the conserva-tion and development of nature with society. When ecological specialists are asked about this, they reply without doubt in a positive way. The NEN started in 1989 the process to take care of nature in an organised way, and facilitated the change to re-establish or restore nature in the long term. However, this long term is until 2018, when the NEN should be completed, and that allows important decisions to be postponed. At the same time the NEN became such an institute with vested interests, that nobody dares to discuss its fundamental principles. In this section we will discuss some aspects about the practical realisation and the ecological effectiveness of the NEN and conclude with a general discussion about the NEN.

3.5.1 Questions about the practical realisation of the NEN

From focus on localisation into content
The starting point of the NEN was not a conflict in society, since already existing reserves (also designated in physical planning) were incorporated in a structure that might offer improved ecological connectivity in future. The new areas and corridors had to be installed in the next 30 years within the matrix of the existing physical planning, and the changes and problems were to be faced in the future. As the most important contribution to these

new areas, agriculture had to disappear from many locations. This process needed time and one solution was to indicate searching areas, where a defined percentage of land had to change into nature. Political solutions would be offered later. This change in land use evoked a lot of resistance from farmers and their representatives. It stimulated discussion about the localisation of the NEN, far more than about its ecological contents.

The task to indicate the exact position of the areas of the NEN had to be performed by the provinces. One observation is that the cooperation between the provinces could have been better. Many provinces developed the NEN within their territory without taking areas and corridors from other provinces as a boundary condition. Here the absence of national coordination is clear. The same problem is valid in the international context: the protection of Natura 2000 species in three international nature reserves on the Dutch frontiers needs more coordination (MNP, 2005: 'Natuurbalans').

Nature outside the NEN
Nature can also be present outside the NEN, for instance in agricultural areas, in towns and villages (Leeuwen, 1995). The hedgerows in sandy areas, or the 360,000 km ditches in Dutch pastures are the habitat of many populations of plants and animals. However, most nature policies and their resources are focussed on the NEN. From that perspective the NEN is a risk for nature outside its borders, including the represented rare and threatened species and habitats.

Farmers involvement
The possibility to incorporate farmers in nature conservation with a specialised agricultural management, or in the activities to maintain the landscape, is one of the strategies to realise the NEN. The voluntary base of this option gives a restriction, another one is the contract with the farmer being valid for six years. Both parties are free to continue or not. The legal incorporation of this non-fixed area in the NEN in the physical planning is impossible. The sustainability of the NEN is thrown into doubt by this policy, since the incorporation of farmers' contributions can be guaranteed only for six years. Till now, their participation has increased. Another aspect of this problem is that individual farmers wish to maintain their type of management, although all surrounding plots might be incorporated in a more heavy kind of nature management, for example with higher groundwater tables. Those farmers have the legal right to continue their choice of management, giving less support to the aims for a stronger nature management.

3.5.2   Questions about the ecological effectiveness

The paradox of goal-oriented national nature
Originally the NEN goals were generally defined as 'nature', perhaps as the location for endangered species, but not in terms of ecosystems (LNV, 2006). This vagueness was solved by the definition of nature target types per physical-geographic region, but a precise aim per site was not defined.

That is a correct way of thinking, since characteristic for nature is that species are free to select their area. Nature is definitely not a garden or zoo. The paradox of this approach is that it remains difficult to evaluate the effectiveness of the NEN on a local level; only at a national scale can be observed whether biodiversity is maintained on a sustainable level, or not. To improve the effectiveness of the NEN precise aims should be allocated to each site. But that is contrary to the very nature of nature.

The aimed a-biotic process
The conditional a-biotic processes, are not defined in a precise way either. It could be expected that precisely here defined landscape-ecological processes could serve as a basic principle in the development or restoration of nature areas. A simple check in the ecosystems in the Netherlands shows that nowadays the river ecosystems including the forelands are managed according to a landscape ecological principle (flooding as a process), within the limits of flood-protection. But the system of brooklets, for example, in the eastern half of the country are not managed in such a way that the infiltration areas are related to the seepage zones of the middle brook valley. The full dynamics of the dune areas with erosion at the seashore and sedimentation by blown-out dunes is never fully applied for reasons of sea defence. It appears that vested interests of owners of areas elsewhere are ruling the desired conditions for nature. As society, we replaced the free natural conditions of the pre-human landscape by a human-organised landscape including the man-controlled environment. Landscape ecology might learn about the perfect conditions for the desired nature but the installation of these aimed processes is not really possible without influencing the surrounding area and its economy. Restoration of nature becomes in this way a real decision for the future of society.

Minor effects of agricultural management
Recently some publications discussed the ecological effects of agricultural management in the NEN (Kleijn et al., 2004). One of their conclusions was that no difference in ecological profits could be proved between areas with restricted and with normal agricultural management. At least this is true for some nesting birds in a six year time period; however, some insect species were stimulated by not mowing the ditch banks. The vegetation was not stimulated to a higher diversity within six years, but other investigations proved that positive development needs contracts with a 25 years period (Wymenga et al., 1996). International research in the effectiveness of agricultural nature management showed that indeed there are positive effects for common species, however, rare and characteristic species cannot be stimulated by this type of management (Kleijn et al., 2006)

Unclear impacts of corridors
The corridors of the NEN should connect ecosystems so that ultimately species can migrate within the landscape to make populations more viable. This assumption can be proved for some species (Opdam, 1987; Mabelis, 1990); however this is not proven for whole ecosystems. The question is if

corridors not only favour migrating species but also the ecosystems or nature target types they connect. New methods for planning connectivity have
been developed by Langevelde et al. (2002).

Doubts about the right position
The input of bio-geographical data in the development of the NEN could be
optimised. Which areas are hotspots of biodiversity and which areas are
inhabited by species to be protected at the national level? The question
whether the NEN is on the right position to protect biodiversity has not yet
been completely answered by research (see Schouten et al., chapter 6,
page 92). The same holds for the question whether essential ecological
processes are present to support the sustainability of the NEN.

3.5.3   Evaluation of the NEN with input from landscape ecology

Science passed by practice
The development of the Dutch NEN is grounded on theories from landscape ecology. The patterns and the processes are formulated on a general level and the first definition of the areas to be incorporated appeared to
be in line with this way of thinking. Although the NEN was postulated without extensive research, we are convinced that expert judgement of many
specialists incorporated the main relations and priorities for areas. However, the implementation of the NEN in society requires compromises. Local people, needed to carry a sustainable network but not always experts
on landscape ecological processes perform this task. They are often concentrated on the realisation of hectares and not on the best ecological location of the NEN. National coordination and concrete aims should be sup-
portive. On the other hand, optimal solutions can usually best be found at
the local level (MNP, 2005: 'Natuurbalans').

Room for ecological processes
Well-defined processes from landscape ecology, such as rainwater infiltration and seepage, are often not incorporated. In the eastern parts of the
country many valleys with reserves along rivulets suffer from this planning
mistake. The eolian process with erosion and deposition of sand in dune
areas is an essential landscape ecological process, but it can be installed
only in a very limited number of locations. It seems that the river ecosystems are the only ecosystem where the dynamic processes can be implemented. A great number of new nature areas are installed in river forelands
and erosion, sedimentation and flooding are allowed in a system that has
always been rich in nutrients. This development is a success story. But in
ecosystems, which should be poor in nutrients, the results are limited and
investments will be needed for many decades (for example Verhagen et al.,
2003).

Large connections lag behind

The other input from landscape ecology, the importance of connectivity and the metapopulation theory, is widely accepted but hardly practiced. At many locations corridors to cross motorways have been installed, but the national corridors have not yet been realised. The actual migration of species has still to start and very isolated and rare populations (for example, Wynhoff, 1998) are not yet supported, since they need a corridor with that special rare ecosystem and not an ordinary forest.

Environmental problems

Ecological and landscape ecological research shows that many environmental problems are still present in nature reserves. In recent decades the general condition in the Netherlands has clearly improved (MNP, 2005: 'Milieubalans'); however, many endangered species and ecosystems need lower levels of environmental disturbance. The nitrogen input is still too high in systems which cannot survive with any input; acid rain is still present in systems with a very restricted buffer capacity; the discharge of buffered groundwater in wetlands is too restricted to restore the balances; desiccation is still a serious problem in many nature reserves (MNP, 2005, 'Natuurbalans').

Nature outside the NEN

Important nature is also present outside the borders of the NEN. It is never tested whether all important values of nature (species and ecosystems) are covered by the NEN, and first indications (for example from Schouten, chapter 6, page 92) inform that areas in the north of the country might be more important to be protected for a number of species. The presence of important populations outside the NEN is most prominent with the Black-tailed godwit, where the international pressure to maintain this population is very high. This bird is nesting in wet parts of the country in many pasturelands with regular agricultural management, whereas the bird-population as a whole has decreased over the last few decades.

Half-way

The 'bottle with nature' in the Netherlands is at the same time half-full and half-empty. We are half-way through the realisation of the NEN and still many problems have to be solved. The general tendency is that during the last decade the decrease in species is slowing down (MNP, 2005, 'Natuurbalans'). Very positive results can be reported. At the same time rare and characteristic ecosystems are still losing species (Algemene Rekenkamer, 2006) and sustainable conditions of landscape ecological processes are insufficiently incorporated. We need another twelve years to reach the final goal.

Fig. 16 Dutch polder with pasture landscape (Photograph A. Barendregt)

3.6   Evaluation

The right concept
Many conclusions can be drawn from this analysis of a complex process; only the most important in the perception of the authors will be enumerated. The primary conclusion is that the NEN, presented in the Nature Policy Plan, created a strong strategy to maintain the presence of biodiversity in the Netherlands. The people involved in the development of the concept of this strategy can be proud that 17 years later this is the case.

Technical problems of actual implementation
However, the realisation of the strategy is a real problem. There is some discussion about what should be performed, but practical problems are the actual dispute. The impact of the NEN in practice appeared to be heavy. Nobody is against nature, however, it should not be in my backyard. The persons who wrote in 1988 that the NEN should be sustainable in 2020, if the acid deposition would be solved before 2000, made a wonderful restriction. We did not solve the main environmental problems in 2007, so the sustainability of the NEN remains in discussion.

Ecological profits not discussed
The social and political discussion up to now about the practical realisation of the NEN seems to be more important than the scientific discussion about the ecological possibilities. It remains striking that a real scientific dispute

about the possible ecological profits has never taken place. This might be so for two reasons: either the concept is perfect, or the discussion about the concept is overruled by the difficulties in the implementation of the NEN. We opt for the second reason; a number of uncertainties about this aspect are enumerated in this chapter.

First of all, the concept of the NEN was not defined in terms of ecosystems or in real conditions, only LNV (2006) defines this in the first time on a national level. This makes it difficult to test the effectiveness of the NEN (Algemene Rekenkamer, 2006). It was already difficult since the conditions in biodiversity in 1990 were not documented as a reference. Second, there was never a real test whether the NEN could serve as a solution to the enumerated problems. Third, the environmental problems are not yet solved at present (for example, acidification and desiccation). Fourth, it has never been tested whether the NEN is located at the locations with optimal site conditions and processes or whether the NEN includes the main hot-spots of characteristic biodiversity (see: chapter 6, page 92).

Natura 2000

The implementation of the Natura 2000 areas did not solve the problems, but it can at least be stated that in 40% of the NEN the Birds and Habitats Directives apply with a strong protection regime. However, in Natura 2000 areas with cultural-historical values are not incorporated. EU nature policy concentrates on populations of selected species and on habitats and not on landscape ecological relations. As Beunen (2006) states "because of the emphasis that is placed on the procedural aspects of decision making, the costs involved have increased, while the substantial goals of the European Birds and Habitats Directives are fading into the background".

The task of landscape ecology

Finally, what about landscape ecology in the NEN strategy? It offered a perfect framework to develop a main blueprint to create a sustainable structure in the ecological relations within the Dutch landscape. However, important landscape ecological relations from the blueprint are not worked out well. The required sustainable conditions to conserve biodiversity are not yet met. There are positive exceptions and the possibility remains to improve the NEN in future.

4 THE NATIONAL ECOLOGICAL NETWORK IN PRACTICE
Experience in land management

Marco Glastra

4.1   Introduction

In the Netherlands, two land management bodies operate at national level, *Staatsbosbeheer* (the State Forestry Service) and *Natuurmonumenten* (the Society for the Preservation of Nature in the Netherlands). In each province there is also a *Provinciaal Landschap* (provincial land management body), and these bodies collaborate under the name *De Landschappen*. This article is written from the perspective of one of these provincial land management bodies, *Stichting Het Utrechts Landschap* (the Utrecht Landscape Foundation). Founded in 1927, this organisation now owns almost 4700 hectares of nature reserve. Land management bodies are crucial to the realisation of the National Ecological Network (NEN). It is they that play a key role in the acquisition of land and the planning and management of nature areas.

NEN: the launch
The launch of the NEN has had far-reaching consequences for the work of land management bodies. Until 1990 the strategy was defensive and the emphasis lay on acquisition and management of existing nature reserves. The objective was to prevent further decline in the natural environment. Campaigns were launched to rescue nature reserves such as Deelerwoud and Geuldal from demise by purchasing them. These efforts were successful, but were at best only able to slow the downward trend.

The advent of the NEN was a turning point for conservation in the Netherlands. It shifted from its defensive position and sought a way forward. Doom-mongering about the decline of nature gave way to visions of black storks, otters and large grazing animals. The prospect of a single interconnected network of nature reserves was outlined. This suddenly provided an offensive agenda for conservation. The concept fits perfectly within the Dutch tradition of delta projects and appeals to the imaginations of policy makers and politicians.

NEN: the concept
The NEN is an interconnected network of nature reserves and consists of 'core areas', 'nature development areas' and 'connection zones'. Key areas of the NEN are the coast, the big rivers and the so-called *natte as* or 'wet axis' – the transition zone between Pleistocene and Holocene deposits in which the larger peat marshes lie.

According to the NEN concept, the core areas were to grow into safe havens for biodiversity, where necessary through the acquisition of directly adjacent agricultural land. In the province of Utrecht the largest of these areas is the Utrechtse Heuvelrug (Utrecht Ridge), at around 20,000 hectares. However, there were a number of missing links in the NEN in crucial places, and here extensive nature development areas were planned, for example in De Venen and Midden-Groningen. These development areas are situated where there is major potential for new nature, and they usually serve to link core areas. In Utrecht they comprise the washland of the rivers Nederijn and Lek, and marshes in the western peat marsh area, which are to be newly developed as part of the wet axis.

The NEN plan was then ultimately to link all existing and new core areas through the creation of ecological connection zones. In the province of Utrecht these run along the streams in the Gelderse Vallei (Gelderland Valley). In total, the NEN provides for an increase in the area of nature reserves of 275,000 hectare – in Utrecht 10,000 hectares. In addition, policy is directed at the conservation or restoration of natural value in areas of farmland that retain their agricultural function.

Creating the vision
Shortly after its launch, the land management bodies produced a stream of proposals for the future of sections of the NEN. Some examples are *Plan Ooievaar* (black stork project) for the big rivers, *Plan Goudplevier* (golden plover project) for the heathland of Drente, and the *Hunzevisie* (Hunze valley project) for a valley in East Drente. Apart from plans for particular areas, there was also attention for a more natural way of managing existing nature reserves, in which grazing animals play a role. Cattle and horses appeared in many reserves, and here and there also beavers and red deer.

Boundaries
The initial NEN still had to be condensed to form a 'net NEN'. Land managers expended a great deal of energy in ensuring that the available hectares were located in the right places. Nevertheless, the final boundary of the NEN is not based purely on ecological reasoning. Social feasibility proved to be at least as important a criterion. As a result of massive resistance from the agricultural sector – and from management bodies that were initially in favour of the plans – the desired continuity of the NEN is far from complete in many places, and areas that are potentially very favourable for natural development, such as the seepage area bordering the Utrechtse Heuvelrug, finally fell by the wayside.
Although the NEN is far from being optimal from an ecological perspective, it constitutes a major step forward and consequently among the land management bodies there is a wide basis of support for its realisation.

4.2   The NEN in 2007

The NEN was launched in 1990 and is to be completed by 2018. In 2007 we are therefore more than halfway there. While it is true that the realisation of the project is behind schedule, much has been achieved. Around 70,000 hectares of agricultural land have been purchased of which 50,000 hectares have been transformed into nature reserves. The acquisition has come about partly through the efforts of the Rural Area Service (DLG) and its predecessors, often within the scope of rural planning projects. Furthermore, the land management bodies have actively purchased land them-selves.

Voluntary sale

Acquisition takes place only on the basis of voluntary sale. For other pro-jects in the national interest, for example in the field of infrastructure or flood defences, it is unthinkable that implementation should be dependent on the attitude of an individual landowner. Where nature policy is con-cerned, for the time being voluntary sale remains a fact. The second half of the NEN may be considerably more awkward to accomplish than the first was, because the remaining landowners have not up until now been pre-pared to sell their land. Price-making plays a role in this. In principle, land is purchased at the agricultural price level. In parts of the Netherlands – among them virtually the whole of the province of Utrecht – land prices are higher as a result of possible urbanisation in the long term. In these areas the realisation of the NEN is impossible or extremely difficult because land-owners adopt a wait-and-see attitude. Here the desired spatial interconnec-tion is far from being accomplished (see *Fig. 17*).

Pace

At the current pace of realisation with the financial resources now available, the NEN will not be complete by 2018. The desire to involve private indi-viduals in the realisation of the NEN has added to the delay. Private con-servation is far from being able to meet the demands of the task in hand, and land management bodies have been sidelined. This situation also ap-plies to large areas of the province of Utrecht.

Spatial Interconnection

Within the NEN, barriers occur in the form of infrastructure, campsites, care institutions and military complexes. The means of the NEN are geared to acquiring agricultural land and not to solving the problems of such barriers. For state infrastructure there is a long-term programme for the defragmen-tation of nature reserves; a number of provinces have comparable pro-grammes for provincial infrastructure. For the other sticking points uncon-ventional solutions must be found. A good example is the Crailo Nature Bridge, where an initiative on the part of land management body *Het Goois Natuurreservaat (*Gooi Nature Reserve) has resulted in the world's longest 'ecoduct' (see *Fig. 18*).

The creation of the NEN is a laborious process around urban areas due to the high land prices and the network's low priority in local government. However, change can be brought about by a proactive attitude on the part of nature organisations such as the Utrecht Landscape Foundation. In 2000 this organisation presented an integral vision for the future of nature in and around the fast-growing city of Amersfoort. The development of new areas of nature was linked to water storage and leisure. A collaboration between local authorities was initiated, which resulted in the creation of plans for new natural areas close to the city. The provincial government allowed for additional hectares to be enclosed within the NEN boundary. As well as the usual financial resources, supplementary funds were obtained from the EU and the Postcode Lottery. The projects are now being implemented and in a short space of time more than 100 hectares of new areas of nature has been created around the edge of the city.

Fig. 17 A Window on the Valley; the development of nature in the outskirts of Amersfoort (Glastra, 2000)

Within the scope of the defragmentation of national infrastructure, *Rijkswaterstaat* (the Directorate General for Public Works and Water Management) has built a variety of ecoducts. The world's longest is in the region of Het Gooi, the 800m long Crailo Nature Bridge, which spans a road, railway line, sports park and in-dustrial area. The nature bridge came about thanks to the efforts of the Gooi Nature Reserve, a land management body. Through active lobbying and fund-raising the organisation was finally able to secure the cooperation of the relevant authorities and raise the €15 million needed. (See also chapter 21, page 405)

Fig. 18 Crailo Nature Bridge[a] *(Woestenburg, et al., 2006)*

Robust connection zones

The need for spatial interconnection has led to an interim adaptation of the NEN: between the large core areas, 'robust connection zones' (in Dutch: robuuste verbindingszones', RVZs) have been introduced. The planned narrow ecological connection zones would never have been able to link the habitats in a functional way. These narrow zones have been extremely difficult to realise because there is little basis of support for them among landowners, who are reluctant to give up parts of their land. A start has now been made on the creation of RVZs between Veluwe, Heuvelrug and Oost-vaardersplassen. In Utrecht work has begun on one of the RVZs, which runs through the Gelderse Vallei and will link the two largest wooded areas in the Netherlands: the Veluwe and the Utrechtse Heuvelrug. The Gelderse Vallei falls within the scope of the Reconstruction Act, so there is not only a

---

[a] see also www.natuurbrug.nl

clear legal framework for its realisation, but also sufficient financial resources.

Planning protection

The advent of the NEN has led not only to the acquisition of new areas of nature, but also to improved planning in the protection of existing ones. A restrictive policy on the construction of new infrastructure or housing applies throughout the NEN. This policy is anchored in the national Spatial Planning Memorandum, regional plans at provincial government level and municipal zoning plans. Exceptions are made only where the interests of society are seriously at stake and the damage must then be compensated for generously. This firm planning basis has greatly contributed to the prevention of further encroachment on nature reserves over the past 15 years.

At the same time, this so-called 'no, unless' principle has also resulted in a certain rigidity. Usually from the perspective of conservation this is far from being a problem. As long as the authorities stick to their guns, project developers come away empty-handed if they attempt to develop housing or commercial areas within the NEN. Where there are sticking points in the NEN, an integral approach can also produce gains for nature. An example of this is the *Hart van de Heuvelrug* (Heart of the Utrechtse Heuvelrug) project. Scattered buildings in this part of the Utrechtse Heuvelrug prevent the area from functioning as an interconnected nature reserve. On the initiative of *Het Utrechts Landschap* (Utrecht Landscape) an exchange is now taking place between 'red' (built up) and 'green' locations, resulting in the defragmentation of the Utrechtse Heuvelrug nature reserve (see *Fig. 19*).
The desirability of this exchange approach is a matter of contention among land management bodies. Some have opted for a 'green boundary' strategy, which leaves no room at all for 'red' development within the NEN, while others have taken a pragmatic line aiming at a net gain for nature.

Basis of support

The launch of the NEN brought a renewed vigour to conservation in the Netherlands. Now we are more than halfway through the project, this vigour has dwindled. The creation of the NEN has become institutionalised and is associated with a great deal of jargon. Apart from the term NEN itself, expressions like 'target species', 'nature target types' and 'semi-natural development' are heard at public information evenings. The vision has faded into the background and there is little interest in the NEN among the general public.

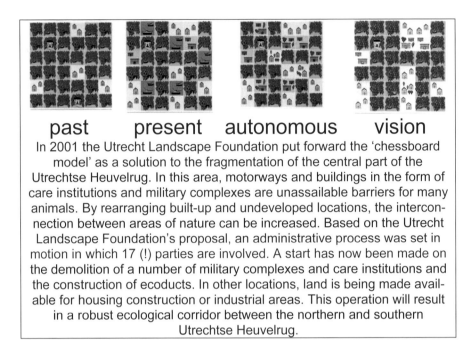

past present autonomous vision

In 2001 the Utrecht Landscape Foundation put forward the 'chessboard
model' as a solution to the fragmentation of the central part of the
Utrechtse Heuvelrug. In this area, motorways and buildings in the form of
care institutions and military complexes are unassailable barriers for many
animals. By rearranging built-up and undeveloped locations, the interconnection between areas of nature can be increased. Based on the Utrecht
Landscape Foundation's proposal, an administrative process was set in
motion in which 17 (!) parties are involved. A start has now been made on
the demolition of a number of military complexes and care institutions and
the construction of ecoducts. In other locations, land is being made available for housing construction or industrial areas. This operation will result
in a robust ecological corridor between the northern and southern
Utrechtse Heuvelrug.

Fig. 19 Hart van de Heuvelrug (Het Utrechts Landschap, 2001)

Thanks to broad governmental support at national and provincial level, the
realisation of the NEN is steadily progressing. But with the most difficult half
of the project still to be completed, further progress is by no means self-
evident. What is more, the implementation also requires support at local
government level. This became apparent in Utrecht during the construction
of an ecoduct in the municipality of Soest. In a referendum, a majority of
residents rejected the plan; the turnout was only just too small for the result
to be valid. For the NEN to be completed, broad public support is extremely
important and this presents the land management bodies with a clear task.
They are better able than anyone else to demonstrate through tangible
results – such as the return of species like the white-tailed eagle and common crane – the NEN's value not only for nature itself, but also for the people who enjoy it.

4.3   Mid-term review

The concept of the NEN is solid. For the conservation of the biodiversity in
the Netherlands it is vital to create large, interconnected nature reserves
and possibilities for exchange to take place between them. This convincing
concept has resulted in a broad basis of governmental support and the
NEN is now embedded in Dutch policy and legislation. For land management bodies it has become the key overarching theme for a great deal of
their activities. It has also provided a strong impulse for collaboration

among them: there is a common agenda to which everyone contributes their share.

The concept dates back to 1990 and needs to grow with the times. Developments in the sphere of climate change and flood defence present new points of departure. With the NEN as an ecological backbone we can anticipate such developments. The story behind the NEN must be continually retold, with the original enthusiasm revived.

5 OPTIMISING THE DUTCH NATIONAL ECOLOGICAL NETWORK

Spatial and environmental conditions for a sustainable conservation of biodiversity

Rien Reijnen; Arjen van Hinsberg; Wim Lammers; Marlies Sanders; Willem Loonen

5.1   Introduction

Background and purpose of study

The establishment of a National Ecological Network (NEN) is by far the most influential concept in Dutch nature policy in the last 15 years (see Barendregt, chapter 3, page 35). The NEN was introduced into policymaking as a 'coherent network of ecosystems that are of national and/or international importance and are in need of sustainable conservation.' (LNV, 1990; Lammers and Van Zadelhoff, 1996).

The NEN can be regarded as a spatial strategy for the conservation, restoration and development of nature and the landscape, which intends to:

1.   enlarge and link nature areas to reduce fragmentation; and
2.   enlarge nature areas to reduce their vulnerability to external influences, spatially separating areas with highly dynamic and less dynamic functions (for example urbanisation and intensive agriculture versus wildlife, low-intensity leisure and drinking water extraction).

After its planned completion in 2018, the NEN is intended to be a coherent network of ecosystems that are of national and/or international importance. All Dutch sites that come under the European Natura 2000 programme are also included in the NEN. By 2018, the NEN is supposed to cover a total of 7.2 million hectares, including 6.5 million ha consisting of large water bodies (the Wadden Sea, Lake IJsselmeer, the delta of the rivers Rhine and Meuse in the south-western province of Zeeland and the Dutch part of the North Sea) and 0.7 million ha of land. The terrestrial part comprises 450,000 ha of existing nature areas (based on the 1990 situation) and 275,000 ha of newly created habitats, including robust ecological corridors that are intended to link nature areas.

In addition to area-based targets, the NEN concept also includes quality targets. In the NEN context, quality has been defined by the Dutch Ministry of Agriculture, Nature and Food Quality (LNV) in terms of 'nature target types', that is, the ecosystem types included in the NEN. These types have been defined in terms of abiotic conditions and the presence of target species. This offers a basis for generic and area-specific environmental policies aimed at creating the required environmental conditions. The nature target types also provide the basis for Dutch national grant schemes for management aiming at the conservation, restoration and development of nature.

Now that the process of implementing the NEN has come about halfway, the Dutch Ministries of Housing, Spatial Planning and the Environment (VROM) and of Agriculture, Nature and Food Quality (LNV) have asked the

Netherlands Environmental Assessment Agency (in Dutch: Milieu- en Natuur Planbureau, MNP) to offer suggestions for further optimisation of the NEN. While the request to identify optimisation options is not a new phenomenon, the reasons for the specific request at this moment in time include the recent decision that the provincial authorities have to specify their ambitions to implement the environmental and spatial conditions that need to be in place to achieve the intended nature target types (LNV, 2004). The Fourth National Environmental Policy Plan (VROM, 2001) indicates that spatial and environmental conditions within the NEN will need to be considerably improved, as fragmentation, eutrophication and acidification are impeding the conservation and restoration of biodiversity in nature areas. The two ministries have therefore asked MNP to assess what environmental and spatial problems need to be solved and how this can be done. This article presents a method to identify ecological problems and to define an approach to solve them, in terms of stages and priorities.

Approach
Our analysis is based on the policy objectives embodied in the recent maps of the NEN (VROM, 2004; VROM, 2005) and its nature target types (LNV, 2004). The first step in our analysis involves identifying the spatial and environmental problems encountered in protecting the nature target types (Section 5.2). A problem is defined as an environmental and/or spatial precondition for a particular nature target type that has not yet been met. The analysis of *environmental problems* focuses on problems relating to water tables and atmospheric nitrogen deposition, as solving these problems is essential for the conservation and restoration of biodiversity in the Netherlands (Lammers et al., 2005; MNP, 2005: 'Natuurblans'). The analysis of *spatial problems* concentrates on the question whether the total area and the spatial configuration of the various nature areas offer sufficient guarantees for the sustainable conservation of the nature target types and the species they accommodate. Sustainable conservation of nature target types ultimately requires both environmental and spatial problems to be solved. Section 0 discusses ways of assessing the urgency of the various problems, based on the principles embodied in the EU's Habitats Directive. Section 5.4 then suggests a method to prioritise problems in various NEN sites in terms of priority, mainly based on the intention to create large-sized nature areas or large systems of smaller ones.

5.2 Identifying problems

5.2.1 Approach: from nature target types to preconditions

The national and provincial governments have set targets for the size and quality of the various types of ecosystem that the NEN is supposed to include by 2018. The provincial authorities have specified these targets in the form of maps showing the intended nature target types. In all, 92 different target types have been defined (Bal et al., 1995, 2001). The Ministry of Agriculture, Nature and Food Quality has clustered these 92 types into 32 'nature targets', divided into three management strategies: large-scale natural habitats, semi-natural habitats and multifunctional habitats (*Fig. 20*).

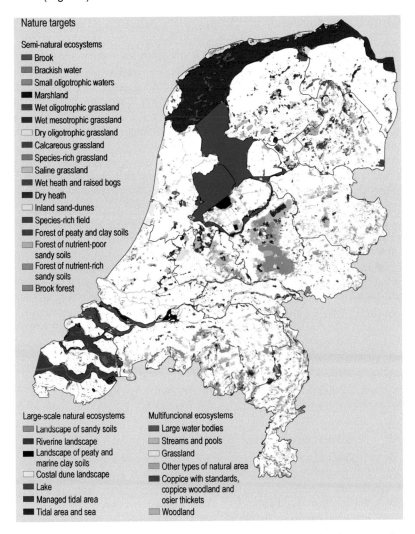

Fig. 20 National map of nature targets. For each area unit, the map shows only the nature target covering the largest surface area. The national nature target map was derived from the 12 provincial nature target type maps. Tweede Kamer, December 2001 (Bal et al., 2001)

Since the aim of the present analysis was to obtain as accurate a spatial overview of the environmental and spatial problems as possible, it was based on the most accurate spatial image of the intended type of habitat that was available, i.e. the map of nature target types. Since the current version of this map still features a number of sites with more than one nature target type, the localisation of the individual nature target types in such sites was further specified, based on current vegetation, soil and hydrology data (Lammers et al., 2005).

The description of the nature target types provides information for each type on aspects like characteristic vegetation types, abiotic conditions, characteristic species (target species), minimum size and management (Bal et al., 2001). Target species are species that are focused on in policymaking, in view of their national and international importance, rarity and decline. They include a total of 236 vertebrates, 260 invertebrates and 546 vascular plant and moss species. The set of target species includes the species in the EU's Birds and Habitats Directive. In addition, there is a clear relationship between the nature target types and the protected habitat types listed in the Habitats Directive.

At local level, the quality of an area accommodating a particular nature target type can be measured by the minimum percentage of target species that are expected to be present. In addition to the local quality requirement of achieving the nature target types, there is also a national policy target, which states that conditions for the sustainable presence of all species that were present in the Netherlands in 1982 (i.e. the indigenous species) must be met (LNV, 2000; LNV, 2004).

The environmental preconditions for the nature target types have been largely derived from those for the plant species or plant communities belonging to these target types (Bal et al., 1995, 2001; Hennekens and Schaminee, 2002). The spatial precondition is based on the area requirements of the target fauna species (Bal et al., 2001).

5.2.2   Environmental problems

The Dutch government's fourth national environmental policy plan (NMP4; VROM, 2001) mentions eutrophication, acidification and desiccation as the most persistent problems impeding the conservation and recovery of biodiversity in the Netherlands. The intensive use of land in the Netherlands has resulted in water table drawdown in large parts of the country, resulting in desiccation problems for many of the originally wet habitats like wet heaths, raised bogs, brook valleys and marshland. *Fig. 21* shows the problems relating to water tables. A water table problem is defined as a situation in which the optimum spring water table of the nature target type for a particular location (STOWA, 2002) does not correspond to the actual water tables as derived from the soil map (Lammers et al., 2005).

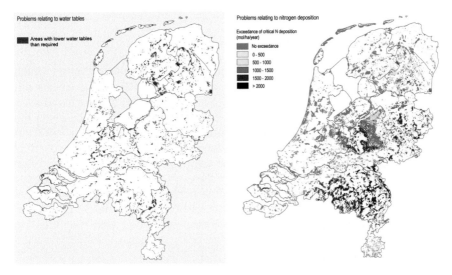

Fig. 21 Problems relating to water tables for the nature target types in the NEN, assessed by comparing environmental requirements and data on current water tables *(Lammers et al., 2005).*

Fig. 22 Problems relating to nitrogen deposition for the nature target types in the NEN, assessed by comparing environmental requirements and data on nitrogen loads *(Lammers et al., 2005).*

Another major problem is the level of atmospheric deposition of acidifying and eutrophying substances. Compared to other parts of Europe, the Netherlands has very high deposition levels of such substances, which affect especially the ecology of oligotrophic sandy habitats like heathlands, raised bogs and forests. These depositions have affected soil conditions like pH and nitrogen availability. As a result, species adapted to oligotrophic habitats have deteriorated and have usually been replaced by more common species. The ecological problems relating to atmospheric deposition can be identified by comparing the critical loads for the various nature target types with the actual deposition levels. A critical load is defined as 'a quantitative estimate of exposure to one or more pollutants below which significant harmful effects on specified sensitive elements of the environment do not occur according to present knowledge' (Nilsson and Grennfelt, 1988). If current deposition levels exceed the critical loads, there is a risk of harmful effects. The greater the difference, the greater the risks and the more harmful the effects (Bobbink et al., 2003). A limited exceedance of the critical loads (approx. 500 mol/ha/yr) can be counteracted, at least temporarily, by management measures. *Fig. 22* shows that critical loads are being exceeded in large parts of the Dutch nature areas. The most serious problems are occurring in the eastern part of the country, where the sandy soils are highly vulnerable. At the same time, this part of the country has the highest agricultural ammonia emission levels.

5.2.3   Spatial problems

A process of fragmentation has led to a situation in which many Dutch terrestrial nature areas are fairly small. For instance, 40% of the forests are less than 500 ha in size, as are 50% of heaths and 70% of marshland sites (MNP, 2003). This has resulted in habitats for many species becoming too small to accommodate sustainable populations (Opdam et al., 1993). In such situations, sustainable presence can only be achieved if the individual habitat patches function as an ecological network (Opdam et al., 1993; Opdam and Wiens, 2002). Opportunities to achieve this are greatly increased by including key patches in the ecological network (Verboom et al., 2001). Key patches are those that harbour a fairly large population which is sustainable in a situation of limited exchange with neighbouring populations.

Our spatial analysis of the nature target types map has focused on the identification of key patches for all target fauna species which breed in the Netherlands (a total of 401 species). The analysis assumed that all key patches are included in an ecological network. This is based on available data on area requirements of characteristic target species for each nature target type. These data were used to assess for what percentage of the characteristic target species the standard for the size of a key patch is currently being exceeded, under optimal environmental conditions. The results of this analysis are shown in *Fig. 23*, for each nature target type.

The figure shows that it is particularly in the larger nature areas that the nature target types include many of the target species, while some of the smaller areas with highly specialised nature target types (such as 'brooks' and 'brackish water') also accommodate a relatively large percentage of the target species. Many of the other nature areas, however, especially the smaller ones, do not allow as many species to form a key patch, and these represent the problem areas.

From the point of view of spreading risks, the sustainable national presence of species in the Netherlands can best be ensured by aiming at a number of key patches distributed over the NEN (Foppen et al., 1998; Opdam, 2002). Based on the work by Foppen et al. (1998) and supplementary expert knowledge, we have distinguished three classes: sustainable, possibly sustainable and non-sustainable. The analysis shows that 6% of the species are not sustainably present (especially birds and large mammals), while 13% are possibly sustainable (including especially species of rare nature target types) and 81% are sustainably present. The non-sustainable and possibly sustainable species include mostly those characteristic of wet heathland, raised bog, forests on rich sandy soils, brooks and marshland.

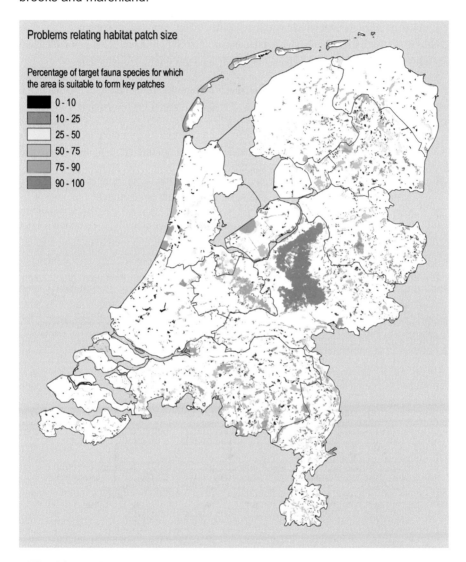

Fig. 23 Local-scale spatial problems. The figure shows the percentage of target fauna species, per nature target location, for which the site is suitable to form key patches. The analysis assumed good environmental conditions and no fragmentation by roads (Lammers et al., 2005).

5.3   Assessing the urgency of interventions

5.3.1   Method to assess urgency

In view of the severity of environmental (*Fig. 21*) and spatial (*Fig. 22*) problems, and their persistent character (VROM, 2002), it is important to assess how urgently the various problems need to be addressed.

Following the Habitat Directive framework (see Pelk, chapter 8, page 119), the conservation status (EC, 2000) of nature target types was ascertained by assessing:

1.   the present *quality* of a protected nature target type in terms of the presence of typical/characteristic species;
2.   the recent *trend* (or specific changes) in the size and distribution of this nature target type or the species that occur in it; and
3.   the *future* of the nature target type/habitat, based on the degree to which ecological soil, water and air conditions are being met or can be met.

The first two of these three aspects (quality and trend) offer a measure of the urgency of interventions. If all three aspects are satisfactory, the Habitats Directive does not require any action to be taken. If quality and trend are satisfactory, but the future conservation is not ensured, for example because of environmental or spatial problems, action is required to ensure sustainable conservation in the longer term (*Fig. 24*).

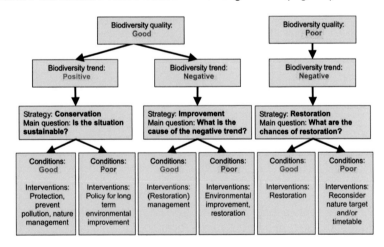

Fig. 24 The urgency of environmental or spatial interventions depends on an assessment of the situation in terms of biodiversity and of environmental, water and spatial conditions (Lammers et al., 2005).

If the quality is still satisfactory but is declining (negative trend) due to environmental or spatial problems, urgent action is required. If the quality is already poor, the potential for improvement should be assessed, though interventions to address environmental and spatial problems are then less urgent. In such cases, one should establish whether short-term interventions are useful and whether the term within which the target must have been achieved could be extended. Whether interventions are undertaken therefore depends not only on the severity of the environmental or spatial problem but also on the consequences for biodiversity.

5.3.2   Nature target types requiring urgent action

Fig. 25 shows the 'conservation status' of nature target types (clustered into nature targets), as well as the environmental and spatial problems (Section 5.2). The table lists only those types with the most severe problems. Nature target types showing a severely negative trend in ecological quality require interventions more urgently than those with a less adverse trend.

The information on trends in ecological quality shows that the nature target types whose problems most urgently need to be addressed are 'wet heath and raised bogs', 'dry, oligotrophic grassland' and 'wet, oligotrophic grassland', as these are subject to environmental problems. In addition, 'wet heathland and raised bogs' are also subject to spatial problems, even if satisfactory environmental conditions are assumed. These problems need to be urgently addressed as these habitats are internationally recognised to be in need of a high level of protection. By comparison, the 'species-rich grassland' target type has less severe environmental problems, and a lower international level of protection.

Nature target types clustered into nature targets	Problem in terms of negative trend in target species over the 1990 to 2004 period (Vonk, 2004)	Environmental problem	Spatial problem	Urgency for policymaking
Wet heath and raised bog	+++	+++	+++	+++
Dry oligotrophic grassland	+++	+++	++	+++
Wet oligotrophic grassland	+++	+++	+	++
Dry heath	++	+++	++	+++
Small oligotrophic waters	+	+++	+	+++
Brook	+	++	+++	+++
Species-rich grassland	+++	+	++	+
Inland sand-dunes	++	+++	++	+
Forest on nutrient-poor sandy soils	++	+++	++	++

Nature target types clustered into nature targets	Problem in terms of negative trend in target species over the 1990 to 2004 period (Vonk, 2004)	Environmental problem	Spatial problem	Urgency for policymaking
Forest on nutrient-rich sandy soils	++	+	+++	++
Marshland	+	++	+++	++
Calcareous grassland	++	++	++	+++
Streams and pools	?	+++	+	+
Coppice with standards, coppice woodland and osier thicket	?	++	+++	+
Other types of natural area	?	+/++	++	+

Fig. 25 Assessment of the 'conservation status' of nature target types subject to environmental and spatial problems. Columns 3 and 4 show the severity of the environmental problems (in terms of atmospheric deposition and water tables) and spatial problems, in 3 classes: moderate (+), severe (++) and very severe (+++). Column 5 shows the urgency of introducing policy to address the problems. The highest urgency is for those target types for which a large percentage (> 30%) of the area in which they occur has a protected status under the EU's Birds and Habitats Directive (>30%), whereas target types for which only 5% of the area has such a status have a low urgency.

5.4   Specification of interventions

5.4.1   Setting priorities within the National Ecological Network

Identifying problems also helps to develop a strategy to improve conservation status. It is obvious that desiccation and acidification require interventions to reduce environmental pressures, for instance by measures taken at the source of the problems, such as reducing groundwater extraction and reducing emissions of acidifying and eutrophying substances. In some cases, habitats have to be enlarged to achieve sustainable spatial conditions. Sometimes, spatial interventions allow both spatial and environmental conditions to be improved, since enlarging nature areas results not only in larger habitat patches for particular species, but also increases the distance between the habitat patches and the locations of polluting activities. It is therefore important to identify sites where interventions would have the greatest effect.

From a national point of view, the obvious choice would be to address the NEN problems in the larger habitat areas, or 'large natural core areas', first. Arguments for this include:

1. Large natural core areas offer suitable habitats for sustainable populations of many species and enhance the sustainability of smaller sites nearby, thereby improving the performance of the NEN as a whole (Verboom et al., 2001).
2. Natural dynamics offer opportunities for the conservation of a rich variety of ecological values. The ecosystem processes required for this can usually only take place in large, continuous habitat areas (Bal et al., 2001).
3. Large natural core areas have relatively shorter borders, which reduces harmful influences from surrounding or enclosed areas (MNP, 2004). Conversely, the internal buffer systems in such large areas mean that environmental preconditions for other functions can become less strict (for example requiring less reduction of emissions as ecosystems are more robust)

Section 5.4.2 defines what is meant by large natural core areas and presents a map showing these areas. Section 0 then discusses interventions to improve the conservation status in large natural core areas. Section 5.4.4 briefly discusses the other types of sites in the NEN (i.e. other than large natural core areas). See the schematic overview in *Fig. 26*.

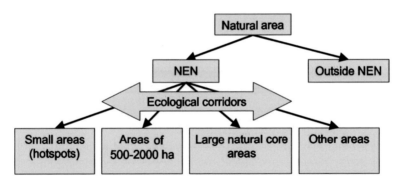

Fig. 26 Nature areas in the Netherlands: a schematic overview
(Lammers et al., 2005).

5.4.2   Large natural core areas: definition and overview

From the perspective of the conservation of species, a large natural core area can be defined as a continuous nature area of sufficient size to allow many species to establish key populations there. There are good chances that species are sustainably present in these areas (see Section 0). Available data on the habitat requirements of various species can be used to calculate how large and continuous a habitat area needs to be for these species. The present study defined large natural core areas as nature areas with a size of more than 2000 ha which can accommodate over 70% of the target fauna species (*Fig. 27*).

A large natural core area does not, however, have to consist of one continuous nature area to accommodate many species; a system of smaller nature areas in close proximity to each other (a so-called mosaic) can function as one large habitat area in an ecological sense. Such areas have been designated 'mosaics of small habitat areas'. Such mosaics can only function as units if the individual small habitat areas within them constitute an ecological network. In such cases, local extinctions can be compensated by new influx from neighbouring areas (Opdam and Wiens, 2002; Opdam et al., 2003). The maximum distance between nature areas that can be overcome differs per species (*Fig. 28*).

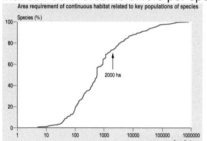 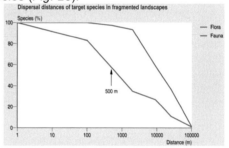

Fig. 27 Relationship between the size of habitat areas and the percentage of target fauna species that can establish large populations in them and so form a key patch, based on the average area requirements of species in the large-scale nature targets of 'landscape of sandy soils', 'riverine landscape', 'landscape of peaty and marine clay soils' and 'coastal dune landscape'. A key patch can accommodate a sustainable population if it is part of an ecological network

(*Data used for the Handboek Natuurdoeltypen (nature target types handbook) by Kalkhoven and Reijnen, 2001; Bal et al., 2001*).

Fig. 28 Distances between nature areas that can be overcome by endangered and seriously endangered plant and fauna species

(*Data from Oostenbrugge et al., 2002*).

The present study defined such mosaics as systems of small habitat areas whose total size is at least 2000 ha, with distances between the small habitat areas of no more than 500 m. The 500 m limit was chosen because many animal species tend to be hampered by gaps of more than 500 m in their habitats, while for many plant species, 500 m may already represent an insurmountable distance. In addition, such a mosaic can only function as an ecological unit if the zones in between the smaller habitat areas are suitable for the migration of animals, especially terrestrial and aquatic animals. In addition, environmental conditions have to be suitable throughout the mosaic area, which is easier to achieve if the area forms a hydrological unit.

A considerable proportion of the terrestrial part of the intended National Ecological Network consists of large continuous habitat areas (390,000 ha) and mosaics of small habitat areas (160,000 ha). These large natural core areas are distributed across the Netherlands (*Fig. 29*).

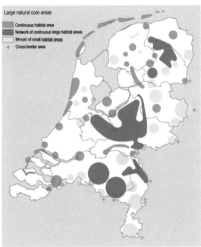 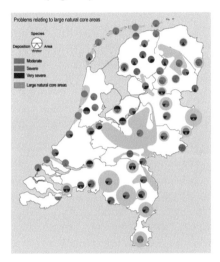

Fig. 29 Schematic map of the 61 large natural core areas. The size of the habitat areas in the border areas with Germany and Belgium is underestimated since many areas extend across national borders (Lammers et al., 2005).	*Fig. 30 Severity of environmental and spatial problems in the large natural core areas. The figure also shows the presence of target species (Lammers et al., 2005).*

Important continuous areas include the dune regions in the west and north of the country, heathlands and peatlands in the north and south, lakes and marshlands in the west, and sandy ridges like the Veluwe, Utrechtse Heuvelrug and Sallandse Heuvelrug regions in the central part of the country. In addition to these large continuous habitat areas, the NEN also includes a number of mosaics of smaller habitat areas in close proximity to each other. Such large mosaic areas, which can function as large ecological units, occur especially in the eastern and southern parts of the country. In all, the large continuous habitat areas and mosaics of small habitat areas cover over 80% of the terrestrial Natura-2000 sites.

5.4.3   Large natural core areas: interventions based on conservation status

The above-mentioned method to set priorities in addressing problems (Section 0) can also be applied to the set of large natural core areas. *Fig. 30* shows which of these areas harbour many species and how severe the environmental and spatial problems are. Information on trends in species could not be taken into account, since these data are not available at local level. The presence of target species is based on distribution data for vascular plants (source: FLORBASE, FLORON). The environmental and spatial problems were derived from the data shown in Section 5.2.

It has become clear that the large natural core areas vary greatly in terms of conservation status. Most of the areas harbour relatively large numbers of target species (plants): about 85% of the NEN locations with the highest plant species densities (floristic hotspots) are covered. The coastal dune areas are particularly rich in target plant species. It was as yet impossible to include data on the presence of fauna species in the analysis (see Schouten and Verweij, chapter 6, page 92).

The large natural core areas also vary greatly in terms of environmental and spatial problems, with the mosaic areas generally showing more severe problems than the large continuous habitat areas. This is because the smaller nature areas that make up these mosaic areas, with their relatively long boundaries, have fewer internal buffer systems, which means that they are more vulnerable to external pollution. As a consequence, strategies to solve problems in these mosaic areas differ from those used in the large continuous habitat areas.

From an ecological point of view, the most effective strategy to address problems in the large continuous habitat areas is by focusing interventions on sites that seriously hamper the opportunities for the area to function as a large unit. These include enclaves that have a disproportionate impact on a large surrounding area (for example because of ammonia emissions, water table drawdown and drainage), as well as fragmentation by infrastructure or a lack of coherent hydrological and/or ecological management. Area-specific management policies could give priority to these locations. This may imply reducing nitrogen emissions by local sources (especially intensive livestock farms and heavily fertilised fields and meadows) within the large continuous habitat areas. A similar approach could be applied to land use forms that require greatly lowered water tables (arable fields, meadows and roads) and are located in large continuous habitat areas where the aim is to establish wetlands. Management policies to combat desiccation could thus concentrate on these areas. If such enclaves are ignored, large areas may not receive the intended degree of protection. It is important to ensure a sound hydrological situation in large continuous habitat areas, since hydrological improvements can also help combat the effects of problems like acidification and eutrophication. Wetland areas need to be protected by preventing local water from draining away or by letting in good quality water from elsewhere. Although water from elsewhere often contains high nitrogen or phosphorus levels, making it unsuitable to solve desiccation problems, it is possible to overcome this by technical measures such as purifying or diverting the water, so that it loses part of its nutrients.

As regards solving spatial problems in large continuous habitat areas, management should aim at actually utilising the area's potential. One option to utilise this potential is by locally increasing the area of particular nature target types (see Section 0 for a discussion of the types of habitat for which this is relevant, see also Pelk, chapter 8, page 119). Environmental conditions can also benefit from the restoration measures (see for example Van Andel and Aronson, 2005). At the same time, local fragmentation problems due to infrastructure within these large continuous habitat areas need to be addressed (see Glastra, chapter 4, page 65). All interventions in such areas should focus on the prioritised nature target types.

Environmental conditions in the mosaics of small habitat areas are often less than ideal, while spatial conditions are, by definition, also under threat. If the land use between the smaller habitat areas included in the network becomes less ecologically favourable, the functioning of the network as a whole will be affected. Improving environmental conditions in large mosaics of smaller habitat areas also requires that land use in the intervening zones is in agreement with the intended nature target types for the area as a whole. This can be achieved by redefining boundaries between planned nature areas or by reducing polluting and/or disturbing activities in the intervening zones. The protection of areas that include wetlands requires the hydrological system to be optimised. This is particularly true for seepage areas and brook valleys. It may be necessary in some of these areas to adapt NEN boundaries to the hydrological system, especially when prioritised nature target types are at risk. However, in view of the current degree of fragmentation of nature in the Netherlands and financial costs, opportunities to create new continuous habitat areas are probably scarce. If creating large natural areas is not possible intervention could focus on improving the integration of the various land-use functions in mosaic areas. This could be tied in with policies for the conservation of landscape structure. Smaller local ecological corridors might perhaps help strengthen the ecological network without the need for large-scale redefining of boundaries.

5.4.4 Wildlife outside the large natural core areas

The large natural core areas of the National Ecological Network are expected to be linked by robust ecological corridors (VROM, 2004, 2005).

In practice, the boundaries of many of these robust ecological corridors have not yet been specifically defined. In the lower parts of the Netherlands, nearly all of the large continuous habitat areas and mosaics of small habitat areas have by now been integrated in the network of proposed robust corridors. This is much less true for the higher parts of the country. In addition to linking the large natural core areas, the proposed robust corridors might also be used to link many of the intervening smaller nature areas (500 – 2000 ha).

Some smaller, more or less isolated nature areas in the NEN are very species-rich. Such hotspots can on the longer term, by definition, only be protected by means of a local approach, as generic interventions to improve environmental quality will not easily achieve the locally required environmental conditions. Hotspots have not yet been sufficiently surveyed at national level, although studies to identify butterfly, bird and floristic hotspots are underway (see chapter 6, page 92). As soon as hotspots have been identified, their conservation status will need to be assessed. Where possible, hotspots should be used as bases to support or colonise nearby large natural core areas. This could involve redefining boundaries of core areas. They could also be included in local or robust ecological corridors and thereby enhancing the ecological potential of all areas thus linked, including the hotspots themselves.

The remaining small habitat areas (smaller than 500 ha) with low biodiversity should be examined to assess their potential. In some cases, it may be doubted whether they belong in the National Ecological Network.

5.5   Conclusions and recommendations

The current state of knowledge confirms that the concept of the Dutch National Ecological Network (NEN) remains a valid spatial strategy for sustainable conservation of biodiversity. Halfway through the planned implementation term of this major project, which started in 1990, it must be said that the spatial coherence and environmental conditions are as yet insufficient to allow the international commitments on conservation of biodiversity to be met.

The remaining environmental and spatial problems are widely distributed across the country. The assessment framework developed for the EU's Habitats Directive can be applied to the nature target types developed for the NEN to specify for each target type and for each site why, where and when interventions are required. This approach offers a basis for setting relevant priorities in further optimising the NEN. The value of this approach will increase if area managers, water boards, municipal and provincial authorities and the national government manage to conclude agreements on the implementation of the NEN and about monitoring.

Large natural core areas include large continuous habitat areas and mosaics of smaller habitat areas. Large continuous habitat areas offer the best opportunities for sustainable protection of habitats for indigenous plants and animals. After the NEN has been completely implemented, 55% of its area will consist of these large continuous habitat areas, such as the coastal dunes, the Veluwe region and the floodplains along the major Dutch rivers. Another 20% of the NEN consists of mosaics of smaller habitat areas situated in close proximity to each other and potentially functioning as one large natural core area. Examples of such mosaics include the Twente, Graafschap en Southern Limburg regions. As the Dutch situation is characterised by severe competition for space, such areas are in great danger of becoming irretrievably lost. If the current qualities and potential of these areas are to be preserved for the future, their spatial coherence will need to be improved, including the hydrological system. This will require additional interventions, which in the short term will especially mean clear spatial planning and in the longer term improving environmental conditions.

6   BIODIVERSITY AND THE DUTCH NATIONAL ECOLOGICAL NETWORK

Marieke A. Schouten; Pita Verwey; Aat Barendregt

6.1 Introduction

6.1.1 Declining biodiversity and its impacts

Modification of the natural environment, either through habitat destruction, fragmentation, pollution or the invasion of exotic species, caused a serious decline in biodiversity over the last few decades (Pimm et al., 1995; Pimm and Raven, 2000). The reduction of biodiversity is expected to have serious implications for ecosystems and society (for example Chappin III et al., 2000; Tilman, 2000). The UN Convention on Biological Diversity (Rio de Janeiro, 1992), ratified by many countries including the Netherlands, therefore urges countries to preserve representative sets of biodiversity. Space, time, money and knowledge constraints force policymakers to search for highly efficient solutions for example to focus on areas with a high conservation value, or the use of indicators for biodiversity. Conservation scientists acknowledged this problem and scientific research over the last decades yielded numerous strategies for site selection.

Dutch biodiversity

Unfortunately, the current trend of biodiversity decline is also valid in the Netherlands (for example Biesmeijer et al., 2006; Tamis et al., 2005; Turnhout et al., 2007). At present the number of wild flora and fauna species in the Netherlands is estimated to be approximately 36.000 (Nieukerken and van Loon, 1995) but basal knowledge about the overall status and spatial configuration of Dutch biodiversity is narrow.

Dutch policy on biodiversity

Various European and international treaties, plans and directives (for example Bird and Habitat Directives) provide a framework for the conservation of biodiversity in the Netherlands. By signing the UN Convention on Biological Diversity, the Dutch government committed itself to preserve a representative set of biodiversity and to contribute to halting biodiversity decline. Furthermore, as part of the Bern Convention, the Netherlands have to ensure that by 2020 conditions will be in place for long term conservation of all species and populations native to the Netherlands occuring in 1982 (LNV, 2000). Policy to implement these goals has been developed. On paper, nature conservation policy in the Netherlands can be subdivided in a spatial or ecosystem oriented policy and the straightforward species protection policy. In practice, the spatially oriented nature policy regarding the maintenance of ecosystems, however, is almost completely elaborated in terms of target species. In order to preserve ecosystem diversity (by means of target biotopes), 1042 target species were selected from different taxonomic groups acoording to their (inter)national importance, rareness or decline in populations (LNV, 2000; Bal et al., 2001). Taken together, these target species represent about 3% of the total number of species occurring in the Netherlands. In addition to the target species, vulnerable and threatened species have been identified (Red List species). The number of these species present in an area is one of the main measures of the performance of Dutch nature conservation policy. The most important instrument of *in situ* biodiversity conservation in the Netherlands is the National Ecological Network, a nation wide network of nature reserves.

Fig. 31 Net National Ecological Network (LNV, 2004)

The National Ecological Network (NEN)
The National Ecological Network is based on existing large nature reserves, such as the coastal strip, the North Sea, the Wadden Sea, and

the forests and heathlands of the Veluwe in the centre of the country. The idea is to expand the area of interconnected nature reserves, thereby increasing the basis for species and promoting exchange between populations. By 2018 the National Ecological Network (NEN) should cover 728.500 ha, about 20% of the total terrestrial area of the Netherlands, which means that currently (January 1st 2005) about 169.000 ha of nature reserve is still to be realised (Rekenkamer, 2006). About 40% of the NEN will contribute to the European NATURA 2000 network.

6.1.2   Hotspots

One of the leading principles in conservation biology is that if one cannot preserve everything, then focus on the areas with a high diversity (Myers et al., 2000; Rcid, 1998). These so-called hotspots can be defined as areas with a large number of species or large numbers of endemic, rare or threatened species (for example Margules et al., 2002; Myers et al., 2000). The GAP analysis approach (Scott et al., 1993) has been developed to identify gaps in existing networks of conservation areas by using information on hotspots of species richness.

Coincidence of different taxonomic groups

However, since basic fine-scaled data on the distribution of most species are lacking, one of the assumptions of this approach is that hotspots would coincide for different taxonomic groups. Numerous studies have been done on whether or not hotspots actually coincide for different taxonomic groups (for example Prendergast et al., 1993; Myers et al., 2000; Tardif and DesGranes, 1998; Harcourt, 2000) and whether hotspots of richness are congruent with hotspots of endemism, rarity or threat (for example Orme et al., 2005). Unfortunately, no univocal answer can be given to both questions.

Keystone species/ taxa, representativeness

Therefore it is difficult to consider for example indicator, umbrella or keystone species/ taxa as surrogates for total biodiversity (for example Williams and Gaston, 1994; Kati et al., 2004; Ricketts et al., 2002). In theory, species' representativeness regarding particular ecosystems is an elegant way of dealing with a lack of knowledge but this theory often does not hold in practice (Andelman and Fagan, 2000). Environmental conditions (for example Dûfrene and Legendre, 1997; Pienkowski et al., 1996) are also being used to represent biodiversity, sometimes in combination with species distributions (for example Carey et al, 1995), resulting in eco-regions that should reflect species distributions and can thus be used to protect biodiversity.

Complementarity

Hereby the focus lies more on the species composition and capturing the environmental variation of an area rather than on species richness alone, as biodiversity cannot be entirely described in terms of species richness. A recent trend in conservation research therefore is the focus on complementarity (for example Vane-Wright et al., 1991; Williams et al., 2005). Complementarity analysis ensures that as many new attributes as possible will be added to an existing network of nature reserves. These attributes can be everything from landscape elements to endemic species. Other conservation issues focus more on population dynamics thriving on the Island theory of MacArthur and Wilson (1967), and deal with size and connectivity of conservation areas. The issue whether it is better to establish several small or single large reserves (SLOSS issue) is still fiercely debated (for example Etienne and Heesterbeek, 2000). Numerous tools have been developed for the establishment of corridors to connect nature reserves.

6.1.3   Beyond target species

Although little is known about the overall status and spatial configuration of Dutch biodiversity, nature conservation policy has been designed and is being implemented. The National Ecological Network ought to conserve biological diversity but, besides target species which represent about 3% of the total number of species occuring in the Netherlands and of which more than 50% are vascular plants, no specific and measurable indicators have been defined for the evaluation of the realisation of the National Ecological Network. Patterns of species richness do not necessarily coincide for different taxonomic groups (for example Prendergast et al., 1993; Myers et al., 2000; Tardif and DesGranes, 1998; Harcourt, 2000) so it is arguable to assume that with these target species the total diversity of the Netherlands would be captured. In earlier studies we examined the spatial organisation of a broad array of taxonomic groups: Syrphidae (hoverflies), Odonata (dragonflies), Orthoptera (grasshoppers), herpetofauna (reptiles and amphibians) and Bryophytes (mosses) in the Netherlands. These groups were selected because they illustrate different ecological functions and detailed, nation-wide data on their distribution are available. In these studies we focused on two important issues in conservation biology, namely, hotspots of species richness and hotspots of uniqueness, regions harbouring species that are 'unique' for that region. In this chapter we will link the results of these previous studies to the current status of the National Ecological Network.

6.2   Approach

6.2.1   Location of hotspots and connections

Although there was a clear concept in 1989, the eventual realisation of the National Ecological Network rather is the product of pragmatism (see Barendregt, chapter 3, page 35). In spite of the fact that the ideas behind the institution, enlargement and connection of existing areas are scientifically broadly accepted (for example MacArthur and Wilson's theory of Island Biogeography), the spatial allocation of the National Ecological Network is not based on thourough scientific research. It is, therefore, not surprising to see that there is increasing evidence for the fact that the National Ecological Network is not optimally allocated (Vereijken et al., 2005; Veling, 1997; Reijen et al., 2007). In this chapter we compare the results of earlier studies on the hotspots of species richness and hotspots of uniqueness based on the species distribution patterns of five taxonomic groups (Odonata, Orthoptera, herpetofauna and Bryophytes) with the map of the National Ecological Network (see *Fig. 31*, Net National Ecological Network, June 2004, Ministry of LNV, division knowledge) and evaluate how much of these areas are currently covered by the National Ecological Network. We consider only the terrestrial part of the National Ecological Network.

6.2.2   Hotspots of species richness

For the hotspots of species richness analysis (Schouten et al, in prep a) a 5x5 km UTM grid of the Netherlands was constructed using ArcView GIS 3.3 (ESRI, USA). Species diversity of each taxonomic group was determined for all 5x5 km square and as the number of species, discarding records that were collected prior to 1965. Only grid squares consisting for more than 50% of terrestrial area located within the Netherlands were taken into account (*n* = 1393). We used three different criteria to establish hotspots. We defined the top 20, 10 and 5% of species richest grid squares for each taxonomic group.

6.2.3   Hotspots of uniqueness

The hotspots of uniqueness (Schouten et al, in prep. c) are based on the uniqueness of species composition. They represent regions that have several characteristic species of different taxonomic groups. Hotspots of uniqueness were defined following a three-step procedure. First we clustered 5x5 km grid squares according to similarity in species composition for each taxonomic group individual, using a numerical classification (TWINSPAN). Then we identified characteristic species for each cluster using the preference index of Carey et al. (1995). Sørensen's similarity index was used to identify corresponding clusters among the different taxonomic groups. Finally we identified regions containing characteristic species for several taxonomic groups.

For both approaches the occurrence data of the individual species for five taxonomic groups were derived form different nationwide species occurrence datasets.

6.3   Results

6.3.1   Species richness

Species richness of taxonomic groups
Correlations between the patterns in species richness of the five groups were not very high (*Fig. 32, Fig. 34*).

	Odonata	Bryophytes	Herpetofauna	Orthoptera
Bryophytes	0,262**	-		
Herpetofauna	0,488**	0,347**	-	
Orthoptera	0,444**	0,228**	0,452**	-
Syrphidae	0,320**	0,282**	0,298**	0,368**

** Correlation is significant at the 0.01 level

Fig. 32 Spearmans' rank correlations for the species richness hotspots of the different taxonomic groups.

The species richness patterns of Orthoptera and bryophytes expressed the lowest correlation (0.228, P< 0.01), while species diversity of herpetofauna turned out to be highly correlated with that of the Odonata and Orthoptera (0,488, P< 0,01 and 0,452, P< 0,01, respectively). The overall patterns of Syrphidae and Bryophytes diverged considerably form those of the other groups. However, the hotspots of species richness seem to be reasonably well covered by the National Ecological Network for all taxonomic groups (36 to 43%, *Fig. 33*).

Hotspots of species richness covered by the NEN
If we use the most rigid criterion (top 5%) we, however, start to see some big differences between the various taxonomic groups.

GRIDS	Bryophytes	Syrphidae	Herpetofauna	Orthoptera	Odonata
Top 20%	36,6%	36,0%	42,6%	40,0%	38,9%
Top 10%	40,2%	32,8%	51,6%	48,9%	41,6%
Top 5%	41,3%	37,1%	56,2%	54,4%	43,7%

Fig. 33 Hotspots of species richness and National Ecological Network. Mean percentage of the grid squares covered by the NEN.

The top 5% grid squares of Syrphidae richness, for example, is poorly represented in the National Ecological Network (37,1% of the total area of these top 5% grid squares is covered by the National Ecological Network) compared to the top 5% grid squares of herpetofauna and Orthoptera richness (56,2% and 54,4%, respectively).

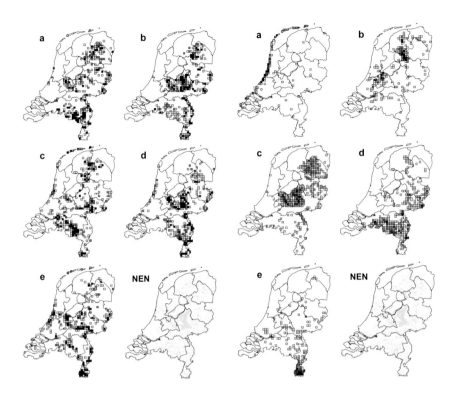

Fig. 34 (a-e). Hotspots of species richness. Spatial patterns of observed species richness of (a) Odonata, (b) Herpetofauna, (c) Bryophytes, (d) Othoptera and (e) Syrphidae. Colors (pink, red, dark red) represent top 20, 10 and 5% of most species rich grid cells respectively.
(Schouten et al, in prep a)

Fig. 35 (a-e). Hotspots of uniqueness. Regionalisation of the Netherlands based on the distribution of species from five taxonomic groups. Colorsrefer to the number of taxonomic groups for which a grid square is identified (the darker the more groups) to the regions: a. DUNE, b. FEN, c. SAND, d, SE and e. LIMB
(Schouten et al, in prep. c)

6.3.2   Uniqueness

Hotspots of uniqueness
In the Netherlands five regions can be distinguished that have a unique species composition for several taxonomic groups (Schouten et al., in prep.

c). These regions can be regarded as hotspots of uniqueness and the combination of these regions comprises the majority of the species of the studied groups, represented in the Netherlands.

	DUNE	SE	SAND	LIMB	FEN
All grid squares	27,9% (82)	29,3% (296)	39,0% (314)	22,2% (126)	24,6% (165)
Core areas: ≥2	32,3% (30)	34,0% (96)	51,8% (157)	19,3% (27)	36,2% (38)
Core areas: ≥3	38,9% (16)	41,9% (28)	65,2% (70)	21,2% (16)	47,6% (14)
Core areas: ≥4	42,3% (8)	42,0% (5)	78,3% (23)	27,0% (7)	-
Core areas: ≥5	-	-	86,5% (6)	-	-

Fig. 36 Hotspots of uniqueness and National Ecological Network. Mean percentage of the grid squares covered by the NEN. Number of grid squares between brackets.

Regions
The first region (*Fig. 35a*) forming a narrow band along the North Sea coast (DUNE), occurs in four of the taxonomic groups but is based on only few characteristic species.
The second region (FEN, *Fig. 35b*) is found in the north and central west parts of the Netherlands and is recognised for three out of the five taxonomic groups. The third region (SAND, *Fig. 35c*) centres on the Pleistocene sand plateau in the central and northern part of the Netherlands and is the only region that is congruent for all five taxonomic groups. The fourth region (SE, *Fig. 35d*) is confined to the south east of the country and is recognised for all taxa except the Orthoptera. Finally, the fifth region (LIMB, *Fig. 35e*), the smallest and most distinct with by far the largest number of characteristic species, is largely concentrated in the southern part of the province Limburg. Taken together, these five regions cover about 40% of the country's surface. Between these hotspots of uniqueness there are large differences among regions in terms of coverage by the National Ecological Network (*Fig. 35*). The SAND region is by far the best-covered region and in particular the LIMB region is poorly covered by the National Ecological Network.

Uniqueness less well represented than species richness
In general we can say that for the selected species groups, the hotspots of uniqueness are less well represented by the National Ecological Network than the hotspots of species richness. This comparison stops to hold true, however, if we focus on the core areas of the uniqueness hotspots. If we only take into account the grid squares that are appointed by many different groups (≥3) the cover percentages are larger. The LIMB region however, remains under represented.

6.4   Discussion

6.4.1   Data used

First of all we have to put our results into perspective with a few comments on the data used. Due to the uncertainties of the species occurrence data used (Schouten et al., in prep. b), results of this analysis have to be treated with care. Ad-hoc data collection produced temporally, geographically and taxonomically biased data. Then, the resolution of our data is, with 5x5 km, very coarse. This resolution is, of course, not suitable to make a very precise evaluation of the current state of the NEN or to make detailed recommendations regarding the hectares still to be added to the NEN. Having said this, we think that we are still able to draw a general biogeographical outline of the Netherlands with these data and can make some brief statements on how well the National Ecological Network contributes to the ultimate goal of the preservation of a representative set of biodiversity for the Netherlands.

6.4.2   Hotspots of species richness

We looked into two dimensions of biodiversity, overall species richness (hotspots of species richness) and uniqueness of the species composition (hotspots of uniqueness). Hotspots of species richness do not coincide perfectly for the different groups. This can have important implications for nature conservation as a larger area may be needed in order to preserve the full range of species. The National Ecological Network seems to be extensive enough to cover most of this variation although it should be noted that the Syrphidae and the Bryophytes, with a species richness pattern strongly deviating from the other groups, are poorly covered by the National Ecological Network.

6.4.3   Hotspots of uniqueness

If we look at the hotspots of uniqueness we see that these patterns differ from the species richness patterns. The FEN region, for example, is not important from a species richness point of view but from an international perspective it is of great importance as many of the species occurring here depend on the Netherlands for their existence. The hotspots of uniqueness, in their widest sense, are except for the SAND region not very well represented within the NEN. When more rigid criteria are applied, coverage improves, although the LIMB region stays underrepresented in the NEN. However, the species from the LIMB region usually are not of international importance as they are species with a southern or alpine distribution, being sub marginal species in the Netherlands. The low coverage of LIMB region by the NEN is also due to the type of landscape. The southern part of Limburg is of course a fragmented landscape with scattered patches of nature in an agricultural matrix and the LIMB region is rich in species because it contains so many ecological gradients. Since the NEN primarily consists of nature areas, coverage will automatically be low. However, these results may be an indication that not only 'true' nature areas are worth preserving , but that landscapes or systems as a whole are also important for the preservation of biodiversity.

6.4.4   NEN focus on vascular plants

These results, although they are quite coarse, indicate that the current position of National Ecological Network is at least subjective. In particular the hotspots of species richness of the syrphidae and Bryophytes are poorly covered by the NEN. This is not a surprising result if one knows that lower taxa are underrepresented in the current biodiversity policy. Vascular plant species fulfil a key function in environmental and conservation research and policy, with more than half of all the target species being vascular plants (LNV, 2000; Bal et al., 2001). However, with approximately 1.450 species, vascular plants form only a small part of the total biodiversity of the Netherlands. There is no reason to assume that cryptobiota or lower taxa of plants are less at risk or of less importance then the well-studied vascular plant, bird and vertebrate species.

6.4.5   NEN focus on rareness and degree of threat

Besides the bias towards vascular plant species, the focus of the current nature conservation policy on the rareness and degree of threat to species implies a risk. Rarity is, of course, a relative concept. What is rare in the Netherlands can be very common in the rest of Europe and more important what is not threatened now can become so in the near future (due to for example the slow impact of habitat fragmentation). We therefore conclude that a better and broader selection of target species is needed. First off all, less popular groups should be better represented, and there should also be more attention for the more 'common' species, thus preventing them from becoming rare species.

6.4.6   Area size and connectivity

As our data are not suitable (for example no information on species abundance available) for in-depth studies on the population dynamics of the species it is difficult to draw conclusions on topics such as area size and connectivity. However, by appointing regions of uniqueness of species composition we showed that there is a clear biogeographical zonation in the Netherlands. Not all species can, potentially, occur everywhere in the Netherlands: biogeography and habitat preferences make that species are restricted to certain geographic locations. Therefore, it might not be necessary to interconnect as many areas as possible by robust connections, as currently seems to be the practice. It could even be a potential risk if non-native species that are currently on the move due to climate change, some of which having the potential to develop into invasive species, could freely migrate northwards. We do however think that areas that are alike from a biogeographical point of view but currently fragmented should be re-connected, as areas should be large enough to sustain healthy populations for many species.

6.5   Recommendations

6.5.1   Target species representing biodiversity

Based on our results, we conclude the target species should give a better representation of biodiversity if we want to successfully maintain and preserve the biological diversity of the Netherlands. Therefore, there should be more attention for less well-known groups such as the majority of insect taxa and lower plant species and for the more 'common' species as well. More research on species from currently underrepresented taxa is necessary in order to get a better understanding of the biological diversity of the Netherlands.

6.5.2   Focus on uniqueness

We do not think it is necessary to interconnect all areas as currently is the practice, but areas that are biogeographically alike should be (re)connected, or enlarged. The focus on areas of uniqueness seems to be more than promising than the hotspot of species richness approach. Species richness hotspots differ for the different groups and for many species groups no distribution information is available. We were, however, able to identify five regions with a distinct species composition for a broad array of taxonomic groups. Fine-scaled studies on these hotspots of uniqueness should be carried out in order to find out what within these regions the hotspots of diversity are and how the quality of the less diverse or degraded areas can be improved.

7   LANDSCAPE IN THE DUTCH 'NATIONAL ECOLOGICAL NETWORK'

Johannes Renes[a]

---

[a] This article is derived from the Dutch version Renes, 2006

7.1 Introduction

The 1989 national Nature Policy Plan ('Natuurbeleidsplan') introduced the National Ecological Network (National Ecological Network NEN, in Dutch: Ecologische Hoofdstructuur), a system of reserves connected by linear ecological zones, which was scheduled to be realised within thirty years. By concentrating the available money on these areas, the plan aimed at a sustainable system of interconnected ecological zones. According to the original plan, the NEN would consist of a variety of different types of areas, including more or less natural ecosystems as well as ecologically rich historic cultural landscapes. Fifteen years onwards and midway in the programme, it is time to reflect as well as to assess the plan's progress and prospects. This article focuses on the impact of the NEN on cultural landscapes.[a]

The NEN has been the single most important issue in Dutch nature conservation in the last fifteen years. Discussions focused on the so-called 'new nature', the most generally used term for what was also referred to as areas that were 'returned to nature' or remoulded into 'real nature'. Landscapes played a minor role in these discussions. In my opinion, this is not justified, in view of the substantial consequences of the NEN for the landscape. The core concern addressed in this article is the question of what fifteen years of NEN have meant for the landscape. Therefore, it is necessary to define the concepts of 'nature' and 'landscape', after which follows a summary of the history of the protection of nature and landscape in the Netherlands, focusing on the distinction between a broad and a narrow definition of nature.[b] This distinction is crucial for a good understanding of the choices made in the design and execution of the NEN.

7.2 The concepts of Nature and Landscape

For many people, the concepts of nature and landscape are almost synonyms designating the land outside the towns. In Dutch popular language it is quite normal to speak of 'going into the midst of nature' when the destination of an outing is in fact an agrarian landscape or a pine forest. In the official Dutch policy, however, the distinction between the two concepts is central. The Department of Agriculture, Nature and Food Quality (in Dutch: LNV, *Ministerie van Landbouw, Natuur en Voedselkwaliteit*), for example, employs a fairly large staff who are responsible for 'nature' and a considerably smaller staff engaged in 'landscape'. Within the Department, the term 'landscape' is used for that

---

[a] The author thanks Mrs Gina Rozario for correcting the English text.
[b] For background information, see for example Keulartz et al., 2000.

part of the environment that is shaped by man, whereas 'nature' denotes the part that develops spontaneously. This distinction is, however, a bit too simplistic, and therefore I will try to define both terms more clearly.

Landscape

The word 'landscape' has a long history and it has acquired numerous meanings. The main distinction is that between territorial and visual meanings. The first type is older. In the oldest known documents, the term 'landscape' is more or less synonymous with 'region': the territorially bound society and its relations with the physical environment. Also the accessory type of government (including the governors themselves) and the whole complex of use rights on the land were designated as 'landscape' (Hidding et al., 2001, p. 9; Olwig, 1996).

During the fifteenth and sixteenth centuries, painters in the Low Countries started to make representations of rural sceneries, which they called 'landscapes' ('*landschappen*' in Dutch). From then on, a landscape could hang on the wall; landscapes became an important theme in painting. Next, the painted object itself was increasingly named 'landscape', which turned 'landscape' into a visual concept that referred to an aesthetic experience, a kind of three-dimensional postcard. Recently, the perception of landscapes was broadened to include other senses: there is a growing pile of publications on 'soundscapes', 'smellscapes' and 'feelscapes'.

It is good to realise that the way we perceive landscapes is strongly influenced by our culture. Many centuries of landscape painting has led to the development of conventions that guide our perceptions. Most landscape paintings, as well as the photographs we take while on holidays, have a standard layout with the horizon occupying on one-third of the surface from the upper or lower end of the photograph, as well as a diagonal, usually from the upper left to the lower right, with a dark foreground, a lighter middle ground and a clear horizon. We rarely realise that our landscape paintings and holiday pictures show the influence of linear perspective that was developed in Europe during the Renaissance. A comparison with landscape paintings from non-European cultures (for example, the indigenous population of Australia) clearly demonstrates the degree to which our perception is culturally biased. The geographer Denis Cosgrove speaks of a European, 'landscape way of seeing' (Cosgrove, 1984).

The different meanings are still in use side by side. In German, for example, the meaning of the word '*Landschaft*', still retains its original meaning, whereas the English 'landscape', on the other hand, is very much a visual term. In English, the original Anglo-Saxon word '*landscipe*' disappeared during the Middle Ages and the present word 'landscape' was introduced by Dutch painters, hence its visual emphasis. In the Netherlands the original, regional, meaning still exists in a fossilised form in the historic name of the province of Drenthe: '*de olde landschap*' (the old landscape). But also among landscape ecologists, who usually define landscape in terms of relationships between the abiotic, biotic and human spheres, the original meaning is still alive.[a] The general public, on the other hand, uses the word 'landscape' usually with a very visual meaning. The similarity shared by all these meanings is that they all have something to do with perception.

For landscape management both groups of meanings are relevant. The visual landscape is, as was explained in the preceding section, a mental construction. This landscape is described mainly in aesthetic terms: beautiful or ugly, large or small scale, open or enclosed. The landscape in the other meaning, as a region together with its visual appearance, can, however, to a certain degree be described more objectively. We can describe and map the elements and structures that together constitutes this landscape.

Nature

The term 'nature' is as complex as 'landscape'. The authoritative Dutch dictionary *Van Dale* provides no fewer than eleven meanings, many of which are about human 'nature'. Popular sayings such as 'the human nature', 'following his nature' and 'nature takes precedence over doctrine' all point to the part of human behaviour that is genetic instead of cultural. The term is also used for the environment, where it again refers to the part that is not man-made. To cite *Van Dale* once again: 'what man sees around him and what is being considered as not yet changed by man, the landscape', whereby the last addition is rather curious.

When speaking about the management of our environment, the term 'nature' is particularly used for the part that exists without human interference. the Netherlands is home to abundant nature in the form of plants and animals that exist and reproduce mainly without human interference. On the other hand, there is no ecosystem in the country that is without human influence; the Dutch landscape is a cultural landscape. This is not unique to the Netherlands; elsewhere on the planet systems free from human interference are also extremely rare.

Ecosystems that were developing after the last Ice Age were already affected by human activities in the form of hunting, some degree of selection and, since ca 7000 BP, agriculture. The beech, a tree that nowadays is dominant in many parts of the country, was only starting its progression in the period that farmers started to influence the landscape

---

[a] An example is the definition by Schroevers (1982, p. 37): 'a landscape is a set of relations that originated from and is maintained by a framework consisting of interconnected spheres of operation and that by its external appearance forms a distinguishable part of the terres-trial space.'

more intensively than the earlier hunters and gatherers. Completely natural landscapes are, therefore, only known from modelling studies while incomplete landscapes can be identified from comparisons with relatively natural systems elsewhere on the world. But at a second glance, these seemingly natural systems prove to be more affected by human action than most people think. The forest of Bialowiezca, probably the most original European forest, has a long history of hunting (Schama, 1995). Recent studies made clear that even the Amazon rain forest must have had during pre-Columbian times a considerably higher population density than nowadays and that large parts must in fact be characterised as secondary forest (which is no argument to treat it less carefully).[a]

Apart from the question of the degree to which humans affected natural ecosystems, we cannot separate nature from society (in this article, I will not join the debate about whether humans are part of nature).[b] There are several arguments supporting this. In the first place, 'nature' is often presented as landscape. Nature photography often shows us nature through landscape pictures. Moreover, perception and assessment of nature is an activity undertaken by humans. Although nature exists without humans, it only receives meaning within the context of human society. Therefore we could even include nature within the domain of the cultural landscape. Without culture, one is unable to distinguish nature. So nature as we perceive it is, to a certain degree, a social construction, the meaning of which is made by society and varies in space and time (See, for example, Castree, 2005).

The way we perceive and value nature says much about ourselves. We perceive nature from our own knowledge, emotions and judgments. The most extreme proof of this position is demonstrated daily on the TV network Animal Planet, which often describes animal behaviour in terms of human behaviour, choices and emotions. These judgments find their way into practical nature management. Anyone who searches the web site of the Department of Agriculture, Nature and Food Quality for 'butterflies' is guided to data on protected animals and nature management. A search for 'caterpillars' links to the Newsletter of the Plant Protection Service of the Netherlands.

7.3 Nature management and landscape: a historical sketch

In practice we use the terms nature and landscape often as opposites, whereby 'landscape' points to the historic and present human influence and 'nature' to what exists mainly without human intervention. This common

---

[a] See, for example, Denevan (1992) and Mann (2006).
[b] Cronon, 1995; Mels, 1999.

parlance has a long history in the Netherlands. Nature protection started during the nineteenth century. A landmark was the protest against the felling of the last 'natural forest', the Beekbergerwoud, during which the author Van Eeden coined the term 'natural monument'. In 1905 this term was adopted by the Society for the Preservation of Natural Monuments (usually known as '*Natuurmonumenten*'), which was founded to protect the Naardermeer (Lake Naarden southeast of Amsterdam), which was threatened by plans to turn it into a waste disposal site. In the following year the new society bought the lake. *Natuurmonumenten* developed into the Dutch equivalent to the English National Trust, although with less emphasis on cultural landscapes. For a long period, conservationists mainly worked on safeguarding natural sites of value by buying up the affected areas. The main organisations were *Natuurmonumenten*, the State Forestry Service ('*Staatsbosbeheer*') and, later, the Provincial Landscapes (provincial offshoots of *Natuurmonumenten*, with the same aims but working on a provincial level). In the course of the 1920s and 1930s these organisations started to buy landed estates, including country houses. During the Crisis of the 1930s, especially *Natuurmonumenten* and a few of the Provincial Landscapes (*Het Geldersch Landschap, Het Utrechts Landschap*) offered a safety net for landed estates when their private owners could no longer maintain them.

Based on a very different motivation, in the beginning of the twentieth century urban planners started to campaign for the protection of agrarian landscapes, usually designated as 'landscape beauty'. From this standpoint, planning instruments were developed to protect areas against urban threats. Perhaps the most important single campaign was the battle against ribbon development, aimed at opposing the encroachment of uncontrolled urban sprawl into beautiful landscapes. The main battlefield was situated at the borders of the river Gein, a small stream southeast of Amsterdam. The opponents of urban sprawl triumphed when the municipality took measures against uncontrolled building activities in the rural area. It was the first step that would ultimately lead to the detailed municipal land-use plans which form the basis of the present Dutch planning system.

New insights caused nature and landscape protectionists to converge and concur. One of these insights was that the protected nature consisted almost completely of cultural landscapes that acquired their characteristics from human activities. Protection proved insufficient to maintain areas such as heathlands: without specific management biological succession that took place ultimately changed the heathlands into forest. Nature protection and physical planning both appeared to be involved in cultural landscapes.

Secondly, it proved necessary – but also complicated – to shield protected areas from external influences. Many reserves were threatened by drainage systems and manuring practices of neighbouring farms. This made it necessary for preservationists to get involved in planning. In 1931 planners and conservationists together founded the *Contact-Commissie inzake Natuurbescherming* (Contact Commission for Nature Protection), renamed ten years later as *Contact-Commissie voor Natuur- en Landschapsbescherming* (Contact Commission for Nature and Landscape Protection), and in 1972 merged with other organisations to form the *Stichting Natuur en Milieu* (Nature and Environment Foundation). For many years the *Contact-Commissie* had a working group in charge of cultural landscapes (Dekker, 1993). The biologist Victor Westhoff, one of the members of that working group, showed how the influence of man could be enriching. This insight laid the basis for a prolonged cooperation between nature and landscape protection. Westhoff (1916-2001) has been very influential in Dutch nature protection for half a century (Windt, 1995).

Consequently a long period followed in which the protection of nature and landscape went hand in hand, albeit with little success. During the postwar period, spatial planning in the rural areas was monopolised by the agricultural sector that, forced by European subsidies, started a systematic destruction of the old cultural landscapes. While the positive effects of land consolidation on agriculture are probably overrated, the negative effects on nature and landscape are enormous and irreparable. Agriculture was necessary to manage the ecologically rich historic landscapes, but refused to do so.

The 1980s were a period of growing consciousness in environmental matters. Parties involved in nature protection chose for a new strategy, in which land was withdrawn from agriculture and given over to natural processes. Pioneering efforts concerned developments in the Oostvaardersplassen, a wetland area in the province of Flevoland, which for a number of years was left alone and that had developed into a rich bird sanctuary. Another booster was the *Ooievaar* plan for the fluviatile region, which proposed to return the floodplains 'to nature'. Execution of the plan was possible because of changes in agriculture, particularly in dairy farming that was bound to fixed production levels after the introduction of quota (1984). Because production per hectare and per cow kept rising, agricultural land was in excess.

In the course of the 1980s studies on the effects of a receding agriculture were published.[a] One of the possibilities that were investigated was a more extensive type of agriculture that would cause less damage to nature and landscape. The chosen development showed a stronger separation of functions (Heyd, 2005, p. 350), in which agriculture abandoned marginal

---

[a] Among others Meeus, 1988; Padding and Scholten, 1988; Wetenschappelijke Raad **voor** het Regeringsbeleid WRR, 1992.

110

lands, such as the floodplains along the major rivers, that could be transformed into 'new' or 'real nature'. The further intensification of agriculture in the remaining parts of the fluviatile region, that was also part of the *Ooievaar* plan, brought a kind of logical coalition between agriculture and nature development. Nature was, just like agriculture, a sectoral interest that was mainly expressed in hectares. Of course this led to a loss of nature and landscape values in the remaining agricultural areas, an outcome that is too often forgotten.

For a number of ecologists this type of nature was the future, for some it even was the only future. Nature managers, who had always fought with farmers for every square inch, saw themselves in command of large surfaces in which they could create their own ideal landscapes. The cultural landscape around the new nature was mainly seen as a worthless agrarian desert.

In 1989, exactly in the period when the pendulum started to swing through in the direction of 'new nature', the national government presented the *Natuurbeleidsplan* (Nature Policy Plan). This plan has had a clear influence on the way the NEN was executed. Had the Nature Policy Plan been published ten years earlier or later, then the plans for the NEN would have contained more landscape and less 'new nature'. In 1999, for example, the influential Belvedere Memorandum, that greatly stimulated the interest in historic landscapes, was published.
Within the sector of nature- and landscape protection the 1990s were dominated by the contrast between 'old' and 'new' nature. In this controversy the term 'old nature' covers the traditional nature reserves, consisting of half-natural ecosystems, as well as the cultural landscapes. During the 1990s, the competition between old and new nature reached many decibels. Many of the advocates of 'new nature' showed little interest or compassion with the historic cultural landscapes.

The differences between both opinions on the point of nature protection are large; it is not an exaggeration to say that we are dealing with different discourses. The advocates of 'old' and those of 'new nature' differ in aims, ethical starting points, sources of inspiration and working methods. With regard to ethical starting points, in nature conservation there has always been a clear vision that puts a strong emphasis on natural ecosystems (Elliott 1997; Palmer, 1997). The main ethical argument for nature protection is that all living creatures, apart from their benefit to humans, have the right to live. From this point of view, many ecologists will conclude that natural landscapes have a higher value than man-made cultural landscapes. Advocates of historical cultural landscapes will find their roots more in the ethics of art, in which a historical landscape as a relic of past human culture is comparable with the paintings of Rembrandt or the medieval cathedrals.

7.4 'New nature' as landscape

When writing on the influence of the NEN on the landscape, a preliminary remark is necessary: 'new nature' is a type of landscape. 'New nature' is akin to nature without human interference, a situation (as has been mentioned earlier) that is only known from (re)constructions. New nature is being constructed within a landscape that has been influenced by humans for thousands of years, often irreversibly (See for example Louwe Kooij-mans, 1997). The use of terms such as 'real', 'authentic' of 'pristine' nature is therefore based on illusions. Moreover, such terminology makes it too easy to forget that new nature is one that is designed. However much the impression is given of undisturbed natural processes, the construction of hills and lakes, the choice of a certain groundwater level and the decision on the density of grazing animals are not the result of the laws of nature but derive from the conceptions of the designers. The result is a designed landscape (See also Bijlsma, 1995). It is a landscape that looks familiar to us: the park-like landscape that perhaps existed at the birth of mankind, which may be one reason for the comforting feelings it brings. It is the type of landscape we find in medieval game parks and that has dominated our gardens and urban parklands since the eighteenth century (therefore sometimes nature development projects are said to follow a 'neo-landscape style'). One of the main pilgrimage sites for designers of new nature is the New Forest, a medieval deer park in England that has been managed in a traditional, but nonetheless rather intensive way.[a] Whoever refuses to see this type of landscape as cultural landscape is disillusioning him/herself and others.

As such, the discussion is back to the situation a century ago, when nature protection was carried out from the basic viewpoint that nature and man were worlds apart. Man had destroyed Paradise (that is, by the way, described in the Bible as a garden) and had to be removed from nature. Apt examples of this way of thinking are the nineteenth-century American National Parks (hence the 'Yellowstone model')[b], from which the original population that had managed these areas for thousands of years was evicted. In a number of African nature parks there is still the tendency to see the original population mainly as a danger to the stock of game. In these examples, social class determines whether one is seen as a hunter or as a poacher.

In the Netherlands the designation of nature reserves is not coupled with human rights violations, but here also the way farmers are kept out sometimes looks a bit overdone. That is especially the case where for example the farmer's own cows are evicted because they are unnatural

---

[a] Renes, 2005, with extensive reference to literature on this theme.
[b] Stevens, 1997.

elements in the landscape, and replaced by imported Scottish Highlander cows. The motivation seems to be a negative view of humans. As a lover of music, architecture and historic landscapes, such a one-sided view of humanity is incomprehensible to this author.

7.5   Landscape and 'old nature'

In all the attention paid to the construction of 'new' nature, it can easily be forgotten that there is still old nature: the old, ecologically rich cultural landscapes that once were the basis of the nature protection movement. Those landscapes are, by the way, situated inside as well as outside the NEN. Many of the most important lie within the NEN, because the development of the NEN started with the landscapes already under protection. Therefore, in its first few years the NEN comprised largely from agrarian cultural landscapes that had been safeguarded during the years because of their natural as well as their landscape values.

It is too easy to regard these old cultural landscapes only as second-rank nature. In the first place particular natural values are connected to areas that have known an enduring and stable type of management (Dirkx et al., 1992). A considerable portion of the animal and plant species on the Red List is tied to cultural landscapes. It is exactly because of the historical dimension of such nature that these ecosystems are extremely difficult to reproduce. The English ecologist G.F. Peterken once commented on the planting of new forests: *"However much one may wish otherwise, few of the new woods will acquire a full representation of woodland [plant] species, and the species that will not be there include some of the most characteristic and attractive"* (Cited by O. Rackham in Journal of Historical Geography 20, 1994, p. 345).

Moreover, (old) cultural landscapes are not only important for their ecological qualities. Next to their ecological values, landscapes hold historical (see for example Etteger, 1999 and Bervaes, 2004) and perception values, and a management based on preserving these qualities will have broader aims than those concerned with the number of species only.

The perception of landscape is strongly dependent on the perceiver. Although there is certainly some commonality in the assessment of landscapes – for example in the earlier mentioned predisposition for half-open, park-like landscapes – such assessment shows huge individual differences. In the perception of landscapes the personal involvement (for example personal memories) and knowledge are important.

Connected with landscape perception is another theme, namely landscape diversity, that is in turn connected to biodiversity. In general it is widely accepted that biodiversity has diminished during the last century. Although the relation between enlargement of scale and diminishing (bio)diversity is less simple than is often supposed – every region develops its own answers to the forces of scale enlargement and globalisation, which leads to new types of diversity – it is difficult to deny that the unifying forces are stronger than the opposing forces. The enormous variation in hedgerows and other field boundaries, with differing types of species and management depending on the region, has disappeared with the introduction of barbed wire.[a] In the same way the diversity in animal and plant breeds has diminished. It is difficult to establish when landscape diversity was at its greatest. Most authors are inclined to compare not with a theoretical or historical optimum but with an earlier situation they know, such as the 1950s or the situation as depicted by the eldest topographical maps.[b] To take one indicator, the total length of hedgerows seems to have reached its maximum around 1920. In this respect the historical geographer Leenders spoke of 'the climax of the enclosed landscape' (Leenders, 1995). Until this period, new reclamations were surrounded by hedges, and it was around 1920 that the fast advance of barbed wire as fencing for demarcation purposes started. Shortly afterwards the first land consolidation projects (*ruilverkavelingen*) began, resulting in ever bigger losses of hedges and other landscape features.

Historical qualities are especially based on time-depth and on historical layers of the landscape. Landscape wears the traces of thousands of years of developments and the many historical relics make it possible to 'read' this history (Barends et al., 2000). One cannot stress often enough that landscape preservationists do not aim at a reconstruction of a former ideal landscape, such as the landscape of the 19th century. It is misguiding to justify far-reaching changes in the landscape with the argument that 'new history' is being made. The prominent nature development advocate Frans Vera once wrote: "Although existing cultural landscapes can be lost in the process, nature development means not a loss of our culture, but a next step" (Vera, 1998). Of course our generation has the duty to add new features to the landscape and to demand high standards for the new additions. But anyone who destroys a historic landscape for the construction of new nature, while adding a new layer to the landscape, at the same time shortens the visible landscape history by several millennia. Historical landscapes differ from (parts of) nature, because they cannot be (re)constructed and they are not replicable. Landscapes can be imitated, but they relate to an old landscape just as a poster relates to a real painting of Van Gogh.

---

[a] For the diversity in management types in the past, see Burny (1999), Burm and Haartsen (2003) and Dirkmaat (2005).
[b] Ecologists use the term 'shifting baseline syndrome' (Pauly, 1995; Sheppard, 1995; this author thanks Dr F. Vera for bringing attention to this notion).

Looking at the old cultural landscapes over the past fifteen years, we have to state that the NEN has meant very little for these landscapes. On the one hand, the efforts and the financial means for nature and landscape were highly concentrated within the NEN, resulting in a continuing degradation of nature and landscape values outside of the NEN. In fact, the changing ideas on land consolidation and, even more, the financial cuts in this activity, have done more for the protection of the surviving historic landscapes than all nature and landscape policies by the different levels of government put together.

International comparisons make clear that the Netherlands has succeeded in its nature policy, but that landscape policy lags far behind (the future will show whether the new National Landscapes are a turning point).[a] The Nature Policy Plan suggested that the old cultural landscapes would have a future within the NEN.[b] The plan contained for example a map that showed substantial overlap of the NEN with the so-called 'Areas with Specific Landscape Values' (*Gebieden met specifieke landschappelijke waarden*). This latter category was, by the way, useless by its very vagueness, comprising areas with geological values (including the entire North Sea), areas with historic values and small-scale landscapes. As usual in Dutch landscape policy, the designated areas show only accidental resemblances with the areas designated in earlier or in more recent reports, controllable criteria were lacking and the amount of money involved was negligible (Renes, 2001).

Hence in practice the landscape part of the NEN has been very disappointing. This is partly because many nature development projects have been executed with little knowledge of, and interest in, landscapes. This often brought unnecessary damage to historic landscape features. Examples are the remodelling of medieval drainage ditches into curving naturally-looking streams. As natural streams they are wrongly situated and the historic landscape has become more difficult to perceive.[c] Another problem occurs in areas that have once been bought by organisations for nature conservation to protect the combined ecological, landscape and historical values and that are now being managed through the use of large grazing animals. It leads to the neglect and sometimes to the destruction of historic landscape features.

---

[a] Aitchison, 1995; Van der Weijden and Middelkoop, 1998.
[b] Nature Management Plan, 1990, pp. 96-97. See also the report on historical cultural landscapes that was produced as part of the preparations for the Nature Management Plan (Haartsen et al., 1989). In general, pleas for the importance of cultural landscapes (for example Mabelis, 1990) for nature conservation and biodiversity received little response during the 1990s.
[c] See for example of such faked natural streams the Vloedgraaf in Central Limburg and the Heigraaf in the east of the province of Utrecht. The word '*graaf*', comes from the verb '*graven*', Dutch for digging.

A more fundamental problem is the shift of attention and funding from 'old nature' and landscape to new nature. The main reason for this is the unequal division of power within the Department of Agriculture, Nature and Food Quality that is responsible for landscape policy. Agriculture has always been the priority. New nature, as a sectoral issue, can relatively easily be combined with intensive agriculture. Care for the landscape is much more complicated, because it makes changes in agricultural practices necessary. While some of the farmers were open to discussion, the Department and the main Farmers Unions were not (Renes, 2001).

A good illustration of the problem is the recent Management Program (*Programma Beheer*) that awards subsidies for results instead of good intentions, which is a positive point. However, there are a number of problems. In the first place, subsidies went from areas to objects. In the past, a manager of a nature reserve had a certain freedom to use the subsidy in his area. The shift of the subsidy to individual objects not only leads to an incredible amount of extra paper work, but also makes it impossible to operate from a broader vision on the future of the reserve. The landscape is lost from sight and faces relegation with regard to priorities. In the second place the possibilities to apply for subsidies are limited to a certain number of objects. These objects seem to be selected for their ecological values. Archaeological objects, buildings, alleys, solitary trees and field walls are excluded. The introduction of the Management Programme therefore means a shift of money from 'landscape' to 'nature'. Subsidies of this kind create their own reality. When only certain types of objects come into consideration for subsidies, these objects will have a larger role in the future landscape.

7.6   Discussion

Landscape is often wrongfully neglected in discussions on the NEN. There are three reasons to ascribe a leading role to the landscape within the NEN. Firstly, as already mentioned, 'new nature' is also a type of landscape that is often as much designed for landscape as for nature. Secondly, the NEN consists of cultural landscapes and important parts of the areas within the NEN are bought and put under protection especially for their historical landscape values. Thirdly, there are the agrarian and urban landscapes outside the NEN that also merit investment in their ecological and landscape qualities.

It is difficult to judge whether the present choices in the execution of the NEN provide more biodiversity than the 'old nature'. According to some, a completely natural system gives the maximal biodiversity, but other experts argue that certain influences of humans have enhanced not only landscape diversity but also biodiversity.[a] There may be a relationship whereby a growing intensity of agrarian use leads to a growing diversity, up to a certain point. Past that point, and diversity diminishes again. An example could be the black-tailed godwit (Limosa limosa), a bird that is characteristic of the pastures in the fenlands. The godwit reached its highest numbers some decades ago, when the intensification of agriculture led to a growing biomass. Since then, the numbers diminished, probably because the ever more intensifying level of agricultural activities affects the living conditions of the bird.

Quite a few animal and plant species are highly dependent on cultural landscapes. Often these are species that would have been very rare in the natural Dutch landscape, but are now part of the Dutch responsibility in an international context.

It is important to realise that the concept of diversity should be judged on different scales. Reconstructing a landscape into new nature will often increase biodiversity within the area involved. But when these nature projects tend to become standardised (hills, lakes, large grazing animals), landscape diversity in the country as a whole can diminish. It is therefore essential to consider where such nature development will take place. *Fig. 37* offers a possible test.

Potential for 'new nature'	Actual values (historical, geological, ecological, perception)	
	HIGH	**LOW**
HIGH	Preservation of cultural landscape and half-natural ecosystems; small-scale nature development within the framework of the historic landscape	Development of large scale 'new nature'
LOW	Preservation of cultural landscape	Development of new landscapes

Fig. 37 Aims for nature and landscape policy based on potential and actual values.

Fig. 37 shows how the execution of the NEN can overcome the present, rather narrow-minded, approach by combining nature aims with an assessment of the value and qualities of the present landscape. For the floodplains in the fluviatile region, this would mean that large-scale construction of new nature would be limited to those areas that have limited

---

[a] Some recent publications hat support this opinion are Dahlstöm (2006) and Heyd (2005, p. 342).

actual values. This would differ from the present policy, in which parts of the floodplains that have not yet been excavated for gravel, sand or loam (and that therefore have the highest historic landscape values) are dug up, whereby the new nature is paid for by the sale of the excavated minerals (Renes, 2003).

Meanwhile, the controversy between old and new nature has died down. Most large organisations, after the hype of new nature was over, went back to a more balanced policy in which the desired future for every individual area is discussed. New nature has become one of a number of possible strategies. There are, however, large differences between individual organisations. The *Geldersch Landschap*, now merged with the regional Trust for Historic Castles (*Vrienden van de Geldersche Kasteelen*) is probably the Dutch organisation that exercises the most balanced relationship between nature and culture. The World Wildlife Fund, on the other hand, is still very much focused on 'wild nature'.

Where historic landscapes have survived in the last fifteen years, it is only to a very limited degree attributable to the merit of the NEN. The diminishing pressure from agriculture seems a much more important factor. In the future, agriculture can even become a partner in landscape management. Ever more farmers realise that landscape management not only gives them access to subsidies, but that they need the support of the larger (urban) population for their own survival.

8 THE NATURA 2000 NETWORK IN THE NETHERLANDS

Setting targets and conservation objectives

Marion Pelk[a]

---

[a] This article is based on the Natura 2000 targets document (LNV, 2006), especially on the summary of this document.

8.1   Introduction

The Natura 2000 targets document (in Dutch: Natura 2000 doelendocument), a policy document of the Dutch Ministry of Agriculture, Nature and
Food Quality (LNV, 2006) explains the targets for the 162 Natura 2000
sites. It forms the framework for the designation decisions and also directs
the Natura 2000 management plans to be drawn up.

Both are important steps in the implementation of the European Natura
2000 network in the Netherlands. Taking as its basis the documents used
previously in selecting and designating the Natura 2000 sites, the Natura
2000 targets document sets out the system used to formulate Natura 2000
targets at both national and site level. The Natura 2000 targets document
also specifies the criteria used to delineate the Natura 2000 sites. For the
bird species, habitat types and other species for which the Netherlands has
European responsibility, conservation status, relative importance in Europe
and main objectives for the Netherlands are determined for the purpose of
formulating the Natura 2000 targets and conservation objectives. If available a common European approach has been chosen (for example for
conservation status).
The Natura 2000 targets document is primarily intended for the parties
responsible for drafting the management plans and for the parties involved
or having an interest in the designation decisions for the Natura 2000 sites.
This does not affect the fact that the document, in connection with the designation decisions, can also play a role in the granting of specific permits.

Before examining the underlying philosophy of the Natura 2000 targets
document, we will start with a brief explanation of the Natura 2000 network.

8.2   Natura 2000

8.2.1   The concept

The European Union has set itself the target of halting the decline in biodiversity by 2010 (Göthenburg, 2003). One of the main instruments by which
this objective is to be achieved is the implementation of the site-specific
parts of the Birds and Habitats Directive. This means setting up a network
of nature areas of European importance: the Natura 2000 network. The
main objective of the network is to safeguard biodiversity in Europe. In this
connection, it has been agreed that the Member States of the European
Union will take all necessary measures to ensure a 'favourable conservation status' of species and habitat types of Community importance. In the
Netherlands, this involves 95 bird species, 51 habitat types and 36 other
species The design and selection of the measures takes account of economic, social and cultural requirements as well as specific regional and

120

local aspects. Clearly, particular importance is assigned to measures to be adopted in the context of safety.

The Netherlands' contribution to the Natura 2000 network is based on 162 sites totalling around a million hectares (two-thirds of which are open water). The Natura 2000 sites are part of the Dutch National Ecological Network (NEN). Nearly 100% of the terrestrial area of Natura 2000 is part of the NEN. About 40% of the terrestrial area of the NEN is also Natura 2000.

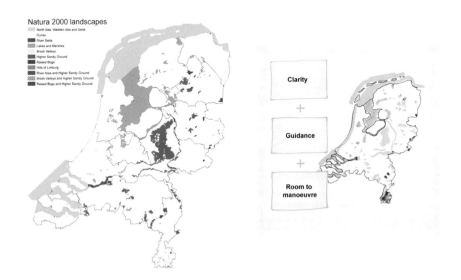

Fig. 38 Natura 2000 landscapes and sites

(LNV, 2006 Natura 2000 targets document)

Fig. 39 Philosophy treefold: Offering clarity, guidance and room to manoeuvre

(LNV, 2006, Natura 2000 targets document)

8.2.2   Instruments used

Because the Natura 2000 sites lie almost entirely within the National Ecological Network, the current or future ecological network measures will also enhance the quality of Natura 2000. Policy and funds for the national ecological netwerk thus also benefit European nature objectives (Natura 2000 targets). Dutch efforts to enforce other European Directives, such as Nitrates Directive and Water Framework Directive will also affect the timeframe for Dutch measures under the Birds and Habitats Directive.

Based on the revised 1998 Natura Conservation Act are three instruments available to realise the set of Natura 2000 targets and conservation objectives of the Natura 2000 sites. These instruments are progressively more specific. They are:

1. designation decisions,
2. management plans
3. and permits.

Designation decisions

A designation decision lays down the conservation objectives and the ex-cact area and delineation of the designated area. It is a formal, legally bind-ing, decision. Designation decisions apply for an indefinite period and are laid down by the Minister of Agriculture, Nature and Food Quality. A desig-nation decision may be revised following a periodic evaluation. In 2006 the Netherlands had formally designated nearly all the conservation sites re-quired under the Birds directive. However these designations do not com-ply with the amended 1998 Nature Conservation Act, as the do not set out the conservation objectives for the areas. The Habitat Directive sites have not yet been formally designated, although they have been notified to the European Commission.

The first tranche of draft decisions has been published for public review early in 2007. Publication of the other designation decisions is expected in the course of 2007. On the basis of the views submitted on the draft desig-nation decisions and the opinions of the provinces, the Minister of Agricul-ture, Nature and Food Quality will adopt and publish the final decisions decisions in the course of 2007 and 2008. Interested parties and organisa-tions which have previously submitted a view will have the opportunity to appeal against the final decision.

Management plans

The drafting and adoption of the Natura 2000 management plans, obliga-tory in the Netherlands, is an important next step. The competent authori-ties (provincial governments, Ministries of Defence, of Transport and Water Works and of Agriculture, Nature and Food Quality) are currently making preparations for the management plan drafting process and interaction with involved or interested parties.

Management plans indicate which policy and management measures are needed to protect and conserve the targeted habitats and species in an area. The plans also describe how this is to be combined with other (exist-ing or planned) land use in the area. Management plans can be used to assess applications for permits under the revised 1998 Nature Conserva-tion Act. Management plans tell site managers, users and stakeholders which activities are prohibited until they have been assessed in acordance with Article 6 of the Habitats Directive and which activities do not require a permit. In other words: the management plan ensures that economic, social and cultural interests are considered when conservation measures are chosen and designed. Management plans runs for six years.

Under the revised 1998 Nature Conservation Act, the above mentioned ministries are the competent authority for areas managed by the national government. The provincial governments are the competent authority for establishing management plans in all other cases. This means that each Natura 2000 site will have more than one competent authority. In principle the provincial governments take precedence, unless more than 50% of the area is managed by the national government. Precedence to ensure that the plan is efficiently drafted and to facilitate consultation with stakeholders.

A manual Management Plan Guide (LNV, 2005) has being written to help those drawing up management plans to translate the programme of requirements into practical measures. Based on the Natura 2000 targets document an update is forseen.

The next three years will be a challenging time for national and provincial authorities charged with drawing up management plans. Plans must be drawn up with 'the field' for all 162 Natura 2000 site. The challenge will be to do this jointly, effectively and efficiently together with the implementation of de Water Framework Directive.

Permits

Permits under the revised 1998 Nature Conservation Act lay down the conditions under which an activity or plan is allowed in or near a Natura 2000 area. Permits are valid for a limited period and are usually issued by the Provincial Executives. In some cases (activities of overriding national importance), a permit is issued by the Minister of Agriculture, Nature and Food Quality. These activities are defined precisely in a General Administrative Order.

8.3 Philosophy of the Natura 2000 targets document

8.3.1 Introduction

The basic philosophy of the Natura 2000 targets document is threefold: clarity, guidance and room to manoeuvre (see *Fig. 39*). One of the underlying ideas in this respect is that the Ministry of Agriculture, Nature and Food Quality has opted to have the details of the conservation objectives, in terms of extent, location and time schedules, worked out in the Natura 2000 management plans. The reasoning behind this is that it is at the level of the Natura 2000 management plans, in interaction with the users concerned and the site managers, that we can best determine where exactly and with what features, to what extent and at what pace the conservation objectives can be achieved[a].

---

[a] This approach is not directly related to the system of setting targets for the NEN, the philosophy is that both systems benefit from each other.

In order to ensure cohesion between the contribution made by individual sites and the contribution made by the Netherlands Natura 2000 network to European biodiversity, a number of choices have been made at the level of the Natura 2000 targets document and the designation decisions. In view of the Netherlands' European responsibility, therefore, this document gives clarity and guidance, where necessary, the further details worked out in the Natura 2000 management plans.

8.3.2   Clarity

The species and habitat types that fall under Natura 2000 must be maintained at or brought up to 'favourable conservation status' at the national level. The Netherlands' contribution to the Natura 2000 network is based on 162 sites totalling around a million hectares (two-thirds of which are open water).

What is required for 'favourable conservation status' is reflected in the Natura 2000 targets. These targets have been set both at national level and for each specific site. The total of all of these targets shows the contribution to be made by the Netherlands to the European network of Natura 2000 sites. A number of choices have been made in this respect, at both national and site level.

Those choices are based on eight guiding principles:

1. Targets should be in harmony with national policy wherever possible, especially on the creation of the National Ecological Network (provided it is in line with our European obligations).
2. Targets should be practically and financially feasible and require minimum effort from the public and the economic sectors and have minimum consequences for them.
3. Existing quality and size should be maintained and, where necessary, improved, both at national level and in each individual site.
4. More effort should be directed at those species and habitat types for which the Netherlands plays a more crucial role, or those whose survival is seriously threatened.
5. Less effort should be expended on species or habitat types where improvement cannot reasonably be expected.
6. Targets should anticipate natural dynamics and climate change, by being able to withstand the test of time.
7. Targets should direct conservation and management efforts in a site, while at the same time leaving scope for a local approach.
8. Targets should be set taking account of existing management budgets.

Examples of choices already made are: for the Haringvliet, conservation objectives are in harmony with the agreements relating to the 'kier' (whereby some of the Haringvliet storm barrier's sluices are left open). For heaths and sand drifts, more effort is needed in the context of management. For fish-eating birds, such as the great crested grebe and little gull, no recovery objective has been formulated at national level as yet. The options for improving the quality of the habitat (sites IJsselmeer and Markermeer and IJmeer) are to be investigated first. For the site Oosterschelde, the conservation objectives have been adjusted in line with the annual decline of the sandflats. In view of the site's vital importance for shellfish-eaters, the options for minimising the decline of the sandflats are being investigated.

Conservation objectives should guide conservation and management efforts in the sites, but also leave scope for a local approach. In formulating conservation objectives at site level, the balance between 'guidance' and 'room to manoeuvre' was therefore an important starting point.

8.3.3   Guidance

As indicated above, details of the conservation objectives, in terms of extent, location and time schedules, are worked out in the Natura 2000 management plans. To that end, further guidance is given with the aid of:

1. core tasks,
2. maintenance or improvement conservation objectives,
3. sense of urgency and
4. credit formulations.

The core tasks indicate the most important contributions that a specific site can make to the Natura 2000 network. They also give an idea of the main buttons that would have to be pressed in order to continue delivering that contribution or to deliver it in the long term. The core tasks are an important aid to focus and the necessary prioritising that may have to be done within the Natura 2000 management plans.

Guidance is given by indicating whether the target is directed solely at maintaining the existing situation or at improving the situation. Based on the Natura 2000 targets at national level and an assessment of the situation in specific sites, it is indicated whether the contribution made by a particular site is sufficient (maintenance target) or whether that site will have to make a greater contribution in the long term (improvement target) in order to achieve the target at national level. In assigning a maintenance or improvement target to specific sites, the principle of strategic localisation was used.

Sense of urgency is used to guide the pace of realisation of the conservation objectives (and the use of necessary measures). In view of the current conservation status at national level and the situation in specific sites, a sense of urgency has been assigned to a number of core tasks. A sense of urgency pertains if an irreparable situation is likely to occur within the next

10 years. A sense of urgency may relate to a problem with water conditions or with land management.

In a limited number of situations, 'credit formulations' have been used. This means that a slight reduction has been permitted for a particular species or habitat type, to the credit of a different species or habitat type. In such a case, the latter is under severe threat and the aim is to expand its habitat or area (in case of habitat types). If there is insufficient scope to achieve both conservation objectives side by side within a particular site, the choice is made in the designation decisions. For example, for a number of species of goose, it has been specified that the size of the foraging area may be re-duced slightly to the credit of, for example, wet alluvial forests or dry grass-lands or the river area. The 'credit formulation' has been applied with as much restraint as possible. This means that, for the other conservation objectives in the further elaboration of the conservation objectives in the management plans, there is sufficient scope in the sites to work out their extent, location and time schedules in more detail. For a number of habitat types and species, 'occurrences' outside the Natura 2000 sites contribute towards the achievement of the national target. For a small number of habitat types and species the decision was taken to formulate 'complementary' conservation objectives. This means, for example, that a target has been formulated for *Molinia* meadows (H6410) in a Birds Directive site. This concerns habitat types and species for which the Netherlands has a special responsibility and which are under threat. One of the reasons is to focus the objective on the Natura 2000 sites. In addition, in a few situations con-servation objectives have been formulated for the development of features that do not yet exist, such as active raise bogs (H7110).

8.3.4   Room to manoevre

An important element of the philosophy of the Natura 2000 targets docu-ment is that the extent, location and time schedules of the conservation objectives are worked out in more detail in the Natura 2000 management plans. There is scope in the management plans for indicating where exactly and at what pace measures must be taken in order to achieve conservation objectives. The starting point is and remains that the conditions for habitat types and species must not deteriorate. This also means that the Natura 2000 management plans state the pace at which a maintenance or im-provement target is to be achieved.

The degree of scope for this more detailed interpretation depends on the nature of the habitat type or species. If the latter places very specific de-mands on the environment, or is difficult to replace or move, there will be less scope. On the other hand, in situations with an active natural dynamic, the system itself 'determines' where, for example, humid dune slacks (H2190) or white dunes (H2120) will occur in a given period. This may also have consequences for the nature and intensity of measures. For example, it is possible to conceive of situations where, as a consequence of an 'improvement target', 'no intervention' is an appropriate measure. The 'room to manoeuvre' approach also creates scope for adjusting the pace of realisation or the nature of the measures in line with developments in understanding and new information, for example.

In the Natura 2000 management plans it is possible to determine, in consultation with interested parties and land managers, which measures will be taken in order to achieve the conservation objectives. In connection with the localisation of conservation objectives, the management plan can also consider where those measures can best be localised. In determining the measures and in localising the improvement conservation objectives in particular in the Natura 2000 management plans, the principle of strategic localisation can also be applied. The principle of strategic localisation was used initially in assigning maintenance or improvement conservation objectives to specific sites.

8.3.5   Strategic localisation

In formulating conservation objectives at site level, account was taken of current quality as well as the options for maintaining or creating a sustainable situation in the long term. Maintenance targets were assigned, for example, if the habitat of a species is already up to standard. Or if it is not possible to improve the ecological requirements any further, or the effort is not counterbalanced by the resulting additional contribution that the site could make towards achieving the Natura 2000 target at national level. The choices were based on the analyses and the site-specific information. For example, the KIWA and EGG report (2005/2006) provides information about opportunities and challenges for habitat types in specific sites and possible ways of overcoming the challenges. Improvement targets were assigned, for example, if the habitat type is not yet up to standard but the necessary measures have already been taken or planned. Improvement targets were also assigned if specific measures are to be planned on the basis of existing policy, for example anti-groundwater depletion policy. The essence of strategic localisation is that improvement targets are assigned primarily to those sites where the maximum benefit (contribution towards achieving Natura 2000 targets at national level) can be achieved with the minimum effort.

8.4   Approach

8.4.1   Introduction

Preparation

The Natura 2000 targets document was first drafted between November 2004 and October 2005 on the basis of discussions involving experts and site managers, available information and expert judgement. From December 2005, consultations were held on the basis of that draft with Dutch provinces and other authorities as well as social organisations, economic sectors and other interested parties.

The Ministries of Agriculture, Nature and Food Quality and of Transport and Water Works also organised joint regional consultations in each of the river basin areas about the Natura 2000 targets and harmonisation with te Water Framework Directive.

In parallel, a quick scan was carried out by KIWA and EGG (2005/2006) and a global cost estimate was undertaken by the Agricultural Economics Research Institute (LEI). The reactions and global cost estimate were also used as input for the final Natura 2000 targets document and the conservation objectives incorporated into the draft designation decisions. The Natura 2000 targets document also has been discussed with the Dutch parliament.

Evaluation dates

The Natura 2000 targets document is based on the best available information at this time. Due to a number of uncertainties about expected developments, such as natural dynamics and climate change, an evaluation is scheduled for the year 2015. If necessary, changes will then be made to conservation objectives and designation decisions. An evaluation of the Water Framework Directive is also scheduled for 2015, which may also influence the Natura 2000 targets. Targets (and conservation objectives) may be recalibrated and, if necessary, adjusted if sites are designated in a subsequent tranche and when the management plans are drawn up.

As at the moment that the Natura 2000 has been published only for the 110 Natura 2000 areas an agreement on the targets and boundaries has been reached with other departments and the provincial governments there is a possiblity to change the document if necessary.

Two process lines

For the preparation of the Natura 2000 targets document a dual approach was taken. One process line focuses on habitat types and species and leads to the targets at national level and also to a picture of the relative importance and conservation status of the habitat types and species for which the Netherlands has responsibility. The second process line leads to conservation objectives at site level. The analyses carried out in connection with this second process line provided important input for the purpose of assigning conservation objectives to specific sites. These steps are explained below. The main results are set out in the next chapter. This chapter concerns the approach taken in formulating the Natura 2000 targets at national level and at site level.

8.4.2   Process line 1: Habitat types and species

More detailed interpretation

Step 1: Define habitat types and species
A concrete product of step 1 are the Natura 2000 profiles (LNV 2007). This document describes the types of vegetation that come under a particular habitat type. A number of definitions have been adjusted compared with an earlier version, for the sake of better coordination with the definition process in the neighbouring countries and the interpretation used in a European context. The profiles also consist of the following elements: profile (description, relative importance), quality (characteristics of good structure and function), ecological requirements, contribution of sites, assessment of national conservation status (including the definition of favourable conservation status).

Assessments

Step 2: Determine relative importance within EU
The extent to which the Netherlands can make a contribution for a particular habitat type or species to favourable conservation status at European level is determined by:

1. the position of the Netherlands within the area of distribution;
2. the extent of occurrence in the Netherlands;
3. the Dutch proportion of the total European area (in the case of habitat types);
4. the Dutch proportion of the European population or the biogeografical population (in the case of species);
5. contribution to ecological variation (this is the case when a habitat type in the Netherlands is clearly richer in species or clearly different from elsewhere in Europe).

Step 3: Assess conservation status
In assessing the conservation status of habitat types and species, the 'traffic light approach' developed in an EU context was adopted. 'Favourable' is green, 'unfavourable – inadequat' is amber and 'unfavourable - bad' is red. In the case of habitat types, the aspects of distribution, size, quality and future prospects were examined. In the case of species the aspects studied were distribution, size of population, habitat and future prospects. The assessment was based on inventory and monitoring data and on best expert judgment. Many experts and site managers took part.

Objectives and targets

Step 4: Define improvement targets
Improvement targets were formulated for all habitat types and species, on the basis of relative importance and conservation status. These targets were then imposed, in addition to the assessment of 'favourable conserva-

tion status', at national level. The most important improvement targets at national level were then put together. They were formulated primarily for habitat types and species for which the Netherlands plays a crucial role and whose conservation status is 'unfavourable - inadequate' or 'unfavourable - bad'. The targets relate to better management or to improving ecological requirements and may concern both individual habitat types or species and cohesive landscapes or systems.

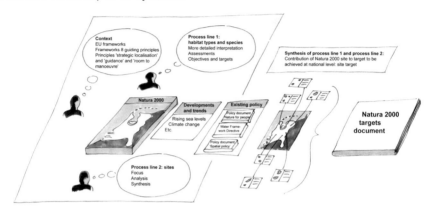

Fig. 40 Natura 2000 targets - Approach
(LNV, 2006, Natura 2000 targets document)

Step 5: Formulate targets at national level

If the conservation status was unfavourable, the targets were formulated in principle in terms of improving quality or extending the area or distribution. If a recovery target was deemed unrealistic, an exception was made. If the conservation status was favourable, the targets were formulated in principle in terms of preserving the current situation. An improvement target may have been formulated to benefit geographical distribution, ecological variation and/or risk spreading (for specific sites).

8.4.3   Process line 2: Sites

Focus

Step 1: Assign sites to Natura 2000 landscape

For the sake of a convenient arrangement, eight Natura 2000 landscapes were identified and individual Natura 2000 sites assigned to them.

Step 2: Formulate core tasks

Core tasks were formulated for the eight Natura 2000 landscapes. This was based on the information from process line 1 about habitat types and species, supplemented by the specific processes for the landscape and specific control options. The core tasks relate primarily to habitat types and species which are under severe threat and/or for which the Netherlands plays an important or crucial role. The core tasks set priorities (including priorities for the management plans) and highlight the similarities and differences between the sites.

Analysis

Step 4: Assess feasibility
For each site and for each central objective, the current situation and the 'seriousness of the objective' were assessed. Feasibility was indicated using the traffic light approach. Green means that the desired situation has already been achieved or will be achieved with measures already planned. Amber means partial feasibility with existing policy. Red means that the desired situation cannot be achieved with existing policy. A sense of urgency was assigned if an irreparable situation is likely to arise within the next 10 years.

Step 5: Assess seriousness of the objective
The 'seriousness of the objective' indicates, in terms of time, money, scale and social acceptance, the size of the disparity between the current situation and the desired situation. This was also indicated in terms of green, amber and red. This analysis was based partly on meetings with experts and managers, a quick scan by the KIWA and EGG focused on water conditions, and information from the Netherlands Environmental Assessment Agency (MNP) and the Dutch Centre for Field Ornothology (SOVON). In addition, the global cost analysis drawn up by the LEI was used for the final formulation of conservation objectives at site level.

Synthesis

Step 6: Formulate site conservation objectives
On the basis of the previous steps, 'focus' and 'analysis', the site conservation objectives were formulated in terms of preservation and/or improvement. For habitat types and species that come under one of the core tasks focused on improvement, conservation objectives were generally formulated in terms of extending and/or improving. A higher target was set in an site if the site has a high potential for certain habitat types or species and relatively little effort is required in order to attain that higher target. In an site with the same potential but where more effort is required, a lower target was set. This made sure that the above-mentioned principles of 'harmony with existing policy wherever possible' and 'practically and financially feasible objectives' were taken into account.

8.5 Natura 2000 targets document in relation to earlier documents

In relation to the frameworks for the formulation of conservation objectives and the designation decisions, it should be pointed out that the Natura 2000 targets document cannot be viewed separately from the earlier documents drawn up for the purpose of selecting and designating the Birds Directive sites in 2000 and the documents drawn up for the purpose of selecting and proposing the Habitats Directive sites (2003-2004). The method used for the selection and designation or registration of sites in documents drawn up for that purpose, such as the Memorandum of Reply Birds Directive (2000), the Justification document (2003) and the List document (2004), is unchanged. The Natura 2000 targets document does state the conditions under which and how conservation objectives are formulated at site level. In principle, conservation objectives are formulated for habitat types and species included in the Natura 2000 databases as submitted to the European Commission in 2004. As a result of the enlargement of the EU to 25 Member States, the species ram's-horn snail was added for the Netherlands. In relation to birds, the lesser white-fronted goose was added on the basis of a decision of the highest Administrative Court (Council of State).

On the basis of new information about the occurrence of habitat types and species, the current assessment of whether a site is making or will be able to make a relevant contribution for a species or habitat type, conservation objectives were formulated for specific sites. This may mean that a species of bird for which a conservation target was formulated in an earlier designation decision is now deemed (on the basis of the criteria applied in 2000) not to require a target any more. The reverse situation may also occur. This means that the criteria were applied using the data from the period 1999-2003 (SOVON and CBS, 2006). The same analysis was applied for habitat types and species. In the case of habitat types, the habitat types for which the Netherlands has responsibility were interpreted in more detail. This, in combination with the assessment of conservation status, led to the proposing of changes to the Natura 2000 databases. The Natura 2000 sites documents state, for each site, if and why changes are to be made to the Natura 2000 database.

8.6 Results

8.6.1 Relative importance

In determining the relative importance of nature in the Netherlands in a European context, we gained a clearer picture of the importance of the Netherlands for habitat types and species for which the Netherlands has taken on the responsibility of protection on a sustainable basis. One of the main ideas underlying the formulation of Natura 2000 targets is that more effort is required for species and habitat types for which the Netherlands plays a crucial role and for species and habitat types that are under severe pressure.

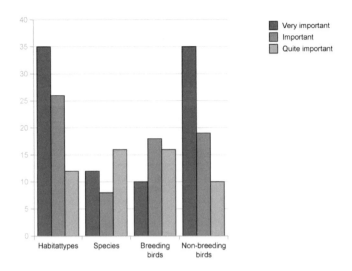

Fig. 41 Relative importance (LNV, 2006, Natura 2000 targets document)

Despite its small area, the Netherlands is of crucial relative importance in a European context for a number of bird species, habitat types, and other species.The *Fig. 41 Relative importance* gives a picture of the distribution of scores between the various categories of relative importance for bird species[a], habitat types and other species. This results in the following picture. For more than half of the non-breeding birds, the Netherlands is of very great importance. The relative importance of the species in Annex II of the Habitats Directive and of the breeding birds is the same: around a quarter scores in the category 'very important'.

8.6.2   Conservation status of habitat types and species

The *Fig. 42 Conservation status* gives a picture of the scores of the various categories of conservation status for the bird species, habitat types, and other species. The figure states the absolute numbers (in the case of habitat types, the numbers of subtypes). To summarise the picture: at present, 12% of habitat types has a favourable conservation status, 54% an 'unfavourable - inadequate' status and 34% an 'unfavourable - bad' status.

---

[a] Some bird species are as well breeding bird as non breeding bird: (95 birdspecies 44 breeding bird and 64 non breeding birds).

133

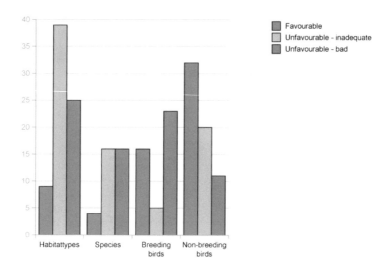

Fig. 42 Conservation status (LNV, 2006, Natura 2000 targets document)

The habitat types relatively often score 'unfavourable inadequate'in relation to quality. In the case of other species, only a minority (12%) has a favourable conservation status, 44% is assessed as 'unfavourable - inadequate' and 44% as 'unfavourable - bad'. In the case of breeding birds, 36% scores 'favourable', 11% 'unfavourable - inadequate' and 52% 'unfavourable - bad'. The score 'unfavourable – bad' for breeding birds coincides to a large extent with the score 'unfavourable – inadequate' for a number of habitat types that form important habitat where improving quality is the main objective. In the case of non-breeding birds, 51% scores 'favourable', 32% 'unfavourable - inadequate' and 17% 'unfavourable - bad'. Non-breeding birds score best, in relative terms, on conservation status.

The picture for each Natura 2000 landscape is different for the bird species, habitat types, and other species. In the landscapes Lakes and Marshes, Higher Sandy Ground and Hills of Limburg, no habitat types score 'favourable'. In the case of species, the proportion of species scoring 'unfavourable - inadequate' is largest in the landscapes Dunes, River Area, Higher Sandy Ground and Brook Valleys. Breeding birds and non-breeding birds show a relatively large proportion scoring 'favourable' in virtually all of the landscapes. A striking result is the relatively large proportion of birds scoring 'unfavourable - bad' in the Natura 2000 landscapes Dunes, Lakes and Marshes and raised bogs. In the case of non-breeding birds, the proportion 'favourable score' in virtually all landscapes is the largest, most clearly in the North Sea, Wadden Sea and Delta, River Area and Lakes and Marshes.

134

8.6.3   Core tasks and sense of urgency

The Natura 2000 targets document includes the main objectives for the Natura 2000 network for the habitat types and species. Those objectives may relate to more appropriate management and/or better arrangement of ecological requirements. Examples include extending damp and wet grasslands and improving quality, improving the quality of sand drifts, open grassy cover in the dunes for breeding birds such as the short-eared owl and improving the quality of the habitat for marsh breeding birds.

The essence lies not in an approach focused on specific habitat types and specific species, but in an approach focused on their mutual cohesion. For the Natura 2000 landscapes, therefore, objectives of landscape cohesion and internal completeness were adopted, which were then elaborated further in core tasks. One of the main objectives at landscape level is preserving sufficient scope for dynamic processes. In dynamic systems in particular, conservation cannot mean that everything stays the same. For example, erosion and sedimentation processes change the size and locations of embryonic shifting dunes from year to year.

For a long-term sustainable result, a coherent approach, as indicated in the core tasks, is necessary. Parts of sites which do not contain any Birds and/or Habitats Directive features also have a specific significance for the sustainable conservation of Natura 2000 habitat types and species. For example, they are necessary in order to create sustainable ecological conditions, for the internal and external cohesion of Natura 2000 sites or the long-term introduction of planned natural features.

The *Fig. 43 Core tasks* summarises the core tasks for two Natura 2000 landscapes.

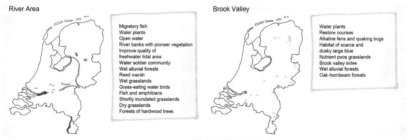

Fig. 43 Core tasks for two Natura 2000 landscapes
(LNV, 2006, Natura 2000 targets document, summary)

A total of 97 core tasks have been assigned to Natura 2000 sites. For a number of core tasks, a sense of urgency has been assigned to specific sites. A sense of urgency was assigned if an irreparable situation is likely to arise within the next 10 years. The assessment is that a core task and the associated conservation objectives will no longer be achievable otherwise.

This means, for example, that specific measures must be taken within the next 10 years in order to bring water conditions up to standard.

Core task	Description	Why	Assigned to sites
1.12	High-tide roosts for birds: Preserve and restore undisturbed high-tide roosts.	Great international importance for migratory birds.	120. Zoommeer; 127. Markiezaat
3.08	Reed marsh: Improve quality and increase size of reed marsh with the associated breeding birds (bittern A021, great reed warbler A298), plus root vole *H1340.	Root vole is a priority. International importance for Dutch subspecies of root vole and for bittern due to large proportion of the population and central location. The River Area has long been of great importance within the Netherlands for threatened marsh birds such as black tern, bittern and great reed warbler.	36. Uiterwaarden Zwarte water and Vecht (, w); 67. Gelderse Poort (, , w); 105. Zouweboezem (w); 112. Biesbosch (w).
5.03	Alkaline fens and quaking bogs: Restore quality and extend area of alkaline fens H7230 and transition mires and quaking bogs (quaking bogs) H7140_A, in mosaic with poor grasslands.	International importance for transition mires and quaking bogs (quaking bogs) in the Atlantic region due to central location and relatively large size. Important at national level for many unusual species and potentially for slender green feather moss. Current area of both habitat types is small.	25. Drentsche Aa-gebied (w); 28. Elperstroomgebied (, w); 45. Springendal and Dal van de Mosbeek (w); 48. Lemselermaten (, w); 52. Boddenbroek (w); 60. Stelkampsveld (w); 65. Binnenveld (, w); 130. Langstraat (w).

Fig. 44 Examples of core tasks

29 Sites have been assigned a sense of urgency in relation to water conditions ( ) (26 land sites and 3 large bodies of water). A water objective (w) was assigned if the water conditions are not up to standard to a greater or lesser extent. 31 sites were assigned a sense of urgency in relation to management ( ). In particular, this concerns the core tasks for grey dunes, reed marsh, dry grasslands and black grouse. The core tasks and the indication of sense of urgency are important to the focus of the Natura 2000 management plans and also to the prioritising of measures.

8.6.4   Natura 2000 targets at national level

The Natura 2000 targets at national level, like those at site level, are formulated in terms of preserving the area and quality of a habitat type, preserving or improving the quality of the habitat of a species, and/or extending the distribution of a species or habitat type. Based on these guiding principles, the Natura 2000 targets were formulated and a number of choices made.

In the case of targets at national level, it was decided to set a high objective for habitat types which have a very unfavourable conservation status and for which the Netherlands has a major responsibility. This relates largely to poor grasslands, the area and quality of which has declined greatly in recent decades. Examples are grey dunes (H2130), *Nardus* grasslands (H6230), *Molinia* meadows (H6410) and alkaline fens (H7230). In various sites, both increasing size and improving quality have been set as conservation objectives. For all landscapes, a recovery objective has been set for the majority of habitat types. This may relate to both increasing size and improving quality, or simply to improving quality.

In the case of species, a recovery objective has been set at national level for more than half of the species. This relates in particular to butterflies, dragonflies and migratory fish as well as to the beaver (H1337) and root vole (H1340). The Natura 2000 landscapes North Sea, Wadden Sea and Delta, River Area, Brook Valleys and Lakes and Marshes in particular are required to make a contribution in this respect. In the case of breeding birds, a conservation objective has been formulated for 23 of the 44 species at national level. The species for which a recovery objective has been set at national level are primarily species which occur in the Natura 2000 landscapes Dunes, Higher Sandy Ground, Raised Bogs and Lakes and Marshes. This concerns marsh birds such as bittern (A021) and little bittern (A022), and dune birds such as wheatear (A277) and short-eared owl (A222) In the case of Higher Sandy Ground and Raised Bogs, it concerns species such as whinchat (A275) and red-backed shrike (A338).

In the case of non-breeding birds, conservation objectives have been formulated for 56 of the 64 species at national level. A recovery target at national level has been formulated for only 9 species. A recovery objective has been formulated for a number of shellfish-eaters in particular. That target is to be achieved by the Natura 2000 landscape North Sea, Wadden Sea and Delta (from the Wadden Sea area). For a number of species such as fish eaters, no national recovery objective has been set for the time being. A recovery objective has been set at national level for the golden plover (A140), black-tailed godwit (A156) and crane (A127). An improvement in the situation must come from outside the Natura 2000 sites.

Target type	Code and description	Natura 2000 target	Explanation
Habitat type	H91F0 Riparian mixed forests of Quercus robur, Ulmus laevis and Ulmus minor, Fraxinus excelsior or Fraxinus angustifolia, along great rivers (Ulmenion minoris)	Increase distribution, increase size and improve quality.	Increase area of habitat type dry hardwood forests at likely sites, preferably near existing sites and adjacent to existing forests. This can be achieved in part by converting cultivated forests at suitable sites. The best opportunities for extension are in sheltered parts of the River Area, where there is no damming up of water, for example in the shelter of railway embankments and bridges. In terms of location, there are good opportunities for extension at sites where the river cuts through a lateral moraine, as in the IJssel valley (Uiterwaarden IJssel (38)), along the Lower Rhine (Uiterwaarden Neder-Rijn (66)) and along the Overijsselse Vecht (Uiterwaarden Zwarte water en Vecht (36)).
Species	H1321 Geoffroy's bat	Preserve size and quality of habitat in order to maintain population.	The only two breeding colonies are in Lillbosch Abbey and the former Mariahoop convent (151). The species overwinters in marlpits in South Limburg: Bemelerberg and Schiepersberg (156), Geuldal (157), Sint Pietersberg and Jekerdal (159) and Savelsbos (160). A large proportion of the habitat (especially the foraging area) is outside Natura 2000. The current summer population comprises 250 – 500 individuals.
Breeding birds	A292 Savi's warbler	Increase size and improve quality of habitat in order to restore a population of at least 5 key populations of 100-400 pairs with a minimum overall total of 2,000 pairs.	Savi's warbler is a breeding bird of the European mainland, found north as far as Estonia. It is most common in Eastern Europe and, due to the decline in numbers in Northwest Europe, the Dutch population is an important and somewhat isolated outpost in the west. With 3% of the EU population, the relative importance is great. The numbers in the Netherlands, but especially the size of the area, have declined, resulting in a very unfavourable conservation status for the aspects of distribution, habitat and future prospects. The aspect of population is still assessed as favourable for the time being. The target set is in accordance with the recovery plan for marsh birds. Over 80% breed in Natura 2000 sites, particularly in Lakes and Marshes (especially in Oostvaarderssplassen (78) (over 25%), Wieden (34) and Weerribben (35)) and around 10. River Area (mainly in % in Biesbosch (112)).
Non breeding birds	A068 Smew	Preserve size and quality of habitat with capacity for a population of 690 birds on average (seasonal average).	The declining tendency in the long term is not significant, due to wide fluctuations. The international trend is positive, however, and the decline in the Netherlands may be a consequence of climate-related shifts in the overwintering areas. In the most important region (IJsselmeer region), the quality of the habitat has deteriorated, though (poor smelt status in IJsselmeer (72) and Markermeer and IJmeer (73)). The future is uncertain due to continuing climate changes. Concentrating this species in the IJsselmeer regio makes the

138

Target type	Code and description	Natura 2000 target	Explanation
			future even more uncertain in view of developments in the fish stock. There is great international responsibility due to the large proportion of the international population that makes its home in the Netherlands (15-25%). No recovery target applies due to the difficulty in controlling suspected causes. The capacity estimate was calculated over 1997-2003, a period following the decline of smelt in Markermeer and IJmeer (73). Within the Natura 2000 network, the sites IJsselmeer (72), Markermeer and IJmeer (73), Veluwerandmeren (76), Oudegaasterbrekken, Fluessen e.o (10) and Alde Feanen (13) make the greatest contribution.

Fig. 45 Examples of Natura 2000 targets at national level

8.6.5   Conservation objectives at site level

In formulating the Natura 2000 targets at site level, a number of standard formulations were used. Some general conservation objectives and conservation objectives for habitat types, species, breeding birds and non-breeding birds have been formulated.

Conservation objectives were formulated, for example, in terms of preserving area and maintaining quality (for habitat types) or increasing the size and/or improving the quality of the habitat with carrying capacity for a population of at least xx pairs. The Reader's Guide to the Natura 2000 sites documents (LNV, 2006) contains a summary of these formulations and, where necessary, states how the explanations of the conservation objectives should be read. It also indicates the situations in which certain formulations are used.

General conservation objectives

1. Preserve the contribution of the Natura 2000 site to biological diversity and favourable conservation status of natural habitats and species within the European Union.
2. Preserve the contribution of the Natura 2000 site to the ecological coherence of the Natura 2000 network both within the Netherlands and within the European Union.
3. Preserve and where necessary restore spatial cohesion with the environment for the purpose of sustainable conservation of the natural habitats and species occurring within the Netherlands.
4. Preserve and where necessary restore the natural characteristics and the coherence of the ecological structure and functions of the site as a whole for all habitat types and species for which conservation targets have been formulated.

5. Preserve or restore site-specific ecological requirements for the sustainable conservation of the habitat types and species for which conservation objectives have been formulated.

Target type	Code and description	Conservation objective	Explanation
Habitat type	*H91E0 Alluvial forests with Alnus glutinosa and Fraxinus excelsior (Alno-Padion, Alnion incanae, Salicion albae)	Expand area and improve quality of wet alluvial forests, riparian forests (subtype C).	The habitat type wet alluvial forests, riparian forests (subtype C) occurs at many sites on the Veluwe, but in most cases it covers only a small area and is of moderate quality. Along the natural streams (Hierdense Beek) and on the transition to the IJssel valley (Middachten) there are larger and better-quality examples. For sustainable preservation of the biological community within the site, it is important to increase both area and quality.
Species	H1614 Creeping marshwort	Preserve area and quality of biotope in order to expand population.	At the moment this concerns one of the larger sites where creeping marshwort occurs in the Netherlands. The biotope of the species has expanded hugely due to natural development and the species has already become established in this new area. Expanding the population of creeping marshwort is necessary in order to preserve the species in this site in a sustainable manner.
Breeding birds	A295 Sedge warbler	Preserve size and quality of the habitat with carrying capacity for a population of at least 220 pairs.	The reedlands in the Groote Wielen are home to one of the key populations of sedge warbler in the Frisian Lakes. From 1993 to 1997, 100-125 pairs were counted each year. For the period 1999-2003, the average number of pairs is estimated at 220. In view of the national favourable conservation status in relation to population size, it is sufficient to preserve the existing situation. The site has sufficient capacity for a key population.
Non breeding birds	A056 Shoveler	Preserve size and quality of the habitat with carrying capacity for a population of 100 birds on average (seasonal average).	Shoveler numbers are of international importance. One of the site's roles is as a foraging area. The data are not yet suitable for a trend analysis. Maintaining the current situation is sufficient because the national conservation status is favourable.

Fig. 46 Examples of conservation objectives

9   ECOLOGICAL NETWORKS ACROSS EUROPE

Rob Jongman and Peter Veen

9.1   Introduction

Land use
In Europe structured development of urban and rural areas, landscape management, water management, road planning, agricultural development and conservation of natural areas have taken place for several hundred years. The main objective was to organise an efficient land use, which could fulfil all required functions in the most efficient way within a country.

Environment
Environmental conservation and environmental management reach far beyond the technical environmental protection such as air and water purification. It also includes functional ecological systems and their variety in spatial forms in their totality. Environmental and landscape planning for safeguarding and development of natural resources are priority issues for national and regional authorities. The notion of "Environment" comprises in its broadest sense the totality of all factors that are of importance for living species and living communities. It refers to the social and psychological environment of man. It is necessary to take natural resources and their mutual relations in consideration in landscape planning. This implicates also a close relationship between the use of natural resources, environmental management and spatial and landscape planning. The objective of spatial planning is to organise functions and space in such a way that it shows the best mutual relationship or, to develop human and natural potentials in a spatial framework in such a way, that all can develop as well as possible (Buchwald and Engelhardt, 1980).

Exploitation
What has characterised the concept of planning in Europe has been the institutionalisation of planning ideals like segregation and functionalism. Its implementation in planning is rooted in changes in social life: growing urbanisation and territorial demands for an increasing population. In planning the ideas of segregation have been pushed towards the extreme edge in the human exploitation of nature. Man's ability to change the forms and functions of the land and its impacts has developed a need for new values in planning decisions and in our concept of nature.

Ecological networks and remaining virgin forests and natural grasslands
In this chapter we deal with the new approaches in Europe for biodiversity conservation. In section 9.2 therefore the role of connectivity and connectedness in the modern fragmented European landscape is discussed. This leads to the conclusion that these features should be included in conservation strategies in the structure of ecological networks (section 9.3). The implementation in relation to the spatial scale of ecological networks is elaborated in section 9.4. The differences between countries in planning are shortly elaborated in section 9.5 and the role of stakeholders is dis-

cussed in section 9.6. As in ecological networks biodiversity conservation is moving outside the reserved areas and claims conservation measures in the wider countryside involvement of other land users and their consent and understanding is essential. A logical consequence is the treatment of public support in section 9.7.

In sections 9.8, 9.9 and 9.10 an illustration is given about aspects of national ecological networks based on field research results for two rare habitat types in Europe: virgin forests in Romania (see *Fig. 47*) and semi-natural and natural grasslands in Bulgaria (see *Fig. 48*).

These habitat types are important in view of remnants of old forests in Europe and hotspots of extensive farming practices. Both habitat types are at this moment under stress of intensification in land use. Virgin forests are under a pressure of cutting high value old trees. Natural grasslands are under a pressure of land abandonment at one side and an on going intensification process at the other side.

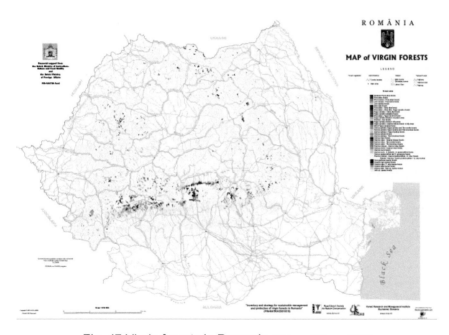

Fig. 47 Virgin forests in Romania (Biris and Veen, 2005)

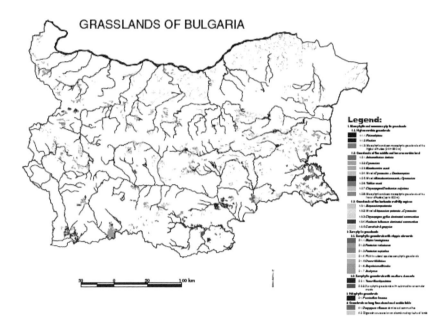

Fig. 48 Semi-natural and natural grasslands in Bulgaria (Meshinev et al., 2005)

9.2   Connectivity and connectedness

Core areas for populations

Migrating species are vulnerable in their lifecycle. They are not all year available to signal the importance of a site as a temporary habitat. European storks (*Ciconia ciconia*) for instance breed in large parts of Europe and they winter in Africa, migrating 10,000 km each season. Species have adapted to the cultural landscapes of Europe, because they were accessible and not hostile. Large areas with good living conditions that are always inhabited are defined as core areas for populations. In good reproductive years species will move from these areas into other – even marginal - sites (Verboom et al, 1991). Area reduction will cause a reduction of the populations that can survive and in this way it is an increased risk of extinction, because dispersal between habitats decreases, causing less exchange of genetic information and less colonisation of empty habitats.

Human impact

Increasing traffic and intensifying agriculture made the European cultural landscape more open on the one hand and more difficult to access on the other. Forests and hedgerows disappeared in intensively used agricultural land, forests became uniform production forests, streams have been straightened and the road-network became asphalted, denser and more intensively used. Last but not least many large and important wetlands have been drained.

Regulation by dispersal and migration

Plants and animals both disperse by wind, water, with help of other species or by own movements. Migration is a specification of dispersal, while it is directed to a certain site. Dispersal is essential in population survival and the functioning of biotopes. However, dispersal can only function if there are 1) sites to disperse from and to 2) means for dispersal. On the one hand animal species will leave a population if living conditions cannot support all individuals and on the other hand species will fill in gaps in populations or sites that are empty. Fluctuations in populations can cause changes in species abundance and species composition of a site. Birth, death, immigration and emigration are the main processes to regulate fluctuations at the population level. Plants depend on other species for their dispersal. However, plant strategies for dispersal are the least known and difficult to detect in practice. Restriction of species dispersal increases the chance of species extinction.

Connectivity and connectedness

The main functional aspect of in the landscape of importance for dispersal and persistence of populations is connectivity and connectedness (Baudry and Merriam 1988) Connectivity is a functional landscape parameter indicating the processes by which sub-populations of organisms are interconnected into a functional demographic unit. Connectedness refers to the structural links between elements of the spatial structure of a landscape and can be described from mappable elements. Structural elements can be different from functional parameters. For some species connectivity is measured in the single distance between sites, for other species it also has to include the structure of the landscape. The connectedness through hedgerows includes the posiblility of corridors and barriers.

Different requirements per species

Routes for species migration consist of zones that are accessible for the species to move from one site to another and back. Due to differences in needs migration and dispersal routes can be manifold, from single wooded banks to small-scale landscapes and from river shores to whole rivers and coastlines. For fish it means that rivers are not blocked by dams and of good water quality. For mammals and amphibians it means that routes are available and that man-made barriers can be crossed. These groups disperse over distances from several metres to hundreds of kilometres. For small mammals ecological corridors can be hedgerows, brooks and all kind of other natural features that offer shelter. Migration is important for grazing animals like red deer (*Cervus elaphus*) and roe deer (*Capreolus capreolus*), for predators like the golden eagle (*Aquila chrysaetos*), the lynx (*Lynx lynx* and *L. pardina*) and the wolf (*Canis lupus*).

9.3   The Structure of Ecological Networks

Core areas, buffer zones and corridors

Ecological networks can be defined as systems of areas of high biodiversity value and their interconnections that make a fragmented natural system coherent to support more biological diversity than in non-connected form. An ecological network is composed of core areas, (usually protected by) buffer zones and (connected through) ecological corridors (Bischoff and Jongman, 1993). Core areas have mostly been identified by traditional nature conservation policies as National Parks or Nature reserves. The insight gained from recent geographical and ecological concepts link this traditional conservation strategy with other land use and integrate nature conservation in general land use policy and spatial planning. In this way ecological corridors and buffer zones are becoming key elements in nature conservation strategy, but also highly discussed elements as they are the landscape elements where many functions coincide.

Biodiversity and sustainability

The meaning and the application of the ecological network concept has changed over the past decade with emphasis shifting from nature protection towards sustainable development for a region as a whole, integrating biodiversity issues. The observed change in thinking originates from the discourse in the international policy arena of the Convention on Biological Diversity, the World Summit on Sustainable Development and the Milennium Development Goals (MCDs), which perceive environment rather as making a contribution to sustainable development than as an intrinsic value to be protected from use. Implementation of these international agendas is increasingly guided by the Ecosystem Approach. This approach can be regarded as a strategy for the management of land, water and living resources that promotes conservation and sustainable use in an equitable way.

Greenbelt Planning

At the heart of the approach is the awareness that, without the effective and sustainable management of ecosystems, there can be no economic development that generates sustainable human and social welfare; equally, without the full engagement of diverse sectors in the economy and society in the management of ecosystems, there can be no effective biodiversity conservation. In that sense, the Ecosystem Approach is a framework for holistic decision-making and action (Bennett, 2004). This shift in emphasis runs parallel with changing paradigms in protected area management that have moved over the years from "strictly nature oriented" to "nature and people oriented" (Phillips, 2003). In some European countries (Portugal) as well as in the USA this approach is named Greenway Planning, integrating local interests with biodiversity conservation and building on the tradition of Greenbelt Planning and Parkway Planning (Jongman and Pungetti, 2004).

Ecological network

The definition of ecological network by Bennett and Wit (2001) and Bennett (2004) is in line with this paradigm shift: *"A coherent system of natural and/or semi-natural landscape elements that is configured and managed with the objective of maintaining or restoring ecological functions as a means to conserve biodiversity while also providing appropriate opportunities for the sustainable use of natural resources".*

Support of stakeholders

One consequence of perceiving an ecological network as a means towards sustainable development is the increasing number and diversity of stakeholders and land use interests that need to be incorporated in the design and should be part and parcel of the management process. In addition it will be evident that the institutionalisation of such a landscape change will greatly benefit from the overall support of the stakeholders. Or as Bennett (2004) puts it: *"No programme of the breath and ambition of an ecological network can achieve results without the active support of local communities and key stakeholders".*

National differences of protection

In Europe many, but not all, important natural areas are protected. Differences in definitions used by countries in Europe can be big and lead to confusion (Jongman et al 2004). Agriculture, forestry and leisure are in some cases allowed, in other cases integral part of the protected area. Traditional land use or land use techniques, especially extensive exploitation of grassland such as transhumance can be a method of management of semi-natural areas. Other categories of protected areas are areas for landscape conservation, nature parks, areas of outstanding natural beauty, etc. These areas can include protected areas for nature conservation. Agriculture, forestry and leisure are more or less limited by rules concerning land use, buildings and environmental protection. Public access is regulated differently. Now through the EU-Species and Habitats Directive (92/43/EEC) some coherency is brought into these developments in Europe. However, national differences will maintain to exist and be taken into account when designing and implementing ecological networks.

Corridors: functional connectivity and physical connectedness

Connectivity and connectedness come together in the concept of ecological corridors. Ecological corridors can be defined functionally to indicate connectivity and as physical structures to indicate connectedness. They can be defined as functional connections enabling dispersal and migration of species that could be subject to local extinction (Bouwma et al 2002). As physical structures they also can be defined as various landscape structures, other than core areas, in size and shape varying from wide to narrow and from meandering to straight structures, which represent links that permeate the landscape, maintaining or re-establishing natural connectivity (Jongman and Troumbis 1995).

Classifications of corridors

In addition to the above classification and according to functionality, corridors can be classified into three or four classes according to the shape that they have: linear, stepping stone and landscape corridors (see *Fig. 49*).

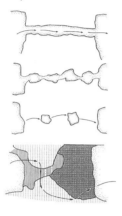 

Fig. 49 Different shapes of corridors: line corridors,

Fig. 50 Fish ladder in one of the headwaters of the Tweed (Scotland) for migration of Salmon
(Photo Rob Jongman)

line corridors with nodes, stepping stone corridor and landscape corridor
(Bloemmen et al, 2004)

As physical structures within an ecological network ecological corridors are multifunctional landscape structures. In Europe ecological corridors are often the result of human intervention in nature: hedgerows, stonewalls, landscapes with small forests, canals and rivers. Others such as coastlines and watercourses are predominantly natural. The nature of ecological corridors and their efficiency in interconnecting remnants and in permeating the landscape depend on the habitat site they originate from and the land use mosaic within which they are embedded in and of which they consist. Their density and spatial arrangement change according to the type of land use. Their connectivity function varies from high to low depending on their spatial arrangement, internal structure and management.

Functions of corridors

Ecological corridors are multifunctional by definition; they have functions for:

1. *Aesthetics*: it makes an area characteristic
2. *Social-psychological well being*: they make an attractive living environment
3. *Education*: they help to understand and experience nature

4.   *Leisure*: nature close to housing
5.   *Ecology*: temporal and permanent habitat and pathways for species.

Ecological corridors are multifunctional in both ecological and societal sense, because they are not the core areas of a nature conservation system but function in the wider landscape. They are also part of 'greenways' that exist in many parts of Europe, sometimes under different names (Haaren and Reich, 2006, Machado et al 1997). They can be as wide as a watershed or as narrow as a trail.

From natural to cultural
They can encompass natural landscape features as well as a variety of human landscape features and are from more natural to more cultural classified as (Florida Greenways Commission, 1994):

1.   landscape linkages, large linear protected areas between large ecosystems including undisturbed rivers;
2.   conservation corridors, less protected and in many cases with leisure functions, often along rivers;
3.   greenbelts, protected natural lands surrounding cities to balance urban and suburban growth;
4.   leisure corridors, linear open spaces with intensive leisure use;
5.   scenic corridors, primarily protected for its scenic quality;
6.   utilitarian corridors, canals, powerlines that have a utilitarian function but serve natural and leisure functions as well;
7.   trails, designated routes for hikers and outdoor leisure having a function as natural corridor as well.

This overview shows the multifunctionality and morphological diversity of greenways and ecological corridors. The more complex a corridor is, the better it can function for different species groups and the more it is multifunctional in an ecological sense.

Positive and negative effects
It must be stated, that corridors also can have negative influence such as the breaking of isolation that is needed for certain species, exposing populations to more competitive species, the possibility of spreading of diseases, exotic species, and weeds, disrupting local adaptations, facilitating spread of fire and abiotic disturbances and disruption of local adaptations (Noss, 1987). Beier and Noss (1998) stipulate that based on empirical research ecological corridors to maintain biodiversity are valuable conservation tools. Not maintaining or re-establishing ecological corridors would mean that mankind neglects the last remnants of natural connectivity and in this way could harm its own nature conservation objectives (Beier and Noss, 1998). Moreover, nowadays practice shows that transport by man are much more important for spreading species and diseases.

Barriers

Finally a network can be hampered by all kind of barriers. Natural barriers do exist at all levels. The Atlantic Ocean is a barrier between America and Europe for most plant and animal species. Mountains and rivers can be barriers for mammals and agricultural roads can already be barriers for insects and spiders. However, much more important are modern barriers for nature, as modern society develops new mechanisms and structures that cannot easily be adapted to by natural species. Canalisation of waterways and the building of motorways however did disturb both the habitat of species as well as their possibility to disperse. Ecoducts and fish ladders can mitigate these barriers (*Fig. 50*).

9.4   Hierarchy of ecological networks

Levels of scale

Ecological networks are effectively implemented at the landscape level; they reflect the complexity of pattern and processes in the landscape. This means that between the Pan European Ecological network and its local application several levels of plans can be developed aiming at decisions and applications for different purposes.

Size of components

The size of network components serves as a criterion of the network hierarchy with four levels (Mander et al, 2003):

1.   mega-scale: very large natural core areas (>10000 km^2),
2.   macro-scale: large natural core areas (>1000 km^2) connected with wide corridors or stepping stone elements (width >10 km);
3.   meso-scale: medium size core areas (10-1000 km^2) and connecting corridors between these areas (width $0,1$-10 km);
4.   micro-scale: small protected habitats, woodlots, wetlands, grassland patches, ponds (<10 km^2) and connecting corridors (width $<0,1$ km).

Mega-scale ecological networks can be considered at global level. The Human Footprint Map (Sanderson et al., 2002) can serve as a base for determining global ecological networks. The macro-scale of ecological networks is represented by macro-regional-level plans such as PEEN maps (Bouwma et al, 2002, Bíro et al, 2006, Jongman et al 2007), the wildlands project (Noss, 1992), or national-level projects within larger countries such as Russia (Sobolev *et al.*, 1995). Most of the projects at this level are used as guiding principles or visions for the future. This macro level can be defined as the (sub) continental level without taking administrative boundaries into account (*Fig. 51*).

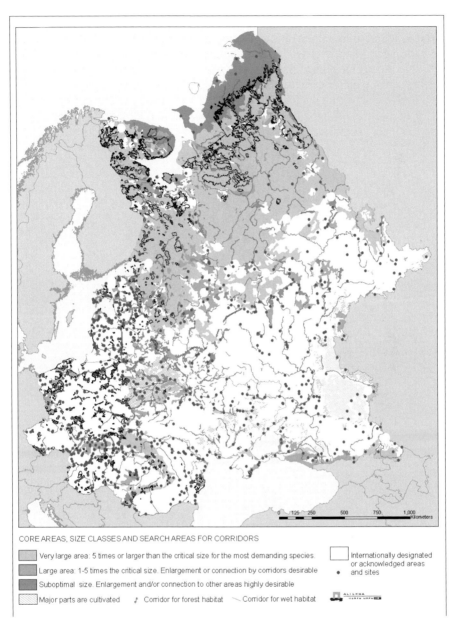

CORE AREAS, SIZE CLASSES AND SEARCH AREAS FOR CORRIDORS

Very large area: 5 times or larger than the critical size for the most demanding species.

Large area: 1-5 times the critical size. Enlargement or connection by corridors desirable

Suboptimal size. Enlargement and/or connection to other areas highly desirable

Major parts are cultivated   ♪ Corridor for forest habitat   Corridor for wet habitat

Internationally designated
or acknowledged areas
and sites

Fig. 51 The Pan European Ecological Network for Central and Easter Europe PEEN-CEE (Bouwma e.a. 2002).

Landscape-level

The landscape-level ecological networks are designed and implemented in a wide spatial scale range, from macro- and meso- to micro-scale projects. At the meso- scalemost significant planning of ecological networks has been carried out (see *Fig. 53* and *Fig. 52*).

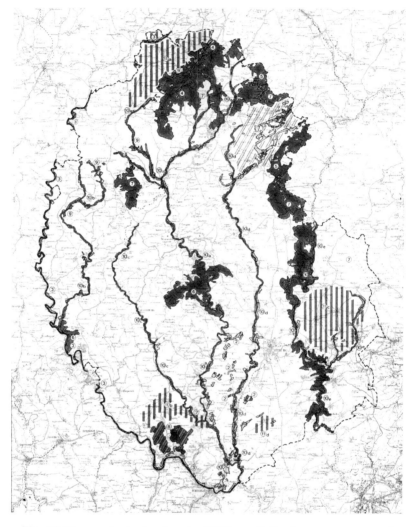

Fig. 52 The ecological network of Bitburg-Prüm (Burckhart et al, 1995)

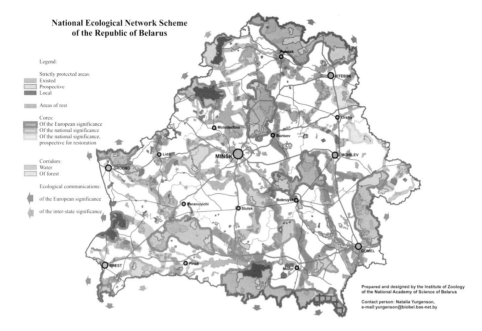

**National Ecological Network Scheme
of the Republic of Belarus**

Legend:

Strictly protected areas:
Existed
Prospective
Local

Areas of rest

Cores:
Of the European significance
Of the national significance
Of the national significance,
prospective for restoration

Corridors:
Water
Of forest

Ecological communications:
of the European significance
of the inter-state significance

Prepared and designed by the Institute of Zoology
of the National Academy of Science of Belarus

Contact person: Natalia Yurgenson,
e-mail:yurgenson@biobel.bas-net.by

Fig. 53 The ecological Network of Belarus
(*Institute for Zoology, Belarussian Academy of Sciences, 2006*)

Micro-scale

Likewise, the most detailed analysis and implementation schemes have
been established at micro-scale (see *Fig. 54* and *Fig. 55*).

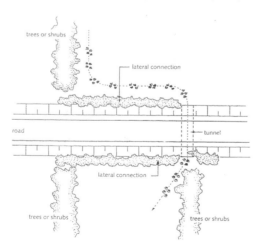

*Fig. 54 Design of a road crossing and landscape structure for a badger
(Meles meles)* (*Bolck et al., 2004*)

153

Fig. 55 Badger tunnel realised in a road project (Photo Rob Jongman)

The challenge of the ecological network approach is to integrate ecological principles, biodiversity, and landscape conservation requirements into spatial planning as well as into implementation.

9.5   Landscape planning and ecological networks

Landscapes fragmented and homogenised for human functions
Important processes in European landscapes are homogenisation and fragmentation of traditional landscapes (Jongman, 2002). New international landscape types develop such as motorway landscapes, leisure landscapes, suburban landscapes, industry landscapes and excavation landscapes (LNV, 1995).

Cultural dynamics disturbing natural dynamics
The European landscape is fragmenting and many species in the small-scale cultural landscapes of Europe are especially sensitive to land use change and changes in landscape structure. The recognition of the existence of fluxes of matter and minerals, population dynamics and genetic exchange on the one hand and compensation of land use that is not compatible with it on the other are the main considerations as arguments for development of ecological networks. Especially administrative borders (national, regional) can be a cause for fragmentation, because plans and priorities are set within administrative borders and mostly not across. Ecological networks require landscape planning across borders.

Different responsibilities and planning organisations
The responsibility for landscape and spatial planning is organised rather differently over Europe and therefore the development of ecological networks is different. In many cases functions and tasks are divided over several ministries and many other agencies depending on the state

organisation. Different views are being developed depending on institutionalisation, scientific tradition and history.

In Germany and Austria landscape planning plays a decisive role as a tool for structuring and maintaining the diversity of the rural areas: its multifunctionality. In other countries nature conservation and landscape planning are strongly integrated (Czech Republic, Slovak Republic) because of the recognition of the relation between them in their cultural landscapes. In countries in southern Europe the need for planning was felt less strongly or at least the execution of planning ideas was less strict. Partly this was, as in Italy is the case due to lack of vertical co-ordination between municipalities, provinces, regions and the national level. In large parts of countries such as Italy discussion is ongoing how culture, nature and other rural functions can be brought together in the same landscape.

Ecological networks in different European countries

Ecological networks have been developed in several European countries such as Czechoslovakia and the Baltic states since the 1970s and 1980s where a strong land use planning tradition had created the institutional environment for allocating functions at the landscape scale. In all Europe habitats were becoming increasingly fragmented due to economic development. The concept of ecological networks is the translation of landscape ecological knowledge on fragmentation processes in the landscapes of Europe and its consequences for populations of natural species. It tries to mitigate the decline of natural species in fragmented landscapes and to overcome the fact that for many natural species the existing nature reserves and national Parks are too small. The concept has become implicit in a variety of international conventions (Ramsar convention, Bern Convention), European directives (Habitats and Birds Directives) and related EU policy implementation (Natura 2000). It has become operational in national and European strategies (National Ecological Networks, the Pan European Ecological Network – PEEN and Pan European Biological and Landscape Diversity Strategy – (PEBLDS).

Aims of ecological networks

The initial aim of the establishment of ecological networks is therefore predominantly protection of nature and biodiversity. Its development is stimulated by science and nature management practice. For example in PEBLDS, the Pan-European Ecological Network aims to ensure that (Rientjes and Roumelioti, 2003):

1. A full range of ecosystems, habitats, species and landscapes of European importance are conserved;
2. Habitats are large enough to guarantee key species a favourable conservation status;
3. There are sufficient opportunities for dispersal and migration of species;
4. Damaged parts of the key environmental systems are restored;

155

5.   The key environmental systems are buffered from threats.

European-level policy initiatives and the growth of international attention for the concept of ecological networks in academic circles, NGOs and in an increasing number of countries in Europe have contributed towards the harmonisation of national concepts into internationally accepted approaches in spatial planning and nature conservation practices.

Top-down planning traditions
Central and Eastern European countries have longstanding experience with biodiversity conservation and ecological networks. Planning approaches however were historically technocratic and top-down. The traditional identification of Nature reserves and national parks did not require involvement of third parties in general as they separated "undisturbed" nature from intensive land uses. However, this division between Nature and other land use appears not long term sustainable. On the one hand there are land use practices that traditionally made use of nature in a sustainable way and contributed to survival of species. Examples are reed cutting in marshland, transhumance in mountain systems and hedgerow planting and management. On the other hand, with changing land use many species cannot survive in the remnants of nature that are left over and intensifying land use, infrastructure and urbanisation threaten their survival. Ecological networks can provide a solution for the problems of intensifying land use and fragmentation. That means however, that new ways have to be found to let populations of species and threatened habitats survive. The urgent need to satisfy EU accession requirements tend to consolidate past practices and to result in the application of blueprint approaches, technical and political solutions rather then locally adapted policies and planning mechanisms, satisfying the interests of multiple stakeholders.

9.6   Stakeholders in ecological network planning at the European and national level

Policy stakeholders
A brief overview is needed on policy stakeholders related to the field of ecological network development in Europe as this is a fast developing area. Moreover, controversies exist around the topic of ecological networks and corridors, both on a political level as well as in research.

Policy arenas
At the European level there are at present four main policy arenas which are concerned with the development of ecological networks as well as a number of NGOs. These are:

1. Convention on Biological Diversity (CBD). During the 7th Conference of Parties (COP) of the Convention on Biological Diversity the relationship between ecological networks and protected areas has been discussed. In the declaration it was stated that the COP invites Parties to consider options, in the context of implementing the programme of work of protected areas, such as ecological networks, ecological corridors, buffer zones and other related approaches in order to follow up the WSSD Plan of Implementation and the conclusions of the Inter-Sessional Meeting on the Multi-Year Programme of Work of the Conference of the Parties up to 2010.
2. Natura 2000 and Habitats Directive (EU). Under the Birds and Habitats Directive the Natura 2000 network will be established that is considered as a European Ecological Network. In article 10 of the Directive the importance of connectivity between the areas is indicated. In the EU until now most attention has been paid to the identification and designation of the Natura 2000 sites itself. The European Environmental Agency has indicated that it will incorporate the research on connectivity between Natura 2000 sites in their work program of 2005. An important aspect of Aricle 10 is that it is subject to subsidiarity – the responsible national and regional authorities are asked to take the initiative – and that it is not obligatory. This urges to negotiate and to reach consent with all parties involved.
3. Policy process of the Pan-European Ecological Network (under PEBLDS, Council of Europe). In 1995, 55 countries endorsed the establishment of the Pan-European Ecological Network (PEEN) as one of the activities to be undertaken within the framework of the Pan-European Biological and Landscape Diversity Strategy (PEBLDS). In order to facilitate the development of the Pan-European Ecological Network a committee of Experts has been established under the auspices of the Council of Europe and ECNC. This Committee meets annually.
4. Alpine Convention. Article 12 of the Alpine Convention underlines the need for connectivity. Four organisations are leading in implementing the Alpine Convention: WWF, ALPARC, CIPRA and ISCAR. In September 2005 a joint workshop was organised to identify the connection areas between the Priority Conservation Areas in the Alps.

Differences between and within countries
In several countries in Europe ecological networks are being planned as part of a legislative task or as a regional or national planning strategy. In a number of countries in Europe legislation has included ecological networks. However in most countries the planning policy or nature conservation/biodiversity policy is leading the development of ecological networks. Moreover federalisation and decentralisation has led to a great variety in approaches. When regional governments have the lead in nature conservation and land management, then usually differences occur within

countries. This means that also coordination between networks is a huge task as there are many approaches and interests.

Comparable approaches and objectives

Despite the many authorities and stakeholders involved in the development of Ecological networks at the national and regional level, the approaches and objectives are rather comparable. There are big differences in the level of detail between plans; but in general most regional plans are well sustained by data and monitoring of change and development. National plans are usually made to develop planning strategies while regional and local plans mostly focus on implementation on the ground. In some regions implementation is well on its way such as in Cheshire County, UK[a].

9.7   Public support for ecological networks in Europe

Ambitions

The plans for the further development of ecological networks in Europe are ambitious. The 5th Ministerial Conference "Environment of Europe" concluded that *"by 2008, all core areas of the Pan-European Ecological Network will be adequately conserved and the Pan European Ecological Network will give guidance to all major national, regional and international land use and planning policies as well as to the operations of relevant economic and financial sectors"*. If possible at all it is obvious that these targets cannot be met without the active cooperation of relevant land use sectors such as agriculture and forestry, and local and regional planning authorities. The Pan-European Ecological Network and other ecological networks will expand beyond the "traditional" domain of nature conservation (protected areas). It will include vast stretches of land over which nature conservation authorities and NGOs have no "jurisdiction". The targets can only be realised in partnerships between the conservation sector (government and NGO) and the various stakeholders involved (ECNC, 2004).

Mutual interests of stakeholders

Partnerships are built on mutual interests. The interests of the conservation sector are believed to be clear: conserving biodiversity. Who are the other partners (stakeholders) and what are their interests? It is argued that the integrity of an ecological network (as landscape mosaic and perceived as part of an integrated regional or national plan) can only be sustained with active support of the "various stakeholders". Generating active stakeholder support for ecological networks has taken many forms.

---

[a] see: http://www.lifeeconet.com and http://www.cheshire.gov.uk/srep

A bottom-up approach in Cheshire
In the case of the "Life ECOnet" project in Cheshire, UK, the approach involved five equally important and co-dependent elements:

1.   Technical development of Geographical Information Systems and the application of landscape ecology principles;
2.   Assessing and influencing land use policy and instruments;
3.   Demonstrating integrated land use management;
4.   Engaging stakeholders;
5.   Dissemination.

Two important principles were embedded in these elements. The approach, and the resulting ecological network, must allow integration of environmental issues with socio-economic functions of the landscape and the acceptance of the landowners and consumers of the landscape. Secondly, the approach must provide an identifiable product on which the varied skills, knowledge and attitudes of stakeholders can focus (James, undated). This means that not only the top down planning approach is important, but that realisation and implementation depend on the bottom up approach of involving stakeholders, both from the field of biodiversity conservation and other sectors of society.

An agenda for public campaigns and meetings in Estonia
In the case of Estonia the approach to gain support took the form of meetings and public campaigns with emphasis placed on (Sepp and Kaasik, 2002):

1.   Multifunctional nature of ecological networks (for example increased environmental health conditions, leisure opportunities);
2.   Conservation of "flagship species" to highlight the importance of biodiversity conservation; and
3.   The accommodation of semi-natural habitats or other "use areas" that allow traditional farming practices in the networks.

NGO-initiatives
 There are many cases, where the initiative did not come from government such as the Yellowstone – to – Yukon (Y2Y) ecological network in Northern America. As most Northern American Greenway plans Y2Y is very much a grassroots initiative enjoying support from a large variety of NGOs and other civil society organisations (360 in total) with the objective to ensure that the eco-region continues to support natural and human communities. In a few states in the USA (Florida, Georgia) the state has embraced these plans into Statewide Greenway Plans based on the integration of biodiversity and civil interest issues (Florida Greenway Commission 1994, Bennet and de Wit, 2001). Comparable grassroots-based plans are developed in Portugal around the cities of Lisbon, Porto and Coimbra (Machado et al 1997). Here the initiative has been a combination of

universities and NGOs. The support from the authorities - and therefore its realisation – is still a difficult process.

Combining other objectives
What these cases have in common is that they focus not only on the conservation of biodiversity but also accommodate the exploitation and consumption of natural resources (Ahern, 2004). Serious efforts are made both to buffer sites of high conservation value from potentially damaging forms of land use and to find ways of reconciling the exploitation of natural resources with biodiversity conservation (Bennett and Wit, 2001).

9.8 The bottom up approach to design: national ecological networks of virgin forests and semi-natural grasslands in Romania and Bulgaria

Mapping virging forests and grasslands
Descriptions of the vegetation and mapping of the locations by observations in the field form the basis to prepare strategies for sustainable conservation and management of virgin forest and semi natural grassland in Romania and Bulgaria. Over the last 10 years, the Royal Dutch Society for Nature Conservation (KNNV) was involved in these projects (results on website www.veenecology.nl). This work was done in close co-operation with local institutions such as forestry research institutes, botanical institutes, universities, governmental bodies and non-governmental organisations. These projects resulted in recommendations for uptake of a list of forests and grasslands in Natura 2000 networks. For the grasslands, special attention was given to the preparation of agri-environmental schemes with focus on sustainable management.

Fragmentation of habitats in Bulgaria and Romania
Information on the degree of fragmentation of habitats in Central and Eastern Europe is scarce. In this presention special attention will be paid to the aspects of fragmentation of the individual habitats in the landscape of Bulgaria and Romania.

Advantages of 'bottom up' approach
Until now, decisions on the implementation of national and European ecological networks, for example the Pan European Ecological Network (PEEN)), are mainly based on a 'top down' approach. The 'bottom up' approach, however, provides much more certainty to reach the aim. This is the effect of starting the process with real field observations. From each observation of the target species or community the exact location is indicated on the map. This makes it possible to :

1. design the network on a high resolution spatial scale;

2. to take into account all relevant ecological factors of the habitat;
3. the positive and negative effects of land use around the habitats can be evaluated by mapping species indicative for environmental stress.

We think that virgin forests and natural grasslands need this type of 'bottom up' assessment because the biodiversity in terms of species richness and habitat development is extremely specific.

9.9 The Romanian national network for forest ecosystems

History and coverage in Romania

In the Neolithic period Romania was covered by extensive forests in the Carpathian mountain range and had extensive forest-steppes and steppes in the lower parts along the Carpathian range (Giurgiu et al, 2001). At this moment most of the forests are located on the Carpathian Mountain range and neighbouring areas (89% in mountainous and hilly regions). In the lower parts of Romania only 11% is covered by forest at present. The total Romania forests cover 6,280,000 ha which means that about 26% of the country is covered with forests (Biris et all, 2006). The state owns 69.8% of the total forest area. Other categories of forest owners are communes, churches, foundations and other public bodies (12.5%), local communities (7.7%) and private owners (10%, all data from 2004).

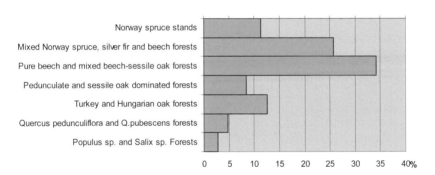

Fig. 56 Main forest types in percentage of total forest area. (Biris et al., 2006)

Beech is the dominant tree species in 34% of all forests and Norway spruce in 26%. All other species cover less than 15% of the forest area in Romania. A relatively high percentage of 69% of the total forest area is naturally regenerated (Biris et al., 2006).

Romanian protection tradition

The protection of forests in Romania has a long tradition (Biris et all, 2006; Giurgiu et all, 2001). Already in the 14th century exists the 'letter of forbid-

161

den forests' in which "branisti" were mentioned where no cutting, hay mow-ing, grazing, hunting and collection of fruits was allowed without the per-mission of the owner. These forests were mainly in the ownership of no-blemen and churches. In the Banat-region, in western Romania, the first protection actions started in 1739 and in Transsylvania regulations were in force from 1781. At the end of the 18th century, several regulations like in Bucovina, Moldavia, Wallachia were in place. However, in the 19th century, after signing the Treaty of Adrianopole (1829) several forests in the plains had been cut to provide land for cereal production. After 1852, Transsylva-nia and Bucovina became a part of the Austrio-Hungarian Empire. This resulted in the development of management plans and the prevention of massive clearings. In the 20th century the influence of France in forest management resulted in a new forest law in 1923 which gave protection to forests in specific locations like slopes and around water bodies.

Communist period
During the Communist period massive cuttings were done because of payment obligations to the Soviet Union. Especially in Bucovina region large territories had been cut for this reason. This explains why beech for-ests are very rare in the area although it is the natural dominating species.

Re-privatisation
After the changes in 1989, the process of re-privatisation started. In 2004 about 10% of all forests were privatised. In the meanwhile the forest legis-lation and nature conservation legislation has been made EU proof. Roma-nia established many new protected areas such as 11 national parks, 5 natural parks, 1 biosphere reserve: in totally, more than 1.2 mill.ha, which means about 5% of the total country.

Forest functions
Forests can have legally authorised special functions in Romania such as:

1. water protection
2. land and soil protection
3. buffering the damage from climate and industrial pollution
4. tourism and leisure
5. scientific research and gene protection.

In total, 50% of all forests have a special function (3.2.mill ha; Biris et al., 2006).

Characterisation of virgin forest complexes
The 'bottom up' project for mapping of virgin forests was initiated by the Royal Dutch Society for Nature Conservation (KNNV) in close collaboration with the Romanian Forest Research and Management Institute (ICAS). During the period 2001-2004 all virgin forest complexes were traced by using a standard form to describe the relevant aspects like structure, age, presence of dead wood, species composition of the different vegetation

layers, historical influence of management, condition of slope, soil and type of boundary. The project's long term goal is the preservation and the sustainable management of virgin forests in Romania. The short term goals include development of a systematic methodology for the investigation of virgin forests, identification of these complexes and preparation of a protection strategy, including an Action Plan. Virgin forests are defined in this project as "natural woodland where tree and shrub species are present in various stages of their life cycle (seedlings, young growth....) and as dead wood (standing and lying) in various stages of decay, with a more or less complex vertical and horizontal structures as a result of natural dynamics. This process enables the natural forest community to exist continuously and without limit in time" (Biris and Veen, 2005).

Virgin forest types:
1. Mixed forests of *Picea abies, Larix decidua, Pinus cembra, Pinus sylvestris*: 46.933 ha
2. *Abies alba* forests and mixed forests dominated by *Abies alba*: 46,645 ha
3. *Fagus sylvatica* forests and mixed forests of *Fagus sylvatica* in mountains: 92,437 ha
4. *Fagus sylvatica* forests and mixed forests of *Fagus sylvatica* in hilly areas: 20,867 ha
5. *Quercus petraea* forests and mixed forests: 3,563 ha
6. *Quercus robur* forests and mixed forests: 578 ha
7. Thermophile Quercus forests with *Q.cerris, Q.frainetto* and mixed forests: 66 ha
8. Xerophyllous oak forests (*Q.pedunculiflora* and *Q.pubescen*s): 66 ha
9. Other forest types (including riparian forests): 6,408 ha
Total virgin forests in Romania: **218,493 ha**

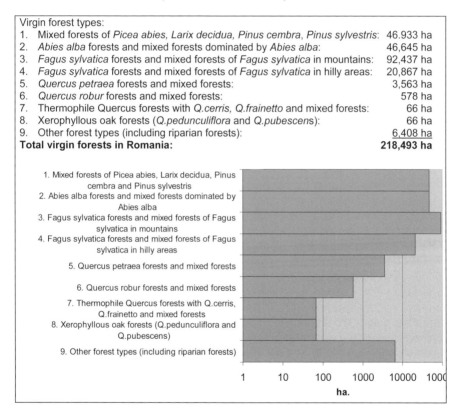

Fig. 57 Presence of virgin forest types in Romania (Biris and Veen, 2005)

Results stored in a GIS-database
The results of the mapping activities are stored in a GIS-database. This database can be used by institutes and ministries for answering specific questions like the location of potential Natura 2000 sites and the type of management of these sites.

As a result of this project a map of virgin forests could be prepared (*Fig. 47*). The diversity in virgin forest types is illustrated in *Fig. 57*.

Virgin forests are important core-areas of the national ecological network because these habitats are of high biodiversity in sense of richness in plant and animal communities and individual species. Due to a long term exis-tence of these forests and the low influence by man, these forests can be observed as one of the richest ecosystems in Europe.

Due to the strong fragmentation impacts, the populations of many species within the virgin forest complexes function as a meta-population and we expect that turn-over rates are high for several species. The results of this 'bottom up dataflow' will be used as a baseline for research within the na-tional monitoring system of forests in Romania. In totally, 3,402 polygons were mapped (*Fig. 58*).

Area-classes	Number of forest polygons
1. 0-1 ha	41
2. 2-100 ha	2,753
3. 101-500 ha	571
4. 501-1000 ha	26
5. 1001-2000 ha	10
6. 2001-5000 ha	1
7. 5001-10.000 ha	0
8. >10.000 ha	0
Total	**3,402**

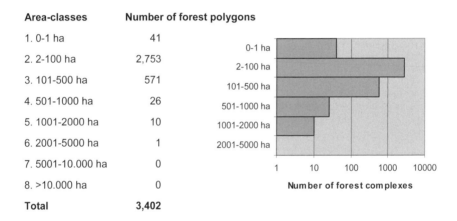

Fig. 58 Division of virgin forest complexes over area-classes
(Biris and Veen, 2005)

This means that the division is skewed and class 2 is dominating the other classes. It is assumed that at least an area of 500 ha is necessary in order to reach complete communities of animals and plants. Of course, the out-come does not exclude that more high class mosaics of individual polygons can be present in reality. The map shows us that on several locations along the slopes of the south-eastern Carpathians virgin forests are present in a real mosaic. This means that fragmentation impacts could be low on these slopes.

Strong fragmentation
The conclusion is that protection of the national ecological network of Ro-mania needs taking care of the negative impacts of isolation due to a strong fragmentation of the rare virgin forest complexes in this country.

9.10   The Bulgarian network of semi-natural grassland ecosystems

Location

Bulgaria is situated in different biogeographical zones: from an Alpine zone in the mountain regions through a Continental zone in the central plains till a Black Sea zone along the Black Sea in the east. Especially the Black Sea zone is important because of its Pontic, South Euxinian and even Mediterranean elements.

Area

Grasslands are important habitats in Bulgaria. It was estimated that in the beginning of the 20th century 1,8 million ha of pastures and hay fields were present in the country. Even in the middle of the 20th century, experts estimated a total grassland area of 1,2 mill ha (Ganchev et all, 1964). According recent CORINE land cover maps it is estimated that 850,000 ha of lowland and hilly land grasslands exist, not taking into account the high-mountain grasslands.

Primary and secondary grasslands

Some part of the grasslands in Bulgaria can be judged as primary grassland. This is mainly high-mountain grassland above 1,800 m. Fragments of steppic grassland complexes are also remnants of autochthonous grassland, especially at the eastern part of the country with low precipitation climatic conditions. All other grasslands can be seen as secondary, basically developed through agricultural practices by farmers.

Cold winter conditions in the mountains

Management of grassland varies in the different parts of the country. The high-mountain grasslands are mainly used for summer grazing. In the past, nomadic Karakachan farmers contributed greatly to the development of pastures. Special Karakachan sheep and horses could survive the cold winter conditions in the mountains. Nevertheless, in winter they migrated by transhumance to the Mediterranean regions. After the Second World War, this transhumance became impossible because of boundaries which could not be passed in the Communist period.

Overgrazing

A specific problem after WW II was overgrazing by sheep and cows. The very rich pastures in the mountains and hills changed in poor *Nardus stricta* grasslands which gave not sufficient food to the cattle.

Reduction of cattle

After the changes in 1990, the total amount of cattle was very strongly reduced until nearly half. The prices for milk and meat are too low for the farmers to provide them with sufficient income. Especially in hilly regions, like the Rodopi Mountains, this caused a rapid change from pastures and hay fields into arable land.

Natural- and semi-natural grasslands

This 'bottom up' project was initiated by the Royal Dutch Society for Nature Conservation (KNNV) in close co-operation with the Institute of Botany within the Academy of Science in Bulgaria (Meshinev et al., 2005). A team of specialists mapped all grasslands during the period 2002-2004. Like the forest project in Romania, grassland data had been collected concerning boundary of polygon, management of grasslands, impacts by man, species composition and special data on the position in the landscape and the type of boundary. The vegetation type in a polygon was determined by using the phyto-sociological method in accordance with Braun Blanquet methodology. This means that of every homogenic polygon a total list of species was made and of each species the abundance was indicated.

Results stored in a GIS-database

The results of the descriptions and the mapping were stored in a GIS-Database. This database can be used by institutes and ministries for answering specific questions like the location of potential Natura 2000 sites and the type of management of these sites.

A map of the natural and semi-natural grasslands is the result of this project (map 2 see *Fig. 48*). The diversity in grassland vegetation is illustrated by *Fig. 59*.

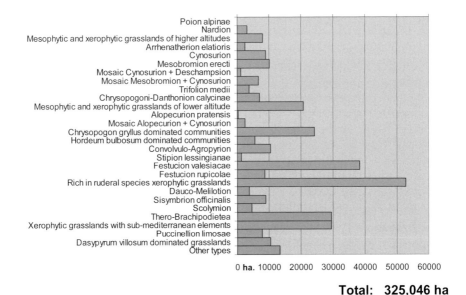

Total: 325,046 ha

Fig. 59 Presence of grassland vegetation types in Bulgaria
(Meshinev et all, 2005)

Natural grasslands are important core-areas of the national ecological network because these habitats are of high biodiversity in sense of richness in plant and animal communities and in species (2,008 plant species were observed, it means 51,5% of the national plant diversity). Due to a long term existence of these grasslands the species richness is very high (Meshinev et all, 2005).

The area-species ratio can be calculated using the database. We also calculated the risks of extinction of every individual grassland site, as a function of the surface. The outcome is presented in *Fig. 60*.

Area-classes	Number of grassland polygons
1. 1-100 ha	1270
2. 101-500 ha	798
3. 501-1000 ha	61
4. 1001-2000 ha	25
5. 2001-5000 ha	8
6. 5001-10.000 ha	1
7. >10.000 ha	0
Total	**3,402**

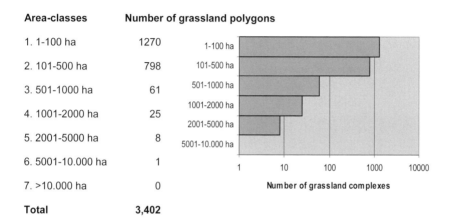

Fig. 60 Division of natural and semi-natural grassland complexes over area-classes (Meshinev et all, 2005)

Strong fragmentation
The fragmentation is very obvious from the fact that the area size classes 1 and 2 are the most dominant ones. The fragmentation impacts are even potentially stronger as is the case of the Romanian virgin forest example because the individual sites are mostly situated in intensively used agricultural fields. There is an urgent need for Bulgaria to develop a strategy for connection of natural grasslands and to make the national ecological network operational for these types of problems.

9.11 Conclusions

Ecological networks are a relatively recent paradigm in nature conservation and ecosystem management. Functions of ecological networks depend on the hierarchy of networks. On different hierarchy levels different applications and management strategies are to be applied. Real implementation can only been carried out at the lowest, regional and local levels. Activities in the research, design and implementation of ecological networks should concentrate on the development of coherent planning and management schemes, while at the higher hierarchical level up to the global scale ecological networks can be used as strategy development tools. This means that upscaling and downscaling of ecological networks is important in the development of a dynamic, modern nature conservation.

Special attention is required for natural ecosystems like virgin forests and (semi-)natural grasslands are remnants of an extensive natural landscape from the past. Due to impacts of man by mainly cutting of natural forests and intensification of agricultural land use, these ecosystems are dispersed and isolated. The identification of these complexes need a 'bottom up data-flow' which includes monitoring of species and habitat types on the meso-scale and micro-scale (see also section 9.4, page 150). Populations of species are limited and turn-over rates can be expected high. The data-base of both project function as a baseline for future monitoring of these populations. The virgin forest complexes in Romania are for 90% included in the new Natura 2000 Network (pers.comm.Biris) and the (semi-)natural grassland complexes make the backbone of Agri-Environmental Schemes in Bulgaria (pers. comm. Meshinev).

10 ECOHYDROLOGY; CONTRIBUTION TO SCIENCE AND PRACTICE

Martin Wassen[a]

---

[a] The author thanks Jos Dekker and Stefan Dekker for helpful comments.

10.1   Introduction

In 1972 the Dutch community of landscape ecologists (WLO) was raised and in 1981 in Veldhoven (the Netherlands) IALE was founded. This book presents an overview of the Dutch contribution to landscape ecology in science as well as policy and planning.

Ecohydrology is a landscape ecological specialisation which brings together concepts, methods and disciplinary knowledge from ecology and hydrology. This chapter defines ecohydrology and reviews the history and state of the art of ecohydrology, both from a Dutch as well as an international perspective. From this chapter it will become clear that in the international context ecohydrological ambitions are more inclusive, aiming at sustainable development in river catchments and recently focusing on global change as well. In the Dutch context the focal point was understanding of spatial relations via water in regional landscapes and the impact of changes in cycling of water and nutrients on ecosystems.

Environmental problems such as desiccation, acidification, and eutrophication were assessed in many Dutch ecohydrological studies. However, explicit attention for the fragmentation problem and how to solve this did not come from ecohydrologists. The Dutch National Ecological Network (NEN) was designed on expert judgement and as such also ecohydrologists were influential, but their advice was limited to the regional scale. At that time ecohydrologists did not care too much about the national scale. Restoring connectivity between fragmented populations was not of prime interest to ecohydrologists and climate change was not yet an issue.

10.2   Ecohydrology or hydroecology?

I always tell my students that a hydrologist can never become a hydroecologist and vice versa: an ecologist can never become an ecohydrologist. Maybe this sounds trivial but the reason for a great deal of confusion is in here. The term 'hydroecology' and 'ecohydrology' both imply research at the interface between the hydrological and biological sciences.

The prefix 'eco' ('hydro') in 'ecohydrology' ('hydroecology') indicates it is a modifier of the word 'hydrology' ('ecology'), and thus, the discipline should be more about hydrology than ecology (and vice versa) (Kundzewic, 2002). However, in practice this rubric has not been applied, as many ecologists refer to ecohydrology and hydrologists refer to hydroecology. This is confusing since ecology and hydrology are quite different disciplines and an ecologist does not necessarily understand anything of hydrology and vice versa. Further contributing to the confusion is that the French term hydroecologie is not a synonym for hydroecology, since it translates as 'aquatic ecology' in English (i.e. the study of freshwater, brackish and marine surface water systems). The use of hydroecologie in francophone pub-lications explains many of the mistaken references to hydroecology in the early 1990s (Hannah et al., 2004). For consistency in this chapter the term ecohydrology will be used throughout since at present it is much more gen-erally used than hydroecology.

Definitions

Despite the loose use of the 'prefix-discipline combination', definitions may reveal the original background of the author. Pedroli, a physical geogra-pher, defined ecohydrology as 'the interdisciplinary study of groundwater hydrology as a component of ecosystems and as a determining factor for the pattern, distribution and development of vegetation' (Pedroli, 1992). His definition not unexpectedly refers to groundwater hydrology as the principal object of study. Me, being a landscape ecologist, once defined it as the landscape ecological study of ecosystems dependent on groundwater and surface water. Hannah et al. (2004) mention that the first clear definition of ecohydrology appeared in a special issue of Vegetatio and relates to wet-lands: 'It states ecohydrology is an application driven discipline and aims at a better understanding of hydrological factors determining the natural de-velopment of wet ecosystems, especially in regard of their functional value for nature protection and restoration (Wassen and Grootjans, 1996)'.

As we will learn later in this chapter, the focus on preservation and restora-tion of wet ecosystems in this definition is resulting from the typical Dutch problems in which human interference in hydrology has frequently led to disappearance and deterioration of ecosystems depending on groundwater and surface water. The definition of Geoff Petts, a physical geographer studying rivers, is: The study of hydrological and ecological processes in rivers and floodplains and the development of models to simulate these interactions (Petts, 1996). Zalewski (2000), a catchment hydrologist, refers to ecohydrology as the study of the functional interrelations between hy-drology and biota at the catchment scale. All these definitions reflect the personal interest of those who defined.

Sometimes this tends to attempting to claim the term for a more specialised part of the field of study. Baird and Wilby (1999) are exceptional in this respect since their definition is that broad that it encompasses a lot more and in my view too much which makes it quite useless. In their definition ecohydrology is the study of the two-way linkage between hydrological processes and plant growth (Baird and Wilby, 1999), thus including all kind of eco-physiological, agricultural and ecological studies on plant growth as long as water is involved. The good thing in their definition is that it includes the feedback of plant growth on hydrology. For instance, by evapotranspi-ration of vegetation or by in-stream vegetation affecting roughness and discharge of streams or through ecosystems altering soil structure and thereby its hydraulic properties.

10.3   A scientific revolution?

In 2001 the international journal Ecohydrology and Hydrobiology was launched[a]. It has been created to reflect the concept of Ecohydrology, which is based on three principles:

1. integrating water and biota at the catchment scale into a Platonian superorganism,
2. enhancement of the absorbing capacity of (evolutionary established resilience and resistance of) the ecosystem against human impact,
3. using ecosystem properties as management tools for biodiversity, water quality and quantity improvement.

Following the need for new solutions in sustainable water management the journal invites especially contributions which provide integrative approaches to aquatic sciences explaining ecological and hydrological processes at a river-basin scale, or propose practical application of this knowledge (Instructions for Authors). These instructions reflect a high ambition of the journals' editors (M. Zalewski and D.M. Harper) with respect to interdisciplinarity and innovative scientific approaches.

New paradigm
It was also Zalewski, who was the first one ever mentioning the new paradigm ecohydrology might be (Zalewski, 2000; Zalewski and Robarts, 2003). In his view the ecohydrology-concept integrates existing fragmented knowledge on hydrological and biological processes at the basin scale into a holistic framework. Following this, he formulated the three principles mentioned above for further progress in the development of the ecohydrology concept: framework, target and methodology. The first principle is the conceptualisation of the catchment as a superorganism in a similar fashion as the Gaia concept of the planet earth (Lovelock, 1995). The second (target) is that the conceptual superorganism can be viewed as a natural state as possessing resistance and resilience to stress. The third principle (methodology i.e. water management) has an obvious link with ecological engineering.

The new paradigm of ecohydrology can thus be seen as the third phase in the development of ecology from a natural history perspective through the understanding of processes to control and manipulation of ecological processes for management of resource quality (Zalewski, 2000). Such a view on ecohydrology could be very useful in attempts to come to a sustainable water and resource management on catchment scale. While hydrologists appear to be actively engaged with the new paradigm (for example Rodriguez-Iturbe, 2000), it has been suggested that biologists are less aware of,

---

[a] see http://www.journal.ecohydro.pl/

or are unconsciously involved in, the ecohydrology and hydroecology revolution (for example Bond, 2003).

Bibliographic survey
Hannah et al. (2003) carried out a bibliographic survey on ecohydrological or hydroecological papers in international journals for the period 1991-2003. On the basis of this survey they concluded that

1. the majority of these papers used the term ecohydrology,
2. there is a gradual increase of such papers in the considered period,
3. the largest number of papers was published in the journal Hydrological Processes,
4. the majority of papers focused upon plant-soil-water interactions,
5. somewhat more than half of the papers resulted from research conducted at university departments of Geography and Environmental Sciences; almost the other half from Bioscience and Ecology departments.

Overall, ecological articles accounted for 59% of publications although these mainly appeared in physical sciences journals. Hydrological papers were fewer (32%) most of them focusing upon hydrology – water resource management. Notably, these water resource management articles often infer ecological implications but contain limited, or no, supporting biological data. Moreover, the bibliographic search showed that there are only few integrative studies in terms of subject matter and/or authors list (i.e. research teams are predominantly composed of groups from either geography or from biology (Hannah et al., 2003)).

Trends
This survey shows a number of interesting trends. Firstly, geographers, hydrologists as well as ecologists may consider their work as ecohydrological and there is still an increasing trend of published ecohydrological studies. However, most of them are biased on plants and there is no convincing integration yet. Apparently, the paradigm shift has not yet occurred and the prophets talking about a scientific revolution did not earn credibility yet. The traditional disciplinary gap between ecology and hydrology still exists. In a way this is surprising since if ecohydrology is considered as a landscape ecological specialisation one would expect that the happy marriage between geography and ecology which landscape ecology is (cf Zonneveld, 1995) would guarantee cross-disciplinary fertilisation.

10.3.1   Ecohydrology in the Netherlands
In the Netherlands ecohydrology started in the 1970s as a much more down to earth approach which can be characterised as learning by doing.

Dutch context
The Netherlands is densely populated and intensive water management takes place in almost every part of the country. Most of its soils have a high

permeability to water flow; hence interferences in hydrology have wide impacts. After the Second World War the Netherlands faced intensification of agriculture with subsequent intensified drainage and redistribution of water from the rivers Rhine and Meuse for irrigation purposes. Groundwater abstraction for drinking water and industry increased dramatically.

Drainage, desiccation and eutrophication

These trends resulted as a side-effect in drainage and eutrophication of nature areas and deterioration of their quality. Vulnerable plant species started to disappear in reserves where distinctly visible environmental causes seemed absent. Later this problem was referred to as desiccation ('verdroging') which in the Dutch context should not be considered literally as physiological shortage of water for growth. The disappearance of species and degradation of plant communities was first noted by vegetation scientists such as Westhoff, Barkman, De Smidt and Van der Maarel. However, only the next generation started to unravel the causes of deterioration. The basis of ecohydrology was then laid by Ab Grootjans (University of Groningen), Geert van Wirdum (National Research Centre for Nature Management; RIN Leersum) and Rolf Kemmers (Winand Staring Centre for Agricultural Soil Science, Wageningen).

Causes of deterioration outside the borders of the nature reserves

They were all biologists (respectively plant ecologists and soil biologist) and for this reason, the roots of ecohydrology are in ecology in the Netherlands. Firstly, they had a hard job in convincing agricultural scientists and water management engineers that the problems of deterioration should be sought outside the borders of the nature reserves. Hydrologists at that time still were busy with supporting agricultural engineering in the rural area like designing effective drainage networks, reclamation schemes and drinking water plants. Simply because hydrologists were not interested the ecologists started to conduct hydrological research themselves. A second reason for ecologists to cross the disciplinary borders was the mere focus on water quantity of the hydrological research at that time. Thirdly, the scale at which hydrological studies were performed (data and models) were too coarse to match with fine-grained vegetation patterns and gradients. Thus, ecologists were forced to gather the hydrological evidence themselves if they would like to convince others that interferences in the regional hydrological system had long distance effects in nature reserves elsewhere.

Hydrologists interested in ecology

Only in the eighties hydrologists started to do ecological relevant research. Especially the so-called hydrological systems analysis of the group of Engelen (Vrije Universiteit Amsterdam) which was built on the theoretical work of Toth (1962), was appealing for ecologists. In this approach groundwater systems were distinguished on the basis of fluxes, water quality and age (Engelen, 1981; Engelen and Jones, 1986). These systems enabled to connect groundwater discharge areas where groundwater is seeping to the surface and nourishing the vegetation to groundwater recharge areas. Thus connecting the flux and water quality of the water feeding the vegetation to soil conditions and land use in the area where the groundwater is recharged by infiltrating rainwater.

Landscape ecologists interested in hydrology

In the late eighties and early nineties there was a sharp increase of ecohydrological studies in the journal Landschap (the scientific journal of the Dutch community of landscape ecologists in the Netherlands (WLO, see Wassen, 1991). The first ecohydrological PhD thesis was completed in 1985 (Grootjans). Then in 1989/1991 seven theses were published in short time (Pedroli, 1989; Koerselman, 1989; Witmer, 1989; Wassen, 1990; Van Wirdum, 1991; Schot, 1991; Everts and de Vries, 1991) illustrating the potentials of this new discipline. In the nineties many more typical ecohydrological theses would follow (a.o. Boeye, 1992; Barendregt, 1993; Kooijman, 1993; Stuyfzand, 1996; De Mars, 1996; Sival, 1997; Van Diggelen, 1998).

Of the fifteen theses mentioned above only three were written by hydrologists (Witmer, Schot, Stuyfzand). The WLO-working group on Ecohydrology was raised in 1988. This active and vivid working group regularly has been organising seminars, workshops and excursions and produced a number of ecohydrological special issues in Landschap (see Meuleman et al., 1992; Van Buuren et al., 1999; Wassen et al., 2001, 2002) and a special issue of Vegetatio (Wassen et al., 1996). Figure 1 gives a scheme of ecohydrological relations and processes on different spatial and temporal scales showing a landscape ecological hierarchy from geology via hydrology to site conditions and vegetation (Mars, 1996). This conceptual model can be regarded as typical for the Dutch approach.

An application-driven interdiscipline

Grootjans et al. (1996a) in their review argue that ecohydrology is an application-driven interdiscipline. They distinguish a number of developments in applied ecohydrology.

Species distribution as a guideline

The first is the use of species distribution as a guideline in ecohydrological surveying. This approach was mainly developed at Groningen University and was initiated by Grootjans (see Grootjans, 1980). It uses the indicator value of plant species and plant communities for judging the environmental conditions at the place they grow. The information on site conditions was inferred from phytosociological literature and mire ecological studies. Subsequent mapping of species distribution on a landscape level resulted in patterns which correlated well to geomorfological and geohydrological information such as (absence of) impervious layers, elevation gradients and calcium richness of groundwater. This generated hypotheses about the ecohydrological relations in a landscape. The thesis of Everts and de Vries (1991) is a prominent example of this approach. Since the method is cheap and covers the whole landscape it was widely applied in land evaluation in

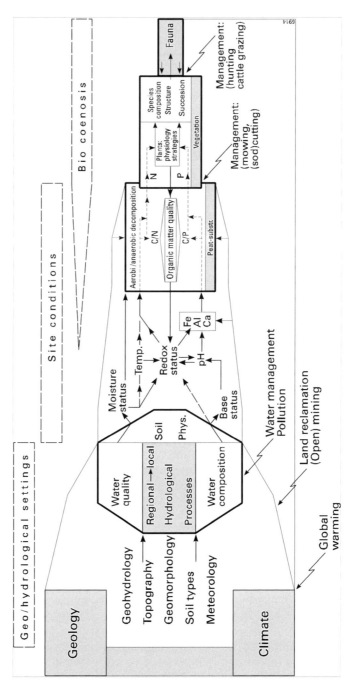

Fig. 61 Schematical representation of ecohydrological relations in land-scape. (Courtesy: De Mars, 1996: pg 134)

the Netherlands. KIWA further developed the method and used it for assessing the potential damage of drinking water abstractions (Jalink and Jansen, 1989; Jalink 1991).

The use of water chemistry

The second approach they distinguish is the use of water chemistry to interpret hydrological systems. In this approach large numbers of water samples were taken and analysed, both groundwater and surface water and the water quality patterns emerging from this were related to vegetation patterns or species distribution on the one hand and to soil and geohydrology on the other hand. In my view this approach is truly overarching cross-cutting ecohydrology, because here ecologists and hydrologists made progress because they cooperated from the start. Groundwater modeling, isotope analyses, calculation of hydro-chemical processes and vegetation dynamics were combined leading to a great added value in the interpretation of dynamic groundwater flow patterns and unraveling ecohydrological relations on different scales. Elegant examples are Wassen et al. (1990), Schot and Wassen (1993), Barendregt et al. (1995), Grootjans et al. (1996b), De Mars and Garritsen (1997). In *Fig. 62* an example of this approach is given in which information of topography, soil, groundwater flow and vegetation are combined in profiles for reconstructing former ecohydrological conditions and predicting potential future developments (Sival and Grootjans, 1996).

The development of ecohydrological models

A third strong application which should be mentioned is the development of ecohydrological models. Although many ecohydrologists doubted their usefulness and reliability at first, looking back we may conclude that ecohydrological models have been very successful. They were applied a lot and were used to evaluate the impact of hydrological interferences and changes in water management on forehand and in this way significantly contributed to sustainable water management (see Witte et al., 1992; Barendregt et al., 1993; Latour and Reiling, 1993; Ertsen et al., 1995).

Usually, in plant ecology three major processes are taken into account, which determine the possibility for a species to grow in a certain place.

1. The environmental conditions (determining if a species can grow potentially),
2. the history and spatial context of a site (availability of diaspores or seeds and connection via dispersal to/from nearby populations determining if a species can reach a site) and
3. interactions among species such as competition, symbiosis, etc, determining if a species can survive and establish for a longer period (cf Van der Maarel, 1976).

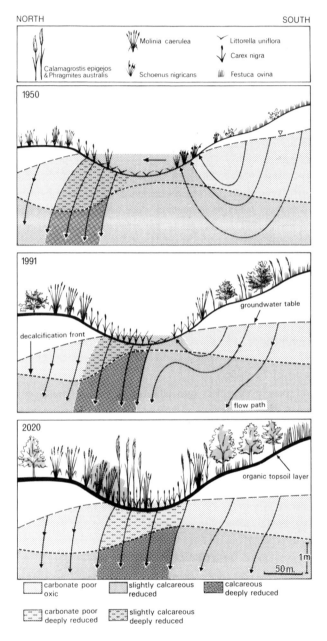

Fig. 62 Reconstruction of ecohydrological conditions round 1950, the situation in 1991 and possible future development (year 2020) of the dune slack Kapenglop on the island of Schiermonnikoog (the Netherlands)
(Sival and Grootjans, 1996)

In ecohydrology Van Wirdum (1979) translated the first one to conditional and operational factors and the second to positional factors. In ecohydrological models mostly only the first category is considered: the site conditions (Klijn and Witte, 1999). Examples of successful ecohydrological models are ICHORS, NTM, DEMNAT, ITORS and MOVE. Apart from ICHORS and ITORS which are empirical models based on large datasets collected in the Netherlands, the others heavily lean on Ellenberg indicator values for moisture, acidity and nutrient richness.

Major differences between the models are: the expert knowledge and field measurements required, the scale level, ecosystem and habitat conditions for which the prediction is made and the number of input variables needed to run the model. Especially the latter requirement seriously hampers the applicability of empirical statistical models such as ICHORS and ITORS since they need defined input values for 28 respectively 39 site conditions. For an overview of ecohydrological models, their characteristics, possibilities and constraints see Olde Venterink and Wassen (1997).

10.3.2   Changing time spirit

In 1993 and 1995 two events took place in the Netherlands which dramatically changed the Dutch attitude towards water. In the slipstream of the discussion on safety also nature could benefit.

Facing flooding disasters

In early February 1995, newspapers and television stations around the world relayed images of the Netherlands facing another major flooding disaster. Fortunately, the reality was less dramatic.

In Limburg province, parts of towns and villages were flooded by the Meuse. But dikes along the Rhine were just strong enough to avoid major flooding. Around Christmas 1993 a similar event took place with flooding in a.o. Germany and the Netherlands, so this early repetition caused consternation: prior to 1993, the last serious river flooding in the Netherlands had been in 1926 (KNAG, 2000). In 1995 the authorities considered the situation that serious that they decided on evacuation. In a few days, some 250,000 people, 300,000 head of cattle, a million pigs, and millions of poultry birds were moved from an area bounded by the Lower Rhine, the Waal and the Meuse. Meanwhile, emergency services and volunteers worked flat out to reinforce the weakest parts of the dikes - and they succeeded. By the time the water level began to fall during the first few days of February, all the dikes had held out, and most of the inhabitants and animals were able to return home within a week of their evacuation.

Once the crisis was over, the province most seriously affected turned out to be Limburg, where the Meuse had flooded much farmland, some villages, and Venlo city centre.

Foreign media portrayed the situation as far more critical than it actually was, even drawing parallels with the great flooding disaster in the south-west of the Netherlands in 1953. The comparison does not bear serious consideration, however, given the vastly different outcome. In 1995, not much land was flooded, no houses, roads or railway lines were destroyed, and - most important of all - there were no deaths. The 1953 floods, by contrast, cost more than 1,800 lives. This does not mean, however, that the danger of a major flooding disaster was absent. And it was this fact that caused shockwaves in the Netherlands. Many Dutch people had believed that, with the completion of the massive Zuyder Zee and Delta projects in recent decades, their country's water defences were after ages of improvement as perfect as to make their country impervious to flooding (KNAG, 2000).

Room for the River

The near-disaster in February 1995 shattered this belief, as the realisation grew that for the Netherlands - situated as it is on the North Sea and the lower reaches of the big rivers - the struggle against the water will never end (KNAG, 2000). This led a.o. to the formulation of a new policy line in 1998 called 'Room for the River' in the 4th National Policy Document for Water Management (in Dutch: 4de Nota Waterhuishouding).

Fig. 63 The theme of the first Dutch so-called 'Natuurbalans' published in 1998 was: Availability of sufficient amounts of high quality water
(Courtesy: Natuurplanbureau, RIVM , Bilthoven pg 13)

In contrast to this we had a number of very dry summers in the late 1990ties and the beginning of the present century. Then, the Netherlands faced water shortages. Especially the amount of freshwater ran short for the two main stakeholders in need of this water: agriculture and nature.

This raised the awareness that fresh water was a valuable resource which we as a society needed to take care of.

10.3.3   Contribution of ecohydrology to nature protection and restoration in the Netherlands

New nature along the rivers

The 'Room for the River' policy initiated a vast number of projects in which the focus was not on reinforcing the dikes but on creating more space along the rivers where regulated floods could top off the river's water levels during high discharges. In these areas often nature was created and eco-hydrologists were advising on how to design these areas in such a way that nature would benefit most (see Pedroli et al., 1996; Jans, 2001; Helmer, 2000; Nienhuis et al., 1998).

NEN

At the same time it was recognised that the realisation of the national ambition to create a National Ecological Network (NEN; LNV, 1990) lagged behind planning (RIVM, 1998, see also contribution Barendregt and Dekker in chapter 3, page 35). Room for the River in this respect was a pragmatic solution to this problem since creation of flood control polders and water retention areas also served nature purposes. The same holds for the implementation of the ideas formulated in 'Plan Ooievaar' (Bruin et al., 1987) which got a significant impulse from the developments following the flood threats of 1993 and 1995. But as a consequence the NEN was shaped differently than originally planned. Away from the rivers the realisation was and still is in many places severely lagging behind planning (RIVM, 2005).

Ecohydrological knowledge put into practice

In a previous section of this chapter already two highlights of applied Dutch ecohydrology were mentioned. The scientific ecohydrological body of knowledge was put into practice in regional analyses of landscapes and via ecohydrological models. Apart from that ecohydrological insights were used in the design, planning and evaluation of the wet parts of the National Ecological Network. Potential areas were identified through regional ecohydrological analyses and these were suggested to be included in the network. Also, in Environmental Impact Assessment ecohydrological knowledge was disseminated mainly via ecohydrologists working at consultancies, RIZA (a research Institute of the Ministry of Traffic and Water Management), the State Forestry Service and the Society for Preservation of Nature (Natuurmonumenten). The latter two are the most important Dutch organisations managing nature reserves.

Local projects
However, most of the practical implementation of ecohydrology was in local projects aiming at preservation of threatened species depending on groundwater (so-called phreatophytes cf. Londo, 1988) in nature reserves. Close cooperation with local managers was an important factor enhancing implementation in restoration projects (see Grootjans et al., 1988; Kemmers and Jansen, 1988; Wassen et al., 1989; Roelofs, 1991; Barendregt et al., 1992; Van Diggelen et al., 2001; De Mars and Garritsen, 1996; Lucassen et al., 2004). Of course, in many of these projects emphasis was laid on the importance of regional groundwater systems and regional water management.

Focus on groundwater
We may conclude that putting into practice ecohydrological knowledge in river marginal wetlands was stimulated very much the last decade, in fact by accidence and not because of a pro-active strategy of ecohydrologists. The concept of the Dutch National Ecological Network was designed and propagated by others and ecohydrologists followed by providing expert-knowledge. Ecohydrologists for a long time did not care too much about the national scale since the largest scale at which there was an operational relation between hydrology and ecosystems seemed to be the scale of regional groundwater systems. Patterns on a larger scale were not yet of prime interest to ecohydrologists. Partly, this was due to the narrow focus of ecohydrologists on groundwater. Connections via surface water were mainly regarded as a threat such as the distribution of Rhine water all over the country. In this sense the attitude towards preservation of populations of endangered species in nature areas was very much focused on maintaining feeding by groundwater and isolating them from surface water. Restoring connectivity between fragmented populations was not recognised as an important issue by ecohydrologists.

10.4   New challenges

10.4.1   Restoration ecology

From description into recovery of ecosystems
In Dutch ecohydrology we see a trend from descriptive research focusing on understanding the reasons for deterioration of wet ecosystems towards studies in which recovery of ecosystems in restoration projects is monitored (see also Grootjans et al., 1996a).

The contribution of Dutch authors to restoration ecology is rapidly growing but the basis for restoration ecology in the Netherlands was laid by Jan Roelofs c.s. (Radboud University Nijmegen). Ironically, Roelofs never called himself an ecohydrologist. An explanation for this might be sought in his focus on soft water lakes, wet acidic grasslands and wet heathlands in which the regional hydrology is not such an obvious key factor as in low-land fens, marshes and non-acidic grasslands. His research is on the edge of hydro-chemistry, plant physiology and plant ecology (Roelofs, 1983; Roelofs et al., 1996; Lamers et al., 1998; Lucassen et al., 2004) and for instance contributed a lot to understanding the processes of internal eutro-phication i.e. eutrophication caused by changing hydro-chemical conditions and subsequently release of nutrients instead of inflow of nutrients from outside.

Monitoring data

Also, the increase of monitoring data from restoration projects led to the definition of success and failure factors for re-establishment of species. Apart from essential physical factors such as proper water table dynamics and feeding by the appropriate water systems it became evident that resto-ration is not always possible. Drainage may lead to irreversible decomposi-tion of peat (Okruszko, 1995) and also the redox-status of peat may irre-versibly change following drainage (Mars and Wassen, 1999). Restoration projects aiming at counterbalancing acidification show that pH and base saturation of the soil are recovering very slowly (Beltman et al., 2001). Also high atmospheric deposition and poor quality of groundwater and surface water may prevent full recovery of nutrient-poor conditions (Bakker and Berendse, 1999).

Restoration prospects

For this reason, Van Diggelen (1998) is more optimistic about restoration prospects of eutrophic floodplains than of mesotrophic fens and fen mead-ows. Finally, re-establishment of the desired species is hampered if they are absent in the actual vegetation and the seed-bank. Regeneration of the vegetation depends in such cases on dispersion possibilities, which are unfavourable for many wetland species in the present-day fragmented landscape (Bakker et al., 1996; Poschlod and Bonn, 1998, Van Diggelen, 1998). Additionally, above and belowground communities influence each other through a variety of direct and indirect interactions (Wardle, 2002).

Succession and soil community composition

Time lags in the response of belowground organisms to change lead to different selection pressures, for example as exerted by above and below-ground herbivores and pathogens (Putten et al., 2001). The strong linkage between succession in vegetation and in soil community composition sug-gests that restoration and conservation of biodiversity in natural vegetation is often too much focused on reducing soil fertility or on introducing above-ground vertebrate grazers (Deyn et al., 2003). The slow recovery of below ground communities might be an important reason for disappointing recov-ery of ecosystems after restoration measures have been taken in former agricultural fields. From the above it seems evident that restoration and creation of new nature can never replace a strategy aiming at preservation of what is still left.

10.4.2   Global ecohydrology

Global issues

Another emerging trend in current ecohydrology is the shift towards global issues. One might say that the godfathers of ecohydrology in the Netherlands Grootjans, Kemmers and Van Wirdum, all ecologists, have left the arena because they feel attracted to restoration ecology more than to ecohydrology. Instead there is a new generation ecohydrologists which is inspired by the ecohydrology perspective advocated by for example Rodriguez-Iturbe. This perspective focuses on climate-soil-vegetation dynamics in regions where water is a controlling factor (see Rodriguez-Iturbe, 2000), Eagleson, 2002 and it roots in the hydrology of land use change (see Van Dijk, 2004). In fact this type of ecohydrology was 'avant la lettre' already practiced in the 1960s in the Netherlands by hydrologists such as Prof. Reinder Feddes (Wageningen University).

Original focus on agriculture or forest hydrology

However, the work of Feddes was very much focused on agriculture and quantitative hydrology and not linked to the Dutch hydro-ecological mainstream as practiced by ecologists. Present-day representatives of this school who have broadened their work to other land use types are for example Sjoerd van der Zee, successor of Feddes, professor at the Soil Physics, Ecohydrology and Groundwater Management Group (WUR) dealing with the physical processes of the unsaturated zone, saturated zone and the interaction between the earth surface, atmosphere and plants; Han Dolman, professor of Ecohydrology at VU Amsterdam working on the interaction of the terrestrial biosphere with the hydrological cycle and atmosphere and the physics of the transfer of water, energy and nutrients from soil to biosphere and atmosphere; Marc Bierkens, professor at Department of Physical Geography, Utrecht University, an expert in stochastic hydrology, catchment hydrology and scaling issues.

Vegetation as a factor in hydrology

They are all hydrologists and in their work the focal point is not vegetation and plant species as an object for preservation but vegetation as a factor influenced by hydrology and climate and feeding back to hydrology and climate. For example, precipitation that falls on the land comes from advected moisture and from local evaporative sources. Vegetation adds to this through transpiration. Such local sources may differ due to micro-scale feedback mechanisms and may have large influence on the magnitude of the local precipitation (see *Fig. 64*).

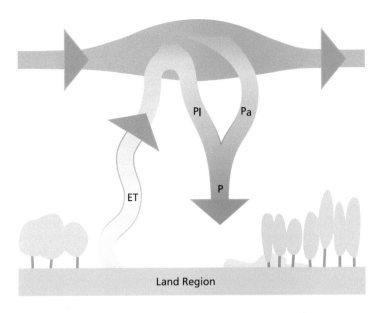

Fig. 64 Macroscale atmospheric water balance. ET = Evapotranspiration, P = total precipitation formed by precipitation derived from local evapotranspiration sources (Pl) and precipitation derived from advected moisture (Pa)
(After Dekker et al., 2007)

Non-linear change
Understanding this complexity helps to understand whether current global change will increase the risk of catastrophic desertification and how such a shift may feed back to the climate system (Dekker et al., 2007). Another example of coupled global and ecosystem change is the current deforestation in the Amazone region. These land use changes affect climate, and may cause increased droughts, because forests contribute to regional precipitation by actively transpiring water. Such coupling may also generate non-linear changes in the climate and ecosystems. Using global change scenarios and coupling land-use, biosphere and atmosphere models may create instruments which predict under what conditions non-linear shifts in climate can be expected and what their effects are on ecosystems worldwide. For example, Valentini et al. (2000) assessed the importance of respiration as the main determinant of the carbon balance in European forests and Ciais et al. (2005) showed that an increase in future drought events could turn temperate ecosystems into carbon sources, contributing to positive carbon-climate feedbacks.

10.5   Conclusion
Ecohydrology in the Netherlands already started in the 1970s. Its roots are in (landscape)ecology, vegetation science and agro-hydrology. The pre-

dominant focus was on groundwater in wet and moist terrestrial ecosystems and aimed at preservation of these ecosystems. Later, surface water was included and restoration and creation of new nature became core business. The Dutch ecohydrological agenda was set by ecologists and hydrologists were from the start only involved indirectly. Major environmental issues Dutch ecohydrologists were focusing on were desiccation, eutrophication and acidification. Fragmentation was for a long time not recognised as an issue by ecohydrologists. For this reason also their contribution to the National Ecological Network was limited.

Internationally ecohydrology started to flourish only 20 years later in the 1990s. In contrast to the Dutch trend, in other countries hydrologists were much more active. This is illustrated by an almost equal share by hydrologists and ecologists in papers in international journals. Ecohydrologists abroad had higher scientific ambitions than the Dutch down to earth approach. They advocated a paradigm shift in ecology and hydrology and aimed at developing a new cross-disciplinary science. Sustainable catchment water management based on integrated models of land use, groundwater and surface water shows how relevant this view on ecohydrology is for approaching complex environmental issues. At present, the Dutch contribution to ecohydology in the international arena is changing. There is a trend of ecologists retracting themselves to (restoration) ecology. The central position of ecologists in ecohydrology is taken over by hydrologists inspired by global issues. The fact that in the Netherlands at present it is mainly hydrologists regarding themselves as ecohydrologist might be a sign that ecohydrology in the Netherlands after thirty years is finally taken up seriously by hydrologists opening up more possibilities for real cooperation. Now, the danger is that ecologists stay apart. Hopefully, ecologists sufficiently keep taking part in defining new ecohydrological questions. The field of Global Change and Ecosystems described above is for sure an ambitious challenge in which also ecologists are needed.

11 HABITAT SCALE AND QUALITY OF FORAGING AREA

determining carrying capacity of Dutch wetlands:
lessons learned from a changing delta

Mennobart van Eerden

11.1   Introduction

Both from a theoretical (foraging theory) and a practical (wetland development and management) point of view, insight is required how animal populations depend on their habitat. Wetlands are considered to be threatened
habitats all over the world (*for example* Finlayson and Moser, 1991). Water
birds such as ducks, geese and swans constitute an eye-catching element,
relatively easy to study and representing a high natural value from the point
of view of nature conservation. The biological importance of a wetland can
therefore be judged by the presence of these water birds. *Vice versa*, many
species of migratory water birds completely depend on the presence of well
functioning wetlands along their flyway. However, little is known about the
scientific background of the relationship between habitat quality and the
number of animals the area can sustain, a concept often referred to as
Carrying Capacity.

The aim of this chapter is twofold: first, to stress the importance of food-
consumer relationships for water birds in wetlands and second, to relate
this information, together with data from literature, into a concept which
combines food exploitation theory in relation to the issue of carrying capacity at a community level. As such the scaling of habitat is an important quality factor, which contributes to carrying capacity in two ways: 1) the overall
food availability and 2) the habitat scale-related dynamics of ecosystem
processes.

11.2   Starting points

An *a priori* set of four lines of thought has been formulated, which is to
serve as proposition statement:

1.   Water bird populations in NW Europe are limited by winter food
      supplies and not simply distributed in relation to the occurrence of
      strictly protected nature reserves; thus food quality and food
      availability have management implications.
2.   Exploitation patterns of staging sites by migrating birds depend on
      patch use at different levels of scale, which relate to

      a.   Geography in relation to migration path (the travel distance -
            food abundance relationship)
      b.   Dietary preferences (limited spectrum per site during time of
            use) explained by food quality
      c.   Food depletion including inter-specific competition in relation to
            a threshold in exploitable food density
      d.   Foraging costs by the consumer leading to different exploita-
            tion thresholds

e.   Other factors not considered here (hunting, disturbance, pollution).

3.   Temporal and geographic exploitation patterns imply an interplay between tradition (return to reliable sites) coupled with nomadic movements to sample alternatives. Habitat scale is important

4.   The creation and maintenance of a network of sustainable staging and stopover sites is necessary in an increasingly urbanised world. The National Ecological Network (NEN) in the Netherlands provides such a network but, in order to meet the desired quality, needs to be evaluated in relation to the points mentioned above.

11.3   The role of food

Regulation of population by food

The question of how animal numbers are regulated has been receiving attention for many decades (Lack, 1954,1966; Wynne Edwards 1962, 1970; Newton 1980; Sinclair, 1989). Food is a necessity for every living creature and feeding ecology has therefore always played an important role in studies, which deal with population regulation. In the earlier studies, the segregation of species was supposed to be the result of competition for a common resource *for example* food. It was not until the 1970s that the few descriptive studies included the first field experiments. So was the long-term study on scottish red grouse *Lagopus lagopus scoticus* one of the first studies to combine the observed cyclic patterns of population development with behavioural features of the animals (Watson and Moss 1979, 1980). The condition of the heather was also taken into account, and it was tried to narrow down the food supply on offer in relation with the demands of the grouse into the profitable parts of their food plants (Miller *et al.,* 1970).

Use of habitat

The question of how a habitat is used also has a long tradition in science. Habitat use models have been constructed to order the observed spatial differences in animal abundance. However, unlike foraging models, habitat research remained descriptive and rarely provided a quantitative prediction about the relation between animal abundance and habitat quality (MacArthur and Levins 1964, MacArthur and Pianka 1966, Schoener 1968).

Individual decision making

More recently, emphasis has been laid on studies, which deal with individual decision-making. The way an animal behaves is considered to be not just pre-adapted to its environment, but, due to a constant selection pressure, in continual dialogue with it, aiming at a maximal achievable fitness. This view has led to a tremendous step forward and has resulted in the rapidly developing concept of foraging theory (Charnov 1976, Stephens and Krebs 1986, Maurer 1996). Based on principles, which originate in economic reasoning, animals were supposed to behave along predictable lines, using information from their environment. Food can be regarded as present in distinct unities, often called patches. Behavioural response to patches of different quality was considered such that the animal should maximise its long-term energy gain and, correspondingly, its fitness (see Ricklefs, 1996 for review).

Metabolisable energy per day (DME).

As in the economy, a well-defined currency plays an important role in assessing the state the animal is in. Nowadays, in foraging ecology the income for the animal is often expressed as metabolisable energy per unit of time, often expressed on a daily basis (DME). On the other hand, by expending energy the animal can afford various activities (locomotion, foraging, reproduction) and may compensate for various costs such as heat loss, feather and tissue renewal). Subsequently, energetic measures are needed to quantify the budget between DME and DEE (Daily Energy Expenditure) of the animals under study (King, 1974, Drent and Daan, 1980, Ricklefs, 1996). Energy budgets may tell whether birds are able to meet their demands from the food supply on offer, or have an income in excess of their timely energy expenditure, which allows them to store fat for migration or later use.

Maximum and Basal Metabolic Rate (BMR)

Important in the context of rate maximisation was the observation that the metabolic machinery sets an upper limit to the amount of energy that can be taken up and extracted from food per day. Upper limits of DME on a sustainable basis were related to the size of animal. Scaled to the level of Basal Metabolic Rate (BMR, the energy consumption of the animal at rest), maximum values of 3 to 6 times BMR (Drent and Daan 1980, Kirkwood 1983), or 7 times BMR (Weiner 1992) were found. Whether or not DME_{max} is tightly proportional to BMR (see discussion in Ricklefs 1996), the fact that per unit of time an upper limit of processable energy exists underlines the importance of time as a constraining factor which dictates the decision making of the animal. Time-related energy budgets play a decisive role in the animal's survival strategy. Sometimes time can be managed actively *for example* when an animal selects the most profitable food items or determines the speed of migration. In other cases the body plan dictates the speed of the process, such as the rate of food processing (gut structure) and the rate of depositing body fat. Thus, beside available energy, also time can be regarded as a major constraining factor in a foraging bird.

Carrying capacity

Carrying capacity, the number of animals that can be sustained by a certain habitat, is intuitively connected to both feeding ecology and habitat quality. No wonder that the concept appeals to policy makers, land-use planners and nature conservationists. Scientifically, however, a great deal of information underlying this idea is still lacking. In studies dealing with waders in estuarine environments, John Goss-Custard has formulated the carrying capacity concept as a testable hypothesis; the carrying capacity of an area has been reached when the addition of one more animal to the habitat under study would cause the death or emigration of another (Goss-Custard, 1985).

11.4   Carrying capacity and habitat scale

Scale

Problems concerning carrying capacity can span the range from stopover sites (10-500 km^2 and more) along an entire flyway for migratory birds (thousands of kilometres) to a single patch of food (0.01-1 m^2) or feeding habitat within a foraging area at a given stopover site (hundreds of metres). Recently, an approach has been in demand to try to scale-up foraging patterns from food patch to habitat level, both with regard to the total amount of food used, and the consequences that limited use of food resources might have on population level (Goss-Custard, 1980, Goss-Custard *et al.,* 1995, 1996, Zwarts and Drent, 1981, Sutherland, 1996). However, remarkably few studies have measured food consumption on a larger scale, such as an entire stopover site of a migratory species, for instance.

Wetlands

As wetlands in the western world can be considered discrete islands in a sea of cultivated habitat, many water birds rely closely upon these sites. Already in the 1960s, the need for protecting wetlands was felt, and a network of protected areas was assembled, aiming at safeguarding the identified flyway populations (for Europe see Atkinson-Willes, 1972, 1976, for North-America see Bellrose and Trudeau, 1988). Migratory water birds form an important group of animals which inhabit these wetlands and, in terms of the values they represent for nature conservation, they are protected by governmental legislation and policy documents (*for example* under the EU Birds Directive, Bonn Convention, IUCN et al., 1985).

Food-consumer relationship

However, the need for studies unravelling the food-consumer relationship remained as only a few sites offered good research facilities and, perhaps more important, no attempts were made by the field biologists to fill this gap. Trying to quantify food-consumer relationships at the level of an entire stopover site may seem an obvious challenge for a fieldworker; in practice it is costly and means a tremendous deal of work. Studies on diet, consumption rates in relation to food density and the numerical response of animals to the food supply on offer need to be confronted with the quantified food stock. In many cases not the absolute food abundance, but the fraction available, or better, attainable for the animal consumers needs to be quantified. For coastal mudflat habitat and waders *Charadriiformes,* this approach has been thoroughly worked out by Leo Zwarts and co-workers (Zwarts, 1996). The waders, almost entirely dependent upon invertebrate prey, possess species-specific prey detection and prey ingestion techniques, whereas the prey consistently delimits predation by an evolutionary play of hide-and-seek.

Long-term studies of water birds

Since 1975 I have taken up the opportunity to work along these lines in several Dutch freshwater habitats. Studies were initiated on a vast range of different species of water birds, being representatives of the different habitat sub-sets within a freshwater ecosystem. Much of the work has been carried out over a long period of time, combining detailed ecological examinations with long-term counting data (10-25 years and more) and periodic habitat assessment (see van Eerden, 1997). This period of time was felt necessary in order to use the change in natural succession of vegetation, benthos and fish as a kind of natural experiment to which the birds were supposed to respond with a change in numbers or a different harvest level. Knowledge about processes and quantified relationships are considered useful, both for our understanding of how to preserve biodiversity in existing natural wetlands and for management measures in man-made wetlands.

Study of newly created wetlands

Some of the newly created wetlands in the Netherlands were chosen as a basis for our research. *Fig. 65* shows the main study sites.

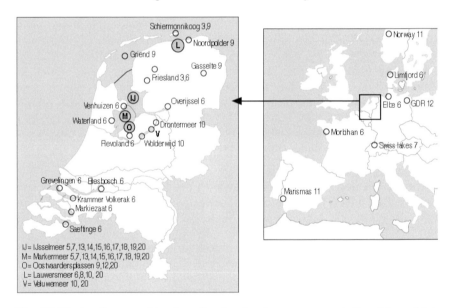

Fig. 65 Study sites used to describe the food-consumer relationships of this study (filled dots). Additional study sites where samples of food were taken are shown as open dots. (Eerden, 1997)

Each main site consisted of a number of typical habitat types. Nowhere disturbance because of leisure or hunting was such that this could have a dominant effect on waterfowl numbers or distribution. Salt-marsh communities and brackish wet meadow systems were studied in lake Lauwersmeer, the former Lauwerszee estuary, which was embanked in 1969. Freshwater meadows and marshes were found at a scale suitable for research in the Oostvaardersplassen, a recently developed wetland in Zuid Flevoland (reclaimed in 1968). The exploitation of submerged macrophytes was studied in the Lauwersmeer and the Borderlakes of Flevoland, created in the late 1950s. The lake IJsselmeer system (closed off from the sea in 1932) was chosen as it presently forms the most important large scale (*c.* 2000 km^2) open freshwater habitat in the Netherlands. Especially benthos and fish consumers are abundant here. All areas have in common that they are, in ecological terms, young of age. The production of food in this phase of natural succession is relatively high and still bears a relation with open, dynamic natural systems of (tidal or seasonally changing watertables) freshwater marshes (Odum and Hoover, 1988).

Two focal points
Two focal points have been set for this study: first, to quantify food-consumer relationships in a bottom-up manner from individual patch use to a community level. Second, to obtain information on the effect of trophic level and habitat scale on these relationships, by combining empirical measures from different species (plant-, invertebrate- and fish-eaters).

11.5   Foraging theory in practice

Daily energy expenditure and foraging rates
An animal requires food in order to balance its daily energy expenditure. As pointed out by Nagy (1987) extended by Nagy and Obst (1991), daily energy expenditure (DEE, kJ day^{-1}) for birds under field conditions scales to body mass as

$$DEE = 10.4 \ M^{0.67}$$

Foraging rates (F, kJ h^{-1}) also relate to body mass as found by Bryant and Westerterp (1980):

$$F = 2.02 \ M^{0.68}$$

The ratio between the two formulas represents the average amount of time a wild bird spends foraging, per day *i.e.* 5 hours. As the slope is nearly the same in both equations, time spent feeding is not related to body mass (Maurer, 1996). On a mass specific basis, smaller birds consume more energy than larger birds, in absolute terms, however, the requirements of large birds are higher.

Food density

Food density is just a simple measure for what is available to a consumer and it can only provide a superficial picture of what potential carrying capacity could be. As pointed out meticulously by Leo Zwarts in his studies on waders *Charadriiformes*, the available fraction of marine benthic prey present on tidal mudflats varies according to species, time of year and even between years (*for example* Zwarts and Wanink, 1993; Zwarts, 1996). The different predators also behave differently with respect to the available fraction, which complicates generalisations about carrying capacity.

Energy expended to obtain food

Water birds in wetlands consist of different ecological groups, which can be roughly divided into herbivores, benthivores and piscivores; energetic density of prey differs (Castro *et al.,* 1989; Karasov, 1996) and also costs for food provisioning vary substantially. Both parameters are considered important determinants of the bird's energy budget (Goldstein, 1989; see Ricklefs, 1996 for review). Water birds show a considerable variation in foraging modes with contrasting energetic costs. Diving under water while pursuing prey and flying are the most costly activities that occur in nature.

Compared to standard metabolic rate, *i.e.* the amount of energy expended when an animal is at rest in the thermoneutral zone, flying and running are most expensive (up to 14 and 12 times BMR respectively, (Brackenbury, 1984). Diving costs may vary between 6-7.5 times BMR for stationary divers in cold water (Leeuw, 1997) and 10 times BMR for pursuit-divers (Nagy et al., 1984). Swimming (2.2 (slow) to 4 (fast) times BMR) and walking at slow speed (2-3 times BMR) are less costly activities for birds (Videler and Nolet, 1990; Nolet *et al.*, 1992; Wooley and Owen, 1978).

A waddling goose or a duck floating on the water's surface requires far less energy than the costs for locomotion by a tern searching for fish in active flight or swimming underwater by a pursuit diver as the great crested grebe hunting for fish prey. Within one foraging guild, foraging costs may also vary. In herbivores, for instance, pecking or straining food from the water (seeds, leaves) is less costly than grubbing for roots or digging pits to extract tubers and stolons. So, it can be hypothesised that food density and quality alone do not determine the level down to which animals may harvest, but that the costs to obtain it also form an important factor.

Foraging models

Fig. 66 shows a qualitative model of the expected relationships between foraging costs, energetic return and patch harvest level.

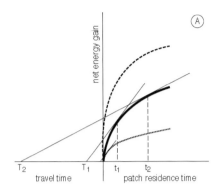

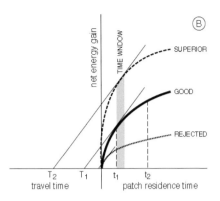

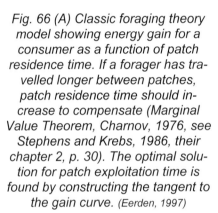

Fig. 66 (A) Classic foraging theory model showing energy gain for a consumer as a function of patch residence time. If a forager has travelled longer between patches, patch residence time should increase to compensate (Marginal Value Theorem, Charnov, 1976, see Stephens and Krebs, 1986, their chapter 2, p. 30). The optimal solution for patch exploitation time is found by constructing the tangent to the gain curve. (Eerden, 1997)

(B) As an extension to the theory now consider a stopover site as a huge food patch and patch residence time fixed, as in a migratory bird with a narrow timetable. It then follows that better sites would attract birds from a longer distance, whilst others are unattractive.
(Eerden, 1997)

It is based on the classic foraging models used to describe patch leaving decisions in relation to time spent travelling between patches (see Stephens and Krebs, 1986). As an extension to these models, foraging efforts and energetic gain for different food types were considered to operate at a higher level than the single food patch. It is assumed that feeding conditions at a stopover site as a whole can be considered a "super patch", as compared to other halting sites along the birds' flyway. Time spent in such a super patch would thus not only depend upon the amount and quality of the food present, but also on the conditions elsewhere. In particular, questions as to timing of migration and duration of stay at a stopover site could be evaluated, if flight distance between stopover sites is considered as travel time between patches.

Analogous to the classic theory, a bird thus would aim to stay longer at a site when the travel distance from the previous haunt was longer (*Fig. 66A*). Also, if the quality of a site on the flyway level in terms of available food increases, this site would attract birds from a greater distance (*Fig. 66B*).

Measuring food density

As stated before, situations with measurable food densities at field scale are rare. In the Netherlands, the struggle against the water has a tradition of at least a thousand years (Schultz, 1992). The more recently reclaimed polders and dammed estuaries provided good research possibilities for the purpose of our study. Often left alone for many years before cultivation could be started, and sometimes designed as nature area, when protection against the sea was the only purpose, natural succession could go on undisturbed. Food for water birds was abundant and comprised often single or a few species. Natural succession caused food density to change dramatically at a rate, which was manageable by the researcher (10-25 years). The complete turnover from one ecosystem to another provided excellent starting conditions to study the effect of a new territory on the numbers of migratory water birds. As such, I consider the biological processes during the making of polders or the damming up of estuaries as large-scale natural experiments, useful for studying the process of colonisation by food organisms and their consumers.

Monitoring and experiments

Impact estimates were determined in the field and taken from literature. The ideal route to study carrying capacity problems is to monitor biological processes in nature, and to carry out experiments both in nature and under controlled (semi) laboratory conditions. Such an integrated study has recently been published on the ecological energetics of wintering diving ducks (Leeuw, 1997).

Carrying capacity at different trophic levels

In this chapter, I am trying to combine issues related to carrying capacity at different trophic levels. Besides obtaining baseline information, the objective is to derive a possible new concept in the study of carrying capacity in a comparative way. Two lines of approach were supposed to be essential for this study.

Foraging costs, patch residence time and patch harvest level

First, to investigate the first order relationship between food availability and response (numerical, functional) or impact (patch harvest level at different levels of scale) by the consumer. The basic idea is that, because costs for foraging differ greatly among the ecological groups under study, as does the energy content of the food, food harvest levels may vary according to the net energy gain per unit of time (gross intake rate minus costs required for foraging, digestion and other losses). *Fig. 67* shows the expected relationships in a graphical way (see also Stephens and Krebs 1988). If foraging costs increase, patch residence time becomes longer and patch harvest level increases. The aim was to test this hypothesis and to derive a conceptual model for carrying capacity issues of natural areas.

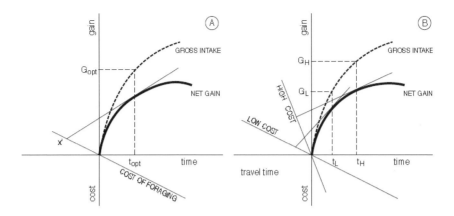

Fig. 67 Model describing optimal patch residence time including foraging costs. In extension to Fig. 66, gross and net intake gain curves have been drawn. Note that the optimal solution is found by constructing the tangent to the net gain curve, whereafter the gross gain is derived. The costs of foraging are subtracted from the net energy gain by rotating the abcissa and drawing the tangent from x^1 instead of x. (A) Energy costs for travel and foraging are the same. (B) Higher travelling costs imply longer patch residence time and a higher energy gain.
(Modified after Stephens and Krebs, 1986)

Herbivore, benthivore and piscivore water birds
Second, by combining results from herbivore, benthivore and piscivore representatives of the group of water birds, all belonging to a few families of birds and often assembled in the same wetlands, I will attempt to con-struct an overall picture of the key parameters that determine carrying ca-pacity with respect to food provisioning in relation with the trophic level. As the scope of this study is wide, it cannot be expected that tailor-made solu-tions can be achieved for all ecological groups as yet. As an example, just the integrated diving duck study mentioned earlier took *c.* 15 man-years (Leeuw, 1997).

Disturbances left aside
In view of the above, it may also be good to state where this chapter is *not* about. Not dealt with are, for instance, the various sources of disturbance (hunting, leisure, traffic), factors that also affect carrying capacity of wet-lands for water birds, and the effects of pollution and pesticides limiting bird numbers.

11.6 The role of habitat scale and completeness

Wintering water birds

Wintering water birds are confronted with rapidly changing conditions while on migration and in their winter territories. Having spent the summer in their northern, mostly undisturbed breeding grounds, they migrate towards the south with their young. The range from which birds reach the Netherlands extends from NE Canada for some waders, via Spitsbergen and Scandinavia eastwards to Russia, far behind the Ural mountains. Two events require special attention during the period outside the breeding area. First, moult of wing feathers imposes in many water birds a constraint on their movement. Being flightless for several weeks (3-5 depending on size of the species), a habitat is selected which must meet conditions of safety and food availability.

Critical surface area for specific functions

Greylag geese (Anser anser) spend the moulting time in reedbeds. Of the many reedbeds along lakes, ponds and canals only some, the most extensive, are used for moult. Mostly they are adjacent to open waterbodies such as IJsselmeer (Vrouwenzand/Steile Bank) or Haringvliet/Biesbosch. In the case of the embankment of Zuid Flevoland, reed marshes larger than 1000 ha developed. They attracted up to 60,000 moulting geese, arriving from central and Eastern Europe (Zijlstra et al., 1991). This example shows that a critical surface area exists for specific functions. Minimising the risk of food shortage, the chance of predation and disturbance are factors contributing to this scale dependent functioning.

Replenishing fat reserves

Second, prior to the period of wintering, fat deposits are usually laid down and this entails special demands on the birds' energy requirements. Stopover sites are often used to replenish fat reserves needed for further migration (Alerstam and Lindström, 1990). As an example of birds using vast areas of open water the black tern (Chlidonias niger) visits the area of Lake IJsselmeer for several weeks in late summer. They hunt for small fish in the upper 5cm of the water column, mainly smelt (Osmerus eperlanus). Concentrations of 25,000-125,000 in autumn are only known from the IJsselmeer; there, the combination of abundant prey in the turbid waters and safe roosts guarantee the possibility of a certain stay over. Birds may fly each day more than 60 km one-way to the foraging area, because of changing wind and turbidity conditions which cause differences in availability of smelt. Probably at smaller lakes, the annual forecast for a successful stay is to little to gather in larger numbers.

Combination of habitats

That not only one type of habitat is involved but the combination of a series of habitats is best illustrated by the example of the great white egret

(Egretta alba) in Oostvaardersplassen. This species bred in very small numbers since 1976 when the water table in this newly created marshland was put up. However, it lasted until the early 2000s before numbers really got up; only after completion of the newly dug pools and inundation of wet grasslands feeding conditions became more successful. So besides wet reedlands for safe breeding, the species needed foraging area in early spring in wet meadows and pools. Here they hunt for small fish, amphibians and voles (Voslamber *et al.*, in prep.). The combination of both habitat types is only occurring in Oostvaardersplassen as yet, and therefore only in this area the species is found breeding in significant numbers (143 pairs in 2006).

11.6.1   Different constraints in different foraging guilds

Digestion capacity

Beside constraints imposed by the environment, mostly because of availability of food, the birds face costs for digestion of their food. This is especially the case for bulk eaters, such as avian herbivores and some diving benthivores, which ingest entire shells. Not only food selection and intake rates, but digestion capacity of the gut sets the limit of food processing in those cases (*for example* Karasov, 1996). Herbivorous water birds, *i.e.* a large group of ducks, geese and swans can only superficially digest plant food. We investigated whether food availability *per se* or the capacity of the gut to digest, could set the major limit to habitat selection and food choice. From field data of digestibility of naturally taken food and an estimated energy return is quantified. Using data from a variety of species, we related apparent digestibility to body size. Among the birds we studied a mass range from 300-12 000 g, which spans the entire range in body mass of this group of birds in the western Palaearctic. It is expected that the smaller species will have to be more selective (and thus be more susceptible to changes in quality of their habitat) in order to meet their daily energy requirements. Some food sources were found not be exploitable because of a too low energy content, others because the intake rate cannot be maintained at the right level.

Die-off

The ultimate effect of low food stocks in relation to harsh weather conditions may have serious consequences. A mass die-off of diving ducks, which took place simultaneously in Switzerland and the Netherlands, was examined. Harsh weather conditions, in connection with body condition of the birds and low food availability were the most important reasons for this to happen (Suter and Van Eerden, 1992).

Plants, bottom-fauna and fish eaters

Food choice and feeding specialisation, in combination with guild-specific constraints have thus important consequences for the ecological energetics of the different waterbird species. Plants, bottom-fauna and fish eaters were chosen as distinct groups in which issues such as foraging behaviour, consumption rates and harvestable fraction were determined. This group was found to be able to harvest food items at rates and to levels greater than any other group (60-90%). Grass leaves, buds, seeds belong to this category which have in common that they are poorly digestible but reasonably easy to obtain.

Benthos feeders

Benthos feeders were studied at lake IJsselmeer and proved to be dependent on a few prey species. The zebra mussel (Dreissena polymorpha) plays a central role in the freshwater lakes nowadays. The levels of predation were less than the herbivores, ranging between 25-60%. The diving costs, due to travel distance (2-5 m) and cold water impose a serious constraint, as well as the presence of Dreissena in clumps.

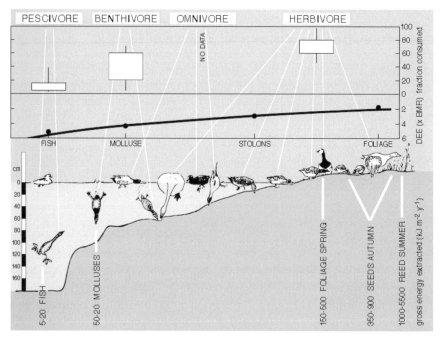

Fig. 68 Schematic cross-section through a freshwater wetland with typical avian consumers. The numbers refer to the energetic uptake and percentage harvest for individual species. The approximated costs for foraging, expressed as multiples of Basal Metabolic Rate (BMR) and the (gross) energetic return of the food taken (kJ g^{-1} DW) are indicated
(Castro et al., 1989)

Prey catchability and harvested fractions
The fish-eaters among the avian consumers consist of the most hetero-geneous group. Like benthos feeders they dive for food, but in most cases a pursuit dive is made because their prey is mobile. In this group specific problems like underwater visibility impose extra constraints on prey catchability. Using data from the large IJsselmeer system, factors such as sex, age and social foraging proved important with respect to the under-standing of feeding ecology in a comparative way. In no other group of water birds is the segregation between the different categories within one species so apparent. Prey exploitation among fish-eaters resulted in the lowest levels of harvested fractions, in most cases 5-20%.

Fig. 68 gives an overview of the different ecological groups of water birds relevant for freshwater wetlands. Beside the schematic positioning along an imaginary cross-section of a wetland, the observed harvested fractions and energetic gain of naturally used food items are indicated.

11.6.2   Changing delta, changing policy directives

Fragmented smaller habitats with more food
To put the current situation of wetland abundance in the Netherlands in a wider perspective, we need to consider the historical situation in the Ne-therlands, starting at least 7000 years ago. Due to climatic change after the past glaciation and reinforced by influences by man, the original realm of sheer endless wetlands finally became narrowed down into well-defined super patches of protected nature areas. The consequences of this large-scale impoverishment caused most likely population declines in most wa-terbird species (Eerden, 1997), where a tentative exploration is made of the quantitative consequences of the continual habitat shift. However, the most impressive changes in land-use have occurred in the last two centuries. Due to agricultural changes, the carrying capacity for wintering herbivorous ducks, geese and swans has greatly improved. It has been suggested that the winter food-bottleneck of this group has been released by the improved quality and quantity of a variety of cultivars providing enormous amounts of seeds, roots and protein-rich leaves (Eerden *et al.*, 2005). As in future the role of agriculture will become less energy and nutrient demanding as a result of the continuous policy to achieve a greater sustainability of the agro business, the situation for water birds may turn worse. Lesser application of artificial fertilisers may lead to lower winter protein content of grass leaves, which affects the grazing of herbivorous water birds such as geese, ducks and swans.

Eutrophication and purification
Moreover the constant eutrophication of the shallow waters has led to a dramatic increase in overall productivity of the water system. This process has caused negative effects such as the occurrence of algal blooms and anoxia with associated fish kills, but like in the agricultural situation, also

facilitated water birds by increased carrying capacity. Since 1980 a generally downward trend in dissolved phosphorous is obvious, resulting in clearer waters with less contaminants, but with a possible lowered productivity as well (although this issue is still under debate by Hosper and Meijer, 1986; Van Liere and Gulati, 1992; Scheffer, 1998). The overall loss of wetland habitat may thus have been temporarily compensated by a higher productivity, a trend that will be broken because of the success of the many purification plants that have been realised meanwhile.

Protection of carrying capacity

The conclusions of these trends do not cause an optimistic mood; now EU induced legislation leads to large-scale protection of the water bodies, the reorganisation of fishery pressure, the extensification of agriculture and the improvement of water quality will lower the carrying capacity of wetlands for water birds. To put it in other words, the overuse by man that unintentionally compensated the loss of wetlands because of drainage and embankment will show the real effect of these measures.

Of course one may argue whether the opposed trend of these policy lines could result in a better perspective for the water birds but this is not an adequate tackle of the problem. Obviously the direction for a naturally increased carrying capacity is a better, although not easily achievable goal. The solution may prove to be the restoration of dynamics (greater role for natural processes which cause cyclic events in succession) and realisation of habitat completeness (linking of complementary habitats, levelling out the effect of of barriers). As such, it is not sufficient to safeguard and connect crucial habitats for water birds alone. This strategy, which forms the basis of the Natura 2000 and NEN network planning, is only adequate if one is able to guarantee the minimum of required dynamics by proper management.

The international liaison

Temporal and geographic exploitation patterns by water birds imply an interplay between tradition (return to reliable sites) coupled with nomadic movements to sample alternatives along the flyway (Alerstam & Lindström, 1990; Lindström, 1995). Many studies have shown the ability of water birds to behave as such. The annual change of climatic conditions such as rainfall, ice and the cycle in temperature and radiation affect the production of young leaves, seeds and fish. Water birds are used to profit at a flyway level, i.e. a global scale of these phenological events. Global warming will alter these events and may cause shifts in vegetation belts and, in turn water bird distribution and numbers, because not all species may be able to adapt easily to this because of geomorphological constraints.of their habitat. Other than many terrestrial ecosystems, the wetlands (ranging from terrestrial marshes to shallow seashores) possess a series of critical features, which explain their specific function as key habitat for water birds. Therefore, the northward shift of ecotones may lead to the change (in many cases deterioration) of the carrying capacity for wintering water birds. This issue will need more attention if one is to judge both the maintenance of the current networks of Natura 2000 and NEN as well as to determine the future direction of development (creation of new sites, policy of species specific goals). It can be easily foreseen that this discussion needs to be addressed at a European (if water birds are concerned a flyway -) level.

11.7   Evaluation and synthesis

11.7.1   Management directives at species level

The factors which determine Carrying Capacity, can be divided into two parts:

1. consumer-based factors: body size, "Bauplan", degree of water contact, trophic level of interaction, and
2. environment-based: food quality and food availability.

The maximum work capacity of a bird and the general time constraint to achieve the necessary income in terms of metabolised energy is considered a major framework for the evaluation (see also Hammond & Diamond, 1997). This implies species-specific and site specific conditions, which form the basis for the suitability of a certain habitat. As for the water birds the main issues can be summarised as follows.

All water birds

For all water birds the basic prerequisite is the availability of safe (predators, disturbance by man) feeding and resting sites of sufficient size and at a reasonable distance from each other. Resting sites are best close (1-5 km) to the feeding grounds, but in some cases distances of 15-40 km might be acceptable, given the quality of the distant food is ascertained. Some water birds are active during daytime, the majority during night time. In relation to disturbance (leisure, traffic, noise, darkness) this issue is relevant. In many cases is reduction of contact with water an important item to reduce heat transfer. Therefore, the existence of sandbanks, mudflats, inundated meadows etc. is of importance as a secondary factor, which makes an area attractive to water birds (see De Vries and Van Eerden, 1995).

Herbivorous water birds

For herbivorous water birds is important the availability of nutrient rich food sources; as protein determines the quality of forage in spring, this requires short swards in the growing phase, for which large herbivores are needed as a management tool (see Vulink and Van Eerden, 1998; Vulink, 2001). As the seasonal growth of foliage causes the deterioration of nitrogen content over time (see Black, Prop and Larson, 2007) and water level determines the start of spring growth, long gradients in micro elevation of the terrain are necessary to sustain foraging ducks, geese and swans for a longer period.

In general, the midwinter habitat of these birds is related to the use of agricultural lands. However, in spring and autumn the availability of wet grasslands is crucial. Retreating water levels by evaporation may guarantee the successive availability of nutrient rich swards; an obvious relation to habitat scale is evident here, as only part of the area functions at a certain moment in time. As not all water birds are grazers, the availability of seeds of pioneer plants (both in freshwater wetlands as coastal sites) is important. Also the true grazers often prefer seeds when available in late summer and autumn (stage of pre-winter fattening up). Pioneer habitat such as mudflats, sandy shores, open patches etc. tend to overgrow with perennial plants, in general less productive.

The dynamics of the abiotic conditions form the most important of the prerequisites for this function, in some cases coupled with extensive grazing by large herbivores (see Vulink and Van Eerden, 1998 for review). Changing water tables and either erosive or constructive phases in micro- succession are needed. Flooding is needed for seed availability again. Finally, the grubbing swans and geese need nutritious (carbohydrates) feed, especially in autumn when they build up fat reserves. Stolons and rhizomes, bulbs or bulbils of perennial marsh plants are important in this respect. These hibernating organs contain energy dense material that is preferred food by the avian grubbers. Like for the seeds, the relation with succession stage of the vegetation is apparent and, again, water table management determines food availability (grubbing needs water, plants with rhizomes rely on water table changes and/or early stages in the succession).

Benthivorous water birds
For benthivorous water birds the availability of abundant (energy density), suitable (size, burying depth, thickness of shell) prey is important. In tidal littoral environments, burying depth of shellfish and worms is important for prey accessibility for waders (see Zwarts, 1996 and Van De Kam *et al.*, 2004 for extensive treatment of this issue). To some extent this also holds for freshwater mudflats in relation to chironomid larvae but generally these prey are easy in comparison with species of the intertidal community. For diving ducks, both freshwater and marine, water depth is of prime importance in relation to the energy expenditure (see De Leeuw, 1997).

For Lake IJsselmeer this implies that less than half of the total community of Dreissena polymorpha is available to the ducks. Siltation, eutrophication and anoxic conditions cause the disappearance of mussel beds. Ice winters, storms and in tidal areas mechanical harvesting of cockles and mussels may deteriorate conditions and cause major setbacks of the population. In cases where water tables can be managed (as in most of the Dutch situation), these arguments may lead to adjusted management in certain cases.

For fish-eating water birds the availability of suitable prey (size, depth) is important. They have to persuade their prey fish, which not only requires diving but also pursuing abilities and is even more energy demanding than in benthivorous diving ducks which go for sessile prey at the bottom. These skilful hunting performance requires a moderate underwater light climate, not too turbid and not too transparent, which in turn is influenced by wind and water current in relation to soil type and water depth. As fish are mobile, they can position themselves in a three-dimensional space. This makes them unpredictable prey, corroborated further by another major factor in the water system and that is predatory fish.

Because predatory fish can govern small prey fish far more efficient than birds can, their presence (or absence) is of major importance to the habitat suitability (read availability of small fish) for fish-eating water birds. As man fishes the fish predators preferentially, their stocks are generally (too) low nowadays. This holds for the open seas, the coastal lagoons and large lake and river systems. Sustainable fishery, allowing a larger standing stock, thus lowers the potential use that birds can make of the system. It is undesirable to promote over-fishing as a management option in favour of fish-eating birds. Because of its effect, however, extra measures will be necessary to safeguard the populations of fish-eating birds under the condition of a larger stock of predatory fish. The fish-eaters are among the water birds, which depend on large open territories, even more than the benthos feeding diving ducks. Because they are active during daylight hours, the chance of disturbance of foraging flocks is larger in this group than in the benthos-eating diving ducks.

11.7.2   Management directives at habitat level

The creation and maintenance of a network of sustainable staging and stopover sites is necessary in an increasingly urbanised world. The challenge is not only to achieve this on paper only by EU legislation like Natura 2000 and at the country level by the NEN, but also to restore the dynamics that originally were thriving the wetland values. Our studies in the IJsselmeer area, Lauwersmeer and delta region, relate to changed wetlands as a result of large-scale flood protection measures. Although they function as important areas for water birds, enforced by law since March 2000 when protective status under Natura 2000 legislation became apparent, their future state in terms of carrying capacity will be dependent on selective management measures:

Open water of large scale is a highly dynamic environment but because of low availability of food for water birds per unit of area requires the preservation of the large scale and associated values (darkness, gradients in food production and food availability) and a sufficient area of undisturbed regions.

Restoring hydrological connectivity of large lakes in the Netherlands (parts of former estuaries) and, where possible also with the marine systems is an option to restore former connectivity and minimise the risk of extinction of food organisms. This is useful for the aquatic organisms such as fishes but also for enforcement of the carrying capacity for water birds.

Natural or restored marshes in the terrestrial component of the wetland range need more direct link to the open water as well as ecological dynamics in order to fulfil their role as food provisioning habitats. The large open lakes in the IJsselmeer area and delta with artificial shores in the form of dikes need a prominent fringe of hydrologically linked wetlands and marshes, in proportion to their scale of open water.

For a more integrated approach of management, the Natura 2000 and NEN networks need to be combined into one system, at least this is useful for the ecological management reasons outlined above

For a proper functioning the relationships with surrounding areas need to be addressed more specifically. For geese, swans and herbivorous ducks for instance, the agricultural connection is clear. By integrated management the damage to crops may be lessened and the overall capacity of the system enlarged.

11.8   Epilogue

Future work will be necessary to arrive at a specific point of focus with respect to to wetland management. As the environmentally based factors are manageable and the species-based only to a very limited extent, directives for nature development should aim at the former. Breaking down the comprehensive issue of the carrying capacity of staging areas into manageable sub-issues, I have come to the following four key topics. These statements, some of which need to be investigated at depth in the field or by modelling, are based on the research outcome in relation to the carrying capacity issue for water birds. Although they especially apply in freshwater wetlands outside the breeding season, their scope partly holds in estuarine and coastal situations year-round.

1. Food production and food quality need to be specific points of attention in relation to the management of wetlands for water birds
2. Foraging costs of the consumer determine necessary energetic return of food, which in turn determines patch harvest levels, and thus carrying capacity is strongly related to foraging guilds. The implication is that fish eating birds need the largest areas, followed by benthos eaters and plant eating birds. Future work should address more specifically in relation to the habitat requirements of individual species, in declining need of priority: fish-, benthos-, plant-eaters.

3. It is possible to scale-up patch use patterns to higher levels (area or zone, stopover site along the flyway) and this allows predictions about carrying capacity of the area. Future management of species should consider this interconnection of sites and should aim at preservation of (parts of) populations rather than flocks of birds associated to certain SPA's.
4. As natural processes such as plant succession, erosion, sedimentation, water currents and water level fluctuation determine the conditions for ecological state, and thus affect carrying capacity, these factors need more attention in future management design. Restoring dynamic forces is especially important in man-made environments.

Although seemingly far away from direct benefits to mankind, these measures might link to initiatives, which are necessary to cope with effects of climate change such as sea level rise. Newly created marshes may contribute to the safety of dikes, which need not be as high as they would be without the adjacent marsh zone. The perception that brackish water in specific areas may be desirable not only from the ecological point of view but also be of help in combating the blue green algae which form a plague in late summer for water leisure is another example. It is clear that the NEN and Natura 2000 networks need to be combined. Not only is scale a relevant parameter for water birds, it should also become an issue at the management level, both in the field as through legislative measures.

12   CONNECTING IS EASY, SEPARATING IS DIFFICULT

Taeke de Jong[a] **T**UDelft

---

[a] The author is grateful for having received valuable suggestions about this text from Chris van Leeuwen, Johan Vos, Lodewijk van Duuren and Sybrand Tjallingii.

12.1   Introduction

This chapter[a] discusses the one-sidedness of an emphasis on ecological connections in nature conservation and spatial planning. It traces back the track of Dutch nature conservation thinking into the typical Dutch ecologist Van Leeuwen stressing separations to restore the balance.

12.2   The emphasis on boundaries apart from areas

12.2.1   Van Eijck

As a student at the Faculty of Architecture in Delft my favourite lectures were those of architect Aldo van Eijck and ecologist Chris van Leeuwen.

Fig. 69 Aldo van Eyck
(Eyck, 1986)

Fig. 70 Chris van Leeuwen
(Schimmel, 1985)

Both emphasised the boundaries between spaces instead of the character of the spaces intself.
'The boundary makes the difference; that's where it happens' they argued.
After all, the task of urban and architectural designers is to draw boundaries. Designers cannot do much more than drawing boundaries to make spaces visible and usable.

---

[a] Based on a lecture for the Dutch-Flemish association of ecology NECOV 2005-01

  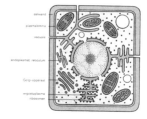

Fig. 71 Van Eyck (1955-1960) Burgerweeshuis (Amsterdam) (Ligtelijn, 1999)	*Fig. 72 Van Eyck (1965) Sonsbeek paviljoen (Arnhem)* (Ligtelijn, 1999)	*Fig. 73 The cell and its membranes* (Vogel; Günter; Angermann and Hartmut, 1970) page 18)

In the seventies Aldo van Eijck could give a lecture without a break for six hours on only a few images from Mali reporting his experiences of Dogon architecture (A.E.v. Eyck, et al., 1968). The Dogon live at a spectacular landscape boundary. Nobody wanted to miss his rare and fascinating lectures and nobody in the overcrowded classroom was bored for one moment by his humorous and furious criticism of Western culture.

12.2.2   Inbetween realms

I remember an image showing the entrance of a hut with thick walls. The entrance had the form of a tree or fungus. So, you could sit in this boundary environment without being forced to choose between inside or outside. You got coolness from the shade or warmth from the sun simply by changing position. Van Eijck called such locations not forcing us to choose 'in between-realms' or 'twin phenomena'. He reproached our culture for forcing choices between false alternatives: "Would you like to breathe in or out?".

Fig. 74 An entrance as a seat: a 'twin phenomenon' or 'in between realm'

12.2.3   Van Leeuwen

The emerging environmental awareness of the seventies made the lectures of Van Leeuwen popular as well. Many remember them. Shortly before his death he attended a conference dedicated to his work (D.J. Joustra, et al., 2004), organised by former students in urbanism and architecture. However, the speeches of that conference showed very different applications, (especially in the field of urban renewal) based on vague interpretations contrasting with Van Leeuwens own usual precision.

Fig. 75 Conference 2004
(D.J. Joustra, et al., 2004)

Fig. 76 Van Leeuwens references
(Ross Ashby, 1957, 1965; Bateson, 1980, 1983)

He knew the outdoor nature like no one else, but at the same time he was an armchair scholar, writer of many dispersed articles and lecture notes (C.G. van Leeuwen, 1971) surprising colleagues and fascinating designers.

12.2.4   Open-closed theory

His 'open-closed theory' (Leeuwen, 1964) was the subject of dispute with his friend and close colleague Westhoff from the University of Nijmegen at the former national institute of nature conservation (RIN). Westhoff, et al. (1975) developed according to Braun-Blanquet (1964) a Dutch syneco-logical system of life communities now elaborated by his successor (Schaminee, et al., 1995) and translated to nature target types (Bal et al., 2001) applied in the actual policy of the Dutch ecological network (NEN). However, that operational approach now loses foundation in the perspective of climate change.

Source:
Fig. 77 Braun Blanquet
(J. Braun-Blanquet, 1964)

Source:
Fig. 78 Westhoff's synecology...
(Westhoff, et al., 1975)

Source:
Fig. 79 ...translated into Dutch nature target types (Bal, et al., 2001)

Van Leeuwen made field inventories himself for many years. Based on that experience he emphasised transitions between such supposed life communities rather than determining the communities themselves (Leeuwen, 1965). Precisely there he saw most rare species, especially if such a transition was spun out along a broad strip (gradient) into an infinite range of unnamed particular environments on a smaller scale. There the ecologically most interesting specialists settled.

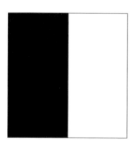

Fig. 80 Limes convergens

Fig. 81 Boundary rich

Fig. 82 Limes divergens (gradient)

12.2.5 Gradient map in national planning

This line of thought was the guideline of the Dutch Second National Policy Document on Spatial Planning (RPD, 1966), by which Van Leeuwen's 'Gradient map' was published (see *Fig. 83*).

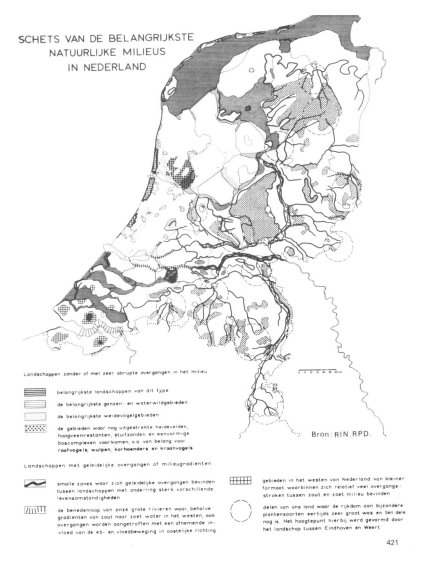

SCHETS VAN DE BELANGRIJKSTE
NATUURLIJKE MILIEUS
IN NEDERLAND

Landschappen zonder of met zeer abrupte overgangen in het milieu

belangrijkste landschappen van dit type

de belangrijkste ganzen- en waterwildgebieden

de belangrijkste weidevogelgebieden

de gebieden waar nog uitgestrekte heidevelden,
hoogveenrestanten, stuifzanden en eenvormige
boscomplexen voorkomen, o.a. van belang voor
roofvogels, wulpen, korhoenders en kraanvogels

Landschappen met geleidelijke overgangen of milieugradienten

smalle zones waar zich geleidelijke overgangen bevinden
tussen landschappen met onderling sterk verschillende
levensomstandigheden

de benedenloop van onze grote rivieren waar, behalve
gradienten van zout naar zoet water in het westen, ook
overgangen worden aangetroffen met een afnemende invloed van de eb- en vloedbeweging in oostelijke richting

gebieden in het westen van Nederland van kleiner
formaat waarbinnen zich relatief veel overgangs-
stroken tussen zout en zoet milieu bevinden

delen van ons land waar de rijkdom aan bijzondere
plantensoorten eertijds zeer groot was en ten dele
nog is. Het hoogtepunt hierbij werd gevormd door
het landschap tussen Eindhoven en Weert

Bron: RIN, RPD.

421

Fig. 83 'Gradient map' in the Dutch Second National Policy Document on Spatial Planning (RPD, 1966)

Citing RPD (1966) :'Gradients are narrow zones with gradual intermediate stages between landscapes with mutual strongly different life circumstances. Examples are contact zones between salt and fresh water environments, between relatively dry and wet areas, between poorly and richly nutricious landscapes and slopes in high areas. Within or directly near these gradual zones one finds a great gradation of environmental types in small compass and as a result a large richness of plant and animal species. To this richness belong nearly all rare plant species in our country.

213

Moreover, here are the regions where in the Netherlands natural edge of wood thickets can develop.
Furthermore, the 'conservative' character of these transitional environments is typical. This assures continued existence of species concerned at these locations, subject to not disturbing the transitional environment fully by changes caused by modern agricultural methods.'

Van Leeuwen surprised colleagues by predicting the square metre where a specific rare plant species could be found. For example I witnessed him when he was already at an advanced age looking around and indicating the place where the Carex pulicaris ('flea sedge', 'vlozegge') should grow. However, the manager of the area never found that species on his territory. The bystanders went on their knees and found the predicted flea sedge. Van Leeuwen did it intuitively, based on 'phenomenal' field knowledge.

12.3   Regulation theory

12.3.1   Relation theory

However, Van Leeuwen could not record that experience in writings otherwise than by sketching a very theoretical framework known as 'relation theory'. That theory is dispersed in many articles and elaborated in different separate directions, always surprising by unexpected relations between 'down to earth' examples. It led to his being made an honorary doctor of the University of Groningen (1974), but the same University published a doctoral thesis judging that theory to be invalid on mathematical grounds (Sloep, 1983). However, the same critique applied also to other ecological theories not studied by Sloep. Opposite that most readers and certainly listeners got the feeling of a crystal-clear and simple framework, relevant to many questions concerning design, spatial planning, urban renewal and nature conservation. At last Van Leeuwen agreed to name his theoretical framework more precisely 'regulation theory', according to his cybernetic references of steering and disturbing.

12.3.2   Spatial and temporal variation

One of the first schemes I remember from Van Leeuwen's lectures shows some basic notions of that theory (see *Fig. 84*). Firstly it shows the possibility of a negative relation between pattern and process in ecosystems in terms of spatial and temporal variation. So, in general difference correlates to stability (often found near vague boundaries), equality to change (often found near sharp boundaries). However, I realised many years later this rule cannot be applied on any level of scale if you take the scale paradox (see *Fig. 205*) into account.

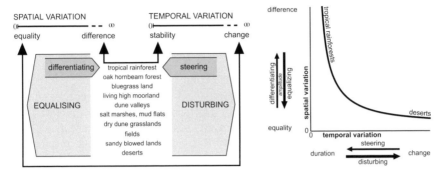

Fig. 84 Spatial and temporal variation in the theories of Van Leeuwen
(author, derived from the lectures of van Leeuwen in 1972)

According to Ross Ashby (1957, 1956) 'equality' is not regarded as the opposite of 'difference' but as its near-zero-value. After all, any imagined difference can always be made more different by adding attributes of difference (for instance difference of place, distance), but it cannot always be made less different. A difference less than the least difference we can observe or imagine is called 'equality'. So, 'difference' and 'change', 'equality' and 'stability' in the scheme are all taken as values of 'variation' (the variable to be distinguished spatially and temporally).

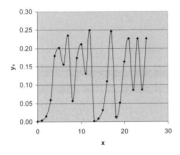

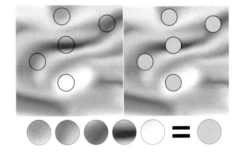

Fig. 85 Chaotic behaviour of
$y_{x+1}=a^*y_x-(a^*y_x)^2$ *where* $y_0=0.001$
and a = 4

Fig. 86 Reduction to the average

To concern equality as a special kind of difference is contrary to the main presuppositions of usual mathematics, the science of equality (you cannot count different categories) and equations. However, chaos equations like $y_{x+1}=a^*y_x-(a^*y_x)^2$ where a>3.6 produce chaotic behaviour even different on different computers using different roundings off (see *Fig. 85*).

The same applies to very small differences of initial values in complex models producing very different results.

The main problem is, mathematical treatment of quantities presupposes qualitative categorisation reducing differences to an 'average' (see *Fig. 86*), tacitly supposed in set theory.

12.3.3   Disturbing and steering

Proceeding that way, Van Leeuwen supposed processes of a second order on both pattern ('process on pattern') and process ('process on process') called 'differentiating' and 'steering' with 'equalising' and 'disturbing' as zero-values (see the grey arrows in left *Fig. 84*). Because these processes are changes as well, they are disturbing and equalising by definition. Stopping a process of disturbing is disturbance as well. Suddenly cleaning a ditch or decreasing the number of grazers could deteriorate the condition of the ecosystem unexpectedly. The consequence of this view appeared to be a recommendation not to change the condition too sudden: clean the ditch or decrease the number of grazers slowly according to the adaptation speed of the system.

So according to Van Leeuwen it is easier to break down differences (equalising) then create them (differentiating) and at the same time it is easier to introduce changes (disturbing) than to guarantee duration (steering). This is a simple verbal expression of the second law of thermodynamics in the perspective of cybernetics. Within that interpretation 'life' is represented as a phenomenon climbing up into local diversity and duration at the cost of global disturbance located elsewhere.

12.3.4   Separation and discontinuity

Second order patterns and processes
Regulation theory became more complicated as soon as Van Leeuwen started to look for a second order of *patterns* as well: 'pattern on pattern' ('structure', ranging from 'separation' causing difference, into its zero value 'connection' causing equality) and 'pattern on process' ('dynamics', gradual ('continuity') or sudden ('discontinuity') changes and stops, causing stability or change). Later I realised distinguishing levels of spatial and temporal scales might simplify the argument and put it into perspective. Perhaps the primary supposition about a negative relation between pattern and process is limited to certain levels of scale explaining exceptions. Perhaps concepts like 'pattern on pattern' are simply a question of scale. 'Difference' is a scale sensitive concept after all (see *Fig. 205*). Moreover, difference, equality, separation and connection are direction-sensitive.

Ligitimate questions
Anyway, many legitimate questions remain. I will summarise some, but not answer them here. The very first question is: "Is this science?". How could you make categories as general as difference and change or separation and connection operational for tests by empirical research? Should you not

distinguish different kinds of difference (for example abiotic, biotic differences, differences observed on different levels of scale) to find mutual relations? What causes what? Are the second order variations dominant? Does separation cause difference or the reverse? How could you imagine separation without difference?

Elaborating these questions you come across fundamental epistemologic questions similar to those I know from the debate about academic design (Jong and Voordt, 2002). They go beyond critics like those of Sloep because equality itself is disputed. Consequently the use of categorisation and classification within categories presupposed in any variable is attacked. The very core of that debate in practice is the question how to generalise solutions of context-sensitive problems bound to specific unique locations and contexts. That question applies to ecology as well, confronted with a confusing diversity of species multiplied by a diversity of specimens and contexts. Management theory also struggles with the inapplicability of reduction into the 'average' (see *Fig. 86*) from empirical science (Riemsdijk, 1999).

From a designer's point of view many design decisions in specific contexts cannot be supported by empirical research aiming at generalisation. "That conclusion does not apply to this specific location!" designers complain. Van Leeuwen's approach offered a terminology directly fitting to design acts par excellence: separating and connecting. It functioned as a great heuristic tool, but many applications fell prey to confusion of scale by lack of scale articulation. Let us now go back to ecological practice.

12.4   Meadowland as a fringe laid out

Shortly before his death Van Leeuwen offered me a clarifying example.
Between meadowland and forest in natural circumstances a fringe emerges through herbivore grazing (see *Fig. 87* and *Fig. 88*).

Herbivores mow with their long necks over the boundary of their reach without treading or manuring (floating head). By doing so, they create prototypes of meadowland. In meadowland (a fringe laid out) without manuring, mowed without treading of note ('hooiland', an alternative etymology of 'Holland') you find species like Serratula Tinctoria ('saw-wort', 'zaagtand') not to be found elsewhere. Species rich steppe grasslands like in the Ukraine and Russia are comparable to meadowlands. Why are there species rich (hundreds per m^2) and species poor (one per ha) grasslands? Instability of a specific temporal scale between dry and wet, cold and warm, fresh and salt seems to be the most important factor.
Such an instability reinforces itself: a dense, solid soil emerges with Plantago major (the tread plant 'common plantain', 'weegbree'). Water remains there, but also flows away easily.

217

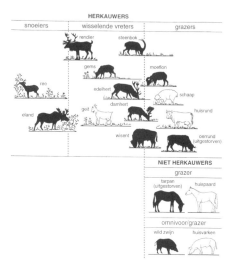

Fig. 87 Metaphors of wilderness
(Vera, 1997)

Fig. 88 Pruners, alternating grubbers,
grazers (Vera, 1997)

That is why even more powerful alternations between wet and dry, cold and warm arise, which cannot to be endured by many plant species. In Moscow dryness is locally suppressed by the fire brigade, again reinforcing disturbance and condensation of the soil. However, a slope stabilises.

In the Netherlands plantago major never grows on a slope, because the contrast between wet and dry is too small. There, other plant species can survive stabilising the environment even further. The Russian species rich steppe has, unlike a desert a stable water balance horizontally and vertically. A desert becomes brackish by evaporation and consequently rising water (ascending moist flow). Salinisation by irrigation is a well known phenomenon. So, a linking between wet-dry, cold-warm, salt-fresh alternation arises there, which does not happen in species rich steppes. Against temporal changes there are stable spatial transitions based on selective separation.

12.5   Selectors and regulators in the landscape

12.5.1   Connection supposes separation

What I would like to bring to the fore is the importance of inaccesibility, isolation, in this case for large mammals. As the concept of ecological networks (ecologische infrastructuur) started its triumphal progress in the Netherlands (D.de Bruin et al., 1987, 'Plan Ooievaar'; primarily based on separation), connections are primarily emphasised.

Fig. 89 The 'Plan Ooievaar'
(Bruin et al., 1987)

Fig. 90 Separation of nature and agriculture: zoning, selection, regulation, 'ecological networks'
(Bruin et al., 1987)

I would like to set against that emphasise, for a while one-sidedly, the importance of separations to arrive at the middle (mi-lieu). The concept of 'structure' (litterally 'brickwork') comprises both separation and connection. Exactly their combination produces particular environments where specialists are at ease. Researching that kind of environment could be named 'structure ecology'. In terms of regulation theory both isolation and connection are a value of separation. Connection is solely a zero value of separation. Connection supposes separation, not the reverse. There are no windows without walls. But there is 'difference in separation', always a combination of separation and connection while separation directs connection.

12.5.2 Selectors and regulators

The first notable combination follows on the 'basic paradox of spatial arrangement' as Van Leeuwen named it: the phenomenon of separation perpendicular to connection.

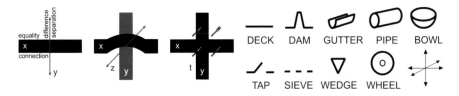

Fig. 91 Basic paradox of spatial arrangement

Fig. 92 Selectors
(Leeuwen, C.G.v. (1979-1980) Ekologie I en II. Beknopte syllabus)

A road is laid out to connect, but perpendicular to that connection it separates. That is painfully felt at crossings. The solution to connecting perpendicularly to the other connection is separating vertically (viaduct) or in time (traffic lights, see *Fig. 91*). However, there are more combinations of separating and connecting. Deck, dam, gutter, pipe and bowl are examples of 'selectors' in one, two, three, four and five directions, selectively connecting into the other directions. That direction-sensitive connection quality cannot be imagined without separation into the other directions. Selectors take care not *everything is going anywhere.*

Taps, lids, valves, wedges and wheels are regulators taking care not *everything is always going somewhere.* Living organisms are complex combinations of selectors and regulators known in technology as mechanisms on different levels of scale (see *Fig. 93*, *Fig. 94* and *Fig. 95*).

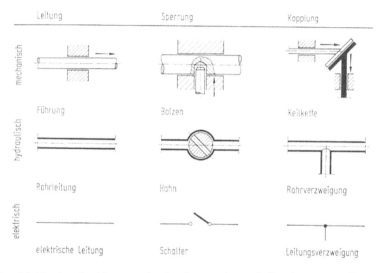

Fig. 93 Mechanical forms of selection and regulation by separating and connecting (Rodenacker , 1970)

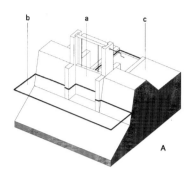

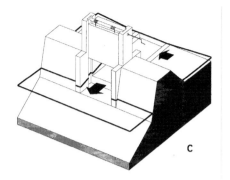

Fig. 94 Sluice closed (Arends, 1994)   *Fig. 95 Sluice open (Arends, 1994)*

12.5.3   Operational and conditional steering

The complex world of selectively separating and connecting occurs right down to the smallest scale of biology: the cell and its membranes (see *Fig. 73*). On that interfaces substances are selected and allowed to make connections with each other. The conditions for specific connections are created primarily by separating substancies that should not be connected (preselection). That already begins with the external membrane separating the inner environment from the entropic outside world. That makes less probable processes possible inside. This range of conditions and the endoplasmatic apparatus necessary to create the right conditions for the right connection is often forgotten in understanding the isolated process of connection operationally (monocausally).

The endless range of conditional functions in the environment seem to require another, perhaps typically ecological way of thinking than the single function with one clear product. Such processes are imitated in systems of retorts and pipes being the armamentarium of chemistry (in Dutch: 'scheikunde', 'skill of separation', not the skill of connection). Madame Curie needed four years to isolate 1/10 gramme of radium from tons of pitchblende. To dissolve sugar in our coffee is a daily activity taking seconds, but separating it afterwards takes much more effort. A heap of manure is easily dispersed, but it takes years to get it out of the ecosystem.

In the same way it is easier to destroy the subtle system of selectors and regulators of a living organism than to rearrange and synthesise it. A violent murder means demolishing separations, starting with those of the skin. Suppose now an ecologically rare location is surrounded by a range of conditional functions we still do not understand completely. Is it wise then to make connections for a few cuddly populations with botanically doubtful functions? Their equalising function in small areas could be that of an elephant in a china cabinet. Other (migrating) animals than grazers do not fit in our small nature reserves, but in vast eutrophic areas elsewhere in the world. There they are needed as mineral transporters comparable with pipelines connecting one sided high productive communities. A much larger number of smaller more rare species of animals needing a smaller area could be supported better by diversification of the botanical foundation. You can wait which superstructure develops thereupon instead of taking the summit of a food web as a target in advance. You should not start building a house with the roof.

12.5.4   Ecological networks

In the doctoral thesis of Van Bohemen (H.D.v. Bohemen, 2004) strikes that the hundreds of millions (!) spent on ecological connections are hardly judged on their ecological effect.

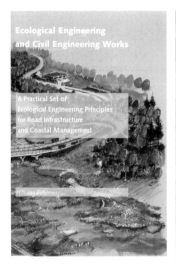

Fig. 96 Technical ecology
(Bohemen, 2004)

Fig. 97 Ecological connections
(Bohemen, 2004)

The argument is: you have to build a wildlife viaduct before you can measure the effect. That phase is now upon us, but it is recognised that just as in epidemiological research cause and effect are difficult to separate. And then we still focus solely on the effect on populations of some species. Which effect the constructed connections show on other species is even more difficult to determine. The deteriorating effect of positive discrimination is well known from hanging on nesting-boxes: other bird species were ousted, insects died out and the plant species having them as postillions d' amour disappeared.

The impact of connections is sometimes demonstrably negative. Examples include the import of alders from Eastern Europe in the seventies or the connection of the Main-Danube canal. The connection of all parts of the

222

world to each other (globalisation) may be the greatest danger. Connecting genetically different races could cause loss of biodiversity. That leads to the subject fascinating me most: levels of scale. At what level of scale connecting is the best strategy, and at what level of scale separating? The best argument for separating areas is the emergence of subspecies, though it takes a lot of time. A crucial question is: are we in the Netherlands in need of other large mammals than grazers if they have better and more sustainable conditions elsewhere? Could not we create in our wet country much more interesting 'ecological conditions' by separation (Tjallingii, 1996, see *Fig. 98* and *Fig. 99*), conditions lacking everywhere else? Holland hooiland!

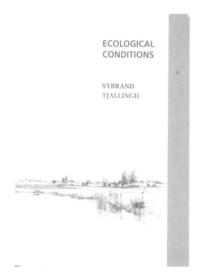

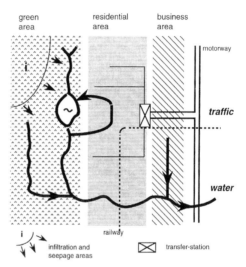

Fig. 98 Ecological conditions
(Tjallingii, 1996)

Fig. 99 Separating dry and wet flows
(Tjallingii, 1996)

A more moderate conclusion is that ecology cannot produce general statements, though politicians would like to seduce you that way.
That is what I learned from the PhD thesis of Mechtild de Jong (2002, see *Fig. 100* and *Fig. 101*).

That methodological problem of scientific generalisation in the context-sensitive relations between one and a half million of species from which we know so little, is something shared by ecology with context-sensitive design (Jong and Voordt, 2002) and management sciences.
The problem of the classical empirical ideal to produce generalising statements (out of bits and pieces, to deduce subsequently from these statements conclusions for specific cases) increases if you realise any species comprises differently reacting individuals. That problem increases even more so, if you realise that any individual arrives in a different context. The urbanist or architect knows the problem only too well.

An ecologist is not invited to copy solutions, but to bring a local field of problems into a common solution by a unique concept. That is not solely an ecological network, but a more complete ecological infrastructure.

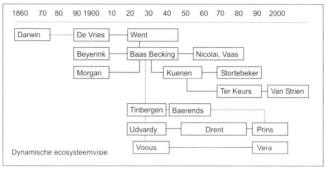

| 1860 | 70 | 80 | 90 1900 | 10 | 20 | 30 | 40 | 50 | 60 | 70 | 80 | 90 | 2000 |

Darwin ⋯⋯ De Vries — Went

Beyerink — Baas Becking — Nicolai, Vaas

Morgan — Kuenen — Stortebeker

Ter Keurs — Van Strien

Tinbergen – Baerends ⋯⋯⋯⋯

Udvardy — Drent — Prins

Dynamische ecosysteemvisie

Voous — Vera

Fig. 100 Separations
in Dutch
ecological thinking
(Jong,
M.D.T.M.d., 2002)

Fig. 101 A genealogy of theories
(Jong, M.D.T.M.d., 2002)

12.6 Conclusion

Generalisation is dangereous, especially if small differences can produce great effects. That is the case in ecology. Biodiversity between species and between specimens within any species is multiplied by the number of contexts they live in. And the physical and social context of any location is different from any other location because every location is unique if only because of its location between the other locations of the Earth's surface.

That diversity is a risk cover for life. But there are different differences. Some of them we call 'equality'. Equality is the basis of expectations. The ecological expectations of our common future are gloomy. However, our imagination covers more than expectations, it opens up possible futures as well as probable ones. The modality of possibility requires an other way of reasoning than probability.

In the advanced technology of pattern recognition the emphasis on similarity shifts into a focus on dissimilarity (Pekalska, 2005). Following that track broadens the view into unexpected, improbable possibilities, opened up by difference. Differences are observable at boundaries. So, it's worth the effort to study boundaries rather than homogeneous areas. They determine the areas, not the reverse. Perhaps it produces cross-border insight.

Town

TOWN

Taeke de Jong

Editorial introduction

The historical contradiction between city and nature is still alive today. However, the appreciation of urban nature is growing, which has resulted in the development of urban ecology. This section gives rather an overview of that field than much 'brand new' insights. Experts will miss many citations, but recent publications like Zoest (2006) give more complete references. Perhaps publications written in other languages than Dutch and English (for example the many German publications in this field) did not yet penetrate sufficiently in the Dutch speaking region.

Dekker (chapter 13, page 230: *Urban ecology, an exploration*) found six different meanings of the term urban ecology in literature, relating to topics ranging from pure ecological science to the design of the ecological city, applying urban ecology in a normative way. Moreover, the object of study, urban nature, is also defined as urban ecology. By these different meanings urban ecology becomes a boundary object. While this gives a common basis of support from different disciplines, it also leads to confusion. Dekker proposes different terms for the different meanings. An important question concerns the relevance of an autonomous theory of urban ecology, alongside other ecological theories. Here opinion is divided. The deciding factor is whether there are sufficient differences between urban and non-urban ecosystems. Most ecologists agree that there are clear differences, and in particular the degree of integration of urban ecosystems is very low. Moreover, as a result of invasion or immigration, cities are extremely rich in alien species. The development of an autonomous theory of urban ecology therefore deserves further discussion.

Denters (chapter 14, page 245: *The urban district, a biogeographical acquisition*) describes the urban area as a mature ecosystem with a specific urban wildlife adapted to locations with constant buzz or surprising tranquility and its development. He makes an inventory of measures to improve its quality. Denters (1994) succeeded in establishing the 'urban district' as an accepted bio-geographical district amongst other districts distinguished in Dutch floristic research. Based on extensive field work firstly in Amsterdam (1994, 1998), later in urban areas all over the Netherlands and Belgium (1999) he published a beautiful field guide for the urban district (2004).

Berg (2001) wrote an essay about human health and nature, establishing that subject as an important policy argument in developing (urban) nature in the Netherlands. She studied the public balance of fear and fascination for nature (2004, 2005), summarising benefits on human health (2005). In her chapter 15 page 259: *Public health* she adds possible disadvantages urban ecologists should take into account discussing urban nature in the context of a municipal policy debate.

Zoest and Melchers (2006) recently wrote a beautiful, very complete and accessible book on life in cities including the human species, its priorities and their impact on other life forms. Its 548 references are not merely scientific references. Much of their contents are explained for the common reader making ecological arguments accessible for public debate. Its first part concerns man, its second part nature and its third part the practice of an urban ecologist, Zoests own task in the municipality of Amsterdam. His chapter 16, page 267: *Driving forces in urban ecology* is a selection of texts form Zoest and Melchers (2006) translated in English with an introduction and conclusion of the author. It gives an impression of the width of the subject, but focuses on the urban-rural gradient on different levels of scale.

Tjallingii, one of the founding fathers of the national WLO and the international IALE, was recently involved in the European COSTII project making an inventory of different urban policies on green urban area. His chapter 17 page 316: *From green belt to green structure* reports some results and developments concluding by recommendations. Drawing on experiences of European cities, the paper confronts two planning strategies for urban development and green areas. Protective greenbelt strategies focus on containment of urban growth and protection of the rural countryside, whereas development oriented green structure strategies perceive green areas as a part of urban quality of life and the identity of the urban landscape. Green structure strategies offer a promising network perspective for growing cities. Supported by road and water network systems, green structures create conditions for urban quality and flexibility.

Wardenaar (chapter 18, page 331: *Landscape architecture and Urban Ecology*) discusses the important role of landscape architecture in Dutch urban areas, often competing urban design by its freedom of form. It starts with a brief history of urban park design concluding recent trends in landscape planning. Landscape ecology plays an increasing role in landscape architecture, but its targets have to be balanced with other green goals in urban area's. Apart from biodiversity the contribution to environmental awareness brings other priorities to the fore. Wardenaar suggests to establish a 'red' ecological network as a complement to the NEN.

Hooimeijer (chapter 19, page 346: *Water city design in waterland*) is an art and culture scientist preparing a PhD on urban design and theory focusing on water cities. In her chapter she stresses the association of Dutch town tradition and the urban functions of water. Cities in a wet territory are connected and defended by water, resulting in a careful water management, readable in the landscape and the town plans like those inspired by Stevin. It gives the country a common memory and cultural heritage to be protected. Hooimeyer tries to make a link between these goals and contemporary priorities like climate change and urban ecology.

Jong (chapter 20 page 372: *Urban ecology, scale and structure*) takes urban ecology in its broadest sense: the distribution and abundance of species (plants, animals *and* humans, their artefacts). He tries to grasp the practice of an urban ecologist in a interdisciplinary framework that can help balancing priorities of different specialists advising urban policy. Scale-articulation seems inevitable to get overview and to bring empirical science, policy and design as the main scheme of human specialisation and difference in language-games together. But within that game of negotiation the biologist has an important message and perhaps a crucial role.

13   URBAN ECOLOGY, AN EXPLORATION

Jos Dekker[a]

[a] This article is based on the abridged Dutch version Dekker, 2006

13.1   Introduction

People use the term urban ecology to refer to very different activities. Even in its ecological sense it is applied in different ways. This article[a] is an exploration of the different interpretations of urban ecology that can be found in the Dutch and international specialist literature.

The exploration begins with a brief history of the attention for nature in the city, and this is followed by an analysis of different definitions associated with different ways of dealing with it. Urban ecology is a social system of people, their organisations and projects, means of communication and so forth. There are a wide variety of approaches to urban nature and the city itself is delineated in different ways. The question as to whether urban ecology deserves to have a separate ecological theory receives conflicting responses. There are parallels with the development of landscape ecology, but also differences.

In this article I do not discuss the wealth and characteristics of urban nature. Neither do I deal with its functions, values and meanings. Government policy is also not considered. Nature receives rather more attention than do people or the environment, and I regularly narrow the focus to the situation in the Netherlands.

13.2   The history of urban nature in the Netherlands

The urban origin of the European appreciation of nature

The present appreciation of nature falls within a long European tradition: the Arcadian tradition. This came about during the Romantic period, although its roots extend far back in Western history. It was well-off city dwellers who wanted to distance themselves from the city as a centre of development in the modern age. Nature in the countryside met their Romantic demands (Koppen, 2002).

The city is therefore the origin of the appreciation of nature. At the same time history sees nature as the antithesis of the city. Nature is seen as being out of place in the city, and for many, city and nature are still the ultimate contradiction (Sukopp, 1990).

City versus nature

A lack of appreciation for urban nature is often found in the Netherlands too, among conservationists and ecologists. In 1937, Jac. P. Thijsse[b] was

---

[a] This article has been adapted from a presentation given at the WLO Symposium on Urban Ecology on 6 April 2006 in Amsterdam. An abridged version appeared in *Landschap* 23(2006)2: 135-144.

[b] Jac. P. Thijsse was an important Dutch conservationist in the first half of the twentieth century. He wrote many books, was co-founder and editor of the journal *De Levende Natuur* (Living Nature), and was co-founder of the *Vereniging tot Behoud van Natuurmonumenten*

disappointed to observe that the Mayor of Amsterdam and aldermen had no eye for natural beauty in the urban environment (Koster, 1939). Half a century later, the Amsterdam alderman Jan Schaeffer is said to have held the same opinion: people who want nature should go and live in Kudelstaart[a].

Thijsse's organisation *Natuurmonumenten*[b] did little better than the Amsterdam local authority. Founded in 1905 in Amsterdam, *Natuurmonumenten* was based there for many years. But the society rarely concerned itself with nature in the city. In a book published to mark its eightieth anniversary, "Ruimte voor natuur" (Room for nature, Gorter, 1986), the city is principally seen as a threat to nature and landscape. The term urban nature is not mentioned. Neither does *Natuurmonumenten* consider nature in and around the city to be one of its core responsibilities for the twenty-first century (Wijffels, 2005).

In 1982 the former Theory Working Group of the Netherlands Society for Landscape Ecology (WLO[c]) produced the book '*Landschapstaal*' (Landscape Language, Schroevers, 1982). In it the city and its nature receive hardly any mention. Ten years later, to mark the twentieth anniversary of its foundation, the WLO held a congress on the past, present and future of landscape ecology (WLO Executive Committee, 1992). The city and urban ecology did not come up for discussion.

According to the first *Nature Outlook* (in Dutch Natuurverkenning, RIVM et al., 1997), compared with the National Ecological Network (NEN), urban areas contain few target species[d] of vascular plants per square kilometre. However, from the same source on which RIVM bases this conclusion (Lemaire et al., 1997) it can also be concluded that three quarters of the target species of vascular plants are found in urban areas, almost as many as in the NEN.

Changing views
Views of nature change, and differing versions of the history of this change have been written. From the seventeenth century the image of the Arcadian landscape was dominant in Europe (Koppen, 2002). An appreciation of the wilderness was a variant on this Arcadian vision. But according to the British historian Simon Schama (1995), in Western history an appreciation of the Arcadian landscape and of the wilderness alternated up until the last century. A constant deepening and broadening of views of nature then took place in the Netherlands (Windt, 1995). At the same time, fluctuating

---

(Society for the Conservation of Nature Reserves). He was involved in the purchase of the first nature reserve in the Netherlands.
[a] Kudelstaart is a village near Amsterdam.
[b] The *Vereniging tot Behoud van Natuurmonumenten* (Society for the Conservation of Nature Reserves), abbreviated to *Natuurmonumenten*, is the oldest Dutch conservation organisation (1905) which purchases and manages nature reserves.
[c] WLO (Netherlands Society for Landscape Ecology) is the Dutch branch of IALE.
[d] Target species are ones that are the focus of Dutch ecological policy because they are internationally significant, rare and/or in decline.

strategies could be observed, partly as a consequence of a number of dilemmas for conservationists (Dekker, 2002). Views on nature are dynamic and this means that the negative attitude to the city and the nature within it may change.

Urban nature
There are indeed positive views on urban nature in the Netherlands. In the introduction to the first issue of *De Levende Natuur* in 1896, the editors – among them Thijsse – write that they want to make living nature 'a source of general pleasure'. They then give a number of examples: 'zoological and botanic gardens, flower markets, florists, … small city gardens, window boxes at the windows of deprived city dwellers, Rover, Felix and the canary, the goldfish bowl, cockchafer sports …' (in Dutch: 'dieren- en planten-tuinen, bloemmarkten, bloemwinkels, … kleine stadstuintjes, bloemen-planken voor de ramen van misdeelde stedelingen, Azor, Minet en de kanarievogel, de goudvischkom, de meikeversport …', Koppen, 2002, p 184).
From 1970 onwards urban nature in the Netherlands enjoyed a growing interest. Prior to this was a long history of development of urban nature, so-called green space planning. Green space planning in towns and cities was developed in the nineteenth century for urban beautification and public health reasons. At the beginning of the twentieth century, garden cities and neighbourhoods were created, later followed by large grassed areas between blocks of flats. In the mid-1970s interest for more natural parks and gardens was aroused by the work of Londo (nature gardens and parks) and Zonderwijk (verges) (Rooijen, 1984). Finally, ecological connection zones in the city became important, as did the concept of the ecological city.
The city is perceived as a 'rock in the meadow' (Windt *et al.* 2003). This image suggests that urban nature is the result of a specific environment. This does not do justice to the social structure of much urban green space in the past 150 years. Urban green space in wealthy residential neighbourhoods is the product of market forces resulting from rich people's demands for a green living environment. It is also an important welfare state amenity for public wellbeing (Rooijen, 1984, see also Van Zoest in chapter 16, page 267).

13.3   Definitions

Urban ecology in specialist literature
Between 1993 and 1997 only 0.4% of the articles in nine leading ecological journals related to cities or urban species (Collins *et al.*, 2000). A personal investigation of the terms 'urban ecology' and '*stadsecologie*' (the Dutch term for urban ecology) using a variety of scientific and other search engines reveals that these terms occur far less commonly than do 'landscape ecology' or the Dutch term '*landschapsecologie*'. A comparable search shows that the Dutch words '*stadsnatuur*' (urban nature) and related terms like '*stadsgroen*' (urban green space) and '*natuur in de stad*' (nature in

towns and cities) occur more frequently than does *'stadsecologie'* (urban ecology). Apparently people are concerned with urban nature, but do not refer to it as urban ecology.

Six meanings

In Dutch and international literature, the term urban ecology has six different meanings. These vary from the purely scientific or sociological to the application of this science (or applied science) in urban design, management and policy. Urban ecology also stands for the object of this science itself: the nature and environment of the city.

Firstly, urban ecology*(ecological science)*. According to Sukopp, a pioneer in this field, urban ecology is 'the study of urban biotopes with ecological methods' (Sukopp, 1990, pp 1-2). Sukopp sees urban ecology primarily as a science (Sukopp, 1998). Many other ecologists take a comparable view.

Secondly, urban ecology is *nature*. According to Tenner *et al.* (1997) it is 'the biological life in the city other than human beings' (p 27). This definition is less widely accepted.

Thirdly, urban ecology is an *outlook or norm*. It is a matter of what is good for nature and the environment (and thus for people). Ecological green space management (Koster, 2000) and the ecological city (Deelstra, 1998) fit into this category. The term urban ecology is often used in this sense and expresses an attitude or ideology that shapes the design, planning and management of the city. Urban ecology is 'consciously treating the city as an ecosystem' (Koning and Tjallingii, 1991). By this they mean that actions are geared to abiotic and biotic effects, such as differentiation processes. Why exactly these processes are more ecological than homogenisation processes, for example, is not explained. The 'ecological city' is principally one that is sustainable and comfortable to live in, with 'factor 20' (less energy, materials and pollution) (Hendriks and Duijvestein, 2002). In its design a variety of sometimes conflicting principles are applied – both higher building density and thinning by more green space and water. Trepl (1995) and Sukopp (1998) expressly distinguish this from urban ecology as science.

Following directly on from this is the fourth definition: urban ecology as *ecological urban development* or sustainable urban design (Sukopp, 2002). According to Tjallingii (1995), urban ecological projects, of which there are already many in the Netherlands, are seeking possibilities to build ecological aspects into urban development. His 'ecopolis' is a collection of strategies (guidelines and concepts) for ecologically responsible urban development. Deelstra considers the design of sustainable cities to be urban ecology (Deelstra, 1998). De Jong (2006) sees urban development as a component of 'scientific ecology'.[a]

The fifth definition of urban ecology is *nature management and policy in towns and cities*. According to Van Zoest (2001), urban ecology, apart from

---

[a] See also his contribution on Urban Ecology in this book.

being a branch of science, is also a policy field, with the objective of promoting nature in and around the city. It can also be a matter of initiatives by
residents' groups (Seiberth, 1981).

The sixth definition, older than the scientific sense of the term, is urban
ecology as a *social science* of the city as ecosystem, known for example as
the Chicago School (Koning and Tjallingii, 1991; Melosi, 2003). This sociological school studied the relationships between people and their urban
environment.

Finally, the term urban ecology is also used in *combinations* of the definitions described above. De Jong sees it on the one hand as a branch of the
biological discipline of ecology, in which people play an important role, and
on the other hand also as design and policy (see his contribution on urban
ecology in this book, chapter 20, page 372).

Niemelä (1999) applies not only a scientific definition but also a broader
one that includes social sciences. Picket *et al.* (1997) and Collins *et al.*
(2000) argue for an integrated approach combining natural and social sciences. Sukopp (1990) does not believe in such an integrated approach,
because there is no common definition and the different methods are difficult to combine.

Boundary object

Because of its widely differing meanings, urban ecology can be seen as a
'boundary object'. This is a scientific sociological term for objects of science
that occur in different overlapping social worlds, meeting specific information needs within them (Star and Griesemer, 1989). A boundary object
must on the one hand be robust enough to be shared, and on the other
hand be flexible enough to be given an autonomous interpretation. A well-
known example of this is biodiversity. Many different approaches go under
this label. This gives the concept strength, but also forms an obstacle to
open discussion (Koppen, 2002).

The advantage of such a boundary object is that it quickly provides a common basis for a shared aim. For urban ecology this can be seen in the major interest in international congresses and the very different subjects that
come up in them (see for example Sukopp and Heiny, 1990; Sukopp *et al.*,
1995; Breuste *et al.*, 1998).

There is a risk that this type of concept may cause confusion or become
meaningless. Sukopp (1995) warns against such problems. Can one for
example regard urban ecology as applying also to city dwellers and their
activities? This is possible according to some people, because city dwellers
'are also organisms', but not according to others (Collins *et al.*, 2000; Picket
et al., 1997). To avoid confusion, separate terms may be used for the different meanings. Urban ecology then refers exclusively to the science of
urban nature and is a branch of biology. Other relevant terms in order of
the different meanings of urban ecology are urban nature or urban ecosystem; nature-friendliness or environmental friendliness; ecological urban
development; urban nature management, or urban environmental policy;

and urban environmentology. This does not then include urban ecology as a pure social science.

13.4   Urban ecology as a social system

The practice of urban ecology
Urban ecology is more than just a definition. It is a social system of ecologists, conservationists, politicians, civil servants, architects and teachers, which could be mapped sociologically. The system is characterised by its institutes, leading figures, organisations and networks, programmes and projects, journals and other means of communication, financial resources, social effects, and so forth, and equally by the theories and methods applied, the outlooks on design, the strategies for protection and policy, and the practices in planning and management. In the specialist literature, however, there is little to be found on the social system of urban ecology.

Organisations
In the Netherlands there is one independent institute for urban ecology, the IIUE (Delft). There have been and still are a number of organisations and projects: the Urban Ecology Platform with a news letter (early 1990s), the WLO working group on Urban Ecology (since 2000), the Ecological City project (TU Delft, 1999-2003), the Urban Ecology research project at international level (Alterra, 2001-2004) and the Alterra research programme Green Metropolises-GIOS (since 2001). Some cities have urban ecology departments, there are a number of green consultancies active in the field and much data are gathered by volunteers, via organisations such as FLORON (for plants).
Organisations at international level include the digital platform ULE (Urban-Landscape Ecology). Other international organisations with activities in this field are or were INTECOL, IALE, the Council of Europe and IUCN (Bohemen, 1995).
According to Trepl, the Man and Biosphere Program, initiated by the UN in 1971, was the start of urban ecology (Trepl, 1995). However, Sukopp sees the start in the tradition of natural history dating from the beginning of the nineteenth century or earlier (Sukopp, 1998, 2002).

Studies
The oldest Dutch study on urban ecology is probably the *Flora Amstelaedamensis* of 1852 (Bolman, 1976), although there were earlier 'urban floras', such as that of Haarlem (1779) (Coesèl, 2001). Probably the first systematic Dutch catalogue of urban nature was 'Vogels van Amsterdam' *(birds of Amsterdam)*, produced by the Dutch Physical Society (NNV) in 1936. According to Reumer (2000), in the Netherlands most has been written about the flora and fauna of Amsterdam. Since 1991 a new series has appeared in this city, in which the first was 'Haring in het IJ, de verborgen dierenwereld in en bij Amsterdam (*Herring in the River IJ, the hidden animal world in and around Amsterdam*, Melchers and Timmermans, 1991).

Internationally the first systematic studies on water, soil, climate and species in cities appeared in the 1970s. The first general works appeared around the time of the Second World War. A new wave appeared from 1989 onwards (Sukopp, 2002).

According to Reumer (2000) the urban ecologist's view is limited. He claims that there are large groups of species of which nothing is known: nematodes, mites, rove beetles, red-tailed bumblebees or springtails – in short, cryptobiotic organisms. The attention for other species is rather unevenly distributed. In German municipalities, for example, vascular plants easily receive the most attention (87% of the municipalities studied), followed by birds (70%), amphibians (66%) and plant communities (62%). In less than 40% of the municipalities was research done on mammals, grasshoppers, lower plants and beetles (Reumer, 2000; after Werner, 1999). Trepl (1995) sees many theoretical questions yet to be answered and hypotheses yet to be examined (see below).

Journals
International journals in this field are *Urban Ecology* and *Urban Ecosystems*, and *Landscape and Urban Planning* also regularly publishes articles on urban ecology. In the Netherlands there is no specialist journal for urban ecology. Dutch articles can be found in *De Levende Natuur* (*Living Nature)*[a], but until recently hardly at all in *Landschap* (*Landscape)*[b].
In *De levende Natuur* (DLN), up until 2006, 51 publications on towns and cities appeared, 19 of which were after 1990, partly through a special issue in 2000. The first DLN article with 'city' as a headword was written by Jac. P. Thijsse in 1899: *The Bee City* (Thijsse, 1899). It was about a bee colony, which he compared to a city. The first article that was really about urban nature was by Pinkhof in 1905, on urban reed land in Amsterdam (Pinkhof, 1905).
In *Landschap* between 1990 and 2005, six articles on urban ecology appeared (Tjallingii, 1991; Peskens, 1995; During and Specken, 1995; Keessen, 1997; Snep, 2003; Windt *et al.*, 2003). In the two 'scientific' articles, ecological knowledge is applied to planning (Tjallingii, 1991; During and Specken, 1995). Both articles concern water in the city.
In 1997 the WLO organised a workshop on 'designing the town-country relationship', from which publications appeared in *Landscape and Urban Planning* (for example Tjallingii, 2000).
Apart from articles there are a variety of publications (Bohemen, 1995): pamphlets, leaflets, reports, illustrated books, symposium reports, theses, studies, grey literature and specific floras and wildlife guides.

---

[a] *De Levende Natuur (Living Nature)* is a journal on research, management and policy in the field of conservation and nature management in the Netherlands and Flanders.
[b] *Landschap (Landscape)* is the journal of the WLO, the Dutch branch of the IALE

Theory
The first Dutch theoretical treatment was probably that of Vink (1980). His book *Landscape Ecology and Land Use* contains a chapter on urban ecology. Vink does not define the term. He sees it as a combination of medical, social and natural sciences.

Theoretical work has principally been carried out in other countries: Germany and the United Kingdom. Publications include *The Ecology of Urban Habitats* (Gilbert, 1989); *Ökologie der Großstadtflora* (Wittig, 1991); *Ökologie der Großstadtfauna* (Klausnitzer, 1993); *Stadtökologie. Ein Fachbuch für Studium und Praxis* (Sukopp and Wittig, 1998) (all referred to in Reumer, 2000). Despite the synthetic nature of these works, there is still much work to be done.

13.5   Multiple approaches

Introductions to the topic often refer to different approaches within urban ecology. The definitions of urban ecology show that these approaches vary from pure scientific research and management to the design of sustainable cities.

From different specialisations
Sometimes the differing approaches relate to specialisations in different aspects of the urban ecosystem – for example, research on water, soil and air quality and the characteristics of certain groups of plants and animals, in relation to their physical features or otherwise. These approaches would appear to be comparatively loosely related, but they can complement each other well. Sukopp (1990) also makes a distinction between historical, structural and functional perspective of research.

From different regions
There are also regional differences. In third world countries more attention is paid to agriculture and forestry in urban areas, as well as to environmental management. In North America attention is paid to hunting species, while in Europe it is chiefly focused on conservation. In the Anglo-Saxon countries there is apparently more attention for urban woodland. In Germany much work is carried out on biotope mapping (Sukopp, 2002).

From different outlooks
Of particular interest are the approaches based on different views or desires in relation to the same aspects. Is nature form, the object of experience or a functional process in which people participate (Tjallingii, 2000)? Is urban nature spontaneous nature or functional green space (Canters *et al.* 2000)? There are many different outlooks on the urban nature that is desirable, ranging from wilderness to design (Gobster, 2001). Van der Windt *et al.* (2003) mention a number of alternative management priorities: improving the quality of surrounding areas versus greening the city, or orientation towards true urban species and environments. They argue for the latter. In their opinion, true urban species (in the Netherlands) are those of

the dry, rocky environments originally found in the southern and/or mountainous regions of Europe, such as cliff-nesting birds, wall plants, and sub-
tropical species – a group of around 50 to 100 species. According to Van
der Meijden (2000) (partly based on Denters and Vreeken, 1998), the
group of true urban plants in the Netherlands easily comprises more than
80 species.

On the other hand one might argue that all species are important, for example for functions such as leisure and health or from the perspective of
education on nature and the environment. Attention for the aesthetics and
functionality of urban nature can give cause for very different priorities in
design and management (Ward Thompson, 2002).

There is a difference of opinion on exotic or alien species. Some urban
ecologists have a clear preference for native species. But from an ecological point of view the aliens are certainly just as interesting (see below).
There are also different approaches to the delineation of the city and the
importance of having an autonomous theory.

13.6 Uncertain urban boundaries

In urban ecology the terms town, city or urban are rarely defined explicitly
(McIntyre *et al.*, 2000). Where this does happen, there are very different
definitions depending on the boundaries applied.

Legal boundaries

Firstly, the city is a *legal* entity, with administrative boundaries. The advantage of this definition is that the responsibility for planning and management
is clear. For the flora and fauna these boundaries would not seem to be
particularly relevant. Furthermore, municipal boundaries often enclose non-
urban areas.

Geographical boundaries

Secondly, the city is characterised *geographically* by number of inhabitants
or density of population, possible in combination with building density.
Many different criteria are applied here. The US Bureau of the Census defines an area as urban if its population is greater than 2500; according to
the UN an urban area has a population of more than 20,000 (McIntyre *et
al.*, 2000). The EU refers to urban areas if the population is more than
2000, while Statistics Netherlands (CBS) refers to population nuclei as
being urban where they have more than 10,000 inhabitants (Vliegen and
Leeuwen, 2005). Drawing geographical borders is evidently not so straightforward. The ecological relevance of these different types of boundaries is
equally unclear (see Zoest, chapter 16, page 267).

Ecological boundaries

Thirdly, boundaries can be defined *ecologically* based on the occurrence of
species or specific environments. If significant changes in species composition or environment occur along the gradient between town centre and outskirts, boundaries might be drawn there. Environmental gradients have

been identified for air pollution, heavy metals in the soil, and temperature, and in relation to these, gradients for soil characteristics in oak woods (McDonnell *et al.*, 1997). Cities have a significantly higher energy use than their surrounding areas do (McIntyre *et al.*; Sukopp, 2004) and they have their own composition of flora (see Denters, chapter 14, page 245). But it is precisely where gradients occur that it is hard to draw clear boundaries (McIntyre *et al.*, 2000).

The extent of the urban area in the Netherlands
The urban environment appears at first sight to be limited in extent. For the Netherlands, figures of 8% (Statistics Netherlands), 12% (MNP, 2006: 'Stedelijk gebied in Nederland') and 15% (Ministry of Housing, Spatial Planning and the Environment) are cited. Of the latter 15%, 8% is made up of publicly accessible sports areas and green space, 1.2% of the Dutch territory. This is not much in comparison to the National Ecological Network (almost 20%) and the agricultural landscape (approximately 50%). Nevertheless, the urban area contains a wealth of habitats, organisms and communities, which is often far greater than that of the urban hinterland (Sukopp, 1998).

The environmental economic perspective
One may also view the urban environment in a wider context. There is private green space, and plants and animals do not restrict themselves to the green areas of the city. Furthermore, towns and cities also have a hinterland that provides natural resources. This approach is one of *environmental economics*. A measure for this is the ecological footprint, the space needed for the production of food, wood, fibre and suchlike, and for the processing of waste, including CO_2 storage. For a resident of Amsterdam, the ecological footprint has been calculated at an average of 4.2 ha (Amsterdam.nl, 2006), more than 200 times greater than the area available to him or her in the city (Amsterdam, 2006). Added up for all its inhabitants, the ecological footprint of Amsterdam is approximately equal to the entire surface area of the Netherlands! A major part of this lies in other countries. The view of the urban ecologist should stretch far into the distance.

The comparability of studies on urban ecology
McIntyre *et al.* (2000) argue that in each study on urban ecology, an interdisciplinary, quantitative and considered definition of the term urban area should be given. Only then can studies be compared.

13.7   A separate theory of urban ecology?

No separate theory
In much literature it is assumed that urban ecology is an element of landscape ecology and that terminology and knowledge derived from landscape ecology can also be applied in the city. According to Niemelä (1999) no separate theory is needed. Urban ecosystems and their specific characteristics (isolation, succession and disturbance) can be studied adequately

using ecological theories such as the theory of island biogeography, the metapopulation theory and the intermediate disturbance hypothesis. In addition Niemelä mentions the human ecosystem model, in which social components are included and for which social theories are necessary.

The theoretical basis of existing ecology is inadequate

Niemelä's proposition is a response to Trepl (1995), whose opinion is quite different. Trepl argues that there is a lack of theoretical basis. Questions for research should not only arise from practice but also from theory. At least a framework is necessary from which questions and possible answers can be derived. He is seeking a 'theory of urban ecology'. A deciding factor in this is whether the particular nature of urban ecosystems differs sufficiently from that of non-urban systems. Only if it does can urban ecology develop into a subdiscipline of ecology with its own theoretical and methodological autonomy, in the same way as limnology. Trepl is not sure whether this is possible. There are certainly differences from most non-urban ecosystems, as many acknowledge[a]. In his search Trepl arrives at a number of questions and hypotheses. These relate to integration, succession and invasion. I will concentrate here on Trepl's former and latter topics and follow his argumentation.

Low degree of integration

According to Trepl, the most important way in which urban ecosystems differ from non-urban ones is their supposed *low degree of integration* (or organisation, or connectivity), which many acknowledge, as a result of the limited amount and low intensity of relations. If this hypothesis is correct, it has consequences for the classification of urban ecosystems. What is then the ecological significance of biotope mapping? Can one still regard the city as an ecosystem in the sense of an ecological functional unit?[b] Is the population approach not more suited to such non-integrated systems than the ecosystem oriented approach? It is even less appropriate here to refer to 'sick' ecosystems than it is in relation to non-urban systems.

High level of immigration and invasion

Another characteristic of urban ecosystems is the *high level of invasion and immigration*. Towns and cities are *extremely rich in species of alien origin*. This is either because they have a low resistance to invasion or because there are a large number of possibilities for species to be transported into urban areas. Habitat conditions are generally favourable everywhere for more species than presently occur. Due to increasing transport there are fewer geographical barriers. The result could be that in the long term every urban area will be inhabited by all the species for which it is suited, wherever they originate from. Locally and regionally this could lead to a vast

---

[a] See also chapter 14 by Denters in this book.
[b] See the chapter 14 by Denters in this book. He refers to the urban area as a fully fledged ecosystem.

increase in species, starting in towns and cities. The major uncertainty here is whether the deciding factor is the low resistance to invasion or the present limitations on transport possibilities. There are very differing opinions on the resistance of urban ecosystems to invasion. One well-known theory is that a wide diversity of species increases the resistance to invasion, but this is controversial. Another equally contentious opinion is that urban ecosystems are highly disturbed, and disturbances increase accessibility. It is about the transport possibilities that we know the least, and there are opposing outlooks on the invasions that can be expected to result from them. According to some the high point has already passed, while others claim it has not yet been reached by far.

Exotic species
By way of illustration, in recent centuries in the Netherlands at least 233 new exotic plant species have become established, 16% of the country's flora. In the last century this increase was constant for each 25 year period (MNP, 2006: 'Plantenexoten in steden'). In towns and cities the percentage is higher. In Amsterdam more than a fifth of the plant species is alien (Denters *et al.*, 1994).[a] Sukopp (1990) estimates the number of 'alien' vascular plants, animals and fungi in town and city centres to be about half of the total number.

Native and alien species, old (archaeophytic) and new (neophytic) aliens do not react in the same way to human influence (Kowarik, 1995). The intermediate disturbance hypothesis, derived from natural ecosystems, does not necessarily hold true for urban areas with many alien species (Sukopp, 2004). Among other things the history of the species plays a role. I see this as an argument for an autonomous theory. Immigrants are probably more interesting for the urban ecologist than are native species.

Trepl principally raises many questions and puts forward many hypotheses. Except from Niemelä there has been hardly any response to his article; it deserves further discussion (see Zoest, chapter 16, page 267).

13.8 Development parallel to landscape ecology

Similarities and differences
Although there are major differences in scope between landscape ecology and urban ecology, there are parallels in their development. Like urban ecology, landscape ecology dates from the 1970s. IALE was founded in 1975. Initially, research in landscape ecology was highly descriptive and focused on structure and spatial patterns. And there too at a certain point during the 1990s the question arose, what is landscape ecology? This led

---

[a] See also the chapter 14 by Denters in this book.

to substantive analyses in journals and proceedings of important areas of study, types of institutes, concepts regarding theories, methods, objects, scale and level, themes, and landscape types (Antrop, 2001).

One of these landscape types is the urban and industrial landscape. In the journals *Landscape Ecology* (1987-1999) and *Landscape and Urban Planning* (1986-1999), respectively only 0.2% and 2.4% of all concepts (in titles, headwords, headlines, headings and captions) relate to urban and industrial landscape. In *Landscape and Urban Planning*, however, a quarter of all concepts from the concept group 'landscape types' relates to urban and industrial landscape (Antrop, 2001).

Phases in the development of Dutch landscape ecology

On the fourth anniversary of the WLO in 1992, Paul Opdam outlined the development of twenty years of landscape ecology in the Netherlands in relation to its application (Opdam, 1992). This development could stand as a model for the development of the scientific field of urban ecology. Summarised briefly, from 1972 onwards the development of landscape ecology went from pattern to process and back again. The first phase was that of environmental mapping, in search of coherence in patterns. It was descriptive research dominated by the demand for data to benefit spatial planning. In the second phase there was a search for spatial processes. Moreover, the research became more theoretical. Hypotheses were developed and tested and simulation models were developed, to an extent independently of application. In the third phase, the new policy on nature in the Nature Policy Plan of 1990, which for that matter made use of a number of theoretical concepts, was an important source for a variety of research questions for which knowledge of pattern and process was necessary.

The mapping phase in Dutch urban ecology

In Dutch urban ecological research the first phase is recognisable, but the second seems to have been skipped, at lease where research itself is concerned. Much has been mapped in towns and cities and this is still taking place for urban spatial planning purposes. Research is being carried out to benefit the development of sustainability, such as research on water management and connection zones. But for the Netherlands I have found no real theoretical studies in the form of theory development, hypothesis, modelling and simulation. This phase could now be launched.

13.9   Conclusions

The historical contradiction between city and nature remains topical. As a result of Romanticism, nature is ill-matched with the city. However, attitudes towards nature can change. There is an increasing appreciation of urban nature, which is expressed in the creation of green space and the development of urban ecology.

In the Netherlands, however, the development of urban ecology as a field of study is limited. The social organisation of urban ecology in the country is fragmented and there is no common organisation or journal.

There are many different definitions of urban ecology in the specialist litera-ture, ranging from the purely scientific to the design of the ecological city, whereby ecological knowledge is applied normatively. Urban ecology also stands for nature in the city.

The different meanings of the term urban ecology make the concept into a boundary object. It thereby gains a common basis of support, but the diver-sity also leads to confusion. Various urban ecologists argue for a clear dis-tinction between the different meanings and their corresponding roles. My proposal is to use different terms for the different meanings: urban nature or urban ecosystem; nature-friendliness or environmental friendliness; eco-logical urban development or sustainable urban design; urban nature man-agement or urban environmental policy; and urban environmentology.

The object of urban ecology – urban nature and the urban environment – is limited in extent compared to the National Ecological Network, for example, yet it often contains a wealth of plant species.

There are many approaches to urban ecology in its different guises. The city is defined in many ways. On the question as to whether a separate theory is required, opinion is divided. The deciding factor is whether urban ecosystems differ sufficiently from non-urban ones, and it is generally ac-knowledged that there are clear differences. The development of an autonomous theory of urban ecology deserves further discussion.

14   THE URBAN DISTRICT, A BIOGEOGRAPHICAL ACQUISITION

Ton Denters[a]

---

[a] This article is derived from the Dutch version Denters, 2006

14.1   Introduction

No other country in Europe is as urbanised as the Netherlands. Stone and asphalt accounts for one fifth of the surface in the Dutch Randstad. Even so, many countries – especially countries in Central Europe – are miles ahead of the Netherlands when it comes to knowledge of urban ecology. Standard authoritative works have been written by, for example, the German authors Sukopp and Wittig (1990; 1993) and the Belgian author Hermy (2005). However, in all fairness, publications on urban ecology and urban wildlife have actually risen significantly in the Netherlands in recent years and have recently been brought together by Van Zoest (2006), thus closing the gap to some extent (see chapter 16, page 267 containing parts of this book and the previous chapter 13, page 230 giving an overview). At present, urban wildlife is attracting the attention of a whole new generation of ecologists who are studying the flora and fauna of cities as a self-contained theme; in other words, with no regard to traditional conventions.

The results are remarkable and demonstrate that cities provide habitats for exceptional, attractive and distinctively urban wildlife. This article looks at the ecological identity of the urban environment which, as a single bio-geographical entity – the urban district – is typified by the specific natural values with which it is entwined. We shall show how these values can help to raise the quality of our cities.

14.2   A mature ecosystem

14.2.1   Flora districts

Flora districts (see *Fig. 102*) reflect the most important dissemination patterns of our wild flora species. For example, we have identified the Dune district, the Fluvial district, and the South-Limburg district, each with its own species. Often, flora districts follow the contours of surface areas, sometimes with climatic boundaries.

This applies likewise to urban districts. The boundaries of an urban district are determined by the temperate city climate and typically urban, largely stony, surfaces. Urban districts are now defined in Dutch flora as: "large stony areas with urban cores throughout the country, where the flora can be both positively and negatively characterised in relation to the more 'natural' districts" (Meijden, 2005). Urban districts are now marked on the map, thereby indicating a shift in ideas about urban nature. The urban district is defined on the basis of its flora, as this is where most of the current knowledge lies, but the actual concept embraces the entire spectrum of urban wildlife.

Fig. 102 Flora districts in the Netherlands and Flanders: Urban district, Haf district, Wadden district, Drente district, Gelderland district, Sub-central European district, South Limburg district, Brabant district, Kempen district, lemish district, Fluvial and estuary district, Dune district (Denters, 2004))

14.2.2　The urban district

The urban district forms a mature ecosystem. It was in the 1990s that I first described the urban district, after studying urban flora (Denters, 1994, 1998). It was already clear at that time that the flora in urban environments was not only exceptional, multi-faceted and diverse but also that it had its very own identity. Urban environments had hitherto received no attention in the conventional system of flora districts; they were simply omitted from the categorisation system.

An urban district consists of areas with 'urban' substrates, mostly stony and disturbed surfaces with strongly fluctuating environmental conditions and an abnormally warm city climate (Kuttler, 1998).

14.2.3　Typical species found in urban districts

Urban districts play host to numerous heat-loving, partly frost-sensitive species, including many naturalising newcomers (neophytes).

Ailanthus altissima	Eragrostis minor	Parietaria judaica
Alnus cordata	Erigeron karvinskianus	Parietaria officinalis
Amaranthus blitum	Erysimum cheiri	Pastinaca sativa urens
Amaranthus deflexus	Euphorbia maculata	Papaver atlanticum
Ambrosia artemisiifolia	Ficus carica	Persicaria capitata
Ammi majus	Fumaria muralis	Petrorhagia saxifraga
Anisantha diandra	Geranium lucidum	Phytolacca esculenta
Anisantha madritensis	Geranium purpureum	Platanus hispanica
Anthemis tinctoria	Geranium rotundifolium	Poa compressa
Antirrhinum majus	Gymnocarpium robertianum	Polycarpon tetraphyllum
Apera interrupta	Herniaria hirsuta	Polypogon monspeliensis
Arabis hirsuta sagittata	Hieracium amplexicaule	Polypogon viridis
Asplenium adiantum-nigrum	Hirschfeldia incana	Portulaca oleracea
Asplenium ruta-muraria	Holosteum umbellatum	Potentilla indica
Asplenium scolopendrium	Impatiens balfourii	Pseudofumaria alba
Asplenium trichomanes	Lactuca virosa	Pseudofumaria lutea
Buddleya davidii	Lagurus ovatus	Rostraria cristata
Campanula poscharskyana	Linaria purpurea	Sagina apetala
Catapodium rigidum	Leonurus cardiaca	Senecio squalidus
Centranthus ruber	Lepidium densiflorum	Setaria pumila
Cerastium pumilum	Lepidium ruderale	Sisymbrium irio
Ceterach officinarum	Lepidium virginicum	Sisymbrium loeselii
Chaenorhinum origanifolium	Leucanthemum paludosum	Sisymbrium orientale
Clinopdoium calamintha	Lobularia maritima	Solanum nigrum schultesii
Conyza bonariensis	Lythrum hyssopifolia	Soleirolia soleirolii
Corispermum intermedium	Meconopsis cambrica	Sorghum halepense
Cymbalaria muralis	Nicandra physalodes	Trachelium caeruleum
Cyrtomium falcatum	Orobanche hederae	Tragopogon dubius
Cystopteris fragilis	Oxalis corniculata	Verbena bonariensis
Digitaria sanguinalis	Panicum miliaceum	Veronica filiformis.
Diplotaxis muralis		

Fig. 103 Urban-dependent plants in the Netherlands (species typical of the urban district): Typical urban, heat-loving plants, strongly dependent on urban biotopes and much scarcer or almost non-existent anywhere elsea)
(realised after Denters, 1998, 2004)

A relatively large number of organisms come originally from dry mountainous and steppe-like regions. They feel at home on streets, walls, roofs, parking lots, ruderal areas, and industry and harbour terrain.

Some species of flora and fauna are 'physically' (wall plants, swifts) or 'historically' (naturalised plants, medicinal plants) bound to the city. Some are such strong city-lovers that they have become dependent on the urban environment. These form the species which are typical of the urban district. There are over ninety different species of urban flora (see *Fig. 103*) The urban district is described on the basis of the flora because it is the most well-known form of wildlife, but the concept embraces all aspects of urban nature.

---

[a]The strength of the ties with the urban environment was determined with the aid of the national flora databank of the Nationaal Herbarium Nederland/Floron.

Amaranthus albus	Epilobium ciliatum	Oenothera species
Amaranthus retroflexus	Eragrostis pilosa	Onopordum acanthium
Arabidopsis arenosa	Erigeron annuus	Rapistrum rugosum
Artemisia absinthium	Galinsoga quadriradiata	Sedum album
Atropa bella-donna	Gnaphalium luteo-album	Sedum reflexum
Ballota nigra meridionalis	Herniaria glabra	Senecio inaequidens
Ceratochloa carinata	Hordeum murinum	Senecio viscosus
Chaenorhinum minus	Hyoscyamus niger	Sisymbrium altissimum
Chelidonium majus	Impatiens parviflora	Solidago gigantea.
Clematis vitalba	Lactucaa	Tanacetum parthenium
Conyza sumatrensis	Linaria repens	Verbascum thapsus
Corrigiola litoralis	Melilotus alba	Verbena officinalis
Datura stramonium	Melilotus officinalis	Vicia villosa
Diplotaxis tenuifolia	Mycelis muralis	Vulpia myuros.

Fig. 104 City-loving plants in the Netherlands: Species which are characteristic of the urban environment, often heat-loving plants which do, however, also appear in some other flora districts (especially the dunes, river areas and South Limburg) (realised after Denters, 1998, 2004)

14.2.4   A rocky habitat

The natural affinity between a large number of species and the urban environment has been neatly described by Reumer (2000): "For most organisms the city is just a habitat. A building is simply an object which, like a rock, can be inhabited. But often, there are subtle differences. Anyone who has been on holiday in the Mediterranean will recall the layers of houses and walls built from natural stone. Some of these walls are scarcely distinguishable from the surrounding natural rock: the same kind of stone, the same gradient, and just as many gaps and crevices. We find all sorts of flora and fauna in these walls: numerous types of fern, other wall plants, wall lizards, funnel-web spiders... To these life forms, a wall is simply an expanse of rock.

That is how the rock dove, the *Columba livia,* sees things as well. The *Columba livia,* the ancestor of our common pigeon, still nests in crevices, regardless of whether they are in a rock somewhere in Spain or in a national monument. A significant proportion of urban wildlife stems from rock flora and fauna." (See *Fig. 105*)

	Conditions	Effects
Climate	Warmer and longer vegetation season	- more heat-loving and frost-sensitive species
Substrate	More polluted surfaces, rich in nutrients, artificial substrates including walls, roofs, paving; drier surfaces due to fast rainwater drainage; cultivated soil with 'alien' plants (gardens, parks, squares)	- more nitrogen- and phosphate-loving species - more tolerant species - surrogate environment for organisms from mountains and rocks - more drought-loving species - favours culture followers - more garden escapes
Disturbance and Dynamics	Frequent disturbance, pedestrian/ mechanical wear and tear, strong environmental fluctuations, steppe-like growth conditions, less competition, more seed transport etc., longer days due to artificial lighting	- more stress-tolerant species and supergeneralists (cosmopolites); especially species from mountainous and steppe-like regions - favours short-living organisms, especially plants with high diaspore production, particularly favourable for wind disseminators - more opportunities for neophytes and adventives

Fig. 105 Conditions and effects of the city on the combination of species compared with non-urban habitats

14.3   Urban wildlife

14.3.1   Surprisingly rich

We have known that there are interesting species among urban flora ever since the nineteenth century. It was then that various guidebooks appeared on the flora of various Dutch cities: Leiden (1840), Utrecht (1843), Amsterdam (1852), Delft (1868) and Den Bosch (1879). In this early period urban flora was studied mostly by botanist-physicians, who were interested in its potential medicinal properties. More recent works have concentrated largely on the surprising diversity (*inter alia*, Pyšek, 1989) and specificity of urban flora (Denters, 1994/1998; Sukopp, 1990). Out of the approximately 1600 species of flora in the Netherlands, two-thirds occur in the cities. The number of endangered and rare species is also remarkably high.

14.3.2   Protected and vulnerable species

There are some 850 different types of taxa in the flora of Amsterdam, including 70 red-list and 32 protected species, which are covered by the flora and fauna legislation (Denters and Vreeken, 1998). The city periphery and long-standing waste ground, in particular, provide an alternative habitat for many species. One notable area is the Amsterdam Westpoort, where, only recently, half a million *Dactylorhiza majalis praetermissa* (marsh orchids) flowered. Even more remarkable is the appearance of *Apium repens* (creeping marshwort) in Vrijbroekpark in Mechelen, Belgium. This endangered species, which was recently added to the Habitats Directive, grows profusely right here on the grass. The only other places where it appears – in extremely sparse quantities – are a few nature reserves.

In some cities in the Veluwezoom (Wageningen) and in Maastricht, the endangered *Thecla betulae* (brown hairstreak) has recently been spotted among the urban greenery, even though it is dwindling in the hinterland – also in nature areas. The city may be buzzing with life, but a surprising atmosphere of calm and stability still reigns in some places (city walls, old city gardens, parks, cemeteries, railway yards, waste ground, etc.). The urban 'oasis' is providing a lifeline for more and more species at a time when rural areas are being used more intensively.

14.3.3   Specifically urban species

The diversity in species usually declines from the city boundaries to the centre (see *Fig. 106*).

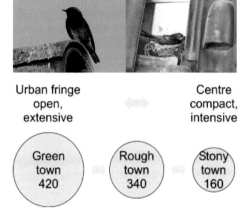

Urban fringe   Centre
open,   compact,
extensive   intensive

Green town 420   Rough town 340   Stony town 160

Artemisia absinthium (absinthe worm-wood)
Asplenium adiantum-nigrum (black spleenwort)
Asplenium scolopendrium (hart's tongue fern)
Asplenium trichomanes (maidenhair spleenwort)
Atropa bella-donna (deadly nightshade)
Ballota nigra meridionalis (black hore-hound)
Catapodium rigidum (fern grass)
Erigeron karvinskianus (Latin American fleabane)
Erysimum cheiri (wallflower)
Lactuca virosa (wild lettuce)
Parietaria judaica (spreading pellitory)
Parietaria officinalis (lichwort)
Pseudofumaria alba (pale corydalis)

Fig. 106 Number of urban species per urban area in Amsterdam,1998

Fig. 107 Some rare species strictly dependent on urban bio-topes

At the same time, the number and density of typically urban species increase. The inner city plays host not only to the most common species but also – and more importantly – to rare species which are strictly dependent on urban biotopes and are therefore explicitly urban (see *Fig. 107*).

14.3.4   Newcomers

Specifically urban species are often heat-lovers and/or sensitive to frost. The number of newcomers (neophytes) is conspicuously high (see *Fig. 108*).

Campanula poscharskyana   1995
Centranthus ruber   1989
Chaenorhinum origanifolium   1990
Cymbalaria muralis   <1644
Erigeron karvinskianus   1993
Pseudofumaria alba   1897
Pseudofumaria lutea   1840
Trachelium caeruleum   2003
Umbilicus rupestris   2006

Fig. 108 Some striking new wall plants in Amsterdam and their first year of naturalisation

Fig. 109 Newcomers to the flora in the Netherlands since 1960. Origin of newcomers (125 species) with percentage per region, 53% of which have settled in urban areas. (Denters, 2004))

Botanical species from all over the world have taken possession of the city and are becoming a regular fixture on the urban scene.

Indeed, new species seem to have been arriving at an accelerating pace in the past three decades. Urban districts are being inundated with new species (on average more than two a year, Denters, 1994), mostly from warm and humid climates. In some cities the flora is no less multicultural than the population.

14.3.5   City-lovers and city-avoiders

Species which are largely dependent on the urban environment are called city-lovers or *urbanophiles (see Fig. 110),* while species which prefer to steer clear of urban areas are called city-avoiders or *urbanophobes.* The third and last category are *urbanoneutral.* Species from this category are found both inside and outside our cities (Wittig *et al.,* 1985). However, more and more of them seem to be migrating to the city. It was the urbanophiles in particular that legitimised the recognition of a separate urban district.

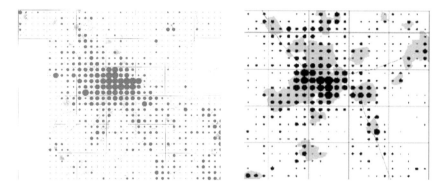

Fig. 110 Dissemination of typical urban plant species in Amsterdam and Eindhoven (Denters and Vreeken, 1998; Rensen-Bronkhorst ed., 1993)

Many city-loving species exhibit a consistent urbanophile pattern, appearing not only in Dutch cities but also in Antwerp, Brussels, Paris, London, Berlin, Warsaw etc. Apart from that there are differences as well as similarities; the composition of species in cities is globally the same but differs in details.

14.3.6   Categories of species

The identification of urban flora turned the spotlight on other urban species as well. We now know that urban wildlife in Amsterdam not only boasts 850 higher plant species, but also 1,100 toadstools, 36 butterflies, 32 dragonflies,16 grasshoppers, 5 cockroaches, 14 ants, 25 woodlice, 50 millipedes and centipedes, 169 migrant nesting birds, 9 amphibians and reptiles, 66 fish, 12 crustaceans and 34 mammals (Halm *et al.,* 2001). The relationships between all these categories and the city are probably comparable with the relationships between higher plants and the city.

14.3.7   Urban bird community

The relationships between all sorts of urban fauna and the city are probably comparable with the relationships between the higher plants and the city. The urban bird community is, at any rate, unique. Various types of birds appear in urban biotopes in densities which are anything from two to 26 times as great as in the rural hinterland (Hermy, 2005). As in the case of the plants, there is a clear trend towards naturalisation. More and more birds are becoming city birds; not just the *Ardea cinerea* (grey heron) but also, for example, the *Accipiter nisus* (sparrowhawk) and the *Falco subbuteo* (hobby).

Another newcomer is the exotic *Psittacula krameri* (rose-ringed parakeet). The first nesting pairs were sighted in the Ruhr valley (Germany) in 1960, after a flock had escaped from captivity. Other sightings were recorded in London, Paris and Brussels. The species also settled in the Netherlands, first in the Hague in 1968, and later in Amsterdam in 1977. A nationwide count which was held in November 2004 revealed that thousands of birds are now flying around in Dutch cities, particularly in Amsterdam (1800), The Hague (3200), Haarlem, Leiden and Rotterdam (300).

14.3.8   Noticeable urban fauna

Needless to say, there are some real city-lovers among other wildlife species as well, the most intriguing examples being perhaps the *Podarcis muralis* (wall lizard), which has established its one and only Dutch stronghold so far in the Hoge Fronten fortifications in Maastricht, and the *Segestria florentina*, an impressive spider who lives in the nooks and crannies of old walls and is found in only a few old harbour towns in the province of Zeeland. Another special city-lover is the *Geopora summeriana,* a fungus which grows under cedars in parks and cemeteries. Sometimes the rare bee-fly *Anthrax anthrax* is observed on sun-drenched walls. In the Netherlands this insect flies primarily in cities. And nowadays the southern oakbush cricket, the *Meconema meridionale,* has also made a home for itself in our inner cities.

14.4   Constant buzz, surprising tranquillity

14.4.1   Intensive and extensive areas

Whereas some species thrive on the constant buzz in urban areas, others find peace and stability. The cities in the Netherlands consist of around 40% greenery and vegetation and there are stretches of waste ground which have lain undisturbed in years.

Constant buzz **-relatively intensive-**	**Surprising tranquillity** **-relatively extensive-**
* stony areas; buildings, paving; especially public space in the inner cities (including shopping areas) and pre-war city districts * green public space in the inner cities; squares, flower beds, parks * new harbour and industry sites; office parks	* historical sites: city walls * parks, old courtyards * cemeteries * old walls; basalt slopes * railway yards * deserted harbour and industry sites and waste (building) sites * corners * roof and wall gardens

Fig. 111 Intensive and extensive urban biotopes

14.4.2   Historical urban greenery

Historical urban greenery is found primarily in parks laid on former city walls, stately residential neighbourhoods, old city gardens and courtyards, and old cemeteries and graveyards. Here we find amongst others *Epipactis helleborine* (helleborine orchid), *Circaea lutetiana* (enchanter's nightshade), *Certhia brachydactyla* (short-toed tree-creeper), *Strix aluco* (tawny owl) and *Celastrina argiolus* (holy blue butterfly). More specifically, urban wildlife appears on disused wasteland where 'historical' species grow, such as *Parietaria officinalis* (lichwort), *Atropa bella-donna* (deadly nightshade), *Hyoscyamus niger* (henbane) and *Artemisia absinthium* (wormwood).

These are plants which in the past – especially the sixteenth century – were brought to the cities because of their medicinal properties and which established themselves on rugged terrain. Stone biotopes are truly urban, particularly old walls, which often harbour very rich vegetation. One need look no farther than Den Bosch (Binnendieze), Amersfoort (Koppelpoort and city canals) and Maastricht (historical city walls) to realise that overgrown walls can add to the quality and feel of a city. Cemeteries have their own attractions, depending on their location, age, use, and management. Besides harbouring a rich hoard of species, these 'resting places' boast various flora of their own, such as *Holosteum umbellatum* (jagged chickweed), *Crassula tillaea* (moss pygmy weed), *Veronica peregrina* (American speedwell) and *Euphorbia maculata* (spotted spurge).

14.4.3   Harbour and industry sites

The constant comings and goings in harbour and industry sites have attracted a permanent and distinct variety of wildlife with many pioneers who owe their survival to the dynamic nature of the area. Typical examples are *Amaranthus retroflexus* (red-root amaranth), *Rapistrum rugosum* (bastard cabbage), *Datura stramonium* (jimsonweed), *Galerida cristata* (crested lark) and *Phoenicurus ochruros* (black redstart), not to mention the 'adventives'; that is, the 'alien species' which are brought in from all parts of the world and break free when the cargo is unloaded.

14.5 The development of urban wildlife

14.5.1 Interesting locations are disappearing

Urban wildlife is resilient; it can take a few knocks. Even so, there is every reason to nurture it more than at present. Density in the cities is having a deep impact on urban wildlife and the way in which it is experienced. More and more building is taking place on deserted and undesignated areas while former harbour and industry sites are being re-designated and urban periphery is being transformed into new business parks.

Measures	Examples
An inventory of exceptional urban wildlife with a view to further development (urban wildlife plan). A policy geared to the recovery and development of specific urban species.	*Department of the Environment:* a hand-book for the protection of threatened wall vegetation; *Amsterdam:* a flora-protection plan for the city; *bird protection:* a protection plan for city birds;
Maintenance and development of important wall vegetation; physical planning protection (protected habitats)	*Amersfoort:* comprehensive protection of flora on city walls / Koppelpoort. *Utrecht:* various projects to transplant wall flora to new/renewed walls. *Amsterdam Zeeburg:* Physical planning protection of exceptional quay walls in the Oostelijk Havengebied.
Re-designate urban wasteland and urban periphery as 'nature discovery areas', which also provide space for specifically urban wildlife.	*Amsterdam:* Klauterbos Zuidoost, Bretten-zone, Oeverlanden Nieuwe Meer, Diemer-park en Diemer Vijfhoek. *Rotterdam:* Rijndamterrein
Create temporary urban wasteland and wildlife play and discovery areas on unused land.	*Rotterdam:* various locations.
A green re-designation / transformation of over-grown, natural old harbour and industry sites.	*Ruhr Valley:* Emscher industrial wildlife park
Stimulate and tolerate exceptional urban and street flora. Set up neighbourhood projects, natural wall gardens and local wildlife gardens.	*Amsterdam:* Buurtnatuurtuin Slatuinen; *Breda:* Willem Merkxtuin; *Groningen:* Buurt-natuurtuin Mauritsstraat
Maintain and restore historical flora, including naturalised plants on fortifications, city walls, historical cemeteries and old city parks	*Maastricht:* Hoge and Lage Fronten, old city walls. *Groningen, Zwolle, Haarlem:* city walls

Fig. 112 More scope for urban wildlife, measures and examples

These new areas with their new lay-outs are often 'neater' than before. Usually, the rest of the open space is tidied up. This takes place – often unnecessarily – at the expense of important wall vegetation, unique urban wasteland flora, and birds such as the *Apus apus* (common swift) and the *Phoenicurus ochruros* (black redstart). Special and exciting spots run the risk of disappearing from our cities – the intriguing stretches of no-man's-land, where you can run into the unexpected, where children love to play, and where nature has a free rein.

14.5.2   Protect, indulge and make space

Modest initiatives are already up and running to keep urban wildlife alive, but not enough attention is being paid to it. This could easily be remedied by urban wildlife development programmes (see *Fig. 112* and *Fig. 113*) which would, involve, amongst others, the protection of exceptional, typically urban species, particularly on the sites where they still occur, as well as the creation of colourful wall vegetation and the establishment of special urban wildlife reserves and nature trails.

Species policy in Amsterdam
Amsterdam is currently working on a plan for the protection of wildlife species. A proposal has been put forward to focus on rare urban flora for which effective measures can be taken (EcoStad/Denters, 2005). The plan covers typically urban, protected species, including some attractive specimens (public species).

	*)	Importance of biotope	Effectiveness of measures	Influence of municipal policy	Final score
		Protection Urgency			
Wall flora	8	+++	+++	+++	**+++**
Street flora	4	+	+	+++	**+**
Flora in urban wasteland	1	+++	++	+++	**+++**
Flora in urban periphery	6	++	++	++	**++**
Flora in parks and public squares	4	++	++	+++	**++**
Curb flora	0	+	++	++	**+**

) = *legally protected species*   **+++** / **++** / **+** urgency: high, fairly high, moderate

Fig. 113 Species-targeting policy in Amsterdam, with priority for typically urban species

14.6   Conclusion

Urban districts have their own identity, with distinct flora and fauna. They host a valuable ecosystem, comprising specifically urban species which are largely dependent on urban biotopes. In the Netherlands, urban areas will only increase, thereby increasing the value of urban wildlife, not least because city residents start to appreciate the nature in their own immediate surroundings. There is every reason from a practical point of view to treat urban wildlife as a mature, fully-fledged system. The idea that it represents second-rate nature is out of date. Anyone who takes an objective look at urban nature will be amazed by its diversity and uniqueness. Our cities harbour wildlife that matters. So, it is important that urban development projects give wildlife the consideration that it deserves and accord priority to 'typically urban' wildlife. This is easily realised. A lot of urban wildlife is completely at home in new developments, but it is vital that city planners, policymakers, developers, urban ecologists, urban nature groups and administrators join forces and work together from the outset.

15 PUBLIC HEALTH

Agnes van den Berg

15.1.1   Introduction

In 1998, the Dutch Ministry of Nature, Agriculture and Food Quality (LNV) launched a campaign titled "Operation Treehut". The aim of this campaign was to give the social values of nature a more prominent position in nature policy. The importance of contact with nature for people's health and well-being was one of the social values that was given high priority in the campaign. In particular, it was assumed that contact with nature would provide an effective means to prevent and decrease diseases and problems that are typical of a stressful urban life-style such as obesity, asthma, chronic stress, heart diseases, and diabetes. However, during a conference on this theme, held in May, 1999, it became clear that there was, at the time, insufficient evidence from rigorous scientific work to support these assumptions.

Almost a decade later, health benefits of nature have become one of the "hottest" issues in Dutch nature policy. Several reviews of the scientific literature have been published, including the Alterra-essay " Van buiten word je beter" (Berg and Berg, 2001) and an influential advice by the Health Council of the Netherlands (Health Council/RMNO, 2004). The general conclusion of the latter review was that there is increasing evidence that contact with nature promotes restoration from stress and mental fatigue. In addition, the Health Council found consistent clues that nature may promote health through other mechanisms, such as stimulation of physical activity and social contacts, encouragement of the development of children, and providing opportunities for personal development and a sense of purpose. The review of the Health Council does not cover more physical pathways by which nature may influence health (for example, air-cleaning effects of plants and trees, life-support functions of ecosystems). However, it is noted that such effects are plausible and should be given consideration in future reviews.

In conjunction with the increasing scientific evidence, there has been a growing interest to put health functions of nature to practice in leisure, therapeutical and other settings. The advisory council for research on spatial planning, nature and the environment (RMNO) has recently listed more than 100 "best practices". These practices vary from the creation of healing gardens near hospitals to the development of "health routes" in leisure areas for people with coronary disease and the transformation of farms into care-centers for mentally handicapped and burned-out persons.

This chapter gives an overview of this changing context and its implications for landscape and urban ecology. It starts with a brief review of research on health impacts of nature in general, followed by a discussion of the values of various ecosystem qualities for human health. It concludes with the implications of these insights for landscape and urban ecology.

15.1.2 Health benefits of nature

People in the Netherlands and other urbanised societies tend to believe that contact with nature provides them with restoration from stress and fatigue and improves their health and well-being. For example, in a nation-wide survey among inhabitants of the Netherlands, 92% of the respondents indicated that they agreed with the statement "a visit to nature gives me a healthy feeling" (Frerichs, 2004). According to the respondents, the primary causes of this healthy feeling were the confrontation with fresh air and the pleasant smell of it (49%), the possibility to cycle, walk or otherwise be physically active in nature (44%) and the relaxing atmosphere and the feeling of "being away" (26%).

In the survey by Frerichs (2004) the concept of nature was defined in a broad way, as "not only woods, moors, lakes, dunes, beaches, rivers, wet-lands, etc., but also green facilities in and around the city. The latter include not only public gardens, parks, and meadows in your nearby living environment, but also nature areas and greenery meant for leisure, such as a cycling route" (p. 6).

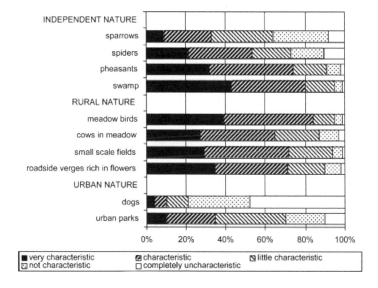

Fig. 114 Degree to which Dutch city dwellers rate elements as characteristic for nature. These results show that lay people have a broad image of nature. (Buijs, Pedroli and Luginbühl, 2006, adapted from Buijs, 2000)

Such broad definitions of nature are very common in social-science research. They reflect the finding that most lay people possess a broad image of nature that includes natural as well as cultural landscapes (Buijs, Pedroli and Luginbühl, 2006). In one study on people's nature images, even rather isolated types of vegetation like flowers along road sides were seen as nature by more than 70% of a sample of Dutch city dwellers (Buijs, 2000).

The widely held belief that contact with 'nature' (in a broad sense) is beneficial for one's health is supported by two large-scale epidemiological studies in the Netherlands (Vries, Verheij, Groenewegen, and Spreeuwenberg, 2004; Maas, Verheij, Groenewegen, De Vries and Spreeuwenberg, 2006). These studies have revealed that the percentage of green space (including urban green, agricultural green, forests, and nature conservation areas) within a 3-kilometre circle from people's home was, on average, positively related to self-perceived health. In both studies the positive link between green space and health was found to be relatively marked among the elderly, housewives and people from lower socio-economic groups. The researchers attribute these findings to the fact that these groups spend a relatively large amount of time in the residential environment.

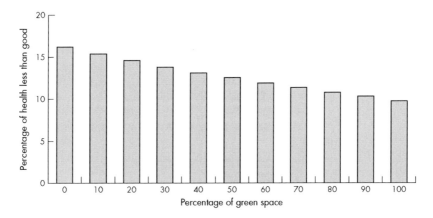

Fig. 115 Relation between amount of green space (in a 3 km radius) and self perceived health (percentage stating their health is less than good)
(Maas et al., 2006)

The results of the Dutch epidemiological studies allow for several interpretations of the mechanisms underlying the relationship between green space and health. These mechanisms may include restorative or stress-reducing influences that result from the contemplation of greenery, health benefits of physical activity (because presence of greenery may stimulate residents to walk or cycle), and better air quality or climate control in greener environments.

Thus far, restorative functions of nature are best supported by scientific evidence. In a recent review, Van den Berg (2005) discusses more than 30 controlled studies that speak to the restorative effects of contact with real and simulated nature. In a typical experiment, healthy volunteers first receive a stress-induction treatment (for example, watching a scary movie or performing a mentally fatiguing task). Next, they are randomly assigned to conditions of viewing or visiting natural versus built environments. Stress is assessed before and after the stress manipulation, and after viewing the natural or built environments. Results of such experiments have consistently shown that stressed individuals who are exposed to natural environments show more positive mood changes, perform better on concentration tasks, are more tolerant to pain, and display more physiological symptoms characteristic of stress recovery than stressed individuals who are exposed to built environments.

Although the benefits of physical activity for people's health are well-known, the importance of nature in establishing these benefits has not yet been directly demonstrated. A recent study by Vreke et al. (2006) showed that, after controlling for influences of socio-economic and ethnographic variables, the percentage of children (ages 4-18) with overweight and obesitas in green neighborhoods was about 15% smaller than in barren neighborhoods. However, because activity levels were not measured in this study, alternative explanations of these findings, for example in terms of different food intake patterns in green and barren neighborhoods, cannot be ruled. Nevertheless, a recent study by De Vries et al. (in press) suggests that it is plausible that the relationship between percentage of green space and overweight found in the study by Vreke et al.(2006) was at least partly caused by higher activity levels of children in green neighborhoods. De Vries et al. (in press) found that the percentage of green space and presence of water in neighborhoods is positively related to children's physical activity levels. Taken together, the findings of these two studies strongly suggest that the presence of nearby nature plays an important role in stimulating children to become more physically active, which may reduce their risk of becoming overweight, and all the health problems that may follow from this condition.

With respect to air quality it has been found that trees and other vegetation can lower local concentrations of particulates and other forms of air pollution by means of their filter function (see Beckett, Freer-Smith and Taylor, 1998, for a review). However, the actual health benefits of such filter functions for residents have not yet been demonstrated. Moreover, the small differences in air pollution between urban and natural environments suggest that the filter function of trees and plants does not have a major influence on air quality at a regional level. The only places where trees and other vegetation may partly account for a positive correlation between nature and health by improving air quality are at local level and directly along busy roads and motorways (for example, Tonneijk and Blom-Zandstra, 2002).

263

15.1.3   Ecosystem qualities and human health

There is no doubt that the degradation of ecosystems may in the long run have serious consequences for the health and survival of the human species. The Millenium Ecosystem Assessment (2005) has shown that approximately 60% of the benefits that the global ecosystem provides to support life on Earth (such as fresh water, clean air and a relatively stable climate) are being degraded or used unsustainably. In the report, scientists warn that harmful consequences of this degradation to human health are already being felt and could grow significantly worse over the next 50 years. Thus, from a global perspective, the quality of ecosystems is extremely important for human health and well-being.

On a personal scale, there is some evidence that ecosystem qualities play an important role in people's perceptions of healthy environments. For example, Ogunseitan (2005) asked a sample of 369 American respondents to rate various environmental characteristics according to their effectiveness in making one feel refreshed or experience restoration. Respondents were also asked to rate their quality of life (including a measure of their physical health) and their current level of restoration. Results revealed four domains of restorative environmental characteristics: ecodiversity (for example, presence of trees, forests, flowers, animals), synesthetic tendency (for example, colors, smells, sounds), familiarity (for example, identifiability, privacy), and cognitive challenge (for example, complexity, mystery). Of these four domains, ecological diversity was most strongly associated with quality of life and the current level of restoration. These findings suggest that individuals tend to believe that some environments, in particular environments with many natural elements, have more to offer in terms of health and well-being than others.

However, there is as yet little evidence from national, regional and local studies to support the notion that health benefits of nature vary as a function of objectively measured ecosystem qualities such as biodiversity, stability, land cover type, degree of organisation, and levels of immigration and invasion. A study conducted in Rome (Bonnes, Carrus, Bonaiuto, Fornara and Passafaro, 2004) found that residential satisfaction towards urban green spaces can be directly linked to the overall quantity/availability of these but are somewhat independent from their overall quality/typology in terms of biodiversity richness. Apparently, the Romans are more concerned for having more green spaces available and less concerned for having green spaces of higher ecological quality. This notion is consistent with results of the Dutch epidemiological studies on relations between urban greenery and health by De Vries et al.(2003) and Maas et al.(2006) which also showed that self-reported health was dependent on the amount of green space but not on the type of greenery.

Type of greenery does seem to matter in physical pathways. In particular, it has been found that conifers are more efficient at absorbing pollutant parti-cles than broadleaved species (Beckett, Freer-Smith and Taylor, 2000). The value of conifers at absorbing pollution comes from a variety of factors including their evergreen habit, their speed of establishment, very high surface areas, and their particular effectiveness at absorbing particles (Beckett, Freer-Smith and Taylor, 1998). In the Netherlands, most conifers are exotic species and typically regarded as a threat to ecosystem function-ing. This illustrates that ecosystem values are not necessarily relevant for, or compliant with, public health values.

By contrast, there are reasons to assume that some characteristics of healthy ecosystems, such as a high biodiversity, may even have adverse effects on public health and well-being. Natural areas with a high degree of biodiversity are typically quite wild and dense. Such areas may not only arouse intensely positive emotions in people, but also intense fears (Koole and Berg, 2005). This fear-evoking capacity of wild nature appears to be a product of evolution; for early humans who had to survive in wild, natural environments that contained many dangers a quick and strong fear reac-tion was crucial for the activation of appropriate defensive actions (Öhman and Mineka, 2001). Especially for individuals who are in a vulnerable posi-tion, for example because they are ill or mentally unstable, confrontations with wild nature may evoke strong (existential) fears and feelings of de-pression and helplessness (cf. Berg and Heijne, 2004). Indeed, evaluations of school field trips and other mandatory nature programs have consistently shown that a small but substantial number of individuals are unable to overcome their fear for wilderness environments and transform it into a positive experience, even after spending prolonged periods of time in these environments (see Bixler et al., 1994, for an overview).

Contact with wild nature may also be unhealthy in a more physical sense. Biodiverse areas often contain many dangerous elements, such as un-tamed large animals that can attack humans, broken trees that can fall on people's heads, and swamps filled with bacteria that may spread conta-gious diseases (Berg, 2004). Of course, these are exactly the same types of dangers that have motivated people throughout history to cultivate "unland" and build cities as safe places to live in. Even in the Netherlands, a country that is often assumed to have tamed nature and banned out all dangers, there is growing concern for such negative health impacts of na-ture (Winsum-Westra and De Boer, 2005). In particular, the occurrence of several accidents with wild cows and horses in newly developed natural areas that are part of the National Ecological Network (see Jongman and Veen, chapter 9, page 141) has stimulated a new awareness of the dan-gerous side of nature.

There appears to be a discrepancy between the perceived dangerousness of nature and the actual risks of getting hurt or killed in nature (Winsum-Westra and De Boer, 2005). Because of our innate tendency to react fearfully to natural threats, the actual risks of contact with nature are generally overestimated. Moreover, the actual impact of natural dangers on physical health is to a large extent dependent on the individual's fitness and coping skills. For people with adequate coping skills, confrontations with natural dangers provide excellent opportunities for improving their mental and physical resilience; for individuals with insufficient coping skills, however, such confrontations may result in injuries and disease.

15.1.4   Implications

In recent years, human health and well-being has become an important criterion for assessing the quality of natural areas and urban ecosystems besides ecological and environmental criteria. Unfortunately, as pointed out in the previous paragraph, it is becoming increasingly clear that human values and ecological values are not interchangeable. The growing recognition of health functions of nature thus seems a mixed blessing for ecologists. On the one hand, it strengthens the case for the importance of nature in society. On the other hand, it weakens the relative importance of biodiverse ecosystems as compared to other types of nature.

How should landscape ecology deal with this changing context? First, there is an urgent need for more research on the health impacts of different types of nature. Landscape ecologist may stimulate this research by asking social scientists to collaborate in their research project. In particular, future research should try to identify health benefits that are specific to contact with wild, biodiverse nature. Most likely, these benefits lie in the domain of personal development and the enhancement of mental and physical resistance. In conducting such research, attention should be paid to individual differences and possible underlying mechanisms, such as coping skills and personality styles.

Furthermore, landscape ecologists should become more aware of the potential conflicts between the health of ecosystems and human health. What is healthy for nature, may not always be healthy for people. Nevertheless, there remain remarkable commonalities in global aims for sustainable ecosystems and their importance for human health and well-being. By being more aware of potential negative health impacts of biodiverse nature on a personal level, the public support for strategies to protect and enhance biodiversity on a global level can be strengthened. The promotion of wild nature, as an answer to the degradation of ecosystems need to be considered in relation to people's personals need for healthy experiences with nature and the suitability of wild nature for meeting that need.

16   DRIVING FORCES IN URBAN ECOLOGY

Johan van Zoest[a]

---

[a] In cooperation with Taeke M. de Jong who selected parts of Van Zoests' recent overview in Dutch: Zoest, J. van; Melchers, M. (2006) Leven in de stad (Utrecht) KNNV-uitgeverij; ISBN-13 978-90-5011-177-5

16.1   Introduction

The domain of urban ecology
Urban ecology can hardly be described as an established and well defined discipline. This is typified by the situation in the Netherlands. There are no research programmes, professorial chairs, or learned journals specific to this field. Furthermore, urban ecologists form a very loose-knit community indeed[a].

It can be argued that there are two main reasons for this state of affairs. One is that urban ecologists have, as yet, failed to encapsulate the distinctiveness of the urban landscape in ecological terms. The ecological patterns and processes encountered in urban environments have a set of common characteristics that are not found elsewhere. These characteristics may be sufficiently different from those displayed by rural and natural environments to merit the creation of a separate discipline (or subdiscipline). The second reason for the present state of affairs may be that the very nature of urban environments means that they cannot be sharply defined. The urban and the rural represent the extremes of a continuum, so any boundaries drawn between the two generally tend to be somewhat fuzzy and arbitrary in nature.

The driving force of human presence
What makes the urban environment unique in an ecological sense is, of course, the omnipresent and all-pervasive influence of humans on the landscape. We shape our cities and towns through planning, design, land use, and management, as well as by introducing and attempting to control various species. The resultant environment exhibits highly distinctive physical, geographical and ecological characteristics. Therefore, if we are to understand the ecology of urban landscapes, we must first comprehend the driving forces behind the urban environment itself. It is the ecological changes along the rural-urban gradient that provide particularly revealing insights into the influence of increasing 'urbanisation' on species diversity, community dynamics and other features of interest to ecologists.

The urban gradient
Luck and Wu (2002) analysed an urban gradient in the Phoenix metropolitan region (Maricopa county, see *Fig. 116*), using spatial landscape metrics according to McGarical and Marks (1995).

---

[a] in a subgroup of the Dutch society of landscape ecological research (WLO)

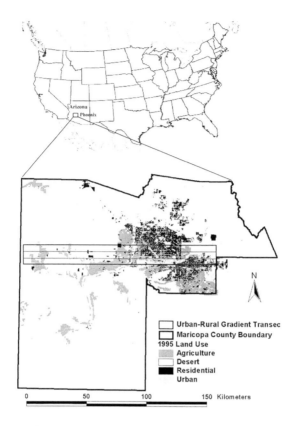

Fig. 116 Phoenix metropolitan region (Maricopa county)
(Luck and Jianguo Wu, 2002)

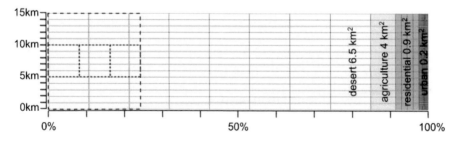

Fig. 117 Mean patch surface of four types of land use in Maricopa county and their proportion of the total area; the 15x15km transect window dottet

Moving along a transect, these authors measured the surface area of patches and other parameters every 5 km, in a window measuring 15 km x15 km (see *Fig. 118*). Relative to the total area of the county, there has been a decrease in the surface area of desert[a], agricultural, residential[b] and urban[c] land use patches. Accordingly, moving along the transect in the direction of the city reveals an increase in the number (per km^2) diversity of patches.

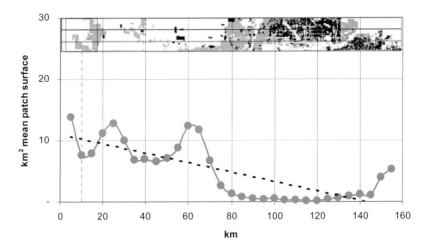

Fig. 118 Window of 15x15km moving along a transect and the mean surface area of patches inside that window (Luck and Jianguo Wu, 2002)

In this chapter we explore the impact of progressive urban fragmentation, diversity, contrast, and of escalating urban dynamics, on the distribution and abundance of plants and animals, and on the human condition.

Questions

We start by briefly reviewing some of the conditions that characterise urban environments (16.2). We then proceed to describe the historical growth of settlements as they transformed into the modern regional cities of today. The effects of this process on public health (i.e. stress-related diseases) are also examined, and emphasis is given to the importance of creating 'good' urban environments (16.3). In 16.4, we turn to the urban–rural gradient, addressing ecological driving forces at several spatio-temporal scales.

---

[a] '**Desert**' including: Vacant, Leisure Open Space, Dedicated or Non-developable Open Space, Rural space.
[b] '**Residential**' including: Large Lot Residential, Small Lot Residential, Medium Density Residential, High Density Residential, Hotel, Motel, or Resort
[c] '**Urban**' including: Neighborhood Retail Center, Community Retail Center, Regional Retail Center, Warehouse/Distribution Center, Industrial, Business Park, Office, Educational, Institutional, Public Facility, Large Assembly Area, Airport

We conclude by positioning urban ecology both as an academic discipline and as a policy-making sector (including normative policy-making).

In doing so, we have attempted to answer two main questions:

1. What is urban ecology (see 16.5)? Is it sufficiently different from other branches of ecology to warrant a subdiscipline of its own?
2. What are the driving forces behind the ecology of urban species, communities and landscapes?

16.2 The urban environment

16.2.1 People, buildings, roads, and abundant energy

Unprecedented environment
In our towns and cities, factors such as high population densities, large concentrations of buildings, and the input of large amounts of energy and materials create environmental conditions that are found nowhere else on Earth. Towns differ from the surrounding rural environment in terms of soil, water management, climate, noise load, and light load.

Risks and opportunities
The risks and opportunities that animals and plants encounter in urban environments vary from one species to another.
From one point of view, life is more stressful for urban animals and plants than for their rural counterparts. They have to cope with high densities of people, buildings, and motor vehicles, in addition to high noise levels, bright lighting at night, numerous dogs and cats, and pollution. Some species are clearly better suited to this environment than others.

From another point of view, life in towns also offers many opportunities. Towns contain many rich sources of food, such as edible litter, gardens with diverse nectar-producing plants, and bodies of water that are well-stocked with fish. For example, Mason (2000) showed that, in Eastern England, urban populations of the blackbird (Turdus merula), the song thrush (Turdus philomelos) and the mistle thrush (Turdus viscivorus) were substantially larger than those in rural areas. These species had substantially more territories in the former environments. Indeed, residential areas may even provide refuge habitats for species that are declining in rural areas. Urban areas sometimes contain micro-habitats that have no equivalent in the natural environment, resulting in completely new combinations of species.

16.2.2   Soil and water management

Urban soils
Soils in towns and cities differ in many respects from those found in agricultural and natural landscapes. According to Pickett (2001) and McDonnell et al. (1997), the following characteristics of urban soils are particularly important in built-up areas:

1.   Heterogeneity. While soils in rural areas form large scale mozaics, small scale variations in the layout and use of urban space mean that urban soils are highly fragmented. These differences are often reflected by the local vegetation.
2.   Compact. Soil compaction (densification) is the result of intensive usage and traffic. Compaction detracts from the soil's ability to sustain life, since rainwater percolates more slowly and the soil is not well aerated. The growth conditions for plant roots are therefore not ideal. The flora and fauna of compacted soils are much poorer than in undisturbed soil, both in terms of the number of species and the number of individual specimens.
3.   Nutrient-rich. Most urban soils are extremely rich in nutrients, due to acid rain, exhaust gases from traffic, cat and dog excrement, irrigation with eutrophied water, and the addition of leaf mould or compost. It is rare to find depleted, impoverished soils in and around the urban region. Soils with only a moderate nutrient content are, however, found on sand and rubble bases, such as those used under water defences and railway yards. The further from the town or city, the more likely it is that the soil will be low in nutrients or even impoverished, and will therefore present the diversity of flora associated with such soils.
4.   Immature. The vast majority of urban soils are relatively young and hence not yet differentiated into various strata. (Soil scientists use the term 'vague soils'.) Older, more mature soils are only to be found in closed or isolated locations, such as private premises, raised (building) land which has remained unused and uncultivated for several decades, the banks of long-established roads and railways, the boundary areas of established sports areas and allotment complexes. In such unused areas with long periods of soil rest, it is common to find the plants and fungi which thrive in a quiet, stable situation.
5.   Contaminated. The majority of urban soils (including the beds of bodies of water) are contaminated to some degree with toxic substances, such as heavy metals, polycyclic aromatic hydrocarbons (PAHs) and pesticides. Although specific discharges do occasionally occur, the main sources of such contaminants are diffuse: precipitation, exhaust gases and the transport of polluted water from another area. Former waste dumps and other historic sources of contamina-

tion (notably former factories and gasworks) are a common feature of the urban areas (see Pouyat and McDonnell 1991).

Development of urban soils

Urban soils develop in a variety of ways. In past centuries, the soils of urban settlements were raised with rubble, refuse, or clay prior to being built upon. In more recent times, land is prepared for construction by raising it with sand. The nature of the soil is determined in part by the top covering layer. For example, gardens, parks, verges, sports fields and allotments are generally given a top layer of organic leaf mould; railway yards and tracksare laid on a layer of coarse gravel, and many old industrial premises are built on clinkers.

Urban soils can be classified into a number of groups based on the methods and materials used to raise or cover the land.

Soil types not found in urban areas

From the foregoing, it is possible to deduce that certain soil types are rarely, if ever, found in urban areas. They include nutrient-poor (impoverished) soils, non-compacted soils and fully matured, stratified soils. Accordingly, the plants, fungi and invertebrates which depend on certain soil qualities will not be found in such areas. They are, however, to be found in remnants of original (open) landscape, in moderately nutrient-rich sand and loam soils, and in and around established woodlands.

Urban water management

Like those in the rural areas, urban water systems comprise a series of waterways, dams and storage areas. Their function is to accommodate inflowing water (rainwater, surface water and ground water) and eventually transport it into a river or to the sea. In the low-lying regions of the Netherlands (the clay and peat-based areas of West-Nederland and the river estuaries), urban water management systems must also provide protection against flooding. In this sense, they do not differ greatly from their rural counterparts.

Percolation

There are some major differences between urban and rural areas in terms of the water balance (see *Fig. 119*). In rural areas, rainwater can percolate into the soil relatively easily, and is transported away via groundwater and surface water. Moreover, a substantial proportion of the water will evaporate, either directly or via plants. In urban areas, rainwater will infiltrate the hard soil to only a limited extent. By far the largest proportion flows into surface water (canals, ditches, etc.) or into the drainage system, from where it is eventually discharged into the surface water via a wastewater treatment plant. Another significant source of water in the urban region is the domestic water supply. After use, domestic wastewater is also transported by the drainage system to the wastewater treatment plants.

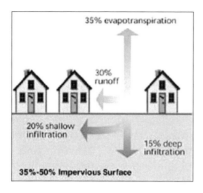 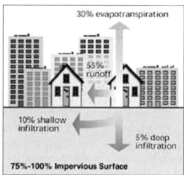

Fig. 119 Urban water management *(Zoest, 2006)*

Drainage

A further difference is that in urban areas, particularly those in the low-lying part of the Netherlands, it is necessary to dewater the soil to a significant depth in order to protect crawlspaces and prevent the various forms of nuisance which can be caused by groundwater. Accordingly, energy must constantly be devoted to pumping away water.

Groundwater flows

The complex morphology of the urban landscape often gives rise to unusual groundwater flows. At the foot of sand-based structures such as the banks of railway lines and roads, and on the borders of areas which have been raised with sand, seepage water will escape and will usually find its way into the seepage ditches. The rainwater which has been filtered through sand in this way is of comparatively good quality. At the base of the slopes of sand structures, it is therefore not unusual to find plant species which thrive on groundwater, such as the branched centaury (*Centaurium pulchellum*), the bee orchid (*Ophrys apifera*) and the common self-heal (*Prunella vulgaris*).

16.2.3   Climate and air quality

Urban climate

The climate of urban areas differs from that of the surrounding rural region in a number of respects (Horbert, 1978).

Wall barley (*Hordeum murinum*) is a type of grass which prefers a warm location. It spreads mainly in the urban areas, and in particular the warmer town centres (see inset: distribution in Amsterdam). Low temperatures slow the development not only of the seeds, but also that of the barbs with which the seeds attach themselves to people and animals in order to 'hitch a lift' elsewhere. According to Davison (1977), wall barley is likely to be suppressed by other grass types in colder areas.

Heat islands
The most notable feature of the urban climate is the formation of 'heat islands'. In the rural areas, much of the sun's energy reaching the earth is
used in evaporating water. Plants assist the evaporation process with their
leaves, which in turn cool the surrounding air. However, the hard surfaces
of the urban areas, such as paving and roofing tiles, create a different
process. Dark materials absorb solar energy which is then re-radiated in
the form of heat. This, in combination with other sources of heat such as
building's heating systems, traffic, industry and the localised 'greenhouse
effect', helps to create heat islands in which the temperature can be up to
8° Celsius higher than in the surrounding area (Gilbert 1989). The size and
shape of heat islands generally follow the contours of the built-up area itself, although according to Sukopp (1998), their exact position is influenced
by the prevailing wind direction and by the presence of any open spaces.

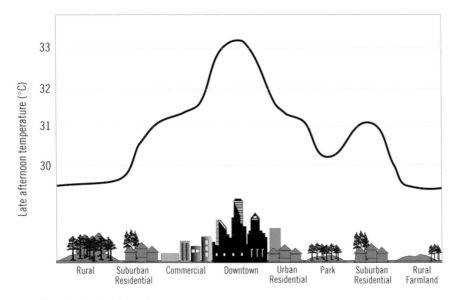

Fig. 120 Heat island (http://adaptation.nrcan.gc.ca/perspective/health_3_e.php).

A study conducted by Franken in Hamburg in 1955 demonstrated that for-
sythia bushes in the urban heat island bloom earlier in the year than those
elsewhere. The contours of the Hamburg heat island are influenced by the
local topography and by the cooler open spaces of the city, such as the
wartime bombsites which still blighted the city at the time of Franken's
study.

Human response
According to Baker *et al.* (2002), the higher average temperatures of the
urban areas also have consequences for humans:

1.   Heat accelerates chemical reactions in the atmosphere, leading to higher concentrations of ozone and hence greater health risks.
2.   In towns and cities in (sub)tropical climates, urban heating results in more 'misery hours' per day, which can have significant social consequences.
3.   Warmer cities consume more energy to power buildings' air-conditioning systems.

Increasing street comfort by incorporating green areas
There is therefore good reason to establish green areas in the warmer urban setting, in order to increase 'street comfort'. Gomez, Tamarit *et al.* (2001) have calculated the exact area of green zones required to produce a more amenable temperature in certain districts of Valencia, based on generally accepted 'comfort indicators'. They state that a temperature reduction of one degree Celsius requires ten hectares of green space to be incorporated in the city layout, while a reduction of two degrees would require fifty hectares, and a reduction of three degrees calls for two hundred hectares.

Growth season
The warmth of the urban climate also has a clear effect on flora and fauna. For insects and other invertebrates, the warmer climate means an extension of the active season. Urban areas also offer plants a longer growth season, i.e. the period in which the average night-time temperature exceeds 5.6° C. In large cities such as London, the difference can be as much as three weeks. The frost-free period in central London is, on average, ten weeks longer than in the outlying regions. Towns and cities are therefore home to a greater number of the plant and animal species which prefer warm locations, many of which originate from more southerly climes. Although the higher temperatures of the city can result in greater heat stress in summer, they also result in less cold stress in winter: a critical factor for many plant species.

Air quality
In addition to 'traditional' air pollution (nitrogen dioxide, ozone, sulphur dioxide and fine particulates), the World Health Organisation (2000) now recognises 'organic' air pollution (for example benzene), inorganic air pollution (for example asbestos) and contaminants in the interior environment (such as radon). Worldwide, the main cause of air pollution is the combustion of fossil fuels and biomass such as wood. In western Europe, motor vehicles and the combustion of gaseous fuels now account for the largest share of air pollution. In the past, most pollution was due to industrial processes and the combustion of coal and high-sulphur-content fuel oils.

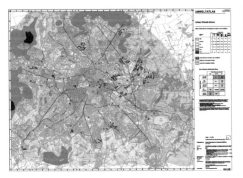

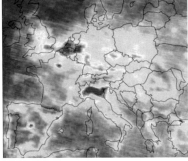

Fig. 121 Urban climate Berlin
(Umweltatlas Berlin[a] cited by Zoest, 2006)

Fig. 122 Air quality (European Space
Agency/IUP Heidelberg [b] cited by Zoest,
2006)

Increasing traffic

Given that a further increase in the traffic on our roads is predicted, emissions – and hence urban air pollution – will also increase. In many cities, over 80% of transport movements now rely on motor vehicles. Projections of traffic growth in western Europe suggest that, based on a 'business as usual' scenario, passenger and freight transport on the roads will have doubled between 1990 and 2010, with a 25% to 35% increase in the number of cars, and a 25% per cent increase in the mileage travelled by car each year. In the countries of Central and Eastern Europe (including Russia), a similar trend will be seen, albeit more gradual.

Improvements

Despite the increase in road traffic, the air quality in most western European cities has actually improved in recent years. During the 1990s, a reduction in the lead content of petrol resulted in a marked reduction in atmospheric lead concentrations. The concentrations of other contaminants also appear to be falling. Nevertheless, some cities continue to suffer from extremely high ozone levels during the summer months, and most continue to exceed the WHO norms for sulphur dioxide, carbon monoxide, nitrogen dioxides and particulates. Statistics published by the EU indicate that approximately 25 million people are exposed to winter smog (concentrations of sulphur dioxide and particulates which exceed the norm values) each year, while 37 million are exposed to (ozone-related) summer smog and almost forty million to concentrations of contaminants in excess of the WHO norms at least once a year. It would seem that such concentrations are a common phenomenon.

---

[a] http://www.stadtentwicklung.berlin.de/umwelt/umweltatlas/karten/pdf/e04_05_2001.pdf
[b] http://www.aecc.be/en/Conservation.html

Plant and animal selection
Air pollution has far-reaching consequences for the flora and fauna of urban areas. On the one hand, it distorts the process of natural selection (resistant or tolerant species are at an advantage in polluted areas), while it also promotes the eutrophication of soil and water. According to Alaimo (2000), the most significant effect is the long-term erosion of the health and vitality of rare and vulnerable plants, while there will also be marked changes in the composition and diversity of the animal and plant populations.

16.2.4   Noise and light

Noise
The average noise level in urban areas is many times higher than that of the agricultural and nature areas. The main sources of urban noise are main roads, airports and flight routes. While it is still possible to enjoy near silence in some rural areas, it is rarely if ever silent in the city.
According to Brumm and Todt (2002), nightingales (*Luscinia megarhynchos*) in Berlin sing louder when there is more ambient noise. This phenomenon is particularly evident during weekday rush hours. Similarly, tree swallow fledglings (*Tachycineta bicolor*) call for food at greater volume when there is more ambient noise (Leonard and Horn 2005). As Oberweger and Goller (2001) point out, singing more loudly expends more energy and is likely to attract predators. It is possible that the selection process in an urban area with considerable noise does not favour acoustically sensitive species such as owls.

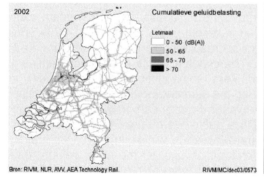
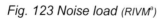

Fig. 123 Noise load *(RIVM[a])*   Fig. 124 Light load*([b]cited by Zoest, 2006)*

---

[a] http://www.rivm.nl/gezondheidenmilieu/themas/geluid/Blootstellinggeluid/
[b] http://www.inquinamentoluminoso.it/worldatlas/pages/fig4.htm

278

Light
In the urban area, street lighting and illuminated buildings account for much 'light pollution', whereby the night sky is never completely dark. Indeed, this is now largely the case throughout the country. Satellite photos taken at night reveal the cities to be islands of light. According to Eisenbeis and Hanel (2003), the 'noise and light climate' of towns may have far-reaching effects in terms of the plant and animal populations, although these effects have been subject to very little research to date.
Some plants and animals will benefit from the longer growth season and the longer average length of the day in the cities (Molenaar, 2002). However, the effects on the plant and animal populations as a whole are difficult to determine.

Attraction and aversion
Some animals are attracted by light (either directly by the light source itself or indirectly by the presence of insects) and are at greater risk than other species of being run over or eaten by predators. Animal species which have a definite aversion to light, such as the Daubenton's bat (M*yotis daubentoni*) are unable to move around with the same ease as they would in complete darkness. Moreover, the areas close to light sources, which would otherwise be suitable for them, become 'off limits'.

Fixation
Some animals can be transfixed by light, unable to move. The rabbit is a familiar example. If caught in a beam of light, the rabbit sees the dark beyond the light as a black wall, and hence as an obstacle.

Disruption of the biological clock
The timing and duration of various animal behaviour during the day is controlled by a 'biological clock', which relies on the natural daily cycle of light and dark. Artificial lighting can disrupt this cycle, whereupon the animals lose sleep and suffer the associated health problems.

Disruption of the biological calendar
Seasonal behaviour such as reproduction, migration and hibernation are accompanied by hormonal and physiological changes. The timing of such changes is precise, with the ratio of day to night playing a crucial role. A disruption of that ratio by artificial light can disrupt an animal's natural biological calendar, whereupon they may breed and give birth to their young when their survival is more difficult.

16.3   Urban dynamics

16.3.1   Urban history

Growth of the urban population
The twentieth century may justifiably be termed the 'century of urbanisation'. This statement is supported by UNFPA (1996), World Resources

Institute (1996), UN Population Division (1999), Brockerhoff (2000), Harrison and Pearce (2001), UN Population Division (2003) and UNFPA (2004). In the economically developed countries, the proportion of the total population living in the urban areas has increased substantially over the past one hundred years. This is due to both autonomous growth and to migration. The trend is set to continue worldwide over the coming decades. Almost the entire growth in the world population projected for the period 2000-2030 will be concentrated in the urban areas. Demographers estimate that the number of city-dwellers will have increased by two billion by the year 2030 (from 2.9 billion to 4.9 billion), a figure which corresponds with the growth in the total world population (from 6.1 billion to 8.1 billion).

The emergence of towns

The concept of the town dates back to the development of agriculture (during the Neolithic Age, 10000 to 5000 BC). The first urban centres were founded in the first agricultural regions – Mesopotamia, Egypt (the Nile Valley), India (Indus Valley) and China (Yangtze Valley). Since then, there have been several waves of urbanisation in various parts of the world. In Europe, the first major urbanisation spurt was during the Roman Age. Some one thousand years later, in the latter middle ages, there was a further urbanisation process. However, urbanisation in neither period can be compared to that brought about by industrialisation in the nineteenth century and the 'technological revolution' of the twentieth century.

Although the first towns and cities are indeed very old in terms of human history, when viewed on the geological-evolutionary timescale they are quite recent innovations. Plants and animals have not had much opportunity to adapt to this new type of landscape.

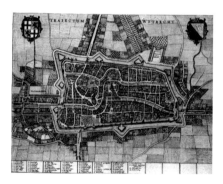

Fig. 125 Utrecht in the Golden Age
(Bleau, 1652)

Fig. 126 Remains in the landscape
(Zoest, 2006)

Dutch towns and cities

The oldest cities in the Netherlands date back to Roman times, i.e. the beginning of the first millennium A.D. Thereafter, we see three main phases of urban design: (I) enclosed towns surrounded by walls or defence works (Middle Ages to approx 1800), (II) the industrial city (1800 to 1945) and (III) the modern city (1945 onwards).

Walled towns (Middle Ages to 1800)

Most European towns have a centre which was established in the Middle Ages. In former times, these towns were not always as compact as they are today. This aspect changed during the development from village to (small) town. Originally, buildings were scattered hither and thither, with many individual houses on one plot of land. As the population grew, the building pattern became more dense – a logical consequence of the fact that most towns (apart from those whose geographic location provided natural defences) were surrounded by walls or other man-made fortifications.

The urban-rural contrast

During this period, there were many differences between town and country. The compact building pattern contrasted with the wide open countryside, the free urban culture with the strongly traditional rural culture, and the diversity of the urban economy with the one-sidedness of the agricultural economy.

Fortifications

The mediaeval town walls were replaced in the fifteenth and (more especially) the sixteenth centuries by earthworks and fortified defences, rendered necessary by the ever greater firing power of the cannon of the day. The system of earthen walls was maintained and adapted until well into the nineteenth century. The urban extensions of the Renaissance era therefore show a clear contrast with the mediaeval town itself. Rather than adopting the former organic street pattern, largely determined by the local topography, the town districts of the sixteenth century onwards were characterised by an orderly grid pattern with closed building blocks. Amsterdam's canal belt district is a prime example.

The defence works also provided the main leisure area for town residents, who would take their constitutional walks here. The walls were planted with trees and bushes, which formed an integral part of the defences. Wood could be used during a siege, thorn bushes would slow down any enemy incursion, while the roots prevented the walls being undermined. In peacetime, the walls provided a sheltered and pleasant promenade.

Growth of the towns

During the eighteenth century, urban growth followed much the same pattern as before, relying on densification and building upwards. Fortifications were either removed or straddled, with new buildings constructed beyond the former town walls. At this time there were very few urban extensions in the Netherlands, since the Dutch economy was floundering. Nevertheless, as Glaudemans (2000) notes, the more prosperous citizens took to spending their summers outside the city at their country retreats.

Industrial towns (1800-1930)

The deteriorating conditions in the towns served to accelerate the exodus of the rich. During the eighteenth and nineteenth centuries, industrialisation drew large numbers of people from the rural areas into practically all of Europe's major cities. The rapid growth of the urban population caused major social and environmental health problems. Although the exact reasons for the growth of the urban population varied from one country to another, most were connected with the introduction of hard capitalist enterprise and industrial production processes. Agricultural crises and evictions forced many farmers and farm labourers to seek alternative employment in the cities. The towns and cities became overcrowded. Many were quite simply dangerous, unhealthy and barely habitable. In Europe and elsewhere, there were three main reactions to the industrial 'horror town': regeneration, the parks movement and the garden cities movement.

Urban regeneration

The French government in particular was of the opinion that a capital city should exude grandeur and beauty, becoming a symbol of the glory of the nation as a whole. Moreover, the authorities feared social tensions and rebellion, the French Revolution of 1789 being still fresh in the mind. To many members of the urban elite, regeneration and 'beautification' seemed an excellent way to render their cities healthier, give them appropriate grandeur, and avoid social upheaval. The old districts which had become over-densified were opened up, broad avenues were driven through, and greenery was added in the form of trees, parks and public gardens, which became the main pillars of the urban regeneration movement.

Haussmann

The prime example of this international trend was the total restructuring of Paris, undertaken between 1853 and 1870 under the supervision of the Prefect of the Seine *Département*, Baron Georges Eugène Haussmann. He ordered many new roads to be pushed through the old, crowded city centre. The original buildings, most of which were by now slums, were demolished to make room for broad, straight avenues and elegant new buildings, most in the classical style.

Haussmann's extensive programme of green amenities was truly revolutionary. He wished to transform Paris into a 'green metropolis'. No fewer than 109,330 trees, all grown in the municipal nurseries, were planted along some two hundred kilometres of avenues. The trees also served to stress the higher social status of a residential area.

Parks

Several new parks were laid, and Paris' two main existing parkland areas – the Bois de Boulogne and Bois de Vincennes – were completely restructured. Each had originally been laid out in the traditional geometric French garden style. Haussman's designers decided to adopt the English approach, with meandering streams, slopes and water features in unexpected places. Winding paths and twisting roads ensured that visitors would find a new and surprising vista at every turn.

Dismantlement of fortifications

The cities' desire for a new look coincided with major changes in the demands of defence. The traditional style of fortifications was no longer necessary, or at least no longer effective as Napoleon's conquest of Europe has shown. It was Napoleon who ordered the demolition of the fortifications in each city he took, and encouraged the creation of public parks in their place. Even where the French were repelled, such defence works came to be regarded as old-fashioned and obsolete, whereupon many European cities converted the former fortifications into green leisure areas.

The parks movement

A second reaction to the blight of the industrial city was the parks movement. Even in the eighteenth century, the increasing prosperity and authority of certain sections of society prompted the nobility to open their gardens as parks for the use of the elite. In Breda, for example, those deemed worthy were given a key to the castle gardens.

The parks movement started in England. The first official acknowledgement of the necessity to provide public parks dates from 1833, when the Select Committee on Public Walks presented its report on the green space available in English cities. The committee concluded that it was the very poor living in the overcrowded city centres who had greatest need of the facilities that parks could offer. The advantages of creating new parks were summarised in terms of physical health, moral and spiritual well-being and political expediency. According to the committee, parks would become the 'lungs' of the city, purifying the air, improving public health, encouraging physical exercise and offering a healthy alternative to the tavern. Moreover, they would elevate the minds of the masses through contact with nature. Because the proposed parks would be accessible to all, regardless of class, they would also relieve social tensions. The classes might even learn from each other. The idea did not meet with immediate approval, but by 1885 the park came to be acknowledged as a standard public amenity and was given a place in every major urban extension project.

Accessibility

Parks were open to all, provided they behaved themselves and observed the moral standards of the day. The idea was that the working class should be educated and elevated to the standing of the more well-heeled citizenry. The park was therefore laid out to provide room for *promenading* –a genteel stroll while engaged in polite conversation – horse riding, taking afternoon tea, listening to music, rowing on the lake, and so forth. There would almost certainly be kiosk selling sheet music and an area set aside to display exotic trees and shrubs. In the nineteenth and early twentieth centuries, parks were also used to commemorate historical events, usually by means of some prominent memorial.

Dutch reserve

In the Netherlands, the parks movement got off to a difficult start, particularly in Amsterdam. The 'powers that be' were indifferent to the idea of public open spaces, almost certainly because most had their own country retreat outside the city. It is notable that Amsterdam's first public park, known simply as 'The Park', was a gift from Napoleon (and hence something that could not be refused!). The first large park in Amsterdam – the Vondelpark – was not laid until 1866 and was intended as compensation for the building-over of the popular Plantage. Sarphati Park followed in 1883. The city's zoological garden, Artis, had been established many years earlier in 1838. This was a private initiative which emulated similar projects in other European capitals. Because much of the funding for large urban greenery projects had to be sought from wealthy private citizens (these were among the first 'public-private projects'), the final result rarely matched the original ambitions.

Socialism

With the increasing influence of socialism (which began to take hold in the Netherlands in the 1890s), public authorities gained greater control over the quality of both housing and public amenities such as parks and greenery. Parks were now laid out by local authorities: the Zuiderpark in The Hague (1908), the Kralingerhout in Rotterdam (1911) and Leijpark in Tilburg (1936) provide good examples. From the 1920s onwards, the design of the parks also offered more space for less genteel forms of leisure, including sports and picnics. From Germany came the concept of the 'people's park', in which sports, entertainment and cultural activities were integrated into both the design and the programme. Such parks offered far more than an edifying constitutional in the open air: there might also be cafés and restaurants, a swimming pool, a sports stadium, sports fields, children's playgrounds, picnic areas, and so on. The Hague's Zuiderpark is an excellent example of this type of park.

Allotments, sports fields, playgrounds

In the late nineteenth century, local authorities also turned their attention to the creation of allotments, sports parks and public playgrounds. The concept of the allotment – a small plot of land on which a household could grown its own vegetables – had existed for many centuries. Now, however, large areas were set aside for such allotments, each administered by an association of its tenants. Similarly, the sports parks were rented to, and administered by, sports associations. The playgrounds movement which had emerged in the nineteenth century now pressed for the creation of play facilities in all major cities. All such developments reflected the growing demand for amenities which were not yet fully available in the burgeoning cities.

Garden city movement

A third reaction to the industrial city which, like the parks movement, had its origins in Britain, was the garden cities movement. In the United Kingdom (more so than elsewhere), industrialisation had led to the idealisation of the rural, outdoor life. For many people, a 'house in the country' was the ultimate ambition. It was, however, an unattainable one. A cottage, let alone a country house, was beyond the reach of all but the rich. According to some idealist thinkers, the solution had to be sought in a transformation of the concept of the city itself.

The garden cities movement was born in 1893, further to a discussion document produced by a society which had been set up to disseminate the ideas of the American writer Edward Bellamy. The concept of the garden city, often credited to the urban planner Ebenezer Howard (1850 -1928), involves a far-reaching integration of the urban and rural environments. This combination of 'the best of both worlds' would provide a response to both the overcrowding of the cities and the isolation of the countryside. The garden cities would be created at some distance from the existing cities, and each would cover an area of some 24 square kilometres. Of this, a sixth would be devoted to housing and other buildings, with the remainder given over to agriculture. Each garden city would be more or less self-sufficient. Moreover, every house would (eventually) be owner-occupied, while all public amenities would be run by the local authority. Once the population had reached approximately 32,000, further growth would be accommodated by the construction of yet another garden city elsewhere.

New towns

Howard's urban model was never implemented precisely as he had proposed (Howard, 1902a and 1902b, 1965). His socialist principles – self-sufficiency and communal living – also failed to gain much ground. Nevertheless, the garden city concept did have a great influence on urban planning throughout the twentieth century. Through the efforts of Howard and his fellow urban planner Sir Raymond Unwin (1863 – 1940), two model garden cities were created: Letchworth Garden City, just north of London, and Welwyn Garden City. The concept was emulated throughout the world. Garden cities or suburbs which reflect Howard's model were later built in the United States, France and the Netherlands (for example Vreewijk in Rotterdam, Vogelbuurt in northern Amsterdam and Rozenhagen in Haarlem). More importantly perhaps, Howard's basic idea of combining town and country has contributed much to the design and development of many of the suburban residential areas built after the Second World War.

Suburban towns (1930 onwards)

In the Netherlands, the Housing Act of 1902 did much to improve the quality of Dutch residential construction. Higher demands were placed not only on the quality of housing, but also on that of the urban extension plans themselves. The entire structure of the urban neighbourhood underwent radical reform. What were the effects?

From garden cities to green districts

Until the 1940s, most construction in all Europe's larger cities was in the traditional closed block pattern. During the 1930s, however, urban planners had begun to embrace modernism, influenced by Le Corbusier and the modernist movement in other areas of art and design. Taking up the ideas of the garden cities movement, the CIAM championed living amid greenery, with plentiful light and air even in the city. CIAM stands for *Congrès Internationaux d'Architecture Moderne*, a movement of 'modern' planners and architects who were chiefly active in two periods: 1930-1934 and 1950-1955. They called for a new design 'language' which differed substantially from that of the garden cities movement. The key features of modernism were high-rise and terraces rather than the closed block, concrete rather than brick, and straight lines without ornamentation. With the car becoming more popular, wider streets were essential, with lines of terraced houses more convenient than the closed block.

Separation of living, work, leisure and traffic

CIAM also called for a strict division between the residential, business, leisure and transport functions. The areas dedicated to each function would be connected by roads and rail. Thus emerged the green suburban residential estate, with a relatively low housing density. Versions of this pattern were to be seen all over the world by the mid-1950s. Some districts incorporated only low-rise housing, while others had high-rise apartment blocks surrounded by greenery and yet others combined the two extremes. Some districts had only straight streets of terraced houses, with occasional courtyards and closes, while others were practically labyrinthine. Others were built around a central 'village green'. This diversity notwithstanding, these suburban districts have many common features. They have much public greenery, many of the houses have a garden, and they have the 'small town' atmosphere which promotes social cohesion.

The finger city

In the early twentieth century, new ideas began to emerge not only about the structure of individual districts but about the form of the town or city as a whole. In order to ensure that green leisure areas would remain within easy rich of all, many urban extension projects opted to extend outwards in a star pattern, with green areas forming 'wedges' between the fingers of the star. The green areas can then be preserved, while further extensions remain possible. Boston, Berlin, Amsterdam and Copenhagen are all examples of this 'finger city' pattern, the result of an integrated vision of urban planning. In most cases, however, this structure takes shape in phases as each urban extension is implemented.

Fringe parks

The contrast between town and country, which was so marked in the pre-industrial city, faded rapidly because urban extensions created suburban residential areas, with business, utilities and leisure areas on the fringes of the urban area. In the more economically important regions, towns and

286

cities converged to form urban networks and regions, such as the Rand-stad, the Ruhr Region, and the greater metropolitan areas of Paris and London. The surrounding semi-rural areas were more emphatically in-cluded in the city's green amenities. From 1960 onwards (and in some instances even earlier), large leisure areas were created on the outskirts of the urban centre to meet the needs of the city-dwellers. Kralingse Bos and Amsterdamse Bos (the latter being laid out in the 1930s) are good exam-ples. In The Hague, the natural dunes area has fulfilled the same function for many centuries.

A changing landscape

"All landscapes are subject to constant change, and that is marvellous to witness", wrote Jac. P. Thijsse in 1942. And if there is one type of land-scape that is even more dynamic than most, it is the city landscape. In the dynamic urban setting, the green structures in and between the cities form the most important arena for nature.

16.3.2 Urban growth

Rapid and different growth in the twentieth century

Fig. 127 Amsterdam 1930 (Zoest, 2006)   *Fig. 128 Amsterdam 1960 (Zoest, 2006)*   *Fig. 129 Amsterdam 1990 (Zoest, 2006)*

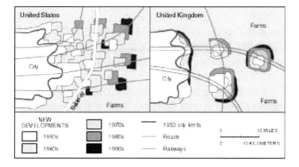

Fig. 130 Urban extension in the US and UK (cited by Zoest, 2006)

Urban boundaries

Where does the city stop and the countryside begin? Geographers and urban planners have formulated different answers to this question. Individual cities are increasingly converging to form regional networks in which town and country overlap. Planners have therefore developed new terms and definitions which treat urban areas as a 'system' rather than as a series of clearly defined settlements. Terms such as 'urban field', 'network city', 'regional city' and 'metropolitan district' refer to extended (and extensive) urban systems which contain several urban centres. There is intensive interaction between those centres in terms of the movement of people, goods and information.

The urban field

The delineation of an urban region is a question of definition and the level of scale applied. It will depend on the patterns and processes one wishes to examine. For example, the exact boundaries of an 'urban field' (a term coined by Friedmann and Miller in 1965) will depend in part on the movements of the city-dwellers during their leisure day-trips. The Los Angeles conurbation is an example of an urban field. In the Dutch situation, the urban field would actually extend across the national borders.

The urban system

At the other end of the spectrum is the term 'urban system', in which one main conurbation dominates all other centres within the system. Here, we may think of London, Paris and Copenhagen.

The urban network

In terms of size and scope, the 'urban network' lies somewhere between these two extremes. According to Batten (1995), such a network develops when two or more cities seek cooperation and complementarity, helped by good and reliable infrastructure for both physical traffic and information. The exact size of the network is a product of people's daily movements for the purposes of work and leisure. This gives rise to the term '*daily* urban system'.

Daily urban system

In practice, the 'daily urban system' is an area with a radius of approximately twenty to thirty kilometres around the main urban centres. An urban network will therefore form a system of (compact) urban centres surrounded by suburban residential areas, business estates and office locations. Beyond these is an even more extensively used 'fringe zone' which forms the transition to the inter-urban and rural areas. The greater the physical connections between the urban centres – whether in terms of infrastructure or convergence – the more the intermediate rural area will become fragmented. A large proportion will then fall under the direct influence of the urban areas, whereupon it can be regarded as a 'fringe' or 'edge' area of the city itself.

The right level of scale

Fig. 131 Conurbation or ...
(Zoest, 2006)

Fig. 132 ...region
(Zoest, 2006)

At the scale of landscapes, and perhaps contrary to expectations, urbanisation actually makes the topographical mosaic more heterogeneous. Viewed from the air, most cities appear as an irregular patchwork with a complex and finely-meshed inner structure.

16.3.3   The urban metabolism

Energy
The metabolism of towns and cities is entirely different to that of agricultural or nature areas in which green plants are the main energy base of the eco-system. The energy system of modern cities is largely dependent on the local import of fossil fuels and food which has been produced elsewhere. Because both the population and the production processes are concentrated in an urban area, the demand for energy is many times greater than in the rural areas, as is the output of that energy in the form of heat, noise and light. Although urban areas cover less than five per cent of the earth's surface, they consume by far the greatest proportion of the available energy (Odum 1997).

Materials
Similarly, cities form 'black holes' into which supplies of resources disappear. Compared to the non-urban areas, cities consume enormous quantities of fuel, food, water, construction material and production resources. Transport and processing of the resultant waste is expensive, and waste management is a problem worldwide. The effects of emissions into the air, soil and water are most immediately visible in terms of the accumulation of nutrients and toxic substances in and around the city. Dissemination in the atmosphere and in water renders these emissions a problem on a continental scale.

The ecological footprint
Calculations show that the urban metabolism is taking a heavy toll on the environment.

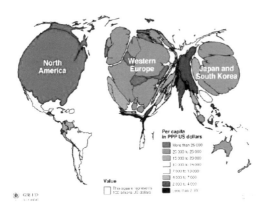

可持續發展
Sustainable Development

Fig. 133 Ecological footprint
(cited by Zoest, 2006)

Fig. 134 Hong Kong
(cited by Zoest, 2006)

A follow-up to a previous study of the urban metabolism of Hong Kong (Southeast Asia) in 1971 shows the rising trends in the consumption of resources and in the production of waste. In theory, Hong Kong needs six hectares per inhabitant to support its economic activities, particularly the marine eco-systems. When fish-farming is included in the equation, a surface area almost two thousand times greater than the built-up area is required. The consumption of food has increased by 20% since 1970, that of water by 40% and that of other resources by 149%. Total atmospheric emissions have risen by 30%, CO_2 emissions by 250%, solid waste by 245% and sewage by 153%. There is a structural overload of land, water and atmospheric systems. Nevertheless, Hong Kong aims to become a 'truly sustainable city' in the twenty-first century (Warren-Rhodes and Koenig, 2001).

16.3.4   Impact

Plants and animals
The cities' use of resources and energy has significant consequences for flora and fauna, partly due to the impact of energy consumption on the urban climate, and partly due to the production of edible waste from which seed-eaters in particular benefit. In most cases, the city is home to a relatively small number of species who are able to take advantage of the ecological characteristics of the urban environment. Emlen's 1974 study of birds in Tucson, Arizona (USA) reveals that just three of the fourteen bird species found in the area - the Inca dove (*Columbina inca*), the house spar-

row (*Passer domesticus*) and the starling (*Sturnis vulgaris*) – account for two thirds of the total bird population and of the biomass produced.

Public health
Epidemiological research by Lucht and Verkleij (2001) suggests that the public health is generally lower in large cities than in smaller towns and the rural areas. Some people are able to escape the city by renting an allotment or space on a caravan or camping site. Others choose to relocate to green suburbs, small towns or villages. Those who cannot do so, however, may eventually suffer some adverse impact in terms of physical and mental health (Maas and Jansen, 2000; Oers, 2002).

Stress-related disorders
As a result of our longer life expectancy and the changes to lifestyle which were introduced during the twentieth century, the most common disorders in the developed countries are the relatively new 'prosperity diseases', such as coronary disease, auto-immune disorders, certain infectious diseases and depression. According to several sources (Cohen and Herbert, 1996: Taylor and Seeman, 1997, the World Health Organisation, 2001), the incidence of these conditions is closely related to stress.

Stress
It has long been known that physical and psychological stress can adversely affect health and well-being (Sternberg, 2000; Keller *et al.* 1994). During the 1930s and 1940s, the world's attention was drawn to this phenomenon by Hans Selye, a Canadian endocrinologist of Austro-Hungarian origin. He discovered that people and animals not only become sick due to infections, bacteria, viruses and other 'tangible' pathogens, but also due to excessive burden on body or mind. He coined the term 'stress', which he defined as "the non-specific response of an organism to external pressures of any kind".

According to Lazarus and Folkman (1984), 'stress' is an umbrella term covering many variables and processes. Physical factors such as excessive noise, heat or cold, chronic illness, natural disasters, lack of food or sleep, and sustained effort can all result in stress. However, psychological factors such as situations which are regarded as particularly unpleasant, intrusive or threatening can also do so. Psychological stress may result from day-to-day concerns such as the pressure of work, or any number of irritations and inconveniences. There are various significant life events, such as the loss of a loved one, divorce, moving house, unemployment or long-term injury, which are likely to result in (serious) stress.

The impact of greenery
To be among plants, trees and water is an effective way to relieve or prevent depression and mental fatigue. According to a report published by the Netherlands Health Council and the Advisory Council for Research on Spatial Planning, Nature and the Environment (RMNO), the physical exercise

such as walking and cycling which outdoor leisure involves can have a beneficial health effect, helping to prevent coronary disease, for example.

The link between greenery and health (see also chapter 15, page 259) is visible not only in the health differences between the urban and rural areas, but also in people's choice of residential location. The Housing Preferences Survey conducted by RIGO in 2003 reveals that the true urban (town centre) environment is still in high demand, but that other settings (semi-rural, village and rural) are also much sought after. Excessive urbanisation, with the physical and social conditions it entails, seems to have strengthened the desire of some demographic groups (such as families with young children) to relocate to a greener setting (RIVM, 2000, 2001; Lucht and Verkleij, 2001; RIGO, 2004, Vries, Hoogerwerf *et al.* 2004). High population density and a shortage of nature, space and tranquillity are, according to Bekke, Dalen *et al.* (2005), also important motives for emigration from the Netherlands, alongside the mentality of the Dutch themselves.

Recovering from mental fatigue

At a certain moment, the mental mechanisms involved in concentration and self-control can become tired, since the energy required for these processes has literally become exhausted. This state is known as 'mental fatigue' and can have various undesirable effects (Baumeister, 1991; Baumeister, Heatherton *et al.* 1994). Mental fatigue can lead to poor cognitive performance. Put simply, you can no longer solve problems with the customary ease, you are less creative and you may become forgetful. Mental fatigue can also cause some people to become more aggressive and less amenable. According to the Health Council and RMNO report (2004), time spent outdoors amid nature is an effective treatment for mental fatigue. Researchers hypothesise that natural environments possess all the qualities and characteristics required to encourage the recovery of the mental mechanisms. The spatial structure of nature areas is complementary to the human cognitive system. Because natural environments are understandable and invite further exploration, they command the attention in an unforced way. Psychologists term this 'soft fascination'. Accordingly, people can function within this type of environment without overburdening the concentration mechanism, which then has the opportunity to recover.

Psychological impact of green areas

Many studies, including those by Talbot and Kaplan (1986), Hazelworth and Wilson (1990), Sebba (1991), Stringer and McAvoy (1992), Fox (1999), Frederickson and Anderson (1999), have identified positive effects of time spent in outdoors (often in the form of tracking or orienteering in a wilderness setting) in terms of autonomy, competence, control, self-confidence, optimism and feelings of (inner) calm. When considering the five mechanisms described above, it becomes clear that time spent amid nature is an excellent means of dealing with stress and trauma, and is likely to encourage integration and personal growth. Experiencing nature has a positive effect on each of the five mechanisms:

Autonomy	Kaplan and Kaplan (1989) state that natural environments enhance feelings of autonomy, control and self-respect because they are psychologically 'low-threshold', i.e. they demand no special skills, while Hartig (1993) states that such environments exert no social pressure.
'Relatedness'	Leisurel activities with a strong social component promote the development of social networks; they help to strengthen friendships and hence feelings of social support (Iso-Ahola and Park, 1996). Most people visiting a nature area do so in small groups (family or friends). Significant social ties are given the opportunity to strengthen.
Competence	Learning to function within a natural environment is largely 'automatic' since it relies upon innate cognitive and emotional skills. This strengthens feelings of competence.
Escape	Provided they offer sufficient contrast with the urban environment and are sufficiently expansive, natural environments suggest – or become – 'another world'. They then provide opportunity for *palliative coping*, i.e. escaping from the causes of stress and trauma (Kaplan and Kaplan,1989: Knopf, 1987).
Relaxation and positive feelings	Experiencing scenes of natural beauty helps to relieve stress and mental fatigue, being linked with positive feelings and improved temperament.

According to the cited researchers, these five components are essential to a sense of meaning and to personal growth. This conclusion is supported by various other studies and tested theories. However, chapter 15, page 259 refers also to negative impacts.

Spiritual Experience Process Funnel model
Based on studies of wilderness experiences, Fox (1999) developed the Spiritual Experience Process Funnel model. This suggests that those who spend time amid nature achieve greater control, autonomy and competence, and become more open to reflection and spiritual contemplation. The effect is brought about not by trekking through the wilderness, but by spending time in any open green environment. Kaplan and Kaplan (1989) conclude from the extensive research into the psychological effects of wilderness experiences that contact with nature brings cognitive calm and the enhanced ability to organise thoughts, consider problems in perspective and establish priorities. The beauty and symbolic significance of the natural world invites one to reflect on the goals and priorities in life.

Urban ecology as a sector of urban planning and management
The Housing Act of 1901 ensured that residential accommodation would henceforth meet certain basic quality requirements. With the help of the disciplines of spatial and urban planning, the government gained greater

control of the growth of the cities, too. Garden suburbs were created as new, healthy residential environments for the working class. Urban extensions adopted the finger model in order to ensure that rural areas remained accessible to city-dwellers. With financial aid from the government, a series of new parks, sports fields, allotments, swimming baths, playgrounds and leisure woodland areas were created. To prevent urban sprawl whereby towns would encroach on the natural areas between them, the 'regional plan' was introduced as an instrument of spatial planning. Planners and nature managers alike were keen to avoid any repetition of the urban development patterns of the nineteenth century, and in the first half of the twentieth century they therefore joined forces to ensure that town and country would be fully complementary.

Standards of green areas changing in time

The development of norms establishing the area of greenery per inhabitant in towns reflects the changing ideas about urban extensions in the Netherlands: from new towns ('bundled deconcentration') with low densities in the sixties, to compact cities to save the open landscape in the seventies and eighties, and a renewed desire for green residential areas in the nineties (see *Fig. 135*).

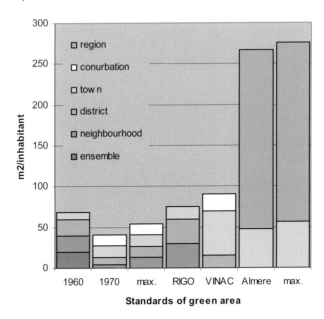

Fig. 135 The development of greenery norms in the Netherlands
(Zoest, 2006)

16.4   The urban-rural gradient

Landscape mosaic

A well-studied urban-rural gradient is offered by the American city of Phoenix, in the Arizona desert (see *Fig. 116*). During the past century, the city has seen substantial growth in both population and area. This process of urbanisation involved the reclamation of desert areas, but went yet further. The growing city has influenced land usage in the wider surrounding area, whereupon the structure of these landscapes has also changed.

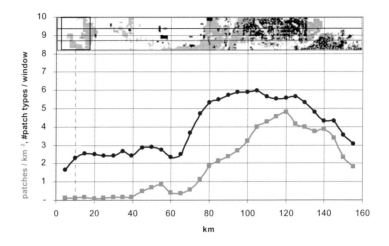

Fig. 136 Patches / km2, number of land use types / window
(*Matthew Luck and Jianguo Wu, 2002*)

By measuring a series of landscape characteristics[a] along a line drawn across the Phoenix metropolitan area, Jenerette and Wu (2001) and Luck and Wu (2002) demonstrate that nearest the centre, the patch size is decreasing and the patch density increasing. This indicates ongoing fragmentation of the landscape. All the features observed indicate marked changes in the topographical structure at 75 and 155 kilometres each side of the urban centre, which exactly corresponds with the urbanisation front of Phoenix on the east-west axis.

---

[a] **Patch richness:** The number of patch types in the landscape; a measure of diversity of patch types.

Class percent of landscape: The proportion of total area occupied by a particular patch type; a measure of dominance of patch types.

Largest patch index: The proportion of total area occupied by the largest patch of a patch type.

Patch density: The number of patches of per 100 ha.

Mean patch size: The area occupied by a particular patch type divided by the number of patches of that type.

Patch size coefficient of variation: Patch size standard deviation divided by the mean patch size; a measure of relative variability.

Landscape shape index: The landscape boundary and total edge within the landscape divide by the total area, adjusted by a constant for a square standard.

Area-weighted Mean Shape Index: A mean patch-based shape index weighted by patch size.

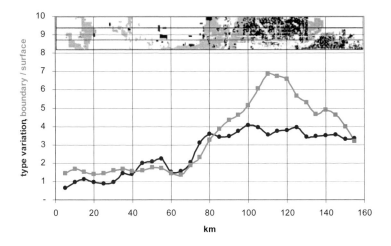

Fig. 137 Land use variation, boundary / surface
(*Matthew Luck and Jianguo Wu, 2002*)

Biotic response

Honnay *et al.* (2002) provides yet another example. They analysed the topographical features of a north-south transect of the Flanders region (Belgium), in 89 blocks of 4×4 kilometres. The higher the degree of urbanisation in a block, the greater the landscape diversity and complexity. In the immediate vicinity of towns and cities in particular, there is a high degree of heterogeneity. There are two possible explanations for this finding. While the agricultural areas have only a limited number of different landscape types, the urban and semi-urban landscapes have a mix of both urban and rural forms of land usage, giving rise to greater topographical diversity. In addition, due to the less efficient and less 'streamlined' structure of the urban fringes, there are many small open areas which provide additional spatial heterogeneity.

The study also established a correlation between the number of plant species found in a particular 4×4 km block and the heterogeneity of the landscape in that block (expressed as the Shannon Diversity Index). The percentage of built-up area is stated for each block divided into quartiles. We then see a clear positive relationship between the percentage of built-up area, the heterogeneity and the number of plant species (see *Fig. 138*).

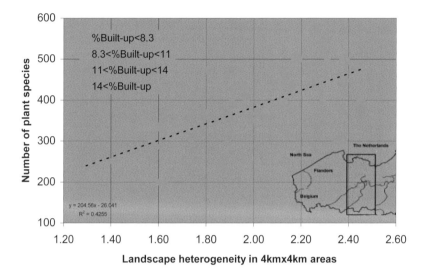

Fig. 138 Number of species and landscape heterogeneity in West-Flanders. (after Honnay et al., 2003)

16.4.1   The city fringe

The zone in which the greatest changes occur is the city fringe. This is the area in which growth is actively taking place and in which the landscape has not yet assumed its final form. The function and structure of the fringe zone are distinctly temporary. It is here that extensive and 'objectionable' functions are generally relocated. Agricultural land use is then supplanted since it is no longer economically viable. Farmers are able to sell their land at substantial profit, often to investors and project developers. The city-dwellers come here to enjoy leisure at weekends. The city fringe is then on the 'waiting list' for urban extension. Once the funding required for construction is in place and there is an economic or social demand for new urban functions, the fringe zone is likely to be the chosen location.

Pryor (1968) defines the urban fringe as "the transition zone in land use, in social and demographic characteristics, which is located between (I) the contiguously built-up urban and suburban areas of the metropolitan centre and (II) the rural hinterland, characterised by chiefly agricultural use." Some researchers choose to divide the urban fringe zone into an 'inner fringe', an 'outer fringe' and the 'urban shadow' which gradually passes into the rural area.

Driving forces within the urban fringe zone

The main cause of the dynamics of the fringe zone is the ongoing demand for space on the part of the urban population – or the speculation of a future demand. Nowhere else is the space claim made by new housing, business areas, amenities and infrastructure so evident as it is in the fringe zone, since building *on* the city is far easier and less expensive than building *in* the city. The dynamics of a growing city affect the fringe zone in many ways.

Space for business areas

A growing city will 'squeeze' business and office accommodation out to the fringe zone. As the city grows, the existing commercial areas find themselves ever deeper within the fabric of the city. Those businesses which depend on good accessibility or require considerable space will prefer to be located outside the centre, in the fringe zone. Garden centres, large furniture retailers, transport companies and suchlike prefer a readily accessible location on the urban periphery.

Space for semi-rural residential environments

People who wish to enjoy a suburban or semi-rural residential environment are also pushed out of the big city sooner or later. The urban centre eventually holds little attraction for them. Provided commuting distances are within reasonable bounds and the available housing not too expensive, many will choose to relocate to a suburban district or to one of the smaller towns and villages around the city itself. Motives for doing so include the need for greater privacy, safety, a larger house, a better environment in which to raise a family, less noise and traffic. In short, they wish to escape the problems associated with a large city, and seek the specific qualities offered by a green residential environment.

Space for leisure

As a town or city grows, so does the requirement for leisure facilities. City-dwellers will wish to visit the city fringe and the rural areas beyond, especially at weekends. Dedicated leisure and nature areas are particularly popular destinations, as is an extensive, well-managed agricultural landscape.

Space for water storage

As a city grows and the area which is built or paved increases, it will also require more space for water storage. It is also essential that the water storage areas are not too far from the urban centre. Where urban extensions are undertaken in the lower-lying regions of the Netherlands, the towns concerned are forced to find additional space for water storage in the immediate vicinity.

Space for allotments, sports fields and other extensive urban functions

When urban boundaries shift, the so-called *extensive* forms of land use – those which offer little or no economic return or which do not make particularly efficient use of the available space – are generally relocated to the new fringe zone. Examples would include allotments, sports parks, campsites and sometimes even cemeteries. Businesses with high environmental impact and utility services which have to be located at some distance from residential areas will either be relocated, or may indeed stand in the way of any extension if relocation proves too expensive. Wastewater treatment plants, power stations and refuse processing plants are functions which cannot be located within the town itself, but which must nevertheless remain reasonably close to it.

Loss of agricultural land
In most cases, an urban extension will lead to the loss of agricultural land in the immediate area. One significant reason for this is that the demand for land increases, whereupon owners are able to command high prices. In today's difficult economic circumstances, many farmers are willing to sell their land. They also realise that land can be subject to a compulsory purchase order. Farmers now avoid investing in land within the 'red shadow' of a town, anticipating future urbanisation. There are also indirect effects; ongoing urbanisation can lead to the fragmentation of a farmer's property, which reduces business efficiency yet further. The tranquillity of the rural setting disappears: farmers are likely to find ramblers on their land. They may suffer vandalism, and will attract criticism of their 'normal' agricultural land usage. Moreover, as the number of former city-dwellers in the agricultural community increases, the balance of political power will shift away from the farmers themselves.

Socio-economic changes in the fringe zone
Along the urban-rural gradient, it is not only the physical landscape which changes but also various socio-economic aspects of land use.
A wide range of functions will be sited in the fringe zone. These functions, while necessary to the city itself, cannot be sited within its built-up area. Land use is generally well-established both in the urban concentration and in the rural areas beyond. Here, there is little reason to make any drastic changes. Around a growing town, however, the fringe zone is constantly in a state of transition from rural to urban uses. Within the inner fringe zone in particular, the majority of functions are temporary in nature. During the process of urbanisation, the temporary functions become more or less permanent ones: housing estates, business parks or industrial areas, for example.

Land ownership
Land ownership arrangements also change as the distance from the urban centre increases. While the land in rural areas and in the urban concentration is owned by 'permanent' owners (such as farmers, housing associations and local authorities), that in the fringe zone is often owned on a short-term basis, having been purchased by speculators. Any change in function is likely to be preceded by a change in ownership. As urbanisation progresses, owners (usually farmers), sell their land to intermediaries such

as project developers, investors, building contractors or real estate agents. Once the land has been developed, it (or the new buildings placed upon it) will then be sold on to the 'end-users': homeowners, housing corporations or public sector organisations.

Leisure
A significant component of the urban-rural gradient is the use of land in the fringe zone for the purposes of leisure. City residents visit the zone to enjoy nature and the landscape. They walk or cycle within a relatively small zone around the urban area itself: a radius of approximately ten kilometres is often cited. They may also visit special attractions or the more unusual landscapes (dunes, woodlands and the coastline) at greater distance, often travelling by car.

Declining social control
The fringe zone is one of the last remaining 'adventurous' areas in daily life, not yet entirely governed by established rules. Accordingly, it can be attractive to those types who prefer to avoid social control and the forces of law and order. Not only can the nature-lover feel like a landowner here, so can criminals, vagrants in their self-built shacks, fly-tippers and various other groups for whom the fringe zone offers the social freedom they seek.

16.4.2   Location-bound biodiversity

The Italian capital Rome has a remarkably large number of insect species. Regular surveys have been held for over 150 years. By 1996, they had identified some 5200 species, in 356 families and 27 orders. This is an extremely high number of different species, particularly when compared with the official figure established for Italy as a whole (approximately 37,000 insect species).

The high number is probably not only a product of the typical mosaic pattern of the urban environment, but also due to the geographic position of Rome midway along the Italian peninsula, with biogeographic influences from both the coast and the Apennine Hills. The insects identified in Rome can be classified in four main groups:

1. Indigenous species on the mixed and dry woodlands, grasslands, marshes and riverbanks (96,5 % of the entomofauna identified to date)
2. Synantropic species (2,3 %)
3. Introduced species (1,2 %)
4. (Locally) extinct species: specialised species (moisture-loving, sand dwellers or salt-loving), the original habitat of which was riverbanks, marshes, brooks and springs. These were not able to withstand the effects of urbanisation such as pollution, dewatering and land reclamation.

According to Zapparoli (1997), Rome's insect life is heavily dependent on a limited number of genetic reservoirs (termed 'hot spots') of insect diversity.

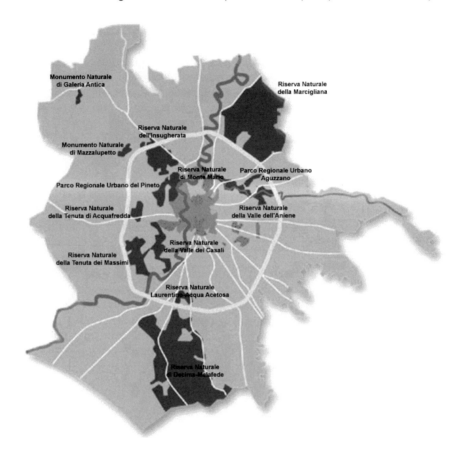

Fig. 139 Core areas: insects in Rome (cited by Zoest, 2006)

The main such 'hot spots' are parks and the historic villa estates such as Villa Pamphili, Villa Ada and Villa Borghese. In addition, the urban and semi-natural green areas such as Pineto, Monte Mario, Acquatraversa and Caffarella form a refuge for many insect species. Archaeological sites such as the Forum Romanum and the Therms of Caracalla also harbour a large number of different species. Some of these areas provide important corridors to the nature and agriculture areas outside the city.

16.4.3   Size-bound biodiversity

The larger the area, the greater the number of species to be found. This is a summary of the 'island theory', a hypothesis for which countless studies conducted since the 1970s have provided collaborative evidence.

There are several possible reasons for larger areas being able to support a greater number of species.

1.   Larger areas are more likely to be found by flora and fauna.
2.   They can include larger (meta-) populations: the larger a metapopulation, the smaller the likelihood that it will die out within a given period.
3.   They offer space for animals which need it, such as top predators and the larger grazers.
4.   They usually have more landscape heterogeneity and a larger range of natural resources. This will attract species with a specialised habitat, which depend on specific types of area. In addition, heterogeneity serves to reduce population fluctuations, which will also increase the (meta-) populations' chances of survival.
5.   They have a greater proportion of 'core' area to fringe area, whereupon those species which are susceptible to negative influences from other habitats (such as predators and disturbance) have a greater survival chance. In large areas of brush or scrub, for example, any intrusion is likely to be confined to the fringes (see below).

Larger animals

The correlation between the size of the area and the number of species to be found there is particularly evident in the case of larger species (mammals, birds, reptiles and amphibians). The more surface area required by a species, the less prevalent that species will be in locations smaller than a given 'cut-off' area. This phenomenon is clearly visible along urban gradients in which the green area gradually becomes more fragmented nearer the city, with only small and isolated areas of green in the urban fringe zone (Suhonen and Jokimaki, 1988); Jokimaki 1999). Many studies of bird populations in urban parks (for example Forman, Reineking *et al.* 2002) confirm the correlation between the size of a park and the number of bird species to be found within it. Smaller parks are unlikely to attract the species which require a certain area in which to live, such as the ground nesters and specialised woodland birds. Studies in Spain and Finland by Fernandez-Juricic and Jokimaki (2001) suggest that an inner-city park of 10 to 35 hectares will have the greatest number of bird species. The size of the park also seeks to be important in terms of the dynamic with which species establish themselves and eventually disappear again. Fernandez-Juricic (2000) found that the larger parks in Madrid had relatively long-established populations of songbirds (such as finches, sparrows and buntings), while the smaller parks had a somewhat faster 'turnover' of species.

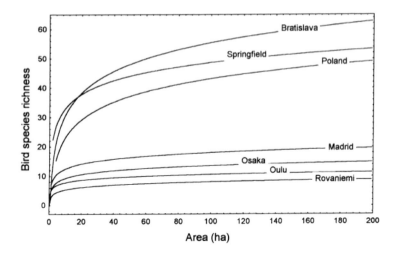

Fig. 140 The correlation between surface area of urban parks and number of bird species to be found in them in Madrid (Spain), Oulu and Rovaniemi (Finland), Osaka (Japan), Springfield (USA), Bratislava (Slovakia) and a number of Polish cities. (cited by Zoest, 2006)

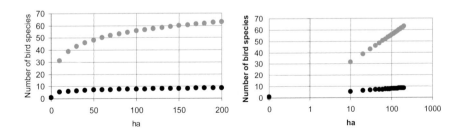

Fig. 141 Island theory: mathematical simulation of maximum and minimum increase in normal and logarithmic representation Fig. 140

The number of species 's' often increases logarithmically with the area 'a' according to s=c+b*ln(a). That means diminishing returns of increasing the area. But, c and b differ substantially between locations. From *Fig. 140* we can derive c=7.2 and b=10.5 for Bratislava and c=3.1 and b=1 for Rovaniemi. Accordingly, if we increase the surface area from 1 to 10 to 100 ha, the number of species will increase substantially from 7 to 31 to 56 in Bratislava, but in Rovaniemi it will increase from 3 to only 5 and then to 8.

Birds
The likelihood of a particular bird species being found in an urban park or wood depends on the size of that park or wood. Each species has its own requirements in terms of surface area. In the Finnish city of Oulu, habitat specialists such as the citrine wagtail (*Motacilla citreola),* the lesser redpoll (*Carduelis cabaret*) and the linnet (*Carduelis canabina*) are found only in the larger parks, while the less 'finicky' species such as the willow warbler (*Phylloscopus trochilus*), great tit (*Parus major*) and finch (*Fringilla coelebs*) are to be seen in even the smallest parks.

Large urban parks not only offer more variation in landscape than their smaller counterparts, they can also accommodate those bird species that require a greater surface area. An analysis was made of the surface area of 54 parks in Oulo was analysed, together with the human activity, habitat and landscape structure in a nine-hectare square surrounding each park. The results of this analysis were then compared to the number of bird species found in the parks. Surface area was found to account for 39 per cent of the variation in the number of species. Of a total of 22 species found in the city, four are not found in parks smaller than 1.5 hectares: the wheatear (*Oenanthe oenanthe*), the linnet (*Carduelis cannabina*), the redpoll and the yellowhammer (*Emberiza citrinella*). The rosefinch (*Carpodacus erythrinus*), garden warbler (*Sylvia borin*) and lesser whitethroat (*Sylvia curruca*) eschew parks smaller than 0.75 hectares. Bird species which thrive in smaller areas were observed closer to the city centre than those which require a larger area.

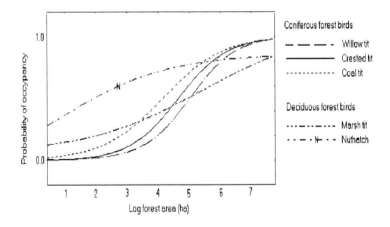

Fig. 142 Forest birds in Stockholm (cited by Zoest, 2006)

Life strategy of birds
Foraging and nesting strategies will also influence a species' choice of habitat. Ground-nesters are rarely found in urban parks, unlike hole-nesting

species which make full and grateful use of the nesting boxes placed in parks. All species of hole-nesters were found in over ten per cent of parks in the study, while only four ground-nesters were observed. Many of the species which build nests at low levels – such as the wheatear, the rosefinch and the yellowhammer – are rarely seen in smaller parks. This may be due to intensive nest predation (for example by crows or cats) or to the lack of suitable bushes and shrubs in such areas.

Buildings
According to Jokimaki's 1999 study, the presence of nearby buildings has a negative impact on three bird species, the willow warbler (*Phylloscopus trochilus*), the hooded crow (*Corvus corone cornix*) and the spotted fly-catcher (*Muscicapa striata*). It may be concluded that the size of the park, the habitat structure of the park *and* the structure of the surrounding urban landscape all help to determine the prevalence of certain species in Oulu's parks.

Boundary-to-central area ratio
Bolger, Scott *et al.* (1997) investigated the response of twenty bird species to an urban-rural gradient in the coastal region of San Diego County, California (USA). The gradient is some 260 km^2 in length and first comprises an extensive, almost contiguous area of scrubland which gradually becomes fragmented, and eventually gives way to a series of isolated habitat locations within the urban matrix. The species studied had clearly differing responses to this gradient. Four species of sparrows favouring open landscapes proved to be the most sensitive to fragmentation and the incursion of the urban fringe into the landscape.

At the other end of the gradient were those species which achieve high densities nearby the urban fringe and in more fragmented areas. In the centre of the gradient were those species which thrive regardless of isolation or the type of habitat. The prevalence of the species which are indeed sensitive to fragmentation or fringe effects was far lower within 200 to 500 metres of the edge of a habitat location. Birds which favour this type of habitat benefit within a range of up to a kilometre from the boundary, depending on the precise species.

Nested distribution
The fragmentation of habitats within the urban matrix gives rise to the 'nested' distribution of species. According to Ganzhorn and Eisenbeift (2001), nested implies that all species which can be found in smaller habitat fragments will also be found in larger ones. One mechanism which accounts for the nested distribution of species is selective extinction and colonisation resulting from the fragmentation of a contiguous landscape (and habitat). The probability of finding a particular species declines in proportion to the degree of fragmentation, as the habitat fragments become ever smaller and more isolated.

A study by Fernandez-Juricic (2002) in Madrid found that the bird populations of 27 urban parks (varying in size from 0.4 to 100 hectares) do indeed
demonstrate a nested pattern, with the larger parks home to all species
which are also to be found in the smaller parks. A similar nested structure
was also found by Natuhara and Imai (1999) in their study of woodland
fragments in Osaka (Japan). This nested pattern is ascribed to differences
in survival strategies, body size, the availability of resources (food, nesting
sites, etc.) and the degree to which the various species prefer a central
woodland location.

Nested distribution in older parks

An interesting conclusion of Fernandez-Juricic's study in Madrid is that the
nested pattern is more often to be seen in the older parks. The degree to
which the parks are isolated from the surrounding area does not appear to
be a relevant factor. The bird populations of these parks are primarily regulated by local factors (size, diversity of habitat, disturbances). Regional
factors appear to be more significant in the case of the more recently established parks, usually on the outer fringes of the city, which are colonised
by birds from the regional species pool.

Non-nested patterns

A non-nested pattern may be expected in situations in which the selective
effects of area and isolation are not dominant, and in which the composition
of the bird population is primarily determined by the influx of species from
elsewhere. In this case, the composition of the regional species pool itself
is a highly relevant factor.

16.4.4   Quality-bound biodiversity

Selection along the gradient

Along the urban gradient, not only does fragmentation increase towards the
city (centre), but the quality of the available habitat will also show significant
variance. The intensity of disturbance will be greater, there will be increased contamination of soil and water with fertilisers and toxic substances, and a greater number of non-indigenous plants and animals will
act as competitors, predators or parasites. The closer to the city, the less
suitable the habitat is likely to be, although there are certain species which
prefer the more urban environment. The quality of the habitat is the main
factor which will determine whether a particular species will establish itself,
particularly in the case of those species which spread well and which can
live even in smaller areas.

Diatoms

An example of the influence of habitat quality on the exact composition of
the wildlife population is provided by a study of diatoms (algae) in streams
and rivers around Melbourne, Australia (Sonneman 2001). Based on the
species of diatoms found living on stones and water plants, the researchers
were able to identify three separate populations, which they named 'east-

ern hinterland', 'western hinterland' and 'metropolitan'. The differences between these three populations are due to both the variation in natural conditions (mineral content of water, basalt geology and annual rainfall) and gradients in the nutrient content of the water (i.e. increased phosphorus, ammonia and nitrogen compounds nearer to Melbourne itself). Although the three populations are similar in terms of diversity, the urban nutrient gradient does have a clear impact on the composition of each population. Diatoms therefore offer a particularly acute indicator of urban influences on flowing water around Melbourne, and especially the addition of nutrients.

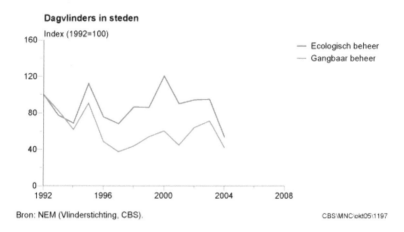

Fig. 143 Habitat quality for butterflies in Dutch urban areas
(cited by Zoest, 2006)

Butterflies

A illustration of species which distribute themselves effectively is provided by Collinge, Prudic *et al.* (2003), who studied the composition and spread of grassland butterflies in the region around Boulder, Colorado (USA). The researchers made an analysis of the species and numbers of individual butterflies to be found on 66 plots (representing four different types of grassland), and also sampled the *quality* of the flora (the number of indigenous plant species in relation to the number of exotic species). The diversity and composition of the butterfly population showed marked variation according to the type of grassland concerned. A significantly greater number of butterflies were counted in areas with long grasses than in those with shorter grasses. The quality of the habitat is also significant. Grassland with few indigenous plant species generally have fewer butterfly species than those with many indigenous plants. The topographical context – the degree of urbanisation in the surrounding landscape – had no relevant effect. The composition and diversity of the butterfly population is thus chiefly determined by the type of grassland and the quality of the habitat it offers, while the degree of fragmentation of the landscape (i.e. the degree of urbanisation) has little impact. The butterfly species in this study will benefit from the maintenance of grassland habitats of high quality.

Paved surface
The high density of people in urban areas places the soil and vegetation under pressure – quite literally. The intensity of human traffic, i.e. the regularity with which the ground is trodden or driven upon, increases the closer one comes to the central, busiest parts of the city. Here, there is a far greater proportion of paved surface, with greenery usually protected by fences. Public green areas are damaged by constant foot traffic (and occasionally motor traffic), while trees are damaged by bicycles parked against them, by errant cars and by mowing machines. In the more densely populated areas of the city, plants (and animals) which are particularly sensitive to such external influences are found only in exceptional circumstances such as on a totally fenced-off site.

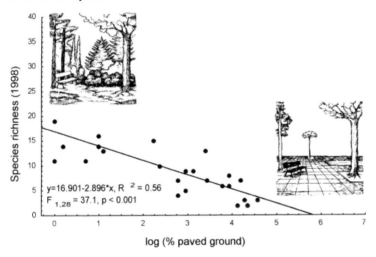

$y = 16.901 - 2.896 \cdot x$, $R^2 = 0.56$
$F_{1,28} = 37.1$, $p < 0.001$

log (% paved ground)

Fig. 144 Habitat quality for birds in terms of % paved surface in Madrid: 'Bird species richness in Madrid is negatively associated with the amount of paved ground within urban parks. This pattern typifies two kinds of urban parks usually sen in several cities. (cited by Zoest, 2006)

Animal predators
In addition to humans, predators such as dogs, cats, crows and seagulls contribute to the overall disturbance in rural areas. As Fernandez-Juricic records (Fernandez-Juricic and Telleria 2000), certain 'disturbance sensitive' species, including ground-nesting birds and mammals, are rarely to be seen in the busier parts of urban areas, where disturbance of one sort or another is so frequent that the animals in question have no opportunity to establish themselves.

Human disturbance

Disturbance by humans has precisely the same effect on animals as that by predators (Gill and Watkinson, 1996). Animals must devote more time to being watchful and alert. They keep their distance from people by fleeing when approached, and seek out the quieter parts of the neighbourhood. According to Fernandez-Juricic and Telleria (2000), they then have far less time to devote to foraging for food and other essential survival activities. The distance to which a human or predator can approach before the animal flees, and hence its sensitivity to disturbance, is a specific characteristic of each species and is termed the 'flight distance'. (Gutzwiller *et al.*, 1998; Blumstein *et al.*, 2003). There are thus species which are more sensitive to disturbance and those which are less so. The flight distance of animals in open landscapes is greater than those in more complex landscapes such as woodlands, while larger animals generally have a greater average flight distance than smaller animals. Within the species-specific sensitivity to disturbance, some variation is possible. Some animals may become accustomed to human activity nearby, whereupon their flight distance will decrease. This is demonstrated by a study conducted by Fernandez-Juricic, Jimenez *et al.* (2001) into the disturbance of birds in urban parks.

Adaptation to human presence

In studies in Madison (USA) and various cities in Finland, researchers such as Knight, Grout *et al.* (1987) and Jokimaki (1999) have determined that bird species can adapt. Species such as the magpie (*Pica pica*), fieldfare (*Turdus pilaris*) have become more accustomed to human presence because they are no longer hunted. This is the reason that several varieties of crow became far more prolific in the urban area during the twentieth century.

Edge specialists

Species which have become fully accustomed to human activity, such as the house sparrow and the urban pigeon, demonstrate a different pattern, with higher densities on the park edges. This is probably due to the better opportunities for foraging (edible waste and deliberate feeding) and the availability of suitable nesting locations in or on nearby buildings. Birds which live on the edges of parks benefit from the abundance of edible waste that is to be found there. Fernandez-Juricic (2001) concludes that urbanisation and urban extension promote the incidence of edge specialists, but that there is only a limited number of specialist species which prefer the central areas of habitats.

Snakes

Differences in sensitivity to disturbance are also to be found among snakes. Burger (2001) noted significant differences between two species: the northern water snake (*Nerodia sipedon*) and the common garter snake (*Thamnophis sirtalis*), both of which are to be found basking in grass alongside the canal paths of the Raritan canal, which runs through an urban green area in New Jersey (USA). Researchers walked along the canal until they sighted a basking snake, observed it from a distance for sixty seconds, and then walked towards it. Northern water snakes were quicker to react to the human approach than the garter snakes. The water snakes' flight distance was also found to increase in proportion to the traffic on the path. Garter snakes, however, showed no variation in flight distance. The conclusion is that it is easier to disturb a water snake than a garter snake. Of those garter snakes observed further than 150 cm from the path, only a few reacted to passing humans, while one in four water snakes did so even when lying between two and three metres from the path.

16.4.5   Layout and maintenance

Impact on nature
For most plants and animals, the quality (and hence suitability) of the habitat offered by urban greenery is dependent on its structure, layout and management. These factors determine whether the habitat requirements of certain species, such as the presence of established trees, water or long grass, are met. Structure and management also determine the heterogeneity of the green area, which will in turn influence its overall ecodiversity. The principles observed when laying out and maintaining a green area will therefore have a significant impact on its ecological value.

Landscape architecture
The principles applied in the 'conventional' management of parks, cemeteries and other urban green areas are derived from the traditions of landscape architecture, formal gardening and forestry. Both the layout and subsequent management are intended to beautify the urban environment, to create 'green oases' (in the case of parks) and to provide space for specific activities such as sport, play and leisure. These objectives also inform the choice of vegetation to be planted and its further maintenance. Alongside practical considerations, including the cost and robustness of the plants, the aesthetic quality of the greenery and the planned leisure amenities are of particular importance.

Ecological maintenance
During the 1970s, a new approach to the management of parks and greenery emerged in many cities of the economically prosperous countries. This new approach was based on ecological principles. In the Netherlands, this development had emerged somewhat earlier, having its roots in the 'instructive' public gardens created by Jac. P. Thijsse. Here too, the main objective is to create something of attractive appearance, but the aesthetic preference is for a more natural, less formally ordered layout, with colour provided by an abundance of flowering plants. Not only is the visitor offered a pleasant and inspiring place in which to stroll, but one which informs and educates, and which enhances the visitor's respect and attention for nature.

Nature-friendly maintenance

The 'nature-friendly' approach to the maintenance of urban greenery is geared towards enhancing its function as a habitat for certain species, and strives to increase overall biodiversity. In most cases, this entails efforts to increase the spatial heterogeneity of the area in question. Gavareski (1976) noted that the bird population in the parks of Seattle (USA) increased not only in proportion to the size of the park, but also according to the degree in which the original heterogeneous vegetation was allowed to remain in place. A large urban park containing the coniferous tree species typical of the northwest coast of America was found to contain a far greater number of bird species than those with cultivated and modified vegetation. The creation of public gardens in some parks had entailed the removal of the natural, indigenous vegetation and of tree stumps and dead branches. As a result, the bird species which prefer dense bushes or scrub are underrepresented in such locations, there being a much higher proportion of the typical urban species.

	Traditional management	Nature-friendly management
Intended landscaping appearance	Dependent on the design style (neo-Romantic to post-modern), but always cultivated and intended to provide an attractive and/or interesting visual appearance..	A natural appearance, with the emphasis on diversity and autonomy.
Suitable for:	Various types of outdoor leisure, sports and games, depending on the type of terrain and the amenities provided.	Simple enjoyment of the surroundings: walking, cycling, sitting or lying in the sun.
Organisation of management activities	Measures are applied to maintain a predetermined overall appearance. A rigid schedule is followed. Responsibility for maintenance activities is frequently outsourced.	The overall appearance is not predetermined to such a great extent (although an area intended specifically for the preservation of native plants is more planned than open meadows). The manager will only intervene if the appearance and/or the ecological quality deviates significantly from the general ambitions. He is therefore proactive, responding to the ongoing development of the area. Activities may be conducted by the manager himself, his staff, or by external specialists.

Fig. 145 Differences between traditional and nature-friendly parks management

There are several types of ecological management, but all reflect one of three main approaches to rural nature management:

The 'heemtuin'
'Heemtuin' is a Dutch term coined by Jac P. Thijsse for an artificial, usually enclosed area, in which nature managers seek to recreate and preserve the typical Dutch landscape of yesteryear. In many cases, native plant species will be accompanied by facilities to attract or support small fauna: beehives, nesting boxes (for birds and bats), brooding pens, artificial hedgerows constructed of dead branches, etc. Such areas reflect the traditional approach to nature protection, as espoused by Victor Westhoff and his contemporaries. It has its roots in a love for the pre-industrial, mainly agricultural Dutch landscape which since the late nineteenth century has been rapidly reduced to just a few remnants.

Parks with nature-friendly management
The 'nature-friendly' approach described above focuses on the heterogeneity and natural dynamic of the park. This may involve planting or removing trees, allowing dead trees to remain in place, establishing herds of cattle or other grazing animals, etc. The approach has its roots in the dynamic style of nature protection in which the ideal is free, self-regulating nature.

New nature
The 'new nature' approach centres around creating various combinations of nature and urban development. It has no precedent and is somewhat experimental in nature. Examples include the nature-friendly management of the hedgerows and ditches surrounding sports fields and allotment complexes, the creation of green banks along urban waterways, and the choice of bushes and trees which will encourage biodiversity. The approach is in keeping with the innovative ways in which nature managers have tried to combine nature with modern agriculture, traffic infrastructure, new water locations, and so on.

The passive function
Ecological management is particularly suitable for green areas which have primarily a passive function, i.e. they are there to be viewed and enjoyed rather than used for more intensive forms of active leisure. They include the quieter sections of parks, natural and farmyard gardens, verges, etc. Many other green areas, especially in the busy urban centres, are rather less suitable for natural vegetation. The crux of the ecological approach is that, within the restrictions of the urban environment, it seeks to maximise the number of trees and plants which can provide a habitat for small fauna (such as birds, bats and insects), or which represent a particular natural value in their own right (such as old indigenous species).

16.5   Conclusion

Urban ecology a separate discipline
Perhaps contrary to expectation, a review of the literature on urban ecology reveals a rich source of academic studies. Although our understanding of urban ecology is still fragmented and sometimes anecdotal, a coherent

picture is emerging of the effects of urbanisation, *via* landscape and environmental parameters, on communities and single species. It is also abundantly clear that the conditions in cities and towns constitute a set of landscapes and habitats not found in any other type environment on the planet. So I think there is a strong case to treat urban ecology as a separate discipline within the broader science of ecology (or landscape ecology if you will), just as tropical rainforests, coral reefs and salt marshes, for example, merit their own communities of research. This is especially so, since world wide most people live in urban environments and there is a growing societal need to promote nature in cities and towns. Urban ecology can help define realistic policy goals and offer the technical knowledge needed to realise these.

Many questions to answer within a frame of general ecological theory
However, our insights in the how's and why's of urban nature are in many respects general at best. With scarce data and general ecological theory to guide us, our ability to answer specific questions posed by policymakers is still limited indeed. Too often, we must resort to educated guesses. For instance, given the sharp decline of house sparrow populations, how important is predation by urban sparrow hawks (Accipiter nisus)? Do populations of native cavity nesting birds like great spotted woodpecker (Dendrocopus major) in urban parks experience competition by rose-ringed parakeets (Psittacula krameri)? What is the role of diseases and parasites in urban habitats, which for many species represent suboptimal environments? A recurrent question concerns the minimum (and maximum) widths for ecological corridors. Why do highly contaminated ruderal terrains, like closed refuse dumps, often contain an extraordinary species diversity? A less practical, but most intriguing question is how well and how fast species adapt to the urban environment. How do the selection processes involved operate?

Handbooks
The touchstone of any science is the ability to describe, explain and predict the phenomena it is interested in. Recognising that urban ecology has made considerable progress in recent years, but still has much road to cover, how should the discipline be advanced?

First, I think there is a need for handbooks and the like, publications which sort and integrate the scattered flow of research results published worldwide every month. Regular updates of the state of the art will help us choose promising courses of research. The publication of Leven in de Stad ('City Life'), by Martin Melchers and myself, is an attempt in this direction aimed at a Dutch audience. And as always in science, we should be especially aware of exceptions, observations that do not fit theory.

Acceptance by politicians and society
Secondly, in order to gain acceptance by politicians and society in general, we must be able to answer simple and specific questions like the ones

mentioned above. We therefore must direct resources to a top 5 research agenda. It is all about organisation and togetherness, ultimately. If researchers and practitioners join forces, write this research agenda and convince politicians of its relevance, money will come. In the Netherlands, the outlook has never been brighter. So what are we waiting for?

Landscape ecology

A last personal remark. To me, landscape ecology is the most comprehensive form of ecology, since it integrates all facets of ecology developed earlier in the 20th century and puts them into a spatial context. However, the 'landscape' in landscape ecology seems to indicate a focus on non-urban environments. Indeed, many landscape ecologists seem to show a marked preference for natural or rural landscapes as research objects, like river systems, woods and fresh water lakes. Although worthy causes, urbanisation makes it difficult and often irrelevant to draw a sharp line between the urban, the rural and the natural. Within metropolitan areas, all of these come together on the urban- rural gradient. So I propose urban ecology should be understood as the ecology of metropolitan areas, that is not just of built up landscapes, but also of the surrounding rural and natural landscapes. One cannot understand the ecology of periurban landscapes without knowledge of the ecology of built up areas and vice versa. Especially in a highly urbanised country as the Netherlands, urban – or metropolitan - ecology has a lot to contribute indeed!

17 FROM GREEN BELT TO GREEN STRUCTURE
green areas and urban development strategies in European cities

Sybrand Tjallingii **T**∪Delft

17.1   Introduction

Urban growth

[a]Urban sprawl is turning many landscapes upside down. From urban islands in a sea of green the situation changes into green fragments in a sea of buildings. In the European context, urban growth is typical in most cities and even the few shrinking cities in Central and Eastern Europe are expected to start growing again once the economic tide can be turned. In the period of industrial revolution in Europe, roughly between 1800 and 1950, urban growth created a rural exodus of poor people moving to the cities where the manufacturing industry provided work. In this period, urban growth took the form of growing central cities with slums and polluting industries.

Urban sprawl

In the last fifty years, however, the nature of urban growth drastically changed. Manufacturing industries left the city and sometimes even the European continent, to give way to a service economy based on unprecedented traffic flows. Urban growth turned to urban sprawl: a mass exodus of middle-class urban citizens from the central cities to the suburbs in the green countryside. The private car connects their houses to working places in the central city or elsewhere in the periphery.

Green belt strategy

Throughout the twentieth century there was a growing concern about the loss of green landscapes and green quality. Increasingly, urbanisation was perceived as a threat to green areas. Planners and politicians reacted by creating green belts around cities. Drawing greenbelt lines on the map, however, does not stop urbanisation. Moreover, green economy also threatens the qualities of the green countryside. Modern farming does no longer produce automatically the favourite landscapes of urban citizens. Modern 'footloose' farms produce for global markets and farmers move away from the spatial constraints of the urban fringe, selling their land for urban development prices. Globalisation has major impacts on the local economy and, therefore, on the local landscape. As a result, the green belt strategy, based on the idea of protecting green areas against the city, is increasingly criticised for being ineffective.

Green structure strategy

An emerging planning alternative is a green structure strategy, based on the idea of developing green spaces as an essential and structural element of the quality of the city. Although the two strategies may not be mutually

---

[a] This article was earlier published in: IFoU 2006 Beijing International Conference, Modernization and Regionalism – Re-inventing the Urban Identity

exclusive at a practical level, they are very different, both in their underlying basic attitudes and in their choice of policy instruments.

The two approaches play a major role in current debates about urbanisation both at the level of the general public, and the media and at the level of politicians and practitioners, struggling to reconcile urban development and green qualities.

COST CII

The urgency of the issues may explain the interest of practitioners to learn from practical experiences in different cities and discuss the underlying strategies. Between 2000 and 2005, researchers and practitioners from 15 European countries focussed on green structure and urban planning in a joint action of European cooperation in the field of scientific and technical research (COST). This COST CII action included field visits and case studies of 9 European cities and numerous detailed studies about ecological, economic, social and planning and design aspects of green areas in the context of urban development. Central to the COST actions is not one systematic comparative research but an exchange of experiences and views that enabled the participants to further explore the ins and outs of the relevant issues.

Leading questions

Drawing on the practical experience and the debates of the COST CII action (Werquin et al. 2005), this paper will discuss some of the leading questions[a]:

1.   What can be learned from experiences with green belt strategies?
2.   What are green structures and how do ecological, social and economic aspects of urban development relate to these structures?
3.   What does a green structure approach mean for design and planning?
4.   What are the conclusions for the future role of green areas in urban planning?

17.2   Two approaches

17.2.1   Green belt

Greater London Plan of 1944

Green belts represent the most restricted form of urban containment policy. Perhaps the best-known green belt is part of the Greater London Plan of

---

[a] This paper is based on my work as a general rapporteur of the COST CII action. A full account of the action's findings is published in Werquin et al. 2005. I am very grateful for the many inspiring field visits and discussions with colleagues in the COST action that have contributed substantially to my understandings and views. It should be stressed that the interpretations and conclusions in this paper are the responsibility of the author.

1944 made by Patrick Abercrombie (*Fig. 146*). Leisure was to be the dominant use in the 'Green Belt Ring' and agriculture in the 'Outer Country Ring' (Turner, 1992). Later, also Berlin, Vienna, Barcelona, Budapest, but also Asian cities like Tokyo, Seoul and Bangkok established greenbelts (Kuhn, 2003; Bengston and Youn 2006). The success of Green belt policies is debatable. In the case of London (Hall, 1996; Turner, 1998), government control did slow down the pace of development in the green belt. However, urban growth jumped over the belt and scattered across much of southeast England.

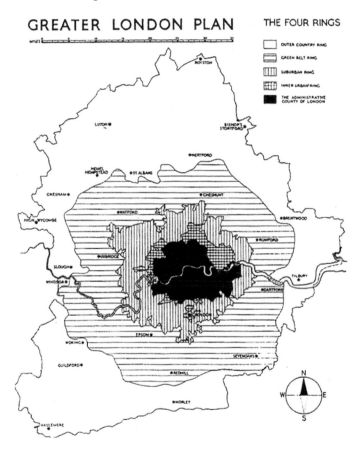

Fig. 146 The Green Belt Ring in the 1944 Greater London Plan (Turner, 1992)

Limited control

In most countries regional and national governments have only limited power to control local preferences for development. A good example is the Dutch policy to keep the Green Heart green (Aykac et al,.2005). This 'green belt' between the cities of Amsterdam, Rotterdam, The Hague and Utrecht is an area with urban growth boundaries clearly set in national and regional policy documents. Yet, building development cannot be stopped. This is partly caused by high suburban pressure from the surrounding cities, but even more by local pressure for developments of the villages inside the Green Heart. Protection is successful in the case of smaller areas with a clear ownership and management situation such as nature conservation areas. But it is impossible to keep larger green belts green if there is a situation with many owners and many different forms of land-use without a clear and promising economic and social perspective for all stakeholders.

The COST Debate on practical and fundamental questions
Green belt strategies are still popular in many cities, sometimes as official policy, sometimes as a strategy for non-governmental organisations criticising official policies for urban development[a]. The strength of the greenbelt idea is its clear and simple message. Its weakness is that in urban practice a full circle green girdle cannot simply contain urban development. The researchers and practitioners participating in the COST action all struggled with the practical and fundamental questions in this debate. Moreover, they were very concerned about the quality of green spaces in the existing city, a field of issues closely related but not addressed by green belt strategies. The emerging concept of green structure offered a better starting point for their discussions.

17.2.2 Green structure

The concept
Although the concept of green structure is rooted in a long history, the term as such dates from the 1980s. In the beginning, as happens with relatively new terms, perhaps even more so if intended to create a common language, there was a lot of confusion about the many possible meanings of green structure. Gradually, however, the COST C11 participants grew towards a shared understanding of green structure as a concept that embodies both a view on the present urban landscape and on the desired future.

Spatial network of green spaces
Green structure links town and country. In a spatial perspective, green structure is more than the sum of green spaces. Speaking of green structure implies drawing attention to the spatial network that links open spaces, public and private gardens, public parks, sports fields, allotment gardens and leisure grounds within the city to the networks of woodlands and river floodplains in the surrounding countryside. Thus green structure

---

[a] 2 European examples are discussed by Kühn (2003). Outside Europe, Seoul and other South Korean cities demonstrate a strong governmental control of greenbelts (Bengston and Youn, 2006). The Greenbelt Alliance in the San Francisco Bay area is a good example of an NGO supporting a green belt strategy (www.greenbelt.org).

highlights the role of greenways for walkers and cyclists and stresses the importance of ecological corridors for wildlife.

Green structure links the past to the future. In a time perspective, green structure expresses a long history and a long term planning policy to make the spatial structure of green spaces a basis for sustainable urban development.

Identity

Sometimes, as in the British Midlands and the German Ruhr Area, green structure is brought back to chaotically developed old industrial areas in order to restore the disrupted system of green valleys. As a result natural and cultural heritage become visible in a new green identity of the urban landscape.

Green structure links different stakeholders. In a decision-making perspective, green structure refers to a green infrastructure that is planned and maintained as a carrier of multifunctional urban development and cannot be claimed by one group of green stakeholders. Green structure is colourful. Green zones along rivers, for example, may perform a multitude of roles, such as: routes for walkers and cyclists, floodplains for water management, ecological corridors for wildlife and attractive edges for residential and commercial development. This implies both multiple funding and multiple use.

17.3   Green structure in perspective

17.3.1   Green structures express identity

Diversity of examples

Typically, cities have their own characteristic green structure. Their green identity is part of the city's history. A few highlights of the field visits of the COST participants to nine European cities may illustrate this point (see *Fig. 147 - Fig. 152*, Green structure, local ecology and variety).

Oslo

Once the power supply of the early industrial city, today the rivers and valleys are the green structure of Oslo, Norway, linking the city's green spaces to the mountains and to the seashore (Fig 2a *Fig. 147*). The river Akerselva is one of these small rivers that brings together many different groups of citizens, organisations, developers and municipal departments in a joint programme for river restoration, urban renewal of industrial heritage and green space development. Oslo shares this change of an historic hydropower river landscape with Sheffield and many European cities that once were the motors of industrial revolution.

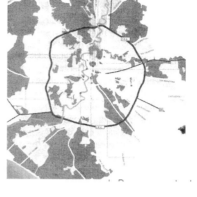

Fig. 147 Oslo, green structure
(Halleux, 2005)

Fig. 148 Rome, green structure
(Halleux, 2005))

Rome

The green structure of Rome, Italy, is closely related to the river Tiber and its tributaries. Some of the valleys, such as the Cafarella valley along the ancient Via Appia, in the southeast, carry a green wedge that leads to the heart of the city. The map in *Fig. 148* shows the big ring road of Rome. The importance of the green structure is symbolised by the construction of a tunnel that allows the ring road to pass without disturbing the quiet green area.

Fig. 149 Warsaw, green structure
(Halleux, 2005))

Fig. 150 Marseilles, green structure
(Halleux, 2005))

Warsaw
The river Vistula plays a key role in the green structure of Warsaw, Poland (*Fig. 149*). Like other cities, including Rome, Warsaw has also a chain of green spaces along the high escarpment of the hills above the river. The parks and gardens on the western escarpment were among the first areas to be reconstructed after the almost complete destruction of the city in the Second World War.

Marseilles
In Mediterranean cities not only mountains and valleys but also water supply aquaducts and canals structure the network of green areas. *Fig. 150* shows the example of Marseilles, France. The small map shows the canal de Marseilles. This canal takes water from the river Durance in the Alps. Its construction not only brought drinking water to the city, but also created conditions for a new green structure of farmland and a range of wealthy houses with gardens (the bastides) built in the previously dry hills now watered by the canal. The bigger map shows how the bastides surround the old city. This illustrates how also private gardens may become part of the green structure. Though most of them are still inaccessible for the general public, they play a role in enhancing biodiversity, in moderating the urban climate and in water management.

17.3.2   An urban ecology perspective

Biodiversity in cities
Biodiversity, climate and water management are important ecological issues for planning and design of green structures (Werquin et al. 2005: 133-206). Until recently biologists did not seem to be very interested in the urban environment as a habitat for wildlife. In recent decades, however, ecological research revealed that cities could be surprisingly rich in species. In many cities, especially in intensively farmed parts of Europe but even in Helsinki, Finland, a city in the forest, the urban area has a higher biodiversity than the surrounding countryside. A green structure approach emphasises networks of ecological corridors that enable plants and animals to move between different core habitat areas and this provides a better chance of survival for vulnerable populations. Oslo and Helsinki and many other cities seek to create better conditions for wildlife restoration and protection through special design and maintenance programmes. An increasing number of cities, such as Munich, Germany and Utrecht, the Netherlands, also create new habitats like wetlands and woodlands and new linkages between these habitats.

Urban comfort
In many ways green structure influences the urban climate and thermal comfort in streets. There is more than shadow and sun. In continental climates, the heat island effect has been closely studied. In summer, higher temperatures and dry and polluted air constrain air quality in down town

areas. In many central European cities, therefore, urban planning promotes green fingers that allow moist and cool air-flow to enter into the central part of these cities. The city of Warsaw is a good example.

Munich
Fig. 151 illustrates how the design of a new development in Munich, Messestadt Riem, creates a large park as a climate corridor to direct the cool and fresh air into the residential development (the small arrows) and into the central city (the bigger arrows). The detailed design of the green corridors is important. In some cases trees seem to slow down air circulation, thus enabling air pollution to accumulate. In windy climates like the coasts of Denmark and the Netherlands, the heat-island effect does not play an important role. Here, green hedges are important as wind breaks in both rural and urban landscapes.

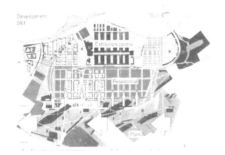

Fig. 151 Munich, new development Riem (Halleux, 2005))   *Fig. 152 Utrecht, new development Leidsche Rijn (Halleux, 2005))*

Water
It is not just that water shapes urban green structure in many ways: green structure may also play an important role in improving water management. Urban growth goes with a dramatic increase of hard surfaces. Rainwater no longer infiltrates into the soil and this causes groundwater tables to fall. Instead the rainwater runs into sewers that are unable to cope with increased peak discharges from paved surfaces. Thus, the sewers have overflows that seriously pollute surface waters. Cities like Utrecht and Munich therefore are disconnecting rainwater from the sewage system, using the urban green structure network simultaneously as a network for infiltration, retention and drainage of rainwater. *Fig. 152* illustrates the design principle for a new development in Utrecht.

17.3.3   A social perspective

Trends of use and perception
How do people use and perceive green spaces and what does it mean for designing and planning the urban landscape? Drawing on research findings

from different countries, the COST action identified some major trends (Werquin, et al. 2005: 217-317). At the level of urban regions sprawl is the trend. Higher and medium income families tend to leave the densely built inner city areas and move to bigger houses with gardens in the urban periphery or in the countryside. A majority of the urban population has more to spend and has a car. Most people have more time for leisure too. As a result car trips to green areas have increased drastically, much more than trips from home to work. These trends may lead to a social segregation of people with more and with less access to green spaces. Many cities, therefore, revitalise public parks and leisure areas in the densely built parts of the city and create greenways for pedestrians and cyclists that connect these parks to other green areas in the urban fringe.

Standards of green space per inhabitant
Since the 1920s, researchers and planners have sought to determine the minimum green surface required for health and leisure in cities (Turner, 1992:366). Although it is difficult to substantiate these figures through research, most cities have adopted minimum green space standards for urban development, varying from 20 - 75 m2 per person. The standards have made it impossible to ignore green spaces in urban planning practice but take no account of quality differences of green spaces and of people (see also *Fig. 135*).

Impact on health and well-being
At an academic level, environmental psychologists and health researchers have found substantial evidence for the positive role of green areas to human health and wellbeing. These studies create a base for accepting that green is healthy, but there is still a gap between this general understanding and knowing how green areas can be designed to make cities more healthy.

Planning and design
At the detailed level of planning and design, however, know-how is developing significantly through a process of learning-by-doing. Some hospitals, for example, try to align garden design with therapy and some schools seek to design school gardens with better conditions for variety in play and activities of children. In search of relevant green space qualities at the district planning level, the city of Stockholm, Sweden used socio-cultural quality concepts that are the basis of so-called sociotopes, outdoor places with special qualities for certain groups of urban citizens.
Questionnaires and interviews with residents, combined with expert information, have led to a map of sociotopes that can act as an interface between public and planners. Here, the focus is on the interactive planning process rather than on the general plan.

Interactive frame
In the absence of general laws with clear cause-and-effect relationships a general plan with goals-and-means is not appropriate for many issues. In

this context, the concept of green structure is useful as a frame that is linked to the identity and ecology of the city and, at the same time, creates conditions for choice and for interactive processes that can make the city a healthier and more agreeable place to live.

17.3.4   An economic perspective

The value of green areas
In an economic perspective many planners and economists assume that the weak position of green spaces in decisions about urban land-use follows from the fact that green areas are not valued in monetary terms (Halleux, 2005).

Contingent valuation
Thus, different techniques have been developed to value green areas to put them on an equal footing with buildings in cost-benefit-analyses. One of these techniques is contingent valuation based on questionnaires that ask people about their willingness to pay for green areas. Critics of the approach argue that the poor are unable to pay and that valuation leads to yes or no decisions about green spaces instead of contributing to an integrated design. Moreover, the long-term conditional value of green spaces for urban development is likely to be underestimated. Although there is much research on the valuation techniques, very little is known about its actual use in real planning situations.

Land price
Other valuations dominate real world developments. Land buying developers tend to jump away from the expensive central areas to the countryside where prices are low. This so-called leap-frogging induces diffuse sprawl and compromises planning efforts aimed at good conditions for services, infrastructure and green spaces. The globalising food production weakens the position of farmers and this only further accelerates these problematic processes. Some cities have adopted land policies to prevent speculation or tax policies to balance private profits and public costs.

Hedonic price theory
Some economists think hedonic price theory may help us to solve these problems. This theory seeks to explain why and how much house buyers are prepared to pay for the attractive environment of a dwelling. Finnish research demonstrated a 5 % higher price for houses with a view onto the forest. In Dutch research, an increase in house price up to 28% was found for houses with a garden facing water, especially if this water was connected with a sizeable lake.
These findings point at the increased value of the edges of green areas. Higher prices will only be paid if, beyond the edge, green stays green. In this perspective, creating long edges, as green structure planning does, is a promising approach.

17.4   Design and planning green structures and green belts

17.4.1   Design concepts

Munich
One of the roots of green structure, as a design concept, is in Munich, where the river Isar is the carrying structure of green spaces. In the late 18th century a three kilometre long part of the Isar floodplain was used for a beautiful park, the Englische Garten. English landscape architects inspired Ludwig von Sckell for his design of the park. For his part, Frederick Law Olmsted went to Munich to get inspired by the Englische Garten for his designs of parks and green girdles, like those in New York and Boston that may be considered as the precursors of the idea of green structure.

Park movement
The idea is also rooted in the history of the public park that developed in the 18th and 19th century in Germany, England and France as many people felt an increasing need to escape from the industrialising dirty and unhealthy cities. Parks had to be breathing places for the metropolis. At the same time, there was a growing awareness that the new parks had to be accessible for every citizen and this led to plans for an interconnected network of green walks and parks cutting through districts for the poor and for the rich. Haussmann and his chief engineer Alphand pioneered with the idea in Paris, but it was Olmsted who brought the approach to maturity with his plans for park systems and parkways in American cities.

Connectors and separators
Abercrombie elaborated the idea in the Greater London Plan of 1944. His ideal was a closely linked park system: "It becomes possible for the town dweller to get from doorstep through an easy flow of open space from garden to park, from park to parkway, from parkway to green wedge and from green wedge to Green Belt." (quoted in Turner, 1992: 368). Obviously, Abercrombie saw no contradiction between green belt and green structure. To him the Green Belt was part of the regional green structure! In the Greater London Plan, the Green Belt was also part of an urbanisation strategy with the New Towns in a central role. Yet, whereas Abercrombie's linked park system was conceived as a network of connections, the Green Belt was primarily aimed at containing urban growth. In a comparison of the Berlin green belt and the Randstad Holland Green Heart, Kühn (2003) points at the double role of open spaces: they act as connectors and as separators. Green belts and the Green Heart are both primarily conceived as separators. Their spatial form is negatively derived from the urban areas they are separating. Green structure strategies, however, are based on connecting processes, of rivers and greenways for example. Their spatial form is often radial. Moreover, as green structure strategies focus on the

quality of green edges, they positively seek to make the contrast of built and green areas an essential part of urban design.

Parkways

There is another important difference between Abercrombie's park system and the green structure approach. Parkways, combining motorway and greenway, play a major role in his view, as in the views of Olmsted and his American followers. We have to realise that at the time of Olmsted and Abercrombie, only a few horseless vehicles made pleasure rides along these parkways. Today, the dynamics of motorised traffic are incompatible with quiet greenways for walkers and cyclists. Parkways can still be elements of an urban design but rather as parts of the road network, not as valuable elements in a green structure. In contemporary urban regions, road infrastructure is the natural ally of dynamic commercial and industrial development, theme parks and big events.

Two networks

Likewise the water network is the natural ally of greenways and green corridors. The two worlds can be characterised as the fast lane and slow lane zones of activities that, together, carry the contrast of urban quality. The two networks of traffic and water, therefore, can act as the carrying infrastructure that creates conditions for both the green and the built structures. Residential areas preferably lie between the two worlds, having good access to both. This two networks approach creates conditions for a true, sustainable integration of green structures in urban planning (Tjallingii, 2005).

17.4.2   Planning strategies

Fixed frame - flexible infill

A network approach of green structures and built structures supported by infrastructure networks of water and traffic systems is different from a masterplan approach for the full surface. Green belt experiences demonstrate that full control of regional development is impossible, even in countries with a top-down institutional planning tradition. Green structure policies, based on the principle of fixed frame - flexible infill seem to be more appropriate in a situation of uncertainty and complexity of regional housing markets and global markets for agricultural and industrial products.

Project based planning

Green belt planning is part of a tradition of protection and permit planning in which the national or regional government sets the limiting conditions and then leans back to wait for local private or public initiatives. In many countries, there is an ongoing debate about the merits of this planning tradition. The importance of formal procedures that include Environmental Impact Assessment and public consultation rounds is clear. But increasingly, governments realise they have to play a more active role as initiators in a design and development planning approach.

This asks for a project based planning process in which green structure planning can be a good starting point (Tjallingii, 2003).

Regional combinations of government sector budgets

At the regional level, the project is to improve a durable frame of traffic and water infrastructure. In green structure planning, the potential synergism between water and green generates promising projects as in river valleys, where many European cities combine flood control schemes with nature protection and leisure projects. Combining government sector budgets in one project is attractive, but the walls between these sectors are often high and the negotiating process may be tough. Non-governmental organisations may be stakeholders in these processes if they can take over management and maintenance of green areas. In some cases the National Lotteries provide a substantial part of their budget. Gambling for sustainability!

Local involvement of private investors

At the local project level, governments seek to involve private investors in the management and maintenance of public green areas. The Oslo Akerselva programme, mentioned above, is one example. In Rome, the punta verde initiative is a promising building-and-investing-in-green-quality approach that involves private investors through long-term contracts with the local government. The local government itself stays responsible for the quality of the main green structure in the agglomeration.

Red-pays-for-green

The discussion of economic aspects of green structure led to a special focus on edge situations. Will it be possible to use part of the extra money house buyers are prepared to pay for the funding of green quality? Several European countries are testing this question in so called red-pays-for-green schemes. The first experiences demonstrate that the option is real. Sometimes there are legal constraints but more important is a well-structured planning process in which the government informs private investors in good time about the rules of the game. As a result, less privileged citizens, who cannot afford to buy houses on the edge, may also benefit from the proximity of green spaces.

17.5   Conclusion

Summing up the observations, interpretations and arguments of this paper leads to the following conclusions.

(1) Urbanisation cannot be stopped.

Green belt strategies and other partial and protective policies, focussing only on containment of urban growth, are doomed to fail. They represent a negative approach to urban development.

Protection of vulnerable green areas remains crucial, but the city is not the enemy. The survival of green areas depends on sustainable support and use by urban citizens. Green qualities should be made part of mature urban development.

(2) Green structure strategies offer a promising alternative.
They represent a positive approach to urban development. They can give identity to the city based on the landscape in which a settlement originated, and its subsequent cultural history. Green structures create a basis for managing ecological, social and economic issues. This spatial frame of open spaces is an attractive element in urban design and creates conditions for choice and for interactive planning processes.

(3) Two networks framework integrates green structures.
A full integration of green structures and built structures requires an adequate infrastructure framework that creates conditions for a zoning of slow lane and fast lane activities. Road networks are good carrying structures for dynamic commercial and industrial activities. Water networks may perform the same role for green area activities.
Residential development typically should be positioned between these two networks.
Motorways cannot be a part of greenways. Good plans for crossings offer a green and blue view to motorists without disturbing pedestrians on the greenway.

(4) The potential synergism with water management can be utilised in projects.
The durable two networks frame should support a project based green area development. This implies a priority for fully utilising the potential synergism between green areas and water management. Water reservoirs, flood control and rainwater storage at different scales offer a variety of opportunities to combine with green structure, from green roofs to national parks in the urban fringe. At the regional level it is crucial to further explore the multifunctional role of farmers as carriers of green structure.

(5) The potential of green edges supports a valuable green structure
Likewise the two networks frame should support a project based built area development. The priority here, is to avoid use of the urban green structure for residential and commercial development. The design and maintenance of public green areas beyond the edge should, at least partly, be made part of the development budget in red-pays-for-green constructions.

Learning-by-doing
These principles are derived from learning-by-doing processes in European cities. New experiences in other contexts may further contribute to this learning process. There are no general answers. The principles are general conceptual tools that may guide designers and planners to find local solutions with local people and local conditions.

18   LANDSCAPE ARCHITECTURE AND URBAN ECOLOGY

The use of Urban-ecological knowledge in landscape architecture

Klaas Jan Wardenaar, Peter Veen VISTA

18.1   Introduction and problem statement

Life in plans

In a publication on the ecology of urban areas a review on the role of landscape architecture cannot be missed. Landscape architects are very much involved in shaping the habitat of living species. The city is no doubt a habitat and in Holland this habitat is becoming more and more important. Most Dutch offices in landscape architecture make use of ecological knowledge in some way or another. The office of Vista landscape and urban design pays special attention to ecology in their designs. Vista was established in 1994 by landscape architect Rik de Visser and ecologist Sjef Jansen en became quickly known for its innovative approach of nature development (Nolan, 1999). Typical for the Vista approach is that nature is not treated a as static entity with a fixed set of species and spatial characteristics, but as a dynamic system with ever changing appearances and often unexpected interactions. Ecological processes are used as an inspiration and as a driving force to create challenging new landscapes, which redefine the relation between program and space. Especially in urban environments, where different programs come together this approach offers interesting solutions, some of which will be shown in this chapter.

This chapter will focus on the question how ecological knowledge can be used in a creative way in landscape design. Basic proposition is that nature is not just one of many functions which has to be accommodated, but a general dimension of environmental quality and spatial development. In 18.2 we will give a short introduction on the historical development of park design in Holland and on recent trends in landscape planning for urban environments. In 18.3 we will elaborate on the underlying goals of urban-ecological planning. A case is made for adding a complementary goal to the classical biodiversity goals, namely stimulating environmental awareness among city inhabitants.

18.2   Landscape architecture in urban areas in Holland

The field of work of landscape architects in urban areas has changed considerably over the years, both in content and in range. Traditionally parks were designed by horticulturists or architects, only later to be taken over by landscape architects. Landscape architects are now concerned with the quality of public space, urban parks, squares and streets, but also with organising and designing leisure landscapes and natural spaces as a context for urban areas and urban development. Many landscape architects are working closely together with urban designers and architects on comprehensive city planning and even laying the basis conditions for urban development. A brief outline of this development will be sketched, as a background for our proposition on possible green goals in urban planning.

18.2.1   Brief history of urban park design

Parks reflect to a great extent the social situation of a certain era. The historical development of parks is a storybook for the different ways nature is used and perceived. A few remarkable stages in urban park design may illustrate this.

Until 1920

In the late 19th, early 20th century parks were seen as the 'lungs of the city'. They were created to improve health of the working class. Trees were supposed to have a positive effect on the rapidly degrading air quality. Parks were designed as romantic oases, introvert and isolated from the 'dirty' city by a green belt. Monumental fences and entrance porches marked the separation from the urban surroundings. Every city has one or a few examples of these parks, in Amsterdam for instance Vondelpark, Sarphatipark, Oosterpark and Westerpark.

Until 1970

From the 1930's on to '60's and '70's parks were basically meant to provide space for Homo Ludens, the 'playing man' (Huizinga,J., 1974). These so-called 'Volksparken' are made up of a network of landscaped paths, which opens up and connects the intermediate spaces with different functions, like sports fields, ponds for playing, private gardens etc. Examples are the Amsterdamse Bos and the Zuiderpark in Rotterdam. In many of these parks the functions have become more and more isolated from public use. Gardens were closed off with water and barbwire, sports fields got fenced. This degraded the parks to a meagre structure of green paths. Recently plans are made to re-integrate private functions and public use, for instance by creating paths that make garden areas accessible.

Until 2000

From the '70's on through the '90's many new parks focused on nature, creating a contrast to the supposedly bad- urban environment. The ecological parks of Le Roy in the seventies of the last century are striking examples. They breathe the atmosphere of ruïns, as if culture is taken over by nature (for example Le Roy et al, 2002). As a provocative comment of today's nature control and wasteful use of resources these parks contain a strong cultural layer. In many other park projects in this period, the ecological program is translated in willfully natural forms, reminiscent of the English Landscape garden, though lacking its visual refinement. These parks seem to ignore the cultural dimension of park design, while on the other hand high maintenance is required to preserve the constructed natural image. Natural dynamics are in fact very much restricted.

From the 1920th on

From the 1990's on a reaction is seen to these natural forms and a more architectural approach to park design emerges. Parks in this period are often designed as green squares, focusing on culture and not, or only partly on nature. Examples are the Museumplein and the Cultuurpark Westergasfabriek in Amsterdam (Koekebakker, 2005; Bouwmeester, 2003) Sometimes the ecological program is aptly integrated in the architectural scheme, forming surprising combinations of nature and culture. Sometimes natural dynamics are used as a way to create new landscapes. This might be recognised as the start of a different attitude towards nature.

Fig. 153 Von-   Fig. 154 Playing in Fig. 155 'Ruïns' in   Fig. 156 Green
delpark entrance the Amsterdamse Mildam by Le Roy square in Culture-
gate   bos   (Nai bookcover)   park Westergas-
(Vanupied.com)   (Vista)   fabriek (Projectbu-
                                                         reau Westerpark)

18.2.2   Recent trends in landscape planning

Along with the changes in content and program of urban green areas, the professional scope of landscape architecture in urban environments has broadened. Apart from traditional architectural and gardening expertise landscape architecture has included more and more adjacent fields of knowledge.

Nature as a subject for planning
The introduction of the National Ecological Network (NEN) as a national policy, following the unexpected and spontaneous outburst of nature in the Oostvaardersplassen in the Flevopolders, made nature a subject for large scale landscape planning. Landscape architects became involved in the implementation of this ecological program, aiming at enlarging and connecting the main nature reserves of the Netherlands. A number of offices saw the cultural challenge of designing the NEN as a new layer in the landscape, expressing today's view on nature. They tried to integrate ecological program and architectural design (for example Feddes et al, 2002). Besides, the emergence of nature development loosened up the generally defensive strategies of nature organisations and stimulated thinking on integration of housing and nature or leisure and nature.

Typical for this development is the study 'Uit de klei getrokken' by Vista. This is a design study for a location in the Haarlemmermeerpolder on how water management could be used to trigger vegetation succession and to create a new landscape with high natural and leisure qualities (Vista, 1996). It was a reaction to the large scale poplar plantings of public green projects like the Randstadgroenstructuur and to the species and pattern oriented ecological planning of the time. Instead 'Uit de klei getrokken' focuses on the natural processes, using water level and grazing as external 'buttons' to steer with.

Fig. 157 a-f. Uit de klei getrokken. Example Fen strategy; possible development in 0, 2, 5, 10, 30 and 80 years after retaining rainwater on the basic slightly undulating clay soil
(Vista, 1996)

Fig. 158 Possible map of the future Haarlemmermeer using different strategies parcelwise
(Vista, 1996)

Not a fixed 'climax' situation in the far future is the goal, but the natural transformations over time. The often unpredictable interplay of ecological factors makes the green area attractive from day one. Subtle differences in the physical basis, underlying even the seemingly uniform landscape of the Haarlemmermeerpolder, are thus expressed in a natural way, with a minimum of artificial design.

Another example of the Vista approach towards ecological planning is the landscape and nature development plan Zanderij Crailoo in Hilversum (1998), situated in a sandy area just east of Amsterdam. This is a plan for nature development on sandy substrates. Transformation of a former sand quarry, later used and fertilised for agricultural use, offered the possibility to make a plan that incorporates a large variety of (ground)water tables and trophic levels of the upper soil. Together with different levels of maintenance a matrix of many different sandy ecotopes could be created, which in time will evolve to different vegetations and fauna communities. The matrix was translated into a largely orthogonal plan, based on the artificial nature of the quarry, with banks and islands all differing in humidity, fertility and maintenance. Along with nature development, erosion of banks and islands will soften the sharp edges of the plan, resulting in a cultural nature garden. A long winding boardwalk allows people to experience this wide diversity of situations, developments and resulting nature. It connects to the 800 m long Crailo Nature Bridge, also designed by Vista, connecting wood and heath area's in the highly urbanised Gooi region, thus creating an ecologically and recreationally rich and continuous landscape (see also chapter 25.3, page 516).

Fig. 159 Nature planning in Zanderij Crailoo (design Vista, photo J.Vlaanderen GNR) *Fig. 160 Scheme of ecotopes created in the Zanderij Crailoo Plan* (Vista, 1998) *Fig. 161 The area after two years development* (Vista, 1998)

Today urban ecology has gained a firm position in many design offices and planning practices. Integration of landscape architecture and ecology seems to have succeeded and results in strengthening both. Mostly however integration of ecology does not go much further than implementing reed beds into a plan. Frequent debate between the fields of landscape architecture and ecology remains necessary to get results that go a little further than the obvious. By doing this, ecological planning can be a cultural intervention, adding a new time layer to the existing landscape.

The city as a landscape
Next to a development of landscape architecture towards ecology, a development towards urban design is visible. On a local scale designs are made for landscape based town extensions or 'green' cities. For this the 'genius loci', the unique characteristics of a place connected with landscape, nature or history, is used to create urban environments with strong identities. This usually goes along with sustainable planning and designing based on the physical situation, in relief, soils, watersystem, ecological infrastructure etc.

Layer appoach
On a regional scale landscape planning and urban planning also become more integrated. Planning on this scale often uses the so-called 'layer approach' to get sustainable solutions. It is based on differences in stability or 'turn over time' of different elements of a certain location. It means that the basic and principally stable physical situation structures the use and development of a region. This so-called first layer consists of relief, soils, water system and ecosystem. It structures further working on infrastructure (second layer) and planning of new functions like housing, working, agriculture, nature etc. (third layer).

Housing in the Bloemendalerpolder
A good example of the integrated approach of urban and landscape development, as advocated by Vista, is the study for a housing project for the Bloemendalerpolder east of Amsterdam. The Bloemendalerpolder is an open peat polder next to the river Vecht and the historical towns of Weesp and Muiden. It borders to the Green Heart of Holland. In this area 2.500 houses have to be build. The proposal of Vista takes the regional identities and the underlying lanscape as a starting point. First there is the river Vecht. It's clay deposits offer firm ground for high-density housing. Next to the river a new 'rivertown' is proposed centred around a new leisure harbour. Without necessarily copying historical forms, the existing and highly estimated rivertowns along the Vecht are an apt inspiration for the urban lay out and atmosphere. Away from the town centre new estates are proposed, another typical housing model of the Vecht area. Secondly there are the wet peatlands in the centre of the polder. These lower soils are less suitable for building and well equipped for water storage. Here low density housing is proposed on land strips and islands in a new wetland, inspired by the nearby nature reserve of the Naardermeer. The new wetland serves as an attractive context for the new houses and offers the recquired water storage. Furthermore is it a substantial stepping stone in the ecological corridor between the marshlands of the Vechtplassen and the IJmeer, the so-called 'Natte As'.

 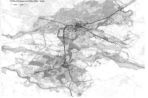 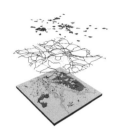

Fig. 162 Urban planning in the Bloemendalerpol- der east of Amsterdam
(design and image Vista)

Fig. 163 Regional plan- ning in 'Knooppunt Arn- hem Nijmegen'
(design and image Vista)

Fig. 164 The layer- approach behind 'Knooppunt Arnhem Ni- jmegen, 1:basis, 2:networks, 3:functions
(Vista, 2005).

Regional Plan for Arnhem-Nijmegen

For the urban network of Arnhem-Nijmegen in the east of Holland Vista made a comprehensive regional plan, which addresses the changing rela- tions between the city and the rural landscape. In this region urban sprawl, infrastructure and green houses have led to fragmentation of the landscape and loss of ecological quality. While the cities have grown, the shortage of leisure facilities and nature reserves has become more apparent. Mean- while increasing river flows and rainwater surpluses pose a serious threat to safety. To counterbalance the urban development the concept of a 're- gional landscape park' is introduced. This regional park will not only give way to leisure, nature and water storage, but also offer a new strong iden- tity to the fragmented landscape. Ecological relations are restored and cul- tural history is made visible. It also steers the urban development. River- fronts are developed to reconnect the cities of Arnhem and Nijmegen to the Rhine delta. For the smaller villages a typology is designed based on their position in the landscape and historical characteristics. This typology offers guidelines for the further development of the villages in a way that the qual- ity of the landscape is enhanced. What is typical for the regional plan for Arnhem-Nijmegen is that urban and rural processes are not treated as separate fields but are combined to give direction and consistency to the ongoing transformation of the region.

Urban identities in a regional context

The landscape architect nowadays not only designs green public space as he always did, but also works as a planner of regional developments, as an inventor of new urban identities and as a creator of new ecology: 'centi- pedes' ready to answer a wide range of questions. It all starts with asking the right questions. In other words, returning to urban ecology: what green goals should we go for in the urban area?

18.3 Green goals in urban area's

Increasing biodiversity is the classical goal for ecological planning, both in rural and urban areas. Much effort is being invested to house rare and demanding species in urban environments. We propose to shift part of the attention towards a complementary and possibly more suitable goal: stimulating environmental awareness among citizens.

18.3.1 Biodiversity

Increasing biodiversity is often considered the most important ecological goal in urban green planning. Oddly it usually aims at creating habitat for species that are typical for rural or natural environments, less so at creating or improving habitat for species of urban environments.

Improving habitat for urban species

In the former chapters of this publication it is stated that the urban environment behaves like a biogeographical district. It has very specific parameters like spatial structure, spatial succession, soil, trophic level, temperature, etc. and houses some very specific species, which usually have their origin in southern, stony environments. In this respect the urban environment should or could be treated as any other biogeographical district, both on a local and on a regional scale.

On a local scale habitat quality is determined by a large number of factors. In short the habitat quality of urban environments is made up by existing urban structures (providing substrate, warmth, shelter, diversity, energy), urban growth (temperate substrate, space, size, import) and urban erosion (wearing of materials, low maintenance and demolition). To improve biodiversity attempts have been undertaken to influence these aspects of habitat quality. Good results have been accomplished with improving habitat for ferns by using cement rich in lime in building embankment walls (ref: several publications of the Museum of Natural History Rotterdam). Also creating hollow traffic noise barriers to provide sleeping facilities for bats, a rare habitat in modern cities, proves to be very effective (Jansen et al., 1997; Klerken et al., 1997; Katoele, 1987) For these kinds of improvements, specific knowledge of habitat requirements and ecological amplitude (see *Fig. 213*) of urban species is needed. According to the former chapters this knowledge is limited. Because of this diversity should be a important factor in planning: if we are not sure how species will react, we better make sure that a maximum of possibilities is offered.

18.3.2 'Red' ecological network

Can ecological planning for urban species also work on a regional scale? Holland has a fine tradition of ecological policy. In 1989 the National Ecological Network (NEN) was introduced. Based on island theory and meta-population dynamics a network of natural areas and nature development areas was defined, connected by ecological zones. Size and quality of the areas and connectivity of the zones determine the viability of species and populations in the urbanised country of Holland.

The National Ecological Network focuses on the biogeographical districts of rural and natural environments. Treating the urban environment as a bio-geographical district (see chapter 14, page 245) requires and justifies think-ing about urban ecology on a regional scale. Improving regional structures of urban environment will undoubtedly improve the vitality of populations of urban species or even improve urban biodiversity. For typically urban spe-cies connection with other urban areas will be positive. In this respect one could add a 'red' ecological structure to the NEN.

Twente
As an example the distribution of the common swift is presented in the Twente region in the east of the Nethernalnds, compared to the ecological main structure and to the urban structure. It clearly shows that the swift, although being a protected animal, has nothing to do with the National Eco-logical Network. Instead it clearly indicates a strong relationship to the lar-ger cities in the region.

Fig. 165 Map of the National Ecological Network in Twente (Vista)   *Fig. 166 Distribution of the common swift (Apus apus) in Twente* (Vista)   *Fig. 167 Urban structure Twente* (Vista)

Rare but typical urban species could be chosen to determine valuable ur-ban areas. For a species like the common swift (*Apus apus*) or the black rat (*Rattus rattus*) this would mainly be the old city centres as core area. Infra-structure like railroads function as lines of dispersal and connection. One could even see urban development as a kind of nature development.

Although logic, it is likely not to be an effective approach, as it decreases the meaning of the green NEN. Rather than adding a new layer to the exist-ing map of the NEN, regional or national 'species protection plans' could be a more effective instrument to maintain rare urban species on a regional scale.

340

Improving habitat for non-urban species in urbanised areas

Next to urban species the urban environment hosts a great number of species of natural and rural environments. Gardens, parks, ponds and leisure areas offer a wide range of habitats. The average density of species in cities is often higher than in intensively used agricultural areas. Water quality is often better. Of course road kills by car traffic, disturbance by light, sound and movement and even cat predation are negative factors. These are inherent to the urban environment, in spite of — often expensive — devices like fauna tunnels, protection screens etc. Therefore ecological planning should focus on species that are less sensitive to the disadvantages of the city. These species may not always be the most rare ones, but they add in a very important way to the living quality of cities and increase the environmental awareness by its inhabitants.

Two projects by Vista can serve as examples of how seemingly unattractive urban environments can offer unexpected circumstances for the creation of new habitats.

Peatland on dioxins, treatment of the landfill Volgermeerpolder. Amsterdam, Holland.

The first example is the cleaning of Holland's largest chemical waste dump Volgermeerpolder, which is now taking place. This area, about 100 hectares just north of Amsterdam, will be totally covered to prevent peoples contact with the waste and to prevent rainwater from flushing the chemicals out. The coverage consists of impermeable foil, soil and water. Slight elevation of the dump compared to surrounding eutrophied peatland, combined with the water tight covering, offered the possibility to create a totally rainwater fed system. By creating a network of small dykes and with that maintaining a thin film of rainwater over most of the polders surface, by introducing a rainwater buffer compartment, and by measurements to prevent eutrophication, a new peatland can emerge. This area is made accessible to the public who can experience the evolution of a water landscape into a new peatland. This refers both to the history and to the future of Waterland.

Fig. 168 Elevation and covering of the chemical waste dump (Vista)   *Fig. 169 Design with the mesh of dikes and waterstorage* (design and image Vista)   *Fig. 170 Possible evolution of the Volgermeerpolder* (Vista)

Wet heathland on the runway of Soesterberg Airforce Base. Soesterberg, Holland.
In the project for the dismantling of the airforce base Soesterberg the question is how to redevelop the area after the airplanes have left. Special attention is needed for the runways. As they were made bomb resistant, they are made of an incredible 6m thick layer of concrete. As this is impossible to remove, plans were made to reuse it. One of these plans consists of creating wet heathland on top. It can be created by retaining rain and local groundwater on the strip by adding edges to the sides and coving the entire strip with poor sand. The water will gently flow over the top of the slightly descending strip. Little dikes along the way can create wetter or dryer parts, thus creating a very interesting park strip. This will add a unique element to the developments and with that strengthen its identity and economic viability.

Fig. 171 Runway of Soesterberg Airbase
(Vista)

Fig. 172. Keeping water on the runway
(Vista, 2006)

Fig. 173 Arial impression of heath strip among developments
(design and image Vista)

These two examples show that in urban projects interesting new habitats can be created. They introduce the element of time and surprise. How long will it take to form a new peat bog on the waste dump of the Volgermeerpolder? What will emerge at the Soesterberg sandstrip, with groundwater running just one foot under the surface? With a well informed 'ecological intuition' the landscape architect can engage in experiments with unknown combinations of environmental factors and create natural 'wonders' in an totally artificial environment. A good understanding of the intricate relationships between environmental conditions and species is required, by the architect himself or by close communication with an ecologist. Thus nature development can really be a cultural thing.

18.3.3   Environmental awareness

An alternative or additional green goal of ecological planning in the city could be suitable to the urban environment. We propose to focus on stimulating environmental awareness among citizens. The city is an area with an extremely high 'hit rate' in the sense of people actually meeting an animal or a plant. If we turn away from the usually negative sound of this towards the positive, this offers a very good possibility to increase people's awareness of nature and their relationship with the environment. It may raise surprise and give insight in some basic ecological principles, most of all the passing of seasons. A single encounter of a child with a strikingly beautiful dragonfly may well raise the child's curiosity and change the basis of its behaviour. This may seem idealistic, but not more so than some of the biodiversity goals.

This goal of awareness leads to a different set of 'goal species'. These can be selected by a combination of limited habitat area requirements, excellent migrating skills, low sensitivity towards disturbance by light or sound, strength and flexibility of populations and a high representativity or *'aaibaarheid'* (cuteness). This aspect of *'aaibaarheid'* has been rightfully criticised from the point of view of biodiversity, but it might be restored in the scope of this goal of awareness.

Focussing on basic ecological principles

First example of this approach is the proposal for the 'Spoorzone' in Amsterdam Slotervaart. Here a park strip is designed, inspired on the old peat and gardening landscape of the Overtoomse veld, which existed till the city rolled over it. In this proposal maximal emphasis was put on creating wet grasslands, peat land reed beds and clear and beautifully vegetated water. Central element in the park strip is a double walkway on a low dyke crossing the numerous ditches by small wooden bridges, based on the ones the gardeners used. The peaty wet ecotopes attract many colorful species, for instance like the dragonfly southern hawker (*Aeshna cyanea*) This species is actually quite common in Holland. Encountering this animal on an everyday walk could raise the curiosity for it and 'burn in' the link of the dragonfly with the wet vegetation.

Source:

Fig. 174 Overtoomse veld mid 20th century
(source Stadsdeel Slotervaart archive)

Source:

Fig. 175 Low lying path through the peatland reedbeds
(design and image Vista)

Source:

Fig. 176 Sudden close encouter

Organising encounters

Another example is the proposal for restructuring the area 'Drijflanen' in Tilburg in the south of the Netherlands. Here a protected nature area is gradually encapsulated by the growing city. The area consists of an old

343

sandy heath land, dense dry forest and a cultivated and enriched creek valley. The viviparous lizard (*Lacerta vivipara*) has been living among the heath until a few years ago, when the area's growing isolation may have caused the lizards disappearance. A plan was made not only to restore the connection with the nearby main population but also to increase the chance of encounters of early walkers with sunbathing animals. Also the rich quality of the riverbed was restored in combination with creating a streamlined bicycle path, for everyday use by inhabitants of new parts of the city. By this the chance of suddenly meeting a grazing roe deer was effectively stimulated instead of avoided.

| *Fig. 177 Green structure Tilburg* (Vista, 2005) | *Fig. 178 Conceptplan Drijflanen Tilburg* (design and image Vista) | *Fig. 179 Boy with lizard* (author) |

Scanning situations and plans like this one comes to the hypothesis that stimulating environmental awareness will work best in a situation where the presence of an animal can be linked to the existence of a natural habitat. While a zoo is entertaining, a little lizard warming itself in the sun on a sandy hill in a heath land park may show us or our children that there is more to life than just ourselves. While the growth of cities will inevitably lead to reduction of natural habitats, stimulating environmental awareness among city inhabitants may well be the only way to get the necessary public support for the biodiversity goals of ecological planning. For landscape architects this approach leads to a different set of design rules. One could state that plans should incorporate surprising encounters, by combining pieces of 'good habitat' with multi-used everyday pedestrian and bicycle routes. It could focus on creating better physical conditions for common species and emphasise their optimal relationship.

18.4   Conclusion

Ecology has gained a firm position in urban planning and landscape architecture over the last years. Environmental and ecological issues have become an integrated part of spatial design on all levels. Yet there are great differences in the way this integration is approached. Nature is often seen as a separate function with it's own ecological program and spatial demands. This leads to planning solutions that tend to focus on limiting human disturbance and maximising opportunities for rare and endangered species.

Opposed to this there is also an approach that sees nature as a general dimension of environmental quality and spatial development. This approach focuses on combining ecological program with other functions and recognises nature development as a cultural intervention. Vista landscape and urban design has been an advocate of this approach since it's founding in 1994, just after the proclamation of the National Ecological Network as a national policy in the Netherlands.

Especially in urban environments a more comprehensive approach to nature is suited, because here many functions come together and the cultural dimension can't be denied. Urban ecology should take into account the wishes of experiencing nature among city inhabitants and search for new ecological possibilities occurring in urbanised environments. Even highly artificial circumstances can result in special ecological qualities, if the right conditions are created. This requires a willingness to engage in 'ecological experiments' and give way to the unexpected, of course based on a firm knowledge of vegetation dynamics and species requirements.

The classical goal of increasing biodiversity may not be the most suitable goal for ecological planning in urban environment. Stimulating environmental awareness among city inhabitants is proposed as a complementary goal, leading to a complete different set of target species and design rules. Main focus is to create opportunities and enlarge chances for encounters with attractive and characteristic species. This may well be the only way to get the necessary public support for the biodiversity goals of ecological planning.

The landscape architect can bridge the gap between the work of ecologists and the work of urban planners, designers and even architects. He can implement knowledge about ecological relationships into designs and guide developments in a way nature is saved or given space to develop. Therefore it is essential that ecologists and landscape architects share their knowledge and work together on environmental issues. Landscape architects should have a basic insight in ecological dynamics to ask the right questions, ecologists should have a general understanding of planning practices to provide the right answers. By thus working together new spatial developments can be effectively used to enhance environmental quality of urban environments as a whole.

19 WATER CITY DESIGN IN WATERLAND

Fransje Hooimeijer **T**∪Delft

19.1   Introduction

There should be a natural abundance of springs and fountains in the town,
or, if there is a deficiency of them, great reservoirs may be established for
the collection of rainwater, such as will not fail when the inhabitants are cut
off from the country by war … For the elements which we use most and
most often for the support of the body contribute most to health, and among
these are water and air. Wherefore, in all wise states, if there is a want of
pure water, and the supply is not all equally good, the drinking water ought
to be separated from that which is used for other purposes.
Aristotle(350 B.C.) Politics, Book 7, Chapter 11 [a]

Deceptive self-evidence of water
The importance of drinking water for cities has always been self evidently
accepted. By contrast, this self evidence makes it non existent in good
times but when there is a shortage, it is a very big issue. Another important
subject this quotation of Aristotle touches on is separating the different
flows of water. Especially in a country like the Netherlands in current times
these flows are matters for great attention of urban ecology and design.

Constant amount, changing distribution
There is an unvarying amount of water on earth: 1,250 million cubic kilometres. It has the capability to appear in various substances and in different
places, but the total amount always stays the same. But 97% of the water
is saltwater and can be found in the oceans. The other 3% is fresh water of
which 87% is frozen around the north and south poles. Of the remaining
non-frozen fresh water, 90% is ground water and 10% is in our rivers,
streams and lakes.

Driving force in Dutch territory
In the Netherlands the presence of an extensive water system has been
crucial in shaping and spatially organising the territory. The fact that the
dependence on knowledge about the laws of the hydrological system and
the technological state of the art had a changing relationship over the centuries is revealed in the changing function of water in the city. The growing
control over water was made possible by improving engineering and motivated by the necessity to protect growing economic values. That is why the
way water is dealt with stands for the way our society has become more
complex over the past centuries. Before 1000 A.D. people simply lived on
the elevated parts in the territory, today advanced technology is used to
prevent threat and inconveniences posed by the water.

---

[a] http://pinkmonkey.com/dl/library1/aris22.pdf

Two trends, three design tasks

Climate change means cities and regions in deltas are faced with three spatial tasks as a consequence of global and local trends: (see *Fig. 180*).

Trends \ Design tasks	Global: Rise in levels of seawater and river water	Locally: Increase in rain and rise in levels of ground water
Coastal defence	Half the country lies below sea level	Subsidence and geological over-turn
Riverbanks	Geological overturn pushes the lands in the east upwards and the western part of the country drops even further below sea level	River flooding in winter
Water storage	Rivers are dry in the summer enabling the seawater to move up the rivers	Tropical rain storms flood the cities

Fig. 180 Trends and design tasks in the Dutch Delta

The design question

In the future it will be necessary to look closely into the water tasks facing Dutch cities on water. As early as in the conference held in honour of the 25th anniversary of the fouding of the WLO in 1997 a landscape ecological agenda was put together headed with the issue of investigation into water systems: attention for water as the carrier of the conditions and processes for the planning and design of natural, agricultural and urban landscapes (Dorp, 1999, p 22).

This contribution will focus on the conditions (function and form) water pro-vides for the design of cities and the meaning of urban ecology. It will be followed by an examination of the spatial qualities of water and the design of public space in considering water. The conclusions will set forward some considerations in designing water cities.

19.2   Cities on a wet territory

19.2.1   Historical functions of water in the Dutch cities

Clarifying history

In order to be able to understand the relationship between water manage-ment and urban planning some notion must be taken from the past. The dominating presence of water and its rules - the hydrological system - has been of significant, maybe even crucial influence on the land use, land-scaping, urban planning, design and organisation of the Dutch territory. Depending on the knowledge and laws of water management and thriving on the development of technical possibilities there has been a shifting attitude towards - and a variety of functions of - water over time. The changing relation between spatial planning and water management can be characterised as: acceptance (till 1300), defence (1300-1500), offence (1500-1800), early manipulation (1800-1890), manipulation (1890-1990) and adaptive manipulation from 1990 (based on Van der Ham, 2002).

The increasing ability to control water and its natural habitat has been supplied by the cumulative enrichment of the fine art of engineering and motivated by the mounting obligation to protect economic goods. The necessity and urge for control brought the development of technology to a level that made the Dutch world famous and is often referred to as the "Dutch fine tradition". The growing complexity of Dutch society is representative of the way the Dutch have treadet the presence of water in their territory. Also, the development of the "fine tradition" in technology is related to the different functions urban water has fulfilled: drinking water, storage, drainage, sewage, military, transport, representation, public space and leisure.

Acceptance until 1000
During the first stage of acceptance people built their houses on higher ground and used the nearby surface water as drinking water, sewage, transport, to drain the production fields and to store rain water. Preparing the wild landscape for agriculture had been done since the nineth century by digging ditches that had a natural discharge on rivers and streams. The system of ditches also formed the property boundaries and produced the necessary organisation of water management: - water boards – as well as a rationally laid out landscape.

Defence: against and with water (1000-1500)
The second stage of defence against water started around 1000 and ended around 1500 with the invention of the water mill. The building of dykes made it possible for cities to be situated at startegic places like the mouth of a river: examples include the birth of the 'dam' cities of Amsterdam and Rotterdam. In addition to the technological possibilities, the foremost motivator of protection was the economic and political circumstances at the time. Cities were granted city rights but there was no centrally organised defence system; each city individually defended its inhabitants and economic goods. Water began to have a two-sided influence on urban planning, as well as the threat it posed, cities start to use water as a means of military protection.

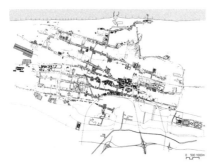

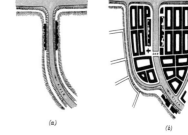

Fig. 181 The Hague on sandy ridges
(Atlas of Dutch Water Cities, 2005)

Fig. 182 Amsterdam on a river estuary
(Burke, The making of Dutch towns, 1960)

Cities of a river estuary

The mouths of rivers guarantee maximum benefit from the water. Dam cities like Amsterdam, Rotterdam, Edam and Schiedam are situated where small peat rivers flow into a larger river. The dyke was the most important condition for urbanism in the polder areas, due to subsidence these areas had become vulnerable to flooding. The dam had an important defensive function and was at the same time the means for discharging river water into a larger body of water. This flow was also used to prevent the harbour from silting up and to keep it available for sea vessels.

The dam as a crucial urban detail

The economic significance of the transport of goods from sea to the hinterland and vice versa became visible through the dam that also formed the economic and physical heart of the city. The sluice could only handle smaller vessels so the goods were taken from the big sea vessels and were sold on the dam where the market was situated. The dam city and the countryside are closely interwoven not only in terms of water management, but also in economics. This was spatially expressed by the construction of public buildings at the dam like the city hall and the church. The economic connection between city and countryside, in their roles as market and production fields, meant that the city had an interest in keeping them dry as well. Larger dyke systems were built to protect the countryside from the washing waters: polders.

Water for military defence and transport

Halfway through the fourteenth century cities were granted city rights. In absence of central government each city was responsible for its own defence. Using water for this purpose made water even more important in the spatial order. As well as the new military function, water became more significant as a mean of transport. After 1630 an extensive system of 'trekvaarten' (water ways with a towpath) was built. In the early Middle Ages the wild and wet territory of the Dutch was hardly passable. Traffic over water was the most efficient, comfortable and fastest mean of transport. Hardly any land routes were built until the end of the eighteenth century, but there existed an extensive transport system over water throughout the whole of the Low Countries. The transport of goods in particular was not easy. The different cities were maintaining their rivalries and kept up certain rights for the transport of goods. Good transport meant having a good position in the transit market and good economic possibilities.

The economic importance of waterways

Jan de Vries considers the 'trekvaart' in his book *The first modern economy* as a modern sign of Dutch development as an economic union. By 'modern' De Vries means that the developments were consciously steered with particular interventions and not just acceped for what they were. In economic relations this meant that investments were made to increase the productivity and strategies were made that could control the process. The construction of the 'trekvaartnetwerk' was crucial for the improvement in the economic situation in the seventeenth century republic. Good transport was available at any season at any time of the day. Because the boats were pulled by horses and therefore did not depend on the wind the service could work with a tight time schedule. Between 1632 and 1667 thirty cities were connected through the 'trekvaart' system for the sum of 5 million guilders. This was a huge amount for the republic; in comparison another grand project was the drainage of six polders in the north for the sum of 10 million guilders. The 'trekvaart' system meant increasing institutionalisation of the republic because it became necessary for cities to work together in using the same time frame (Vries, 1997, p. 35).

Offensive water management: (1500-1800)

The third 'offensive' stage, steering on the circumstances of water, made its entry in the sixteenth century. The first sign of this period is the building of the very first mill in 1408. Instead of natural discharge the water can be pumped to higher situated canals to drain the polders. The control of river water was organised by connecting all the dykes, providing a closed system. Before, each region had an individually organised protection system. An important effect of the possibility of controlling the water is that the water system can represent something: the ability of control. An exceptional example to the new function of urban water as representation is the ideal city of Simon Stevin (1548-1620), an important military engineer, as he described in 'Designing Cities' (1649), which was published posthumously. The design and organisation of the city is based on his experience with designing military camps and fortifications together with the laws of water management. The powerful regents and merchants in Amsterdam used the same principles as Stevin in the design of the first 'droogmakerij' (a polder that is pumped dry) the Beemster (1608-1612) and the seventeenth century city expansion of Amsterdam that was realised in the same period.

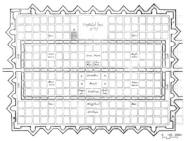

Fig. 183 Wind
mills for water
management
(Meyer 2005)

Fig. 184 Ideal City Simon Stevin
(Heuvel, 2005)

Fig. 185 Simon
Stevin *(Heuvel, 2005)*

Amsterdam Ring of Canals

The Amsterdam canals are good examples of a city expansion that is harmoniously built with the use of the characteristics of the territory and – motivated by the fight against water - expresses a strong feeling of citizenship and powerful tradition. Preparing the wet lands with a high ground water table around 1600 was done by raising the area with the mud that came from the canals. These had a double function: drainage and providing the necessary material to prepare the site. The plan of the three canals was realised with a strict orchestration. The Herengracht, Keizersgracht and Prinsengracht were connected by smaller canals and streets that connected existing roads of the inner city. Along the main canals, rules governed the spatial dimensions in developing the type of buildings and the use of the lots. This is where the royal houses and office buildings of the regents and merchants are situated, often with garden. The row of trees along the canal was supposed to represent unity in the street, as if all citizens along the canal were equal. Labourers' dwellings and industries that produced dirt, smell or noise were forbidden. For them another expansion was built. The Jordaan does not express the finance and effort that was taken to realise the three canals. The polder ditches were used as the ground plan to order the streets and the building blocks.

Blue print

The collective watch out (for a proper city expansion) and the realisation of an ensemble that meets (even though limited) public interest is a very interesting left over from the economy of the Middle Ages. The fact that it was actually built can be attributed to the commitment of private enterprise and to the fact that it was a blue-print of society at that time. The city was not built according to the baroque ideal of a capital city, but instead the principles of the organic life structure (Wagenaar, 1993, p. 9-12).
It was *'(…) a miracle of spaciousness, compactness, intelligible order. It accepted all that was valid in baroque planning, with just sufficient variation*

in the individual unities, combined with the rich tracery of trees boarding the canals, to take the curse off the military regimentation of baroque classicism. The successful breaks in the direction of the spider- web plan keep the distant vista from being empty and oppressive.' (Mumford, 1961)

Dutch Renaissance
The seventeenth century was a Golden Age for the Republic, when cities flourished through economic growth, and the period in which polder expansions were built. The cities stepped off their "dry core" and, under "strict control," raised and drained their expansions. The characteristic of modern civil society is that the future is consciously planned on the basis of rationality, mutual consultations, and decision-making. Political independence was accompanied by a flourishing of science, technology, and art: the Dutch Renaissance. The expression of this flourishing in regard to urban design was Simon Stevin's design, based on existing size and structure principles of agricultural engineering and urban design. The perspectives of water management, derived from the pattern of the polders, were directly applied in his city. Simon Stevin is the first example par excellence of the capability to integrate hydraulics an urban vision: urban engineer, technically trained and a self-taught urban creative thinker, he marks the excellence that mismatching boundaries can produce.

The ideal City of Simon Stevin as a scheme
The ideal city of Simon Stevin was not built, but used as a scheme. In a period of 40 years the VOC built and rebuilt 130 colonial cities all over the world apparently according to Stevins scheme (Oers, 2000). A scheme of the same conceptual calibre is the *grachtengordel* (ring of canals) in Amsterdam (build around the 1620s). It is an integral design of land restructuring, surveying, and water management resulting from the cooperation of the merchants for whom the city expansion was built, the city carpenter Hendrik Jackobzn Staets and the surveyor Lucas Jansz. Sinck, who drew up the plans. Amsterdam did not try to follow the idealistic view of a capitalist city as other European cities did, but implemented this design as a blueprint of social and economic life, making use of technological possibilities (Wagenaar 1993, pp. 9-12.). Hoeven and Louwe (1985) show the influence of a Stevin scheme by a morphological analysis of Amsterdam.

Polder politics
An important characteristic of both the Dutch Renaissance and polder cities is the result of the strict control: the absence of any idealistic expression. Vision and beauty have to be paid for, and the entire budget was used preparing grounds for city expansion. The individualistic citizens did not see the need for an urban composition based on central authority, like the monumental plans in other European cities, simply because it did not exist in the Dutch society (Wit, 2003 p. 11). The Dutch tradition of always trying to reach group consensus, the so called 'polder politics,' is directly related to the fight against water. A count can own a lot of land but it is useless when there are no farmers working for him to keep it dry. This dependency

is something the farmers are well aware of and they make sure that the count on the other side is aware of their interests (Lendering, 2005).

Early manipulation 1800-1890
The phase starting in the beginning of the nineteenth century, after the French occupation, is characterised by an exploding population growth and industrialisation through the transformation from hand to machine labour, made possible by the introduction of the steam engine. This altered strongly the forces shaping the (urban) landscape (Ham 2002). Also, it changed hydraulic technology. It became easier to control water, make it do things it would not naturally do: i.e., to manipulate it. Civil engineering became a more defined discipline (with clear tasks, education, and a discourse) at the end of the eighteenth century.

Demolishing fortresses for urban extensions and parks in the 19th century
Even though the obligation to hold on to the fortifications was only repealed in the Fortification Law (1874) many cities broke down their ramparts around the 1850s. The moats that usually accompanied the ramparts became part of the desperately needed public space. Finally the dense cities could expand and urban plans for open space and spatially set up dwelling areas were drawn. Two important examples are to be found in Utrecht and Rotterdam, both designed by the landscape architects family Zocher.

Source:

Fig. 186 Waterproject in Rotterdam van W.N. Rose (1854) (Municipal Archive Rotterdam)   *Fig. 187 Heemraadsingel as third annual ring (Municipal Archive Rotterdam)*   *Fig. 188 The singels in Utrecht van J.D. Zocher (Municipal Archive Utrecht)*

In Utrecht they redesigned the old moat system to a system of 'singels', canals in a park setting with grass alongside and the water often gently rippling. In Rotterdam they also designed a singel system, not in the situation of a deconstructed fortification system but as the first controlled expansion of Rotterdam.

Urban sewage
The representation through water stands out against one of the oldest functions of surface water and still present during the third stage control: sewage. Rubbish was thrown in urban water until the beginning of the twentieth century. Most cities built an underground sewage system around 1890, the

only solution against recurring cholera epidemics that were caused by the bad hygienic situation in the city. In Rotterdam it had been forbidden to throw anything in urban water since 1719, but this did not have the desired effect. It was not possible to clean the water by flushing with river water because the water management was controlled by the water board. Their job was to maintain the same groundwater table, needed for agriculture, so flushing water in their minds was useless.

Rotterdam Water Project 1854
In 1854 W.N. Rose, Town Architect to Rotterdam, and J.A. Scholten, works surveyor of the Schieland Water Control Board, designed the Water Project to adress the town's troublesome water-resources management and structured its expansion into the outlying polders. This project, the first large-scale nineteenth-century up scaling operation in Rotterdam, is comprised of five 'singels', gently winding watercourses accompanied by tree-lined avenues. Westersingel, Spoorsingel, Noordsingel, Crooswijksesingel and Boezemsingel together envelop the original triangular settlement like growth rings in a tree. The landscape architects J.D. Zocher Jr. and L.P. Zocher embedded these singels in a landscape-style setting, giving the Water Project the added resonance of a major urban embellishment program. It was the first time that a laid out water structure fulfilled the function of public space (Hooimeijer and Kamphuis, 2001).

Manipulation 1890-1990
The last stage of manipulation came around 1890 after a long period of control and the fine tuning of the technology of water management. The invention of the conduct engine opened a world of new possibilities and we can see an increase in scale concerning preparing sites for building. By using the technique of flushing the sand from a lake onto the expansion area the preparing of building sites became much cheaper. This slice of sand covered the original landscaping and made it difficult to grow plants and trees. At the same time the claim for public space became greater at the beginning of the twentieth century. Under influence of Health Boards and the Garden City Movement increasing awareness about the necessity of nature in the urban context was established. Water was, as well as being necessary also a popular element in these green structures. The popularity grew a decade later when the building of leisure areas took flight.

Water leisure
At the beginning of the twentieth century the open water systems become more attractive for leisure. Together with the more free time the Health Boards emphasised the necessity of leisure as part of civilisation. Also hygienic aspects of water play an important role in this. Already from the 1850s swimming pools were built to enforce personal hygiene. Not all houses then had a bathroom, let alone a bath or shower.

Garden city movement
The garden city movement (Howard, Unwin) around 1900 put forward the claim for public space, greenery and entertainment in the cities with more strongly. In many development plans - for example of Amsterdam - the method of Dr. Martin Wagner and his Städtische Freiflächenpolitik (1915) is used. In his view sport and leisure needs 4 m^2 per inhabitant, and 13m^2 for woods near the city, 65% of the necessary leisure space. Particularly in the development of the parks in which sport is important, we can see water as a structural element.

Recent requirements for urban water storage
At present (1990 until now) we are in the of adaptive manipulation stage shifting towards an attitude where the rules of water have to be accepted and the technical heritage to be dealt with. The problems are becoming greater and despite the "fine tradition" things are not under control. In spring and autumn rivers flood and tropical rainstorms occur much more often than the civil engineers and their systems were designed. The rivers have to be given back their original bedding and in recent legislation it is obliged in new expansion areas to bring in a certain percentage of surface water for the storage of rainwater. Surface water always has had this function but it was never build for this purpose specifically until now. The reason for this is that the high technology water systems built during the manipulative stage are less flexible. The system uses a fixed capacity and is not able to anticipate peak situations. It is important to have a flexible system that can carry a lot of water and has storage for dryer periods.

19.2.2   Position and profile in existing urban areas
Design of water?
All pro's and con's of living in a wet territory define not only the physical but also the organisational structures. During the acceptance, defence and offence phases, the planning of urban water was in the hands of water boards together with the municipalities and the design done by civil engineers. The profession of civil engineer was finding more than technological solutions for construction problems, as described above. Simon Stevin also designed cities and was concerned with urban concept and form. The establishment of the profession of urban design in the beginning of the twentieth century changed this. We recognise a secularisation of professions in which each profession was given a certain aspect of urban planning. During the manipulative period the civil engineer became the designated professional in dealing with water management because of his excellence in "fine tradition". In designing cities water management was no longer a designing instrument, simply because it was taken out of the portfolio of the urban designer. They did not have to worry about dry feet in planning new cities; the civil engineer armed with prosperity of the "fine tradition" made it all possible (Heeling, 2003).

Integrating water systems in the urban composition

The different functions of urban water (storage, drainage, sewage, military, transport, representation, public space, leisure) are crucial in the design and the positioning of surface water in the urban context. The function of the waterway is determined the public accessibility of the water, if it is steering or adapting to the urban structure and if it is separated or integrated in other systems. The water system can be part of the composition of the main structure of the urban design, or form a secondary or substructure in the city plan. Also the water can be used to boarder the city area.

Canal, singel and dyke

In the Netherlands the vocabulary in cross sections of water systems contains only two examples: canal and singel. These cross sections are defined by their original function, a stone quay (canal) belongs to the function of transport and gently winding watercourses accompanied by tree-lined avenues (singel) are part of the rural country and public space. Also part of the vocabulary of the urban water system is the dike. They are not always as evident because they are hidden by buildings, but can be recognised by differences in height or the name of a street: dyke or Hoogstraat, meaning high street.

Crossing locations

The particular design of water systems can be an important structuring component of the urban composition or more secondary and integrated with other structures. a logical consequence of the fact that most villages, later cities, were located on the crossing of a land and waterway, was that natural waterways took a central position in the compositions of cities. For example cities along the river like Rotterdam (Maas and Rotte) and Amsterdam (Amstel and IJ) or cities on the coast like Vlissingen en Scheveningen. The waterfronts often are important public spaces where the arrival of goods and people from far away places attract urban functions. In these traditional cities the different elements (buildings, traffic and public space) are orchestrated in a unity (the buildings close in the public space). In the Netherlands due to the necessity of drainage and transport, surface water was until the manipulation stage seen as an instrument to expand the city, as important as streets and squares.

Internal canals

City water that was designed specifically for transport usually is a part of the main structure of the urban plan, like for example the Grachtengordel. Moreover, streets, squares and canals were for a long time the three elements that gave form to urban expansion. In the publication of the State no. 158 of June 22 in 1901 an important change was made to the stipulations concerning expanding cities. Before that time the three important elements that gave form to expansions were street, square and park. The stipulation was that where these elements should be situated it was prohibited to build houses. In this publication no. 158 they changed the park into canal because they considered that with the expansion of most Dutch cities, and

Amsterdam specifically, the digging of canals to be of more importance than parks.

External 'Singel': belt, canal and boulevard; 'annual rings'
From the 1850s the fortification cities were dismantled and between the old and the new city usually a green area was built with the use of the old moat as a singel. This new encircling structure served as an 'annual ring' in the city pattern. In the twentieth century this space was usually taken up by infrastructure.

Even though Rotterdam was not a fortification city, it did use an encircling park with a 'singel' structure to shape its first expansion. The Water project is like an echo from the first annual ring: the old city centre with its a triangular shape. The term singel is used in the meaning of boulevard, it is a walkway around the city. The third annual ring is the Heemraadsingel, a 'singel' structure that works the same way as the Water Project and which was build 50 years later. It was an improved version of the Water Project because according to the new developments more leisure space, was put in by enlarging the profile of the singel. Also a park was attached to the structure to make a significant green structure in the densely populated area. This was possible because the main infrastructural function (like the water Project had) was included in other structures and not along the Heemraadsingel. The main infrastructure was provided by the Mathenesselaan and the s' Gravendijkwal. The independence of green structures at the beginning of the twentieth century meant that they were used as separate composition elements in the city pattern.

Technological prosperity
The use of steam-driven pumping stations for the city expansion of Rotterdam is one of the first examples of a technological approach to urban waters, which would become common a hundred years later. The most important input for this development was technological progress in combination with an explosive growth of urbanisation. Hygienic problems, caused by the city's water were slowly but surely of influence to the spatial effect of water management, due in part to the progressing development of the steam engine and later of the internal combustion engine.
The construction of the sewage system or network and the drinking water supply created a division between the systems required for groundwater level control, wastewater discharge, and drinking water supply. A large extent of the urban water system was constructed underground. From the end of the nineteenth century, traffic and transport by water was largely replaced by transport by train, tram, and motor vehicles. This led to the filling in of a number of canals and singels (greenbelts with a waterway incorporated), resulting in a drastic decrease of surface water (Vries 1996). The water structure remained important for storage and drainage, but was no longer used as a design instrument.

Fig. 189 Pendrecht (Rotterdam)
(Heeling; Meyer et al., 2002)

Fig. 190 Pendrecht (Rotterdam) (Mu-
nicipal Archives Rotterdam)

Public space

Whilst the functional meaning of water increasingly receded, 'nature' in the city gained proportional interest. Around 1900, influenced by health commissions established to protect the hygienic interests of the city, the demand for more public green spaces increased. Besides traffic, buildings, and water, a new element in the city structure was introduced: public areas. At the same time, structures of buildings, traffic and the combination of water and green spaces were separated. These structures coincide in traditional cities, such as in the Amsterdam ring of canals, which has one single main structure containing all the elements.

In the beginning of the twentieth century it appeared as if water and green spaces could only guarantee their right to space when they were combined. Water is part of the green structure, and the green spaces acquired rights as a public area as a component of water. At first, the combined structures of water and green areas were of importance and formed the backbone of the city design in combination with the structure of traffic. In Plan Zuid (1915) in Amsterdam and in Blijdorp (1931) in Rotterdam, the structure of public areas and water (on which people can also sail in Amsterdam) follows a displaced shadow of the traffic structure (Kamphuis and Hooimeijer 1997). Then, as well as now, the urban composition became a derivative of the traffic structure and accessibility principles due to the importance of motor vehicle traffic.

Urban extensions after the Second World War

At the beginning of the twentieth century urban development became an autonomous discipline and the tasks were divided. Civil engineers solved the water problem and offered the urban designers the possibility of designing a plan based on this. This was graciously accepted, especially during late post-war building. Technological progress, such as improved pumps and calculation methods made the preparation of a larger site possible by raising it with sand. This meant that in combination with an underground drainage system, significantly less surface water was needed. At present, water is considered to be a waste product, and is situated alongside the outskirts of districts, integrated into the infrastructure or the green space system. The water system as designed by civil engineers cannot be recognised as such, since underground pipelines alternate with the surface water. Moreover, the sand package provides the urban designers with a tabula rasa on which each required urban design can be realised without any concern for the water system. Where in cities up to 1940 the total surface of the city contained 12%-15% of water, in post-war city expansions, this percentage was often reduced to less than 5%.

The come-back of water as public space
From the seventies, various plans to re-establish previous waterways were developed and realised. For example, in Utrecht part of an old outer canal was once again excavated after it had been filled in for traffic reasons (Vries 1996).
In contrast to the sober singels of the reconstruction of the Netherlands, the natural character of the green belt and water suited the residential areas built in the seventies perfectly. Residences were located in courtyards and encircled by green structures with singels. In the water structure design and profile, the natural character of the water was well used. In this period, nature-friendly banks came into fashion, which gave the singels an entirely different appearance. Influenced by books such as "Silent Spring" (1962) by the author Rachel Louise Carson, more attention was given to ecology and nature. She describes how the use of modern herbicides resulted in over cropping and deformation of the plant world. Her book inspired designers to use nature and city more as an integral issue instead of looking at two separate worlds (Steenhuis and Hooimeijer 2003).

Watertoets
In the last two decades there was a sense of decline, but eventually water made a comeback in cities. In the eighties, as a result of economic recession, attention to public areas decreased, which was of great influence to the application of the singels as a natural element. The positioning of public areas in the urban development plan and the profile of the singels were quickly reduced to being functional and virtually maintenance free. There were no financial means available for building a high-quality public area, so the design was based on a minimum of maintenance costs.
In the nineties, the rediscovering of water as an element of urban development composition coincided with the effects of the changing climate. The most recent law, the Watertoets (2003, Water Test, see also 21.5.3), compels new expansion districts to comply with certain hydraulic conditions, to be reviewed by the water board. This is also how the regional scale of water control is organised, as here the law of communicating vessels applies: if one interferes with part of the system, this will influence the complete system. For this reason many municipalities have created a water plan,

which also maps the hydraulics in the existing city, sets guidelines, and defines future spatial issues.

Storage

With regard to polder cities the emphasis is on maintenance and storage instead of the transferring of rainwater. For example, temporary water reservoirs in public areas are used for polder cities, in the form of wadis or by using roofs which can temporarily retain water (such as grass roofs), reservoirs to collect rainwater for use as toilet water, more gardens and ponds instead of tiled terraces, and fully built inner areas and more surface water in the district. The sewage and rainwater are to be collected in a 'separate system'. Rainwater does not have the same level of contamination as sewage water and therefore does not require the same purification. Moreover, the sewage system cannot process the rainwater in a heavy storm.

IJburg

In many new development districts, surface water is given a structural role, as in the old days. In IJburg, the water serves as a backyard, positioned as a division between private and public areas. This natural boundary makes living next to the water as a natural fencing immensely popular, as it is often referred to as 'real estate water', since the water brings in money.

Fig. 191 A usual VINEX-extension
(Palmboom van den Bout)

Fig. 192 Airstrip Ypenburg

19.3   Urban ecology

Drainage as primary spatial order

As welle as the strong functional and spatial connection of water in the Dutch cities as described above, there is also a strong base in urban ecology. More over urban ecology has been very decisive in the building of water structures in the past.

The Dutch situation is exemplary because there is no natural landscape; the whole Dutch landscape is made by human hands, in which the water system was already constructed by digging ditches to drain the wet territory. After subsidence occurred as a result of the drainage a pomp system with mills was supplied that could manage the water system even better. The steam engine and the induction engine even enlarged this pumping capacity. In the current situation the rural and urban water system in the west of the Netherlands is pumped (see *Fig. 194*).

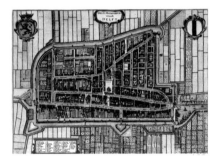

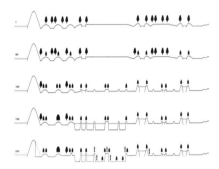

Fig. 193 Agricultural allotment recognisable in Delft (Municipal Archive Delft)

Fig. 194 Increasing pumping capacity caused subsiding over the centuries (Meyer, Hoekstra, et al., 2000)

Ecosystems caused by man

Like the Dutch polder landscape the cities also have a culture bound ecosystem: structured communities consisting of different types of organisms (human, animal, plants) that have a certain, also structured, relationship with a-biotic environmental factors: climate, soil, water and relief.

Ecosystems are open and conditioned systems the key variables of which can be controlled by humans. The water cycle is a subsystem of the ecosystem. Because of the influence of humans there is an increasing variation, fragmentation, mobility and mechanisation. The influence of humans on the Dutch ecosystem in rural and urban areas is extremely large.

Since the oil crisis in the 1970s there has been an overhaul in the approach to ecosystems. Ecologists have become players in the spatial planning (Leeuwen 1965, 1966 p 17). This is also the explanation for the return of water in the design of cities as described above.

Natural conditions, human impacts

The human influence is dominant in the Dutch landscape and even more so in the Dutch urban ecology. Humans choose a location on grounds of its possibilities (conditions see *Fig. 195*, 1-10) and consequently they are directly influence the nature and size of the artificial attributes of the landscape (houses, streets see *Fig. 195*), on processes that occur, the development of natural components and even the climate (see 12, 13, 14 in *Fig. 195*). Rhis is all followed by the various second, third and fourth order of effects (15-20 in *Fig. 195*) that are object of investigation in the design of cities.

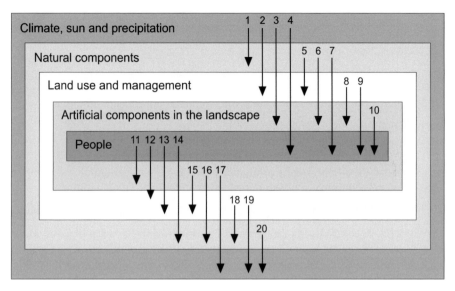

Fig. 195 Relations to be studied in urban design as condition (1-10)and as cause (11-20)

Towns in an urban landscape

The city is an open system, exchanging energy and mass with its environment. The system 'city' has a variety of subsystems: centre, industry, suburbs, parks and traffic. At the base of the city is the natural a-biotic landscape like water and relief. Urban landscapes are superimposed on this. Compared with most rural landscapes the system 'city' is characterised by the large number and scale of artificial elements.

But also the landscape as a whole plays an important part in the city. The subsystems of the city are partly connected to the original landscape on which the city is built, and often connected to the settlement history of the city (Vink 1980, p. 88-89). Especially water plays an important role in Dutch cities as we have seen in the former paragraphs.

Climate change and urban heat island

Climate change is very significant for the Netherlands because it is a strong environmental factor in the landscape ecology and consequently the urban ecology. The following elements of the city are obviously different with the natural ecosystems and will have greater effects when climate change occurs:

1. Stone buildings and streets hold on to heat longer.
2. Urban functions produce heat.
3. Artificial water system of ditches, drains and sewers produces more heat because this water is not used for evapotranspiration (sum of

evaporation and plant transpiration) which cools the environment down.

4. Pollution reflects the sun and keeps in the city's heat.

The city is a heat island and differs greatly with the natural surface in the reflection of radiation, specific heat and the heat colliding coefficient.

Urban drainage requirements

Because of the swift discharge of rain in the countryside the humidity is lower. Temperatures are 0,3 to 0,5 °C higher in the city compared to the countryside. The city's heat production causes a low pressure area and a continuous air stream from the countryside to the city (Vink 1980, p. 90-91). The hydrological demands of the city are:

1. The drainage in the city should be more controlled; the ground water level needs to be maintained because of the foundation of houses.
2. The use of surface water for storage and buffer is recommended.
3. In order to maintain a high level of hygiene an extra drainage is necessary.
4. Raised building sites (for stabilisation and drainage) need sufficient drainage.
5. The plan for drainage should be adapted to the buildings.
6. Before expanding the city a position should be adopted with regard to exceeding groundwater levels; architects, economists and hydrologists should be concerned about this together.

Water management units

The water system in the polder landscape and in the city is built up out of different water steps or compartments surrounded by dikes and discharges via a pump or natural discharge. These units or compartments are laid out in the maps of the water state (see in *Fig. 196*). A discharge unit is an area, separated form other units by dykes, with a system of waters discharging on surrounding waters connected with the greater water system (river, sea). The ground water level, and the fluctuation in the different seasons, is an essential aspect of these units and their landscape ecology.

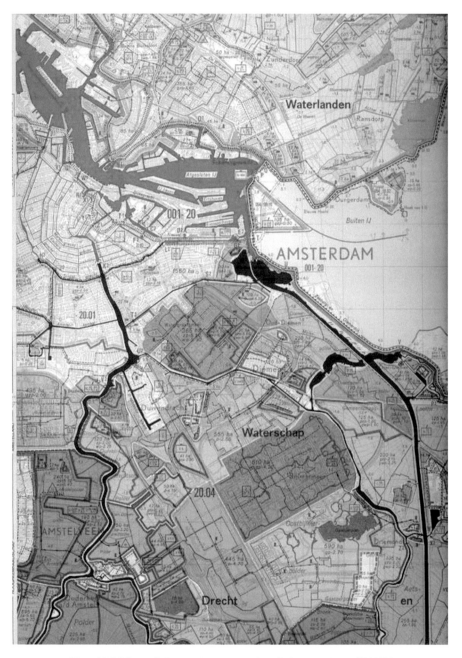

Fig. 196 Map of the water state, detail Amsterdam (RWS)

Disturbed water balance

The easiest way to characterise the ground water levels is through its steps. Water management in urban areas has a water balance, a ground water level that differs from the one in the countryside. Due to the lack of vegetation evaporation is lower in the city. Storage of water in the ground can only be used partly and the same goes for the outlet waterway function of larger and smaller ditches, forming an important buffer in the country-side. Rainwater forms pools on the pavement of streets and squares and two thirds of the rain infiltrates in gardens, pavements and parks. Because of this high trees especially suffer from drought faster (Dorp 1999, p. 30).

Because of ground improvement and drainage the water balance in the ground is disturbed. By improving the profile, the top layer and equalisation the original rural water balance disappears and a new mechanical, inflexi-ble water system comes into place (Dorp 1999, p. 107).

Soil in urban extensions

The making of landscape ecological soil maps is very important due to the following aspects:

1. Building new expansions on raised grounds (with sand) has positive effects on the technical side of the project (no foundation piles, plumbing is above ground water level and does not freeze in winter, no irregular subsidence) but it has also negative effects: lost of origi-nal landscape identity, bad conditions for growing vegetation with high costs for improvement.
2. Preserving the original soil structure will also preserve the identity of the area, with small adjustments, and the costs for the building of ur-ban green areas are reduced. For example Craneveld in Arnhem and the village Son near Eindhoven. In many cases this leads to a greater diversity of life forms and a higher standard of living condi-tions. At the same time costs of building and maintenance can be re-duced.
3. Making use of the hydrological situation: special attention for boundaries (wet-dry, high-low).
4. Soil suitability can be shown: possibilities and limitations of a given situation: a) soil map, b) soil landscape map, c) ground water steps, d) carrying capacity, e) soil suitability for wood, f) the same for sports areas, g) also for buildings without pile foundations (Dorp 1999. p. 92-93).

The importance of soil maps is based on the fact that the structures of Dutch cities are derived from the landscape structures, as described above about dam cities.

Haarlem

A nice example is Haarlem that started around 3000 BC on a sand ridge along the dune coast. The longitude shape of the sand ridge was taken over by the old centre of the village. These sand ridges are decalcified sand soils, in which a peat layer of maximum 50 cm can occur. On the most elevated parts of the sand ridge the roads were situated. The surface of the sand ridge of Haarlem is about 1,50 to 2,50 metres above mean sea level with a ground water level of about 60 cm beneath mean sea level. On both sides of the sand ridge are old beaches can be found overgrown with peat that is overlaid by a layer of clay. These clay-peat-sand soils are made into artificial building sites by raising them with sand.

Sand supply
Raising grounds used to be done with vehicles, a technique for flushing sand over the area with pumps and water was invented. Both these methods have had very different ecological results. Especially the dense layer of sand that is the result of flushing (the water cements the sand grains together) has bad results for the fertility of vegetation. The sand layered by vehicles is much less dense (not a perfect grain fit). There is an obvious series of soil capacity for the growing of trees:

1. sand ridges (well drained, large pores volume, easy for roots, easy growth)
2. raised by vehicle (heterogenic, sometimes good and sometimes bad growth)
3. raised with water (excellent top layer of 23 cm, not easy for roots, damage by draught) (Dorp 1999. p. 94-95).

Soil as a basis for urban extension
In certain clear cases the soil map is enough to be able to see the suitability for urban design. This was especially the case 30 years ago when the Dutch soil maps were readable by laymen. With the increase of knowledge about soils it became more difficult, but not always impossible.
The legend on landscape ecology maps usually has a symbol for 'other diagnostic characteristics' that contains facts like texture, drainage, slope and stability. In making the structural plan for the Haagse Beemden in Breda a beautiful connection to the original characteristics of the original landscape was made by the use of good maps.

Strategy
Koning and Tjallingii (1991) formulated the following fields of attention in considering urban ecology:

1. **Flows**
    Water, energy and goods come into the city and leave the city as products or waste. From an ecological perspective it is crucial to handle these flows sufficiently and orientated on circular courses, and to avoid wasting goods. Also flows of organisms, plants and animals need to be considered in this.
2. **Areas**
    There must be spaces where nature can be experienced. These

spaces must be made an integral part of the living and working conditions to make them more attractive, also they must provide places for leisure and places for plants and animals.

3. **Commitment**
    Attention for the relationship between human and nature as part of their daily city experience. How does he treat nature and how is it experienced?

These aspects offer conditions for a good ecological circumstance and can be used to design the physical structures. Examples are water flows and water treatment in parks connected by green areas to the boarder of the city (Dorp et al. 1999).

19.4   Spatial quality of water in the city

19.4.1   The value of experience of water in the city

Perceptive value of natural water

Apart from being a technical necessity, water in the Dutch cities is also a highly valued spatial quality. The experience of water is determined by the quality of the water. The more clean and natural it looks the more it is experienced as pleasant. The main causes of disturbances are dirt, stench and murky water. Secondly are duck weed, detergent and strange colours. A bad water quality brings up unsafe feelings: it is worse to fall into dirty water then when it is clean.

The wish to make water in the city more natural by the use of nature friendly water bank, where plant scan grow and animals can live and mate, is part of wishing to see water as a part of nature in the city.

Identity, leisure and separation functions

The user value of open water is an important aspect that differs from city to city. The water can be an aspect of the identity of the city as a water city, like Venice, Bruges, St Petersburg and Hamburg. Or Amsterdam that is a city of canals, and always has been. The water is extensively used for living on houseboats and going about in little boats. Some cities even use the water as an extensive public transportation line, like Venice and Rotterdam. The value of use has an interesting history as described in one of the previous sections. The most important value is now formed by the use of water as a separating entity between private and public space. It forms a comfortable buffer with peace and nature, an unused open space, claimed by nobody. The view over water is well appreciated and living along the water is even valued in the price of real estate.

Fig. 197 Venice (Fransje Hooimeijer)   *Fig. 198 Houseboats Amsterdam*

(Meyer, 2002)

Alternative water management
The importance of the quality of water, as well as the quantity, is very much object of study. Most city water is too dirty and the maintenance is not as it should be. "This is possible with combining quality with the design of the open water. People have attention and respect for those things valuable to them and things that improve their living conditions. Qualities that can be experienced are more successful than orders and prohibitions." (Pötz and Bleuze, 1999). Thus, by giving the water a prominent position in the city environment more care will be given to the quality of the water.
Physical qualities of water threaten to disappear and a process of awareness is needed to bring better water and living conditions back to the city. The following technical measures are available:

1.   Store and clean the water
2.   Make the water system visible
3.   Install a circular course

Public awareness of water quality
Making the water more visible by making the discharge canals part of public space has been done in Leidsche Rijn in Utrecht. Infiltration areas where rainwater is infiltrated visibly through ditches is done in Ruwenbosch, Enschede and reuse of water by making water storage areas and treating the water to be used in toilets to save expensive drinking water can be seen in Poldendrift in Arnhem.
It turns out that commitment results in more knowledge, houseboat dwellers in Amsterdam are better informed about the water circumstances than other people in Amsterdam.
The experience of cleaner water is supported by the presence of animals in and on the water. Causes of pollution are found in industries and lack of flushing the canals with cleaner water. Consequently less dirt will be thrown in the water, especially if it is cleaned more often (Noort and Aarts, 2000).

Factors in spatial quality of water
The overall conclusion is that the spatial quality of water is determined by the following:

1. water is public space with a high value of attraction
2. water offers nature, plants and animals that enhance the overall water experiences
3. water offers peace, not much can happen on the water that will be a disturbance (like for example a motorway)
4. water often has a high historical value
5. water belongs to the identity of the city and its inhabitants.

19.4.2   Design of public space

Quality
The current water tasks include the improvement of quality and dealing with increasing quantity of water in the city. The design of public space has to deal with both aspects. Designers have the opportunity to design waterways with natural friendly banks to improve the quality of the water. It is also better for streets and waterways to be designed in such a way that they are not too close to one another. Cars have a great polluting effect on open water. Also the flow of the water can be regulated by ensuring that polluting factors are located towards the final part of the area where the water flows, rather than at the beginning.

Quantity
Measures to deal with the increasing quantity of water are also part of the design of the urban area and the public space. An urban area has to deal with five types of water: rain, drinking water, surface water, groundwater and waste water. Also there are three sources: rain, drinking water and surface water. Rainwater is discharged through infiltration (44%), evaporation (33%) and flowing in the surface water (22%). As well as these 'natural' ways of discharge there is a drainage system for rainwater, usually combined with the sewer system that also discharges waste water.

Water balance
The designer has a great influence in designing a public space that properly deals with the natural ways of discharge infiltration, evaporation and flow. This balance can be changed when for example larger areas of surface water are implemented in the design. The percentage of flow will rise so infiltration and evaporation go down. In an area with a large pressure on space and a high density a designer can for example make sure that the parking lots are half paved so that they serve as an infiltration field and make a rule for buildings that they should catch rainwater for the use of the toilet. In that case the percentage of surface water can be reduced significantly.

The choice for the design of the surface of the urban area, in relation to the amount of surface water and elements that can increase evaporation is an important design aspect of an adequate water system.

19.5   Conclusions

The water city in water land is highly dependent on the rules of the water as a part of landscape and urban ecology, especially in the Netherlands and especially since climate change is a very important strong environmental factor.

We have seen that during the phase of acceptance and defence (- 1500) there was a great coherence between humans' actions and the logics of the water system, based on a flexible (mental and physical) attitude. During the offensive phase technological knowledge was developed in coherence with the knowledge of the capriciousness of the territory. This was very lucrative and a condition for economic prosperity. It can show us now that specialisation of different fields of knowledge is not always the ultimate solution.

The reintroduction of water in the city came with the upcoming importance of landscape (and city) ecology in the 1970s. A crucial aspect are the maps used to be able to determine the capacity of an area for urban use. It is a base on which all the other aspects of the urban design should be built.

In the early manipulative phase we can see the example of integrating different urban tasks in one plan. Water should not be handled as a separate problem, but as a task that also brings quality and conditions for a better living environment. As we have seen water is a much appreciated element of public space and with good design and maintenance it is a gift to the city. The designers have ignored the water task during the phase of manipulation. The great post war building task was done on sand layers. All the technical aspects were dealt with by the engineer, and was not something that concerned the urban designer. Since the design of the public space is crucial in dealing with the water task these two disciplines need to come together again. More attention to water and historic preservation had done a lot of good for the water task in the phase of adaptive manipulation. Projects are approached from different angles and with attention to the original structures of the landscape.

20 URBAN ECOLOGY, SCALE AND STRUCTURE

Taeke de Jong[a] T U Delft

---

[a] With acknowledgements to Drs. Johan Vos, urban ecologist of the Municipality of Zoetermeer.

20.1   Introduction

Beyond understanding
Urban ecology does not stop by understanding alone.
Through its representatives, urban ecologists in service of municipalities or advising urban institutions, it is part of the regional system itself.
It produces images of probable, possible and desirable futures from a biological point of view, influencing the development of towns and infrastructure covering the landscape of and around cities. It takes part in design, planning and decision making.
This chapter tries to discover and develop operational concepts within these contexts.

Urban ecology in an urban context
In practice the urban ecologist meets stakeholders, politicians, designers and specialists experienced in the fields of management, culture, economy, technology, demography and physical planning. They all do have their own jargon and use different language games to convince. Moreover, they got their experience from different levels of scale, inclined to generalise their conclusions into other times, locations and levels of scale. The national administrator thinks national experience is valid on any municipal level, the economist thinks an earlier researched local trend will be regionally reliable as well, the designer thinks about the area as if it were a building and so on. But in practice the discussion is always about one context sensitive case, on a determined level of scale, with many stakeholders, each with their own field of problems and aims to be balanced with facts and imaginations in a project. In this chapter I suppose this behaviour of one particular and dominant species *belongs* to urban ecology. How to clarify this web of desires, information and imagination determining the final realisation of projects? The urban ecologist needs to know that peculiar way of human behaviour as far as it is useful for her or his practice, because it dominates the system to be advised.

Three aspects of practice
This chapter tries to unravel three coherent aspects of practice: the language game by which one speaks (talking about desirable, probable or possible futures), the ways of thinking about solutions (operational or conditional) and the levels of scale to distinguish to make proper conclusions. It does so from an ecological viewpoint: the distribution and abundance of individuals, communities, systems, subject of physical policy and planning as well. After clarifying that limited viewpoint, it starts with the levels of scale to be distinguished being administratively, scientifically, and technically relevant as a condition to keep a clear view on the other aspects.

Three ecological concepts relevant for any specialist
Then it chooses a technical viewpoint nearest to practice: how could we operationally imagine the actual and future urban distribution of individuals,

communities and systems; how could we guarantee their efficient coherence if realised and how to represent that in legends of a drawing, variables of research, items on the political agenda? That will not be elaborated in detail, but just enough to go back to three ecological concepts surprisingly recognisable and convincing for other specialists: ecological tolerance, the technical-ecological definition of environments and the distinction of conditional or operational thinking.

The message of ecology
At that time the time is ripe to come back to the communication of an ecologist with specialists in the project team: how to show them the importance of context, the object of any ecological study by definition. It will redefine the role of an urban ecologist in a planning team and we can conclude that ecology has a message for them.

20.2 Urban ecology

Dynamics and diversity
Urban ecology as a scientific domain of study is a subdiscipline of landscape ecology, the science of individuals, communities and systems of plants, animals *and* humans, their artifacts related to flows of matter and energy on a regional level of scale, be it with more attention to the dominance of humans. The dynamic landscapes within and around towns and conurbations have special characteristics justifying a sub-specialism. The dynamics of human activities and artifacts increase from natural outskirts into rural, built-up and city areas, gradually or with great local contrast (see chapter 16, page 267). However, along the same urban gradient spatial fragmentation and diversity of land use increase (see *Fig. 118*). If biodiversity decreases or increases along that line depends on the level of scale. In natural circumstances diversity is often connected to low dynamics. That may be an important directly observable difference compared to usual ecology.

20.2.1 State of dispersion and density

Density, dispersion and form
According to some internationally authoritative text books (Andrewartha, 1961; Krebs, 1994; Begon, Harper et al., 1996) ecology is at least the science of distribution and abundance of organisms.

Fig. *199* shows that different states of distribution (or supposed to be less conciously intended: 'dispersion') in the case of the same density: 100 spots on the same surface. So, abundance (in a standard surface or volume called density) and state of dispersion are different concepts. In the case of the same density different states of dispersion are possible after all. Fig. *199* also shows the designers' concept of form or shape supposing state of distribution.

Urban ecology adds the distribution of human artifacts (including human institutions) as an important factor. It studies the density, distribution and mobility (distribution in time) of urban components (form and morphogenesis), understood by their separations, connections and operations (structure and organisation), their use and impact (function, eufunction and dysfunction included).

Change of distribution like *processes* of accumulation (concentration) or dispersion (deconcentration) on different time-scales could be named as morphogenesis even if there is not yet change of abundance (growth or decline). The description of form and morphogenesis (morphology) still could abandon causal explanation by supposing separations and connections (structure) or use by any species (function). That is why form can be primarily the object of design (studying possibilities apart form probabilities).

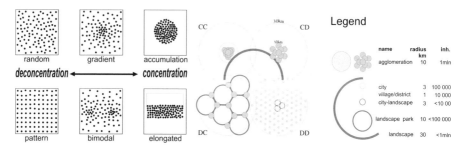

Fig. 199 States of dispersion of   Fig. 200 One million people in two states of
    100 specimens in the same   dispersion on two levels of scale
    density on one level of scale   ('accords' $C_{30km}C_{10km}$; $C_{30km}D_{10km}$; $D_{30km}C_{10km}$
                                and $D_{30km}D_{10km}$)

Fig. 200 shows that morphological categories like concentration and deconcentration could have different meanings on different levels of scale. There are states of dispersion to be described as concentrated in a radius of 30km (C_{30km}), but in the same time deconcentrated in a radius of 10km (D_{10km}) and the reverse ($D_{30km}C_{10km}$).

20.2.2   Urban ecology and urbanism

Urban ecology is ecology of the 'urban district' as a biogeographical unit amongst other biogeographical units (Denters, 1999 and chapter 14, page 245). It is the science, design or policy of distribution and abundance of species in case of a relatively high abundance of people (and their artefacts) in different variants of distribution (urban and rural area). Urbanism is a kind of autecology, studying the optimal living conditions of the human species, its states of dispersion and those of its artefacts.

20.2.3 Different kinds of ecology

From the PHD thesis of Mechtild de Jong (2002) I conclude six kinds of ecology in the Netherlands in the previous century competing for governmental assignments (see *Fig. 201*).

	abiotic	biotic	Dutch university
environmentology	environment	society	
autecology	habitat	population	Wageningen
synecology	biotope	biocoenosis	Nijmegen/Wageningen
cybernetic ecology	abiotic variety	biotic variety	Delft
system dynamic ecology	ecotope	ecological group	Leiden
chaos ecology	opportunities	survival strategies	

Fig. 201 Kinds of ecology in decreasing anthropocentrism

In the present-day Dutch target species policy, supported by the European Birds and Habitat Directive, synecology now is most successful. However, climate change changes its description of biotopes, biocoenoses, their distribution and abundance. Cybernetic ecology focuses on steering mechanisms (separations and connections, structure), their distribution and abundance, suitable for *creating* or restoring living conditions (technical ecology).

20.3 Scale and size

Urban dynamics on different levels of scale

The change of urban landscapes on different time scales (see *Fig. 202*) cannot be understood without a selective study of human dynamics as the primary driving force behind it.

within last	changes in urban areas	within last	changes in urban areas
millenium	Mediaeval, Industrial, Modern towns	week	alternating work and weekend
century	economic development	day	intensity of use, transport
decade	groundworks, building activities	hour	sunlight and precipitation
year	seasons	minute	human activity
month	migrations, flowering periods, trade		

Fig. 202 Urban dynamics on different time scales

Apart from these time scales, already mentioned by Van Zoest (chapter 16, page 267), this chapter focuses on spatial scales to find a sound structure of the discipline (paragraph 20.3).

States of dispersion on different levels of scale

Urban ecology studies 'states of dispersion' of species and artefacts on different levels of scale in the same time. Doing that systematically (see

Jong and Paasman, 1998; *Fig. 200*), you can compare designs mutually, like variants for Randstad (Jong and Achterberg, 1996) or judge local spatial visions ecologically based on rareness expressed in kilometers and replaceability in years (method Joosten, 1992, applied in Jong, 2001; *Fig. 203*).

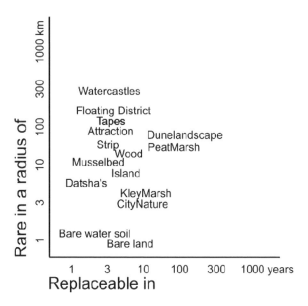

Fig. 203 Ecological evaluation of neighbourhoods named by designers (r=300m) according method Joosten (1992)

The scale of a drawing can be named simply by a nominal radius R from the range {…1, 3, 10, 30 …m} globally encompassing its 'frame' in (supposed) reality, and the smallest drawn detail (*granule*), named by a nominal radius r from the same range.

The distance between frame and granule determines the *resolution* of the drawing. If that distance is small, designers speak about a sketch design, and if it is large about a blue print. Any scale (combination of frame and granule) presupposes a specific legend, a specific vocabulary of the drawing and consequently technically, scientifically and from a viewpoint of government and management a different approach.

20.3.1   Technically

For example, if you aim for diversity in vegetation on different levels of scale, on every level of scale there are different technical means at your disposal (see *Fig. 204*).

Operational variety conditions for vegetation	in a radius of approximately
elevation, soil	30km
soil, water management	10km
seepage, drainage, water level, urban opening up	3km
urban lay-out	1km
allotment (dispersion of greenery)	300m
pavement, treading, pet manuring, minerals	100m
difference in height, mowing, disturbance	30m
solar exposition, elevation	10m

scope

'difference'

'equality'

The radius should be interpreted elastically between adjacent radiuses.
The last four levels of scale hide from the usual view of observations per km^2.

Fig. 204 Operational variation per level of scale (Jong, 2000)

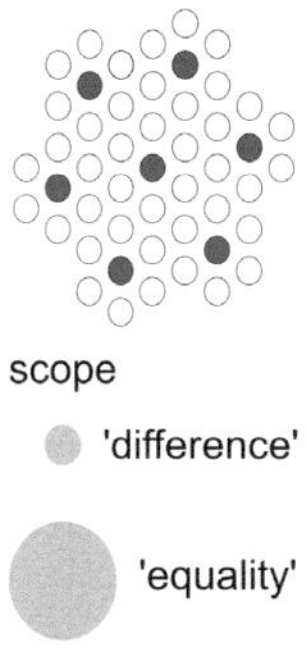

Fig. 205 Scale paradox

However, what causes diversity on one level, could cause homogeneity on another level of scale (scale paradox, see *Fig. 205*).

20.3.2   Scientifically

All ecologies from *Fig. 201* are scientifically meaningful if arranged to the scale of their most appropriate application (see *Fig. 206*). So, microbiology applies on levels of scale and size measurable in micrometers. Chaos ecology stressing individual opportunities and survival strategies or biology stressing cooperation and competition of specialised functions (organisms or organs) applies on levels measurable in millimetres, and so on.

Fig. 206 shows a preliminary distinction of levels of scale and ecologies supposed to be most appropriate on each level of scale.

nominal	abiotic	biotic	ecology
kilometres radius			
10000	earth	biomen	environmental ecology
1000	continent	areas of vegetation	
100	geomorfological unit	plant-geographical or flora-districts	
10	landscape	formations	landscape ecology
metres			
1000	hydrological unit, biotope	communities	synecology
100	soil complex, ecotope	ecological groups	system dynamic
10	soil unit and transition	symbiosis	cybernetic
millimetres			
1000	soil structure and ~profile	individual survival strategies	chaos ecology
100	coarse gravel	specialisation	biology
10	gravel	integration	
1	coarse sand 0,21-2	differentiation	
micrometres (μ)			
100	fine sand 50-210	multi-celled organisms	micro biology
10	silt 2-50	single-celled organisms	
1	clay parts < 2	bacteria	
0,1	molecule	virus	

*Fig. 206 Ecologies arranged to their primary supposed range of scale
(Jong, 2002)*

20.3.3   Administratively

That does not mean these ecologies have to be limited to their primary level of scale as presented in *Fig. 206*. Present-day Dutch nature conservation policy has a synecological origin (Bal et al., 2001) but it focuses on different levels of scale by that approach (see *Fig. 207*).

In *Fig. 206* it is supposed synecology has its most appropriate application on areas of 1km radius approximately, but it claims to offer tools of nature conservation at 3km, 300m and 10m as well.

	Main group 1	Main group 2	Main group 3	Main group 4
Name	almost-naturally	supervised-naturally	half-naturally	multifunctional
Radius	**3km**	**1km**	**300m**	**100m**
STRATEGY				
spatial scale	Landscape > thousands of ha.	Landscape > 500 ha.	ecotope/mosaic to approx. 100 ha.	ecotope mostly a few ha.
· location	mostly process-determined	process and pattern-determined	process-, pattern- and species-determined	pattern- and species-determined
· processes	not directed	directed integrally	directed in detail	directed in detail
· patterns	not established	not established	established, perhaps a cyclical succession	established
directing variables	none	process-focused on landscape level	process- and pattern-focused up to ecotope level	process- and especially pattern-focused up to ecotope level

Fig. 207 The levels of scale in Dutch synecological nature conservation policy (Bal et al., 1995)

From a viewpoint of local government according to *Fig. 203* a different approach is imaginable. A municipality could focus its policy on a specific scale of rareness (identity). For example focusing on global (R=10 000km), European (R=3000km, tables of Flora- and fauna legislation), national (R=300km, Dutch 'Red List' species), provincial (R=100km), regional (R=30km) or local (R=10km) rareness, are different policies. Large cities could focus on global identity, small ones on a regional identity.

20.4   Structure

20.4.1   Coherence in state of dispersion

Technical ecology concerns determining states of dispersion of legend units by *design*. 'Form' presupposes that the state of dispersion is arranged into lines and surfaces. So, a designer or modeller draws the shape of any legend unit relevantly distinguished on the level of scale at hand (see Fig. *208* and Fig. *209*). These influence other legend units amidst of which they are supposed to function in reality. By doing so he or she presupposes coherence (structure) of legend units comparable with symbiosis.

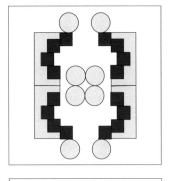

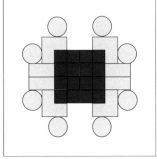

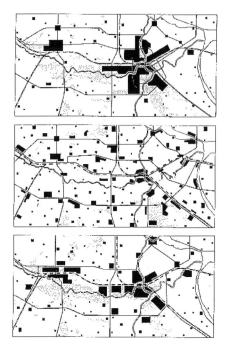

Fig. 208 States of dispersion R=100m

Fig. 209 Accumulation, sprawl and 'bundeled deconcentration' R=30km
(RPD, 1966)

So, technical ecology studies legend units (categories or combinations of categories, types) in urban and landscape design, their possible states of dispersion and its impact on their environment.

20.4.2 The state of dispersion of green areas related to human presence

For example, if you give the green surface in a neighbourhood, district or town the same size of radius as the average walking distance to reach it ('standard green structure' as I will call it), then in each case approximately 10% of the surface is green (see Fig. 210).

Open area	within	radius
• Landscape	100km	30km
• Landscape park	30km	10km
• Urban landscape	10km	3km
⌣ Town park	3km	1km
◯ District park	1km	300m
◦ Neighbourhood park	300m	100m
Ensemble green	100m	30m

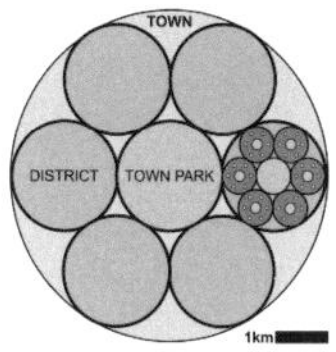

Fig. 210 'Standard green structure'

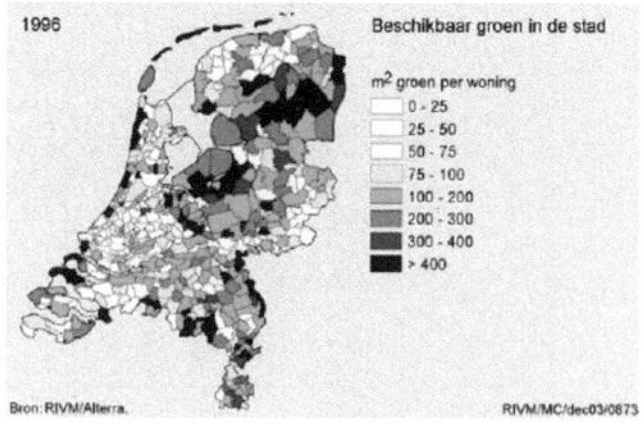

Fig. 211 m^2 Green per dwelling (RIVM, 2003)

In a town (R=3km, surface 30km^2) of 100 000 people that is 30m^2 /inh. town park.

In a district (R=1km, surface 3km^2) of 10 000 people that is again 30m^2 /inh.district park.

In a neighbourhood (R=300m, surface 0.3km^2) of 1000 people that is again 30m^2/inh. neighbourhood park.

If that approach is continued to the fragmented green, any inhabitant would have 120m^2 green at their disposal or approximately 300m^2/dwelling. In Dutch terms that is a maximum (see Fig. 211), but also an easy to handle target standard. The usual standards changing in time (see *Fig. 135*) could be expressed in percentage of this standard green structure.

This approach could include at least 6% water (Hoogheemraadschap van Rijnland, 2002) in urban areas according to the obligatory 'Water test' (Watertoets[a]), to be combined with wet nature (Bureau Stroming, 2006).

20.4.3   Large green areas far away or smaller ones close-by

Now you can work out how much any town deviates from that standard and which level of scale has a relative (dis)advantage. It raises the fundamental question if a town prefers more large green (easy to maintain) at great distance (little public support) or more small green at small distance deviating from the standard. It determines largely the green identity of an otherwise equal amount of green surface per inhabitant.

20.4.4   Boundaries between legend units

On the boundary of legend units mutual effects cause different questions. Formulating these questions depends on the definition of the legend units. If you indicate built-up areas by 'red' (R), paved areas by 'black' (B), unpaved by 'green' (G) and water by 'blue' (B), then *Fig. 212* gives an example of questions under discussion.

---

[a] http://www.watertoets.net/

radius in m	Red/Green	Red/Blue	Red/Black	Green/Black	Blue/Black	Green/Blue
30 000	national dispersion?	building on the beach?		highway landscaping	IJsselmeer Dam	Dutch waterland
10 000	Green Heart?		mainports			Casco-concept
3 000	buffer zones?				Tjeukemeer	2 networks strategy
1 000	town park?	Dream of the real estate agent	noise nuisance		harbours	bank recreation
300	district park?				boulevards	
100	neighbourhood park?		opening up			
30	fragmented green?	drainage		shoulder management		taluds
10	court or garden?				quays	
3	ornamental green?	water cities	building line margin			
1						campsheeting

Fig. 212 Questions at the boundary of legend units

These are well-known questions of spatial lay-out and maintenance on different levels of scale: separating or connecting red and green, red and blue and so on. Mixing means separating and connecting on a lower level of scale, segregation on a higher level of scale. So mixing is not the same as connecting, segregation not the same as separating.

Boundaries (separating *and* connecting zones) could be sharp (limes convergens) and vague (limes divergens), taking up little or a lot of space. If a boundary area takes up a lot of space, it could be indicated by a separate legend unit. In these 'in between areas' (mi-lieus) the more homogeneous properties of adjacent areas vary as a transitional stage (gradient) from one legend unit into the other. A transitional stage offers more specific possibilities of settlement for species and artifacts with a specific ecological tolerance of a varying 'factor of settlement'.

However, any legend unit could be interpreted as 'milieu', as a transitional stage between two adjacent legend units. Such a transition can be operationalised for each direction between adjacent legend units as 'environmental variables'. By such a transition of properties within an originally homogeneous legend unit more specific chances emerge for species and artifacts.

20.5 Ecological tolerance

20.5.1 Inbetween areas

The curve of ecological tolerance shows the chance of survival as a result of values of an environmental variable between 'too little' and 'too much' (see *Fig. 213*).

Suppose for example that water is present as an environmental variable for plants. In that case the chance of survival of a specimen, population or species varies between drying out and drowning.

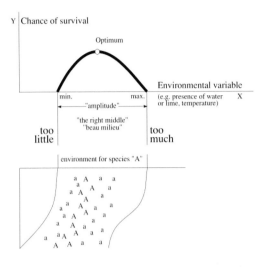

Such an effect can be observed in the field as shown in *Fig. 213* below. Suppose in that figure a slope is imagined from high and dry to low and wet. Species A will survive best near its optimum. On the line of optimal moisture we will find healthy specimens (A). Somewhat higher marginal specimens (a) still grow, but even further from the optimum it is too dry for that species, or in the opposite direction, too wet.

Fig. 213 Ecological tolerance in theory and practice

However, the marginal specimens are important for the survival of the species as a whole. Suppose for example it is raining for a long period. The low-standing marginal specimens will drown, the healthy specimens will become marginal and the originally marginal specimens standing high and dry will begin to flourish. In a long period of drought the opposite will occur.

Levelling out the water level for agricultural purposes to the favour of one chosen food vegetation implies not only loss of other species but also an enlarged risk for the vegetation. Long lasting dryness or precipitation destroys the whole yield. So, diversity of environmental properties and species spreads that risk.

20.5.2   Diversity as a risk-cover for life

In general you can suppose diversity to be a risk-cover for life (Londo, 1997).

In the diversity of life forms there have always existed species (or within a species a deviating specimen) surviving a change of circumstances. 'Survival of the fittest' supposes a diversity from which in changing circumstances the environment selects. A decrease in (bio)diversity undermines the resistance of life against catastrophes. From the approximately 1.8 million species we know (probably a small part of what exists), we have lost already approximately 100 000. So, not only do we cause ecological disasters, we also undermine the resistance of life against these disasters.

However, within this impression of things a less friendly lesson is hidden raising ethical questions. A reserve of marginal specimens offers a favourable effect on the chance of survival of a population. And one level of scale higher: marginal populations could save the species in a changing environment. But could we accept marginal human populations?

20.5.3   Health

So, health is a scale sensitive concept. It is also a concept bound to the individual: the ecological tolerance differs even per individual. What is good for one may be bad for the other. In that way an environmental variable (for example medication) could show a different effect on different people. Even if a medicine turns out to cure a symptom in 99% of cases it can cause rare other symptoms, different to those caused in all other individuals exposed to this environmental variable. The individual side effects do not have to reach the statistical mass to appear in the instruction leaflet indicating the used medicine as a cause. However for all individuals together they may raise many new misunderstood symptoms, each too rare to prove the cause. Within current medical practice these sympoms are suppressed by medicines (with new individual side effects as a possible consequence). So, it is imaginable that medical treatment ends up in a negative cycle causing its own work (Jong, 2005).

20.5.4   The impact of operational and conditional treatment

McKeown (1976) argued the increased life expectancy in England and Wales in the beginning of the 20th century caused by decreased infection illness could not be caused by medical treatment, but by improved hygiene (sewage, drinking water supply, housing, working conditions) and food. It pleads for a more conditional than an operational approach to health (see pag. 389). However, Mackenbach (2004) made plausible that at least 20% of the improvement in life expectancy could be attributed to medical care.
Public perception of medical treatment is more positive through its immediate impact. The eventually negative side effects afterwards are not demonstrable in many cases and thereby neglected, but together they could have a substantial negative effect (80%?) on the average life expectancy as a whole.

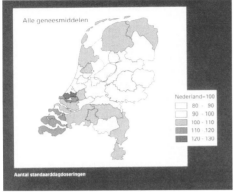
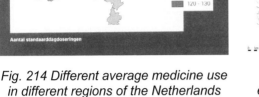
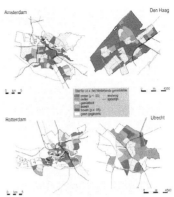

Fig. 214 Different average medicine use in different regions of the Netherlands (Batenburg-Eddes et al., 2002)

Fig. 215 Different life expectancy in cities. (RIVM)

Risk perception rests on incomplete knowledge, uncertainty and fear, aggravated by one sided news coverage about extreme cases, exploited by politicians and insurance companies ("We sell fear").

20.5.5   Generalisation clouded by context factors

The epidemiological research should give evidence, but struggles with enormeous methodological problems. The excess of environmental variables not to be eliminated clouds every supposed causal relation. And like medicine research and toxicology it struggles with reliability of statistics applied on heterogeneous sets.

For example from Fig. 214 and *Fig. 215* you hardly can conclude causal relations (income?, stress?, green in the neighbourhood?). Moreover any organism has by nature many compensating feed-back mechanisms, especially if it can adapt to its environment or move into environments more suitable for its tolerances. In that case reverse causality has to be supposed: favoured or marginal groups gather in specific environments (the 'Gooi' in Fig. 214, or deprived neighbourhoods in *Fig. 215*) with preventive or just illness inducing conditions. The municipal health service (GGD) of Rotterdam in the 1970s found concentrations of schizophrenia in the inner city. What is cause and effect?
Moreover, after some time excessive hygiene or stress could decrease resistance.

With so many uncertainties an urban ecologist or urbanist could only learn from nature to reduce risks through the creation of diversity of environments.
Standardisation into averages (see *Fig. 86*) is not always the most productive way.

20.6   Definitions of environment

Research, design and policy
The probability, possibility and desirability of these aspects (modality) represent scientific, technical and political viewpoints respectively. As soon as we accept decreasing biodiversity and human health as a problem (probable but undesirable), urban ecology becomes 'environmentology' aiming at possible 'sustainable developments' (keeping possibilities for future human generations). After all, any defined 'environmental problem' could be reduced to decreasing biodiversity and human health.

20.6.1   Empirical and technical definitions
'Environment' according to the usual Dutch definition of Udo de Haes (Boersema, C. Peereboom, et al., 1991; see *Fig. 216*) is 'the physical, non-living surroundings of society in reciprocal relationship'. However, for technical operationalisation people belong to these surroundings. Moreover, I cannot imagine people without environment, but environment we can imagine without people (for example other planets). That is why the technical definition we have chosen reads 'the set of conditions for life' (Duijvestein, Jong, et al., 1993; see *Fig. 217*).

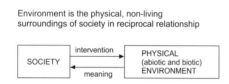

Environment is the physical, non-living
surroundings of society in reciprocal relationship

Environment is the set of conditions for life

	intervention	PHYSICAL
SOCIETY	→	(abiotic and biotic)
	meaning	ENVIRONMENT

Environ-
ment

People

accomodation

adaptation

Fig. 216 Environment according to
Udo de Haes

Fig. 217 Environment in a
technical sense

Social science and geography may stress causal relations (based on probability, the arrows in *Fig. 216* and *Fig. 217*) while technical science cannot abandon conditional relations (based on possibility, such as 'no people without environment', 'no economy without dikes', 'no land without pumping stations'). Conditions are often taken for granted by empirical science as 'ceteris paribus' suppositions, but in technology they are the essence of 'environment'.

20.6.2   Distinguishing many types of environment
By that general technical definition you can define different meanings of 'environment' by substitution of 'conditions' and 'life' more specifically (see fig. *Fig. 218*).

conditions for	human	animal	plant	life
managerial				
cultural	A			
economic				
technical			B	
ecological				
mass/time/spatial				

Fig. 218 Substitution possibilities in defining environment

For example, cultural conditions for human life (A in *Fig. 218*) are a kind of environment ('cultural environment') different to technical conditions for plant life B (like dikes in the Netherlands). They even can be contrary to each other. Conditions can be operationalised by environmental variables taking different values on different locations. Settlement factors for species, populations or their artifacts consist of these values. 'Environmental problems' are simply defined as 'lacking conditions for life'.

20.6.3   Conditions for diversity beyond empirical expectation

It is not easy to recover all tolerable conditions for any life form. There are many species, especially rare ones with small tolerances of which nowhere near all ecological conditions are known. Specific conditions are often forgotten so that target species do not appear. However, within the urban environment you can vary conditions infinitely ('environmental differentiation'; Jong, 1978), so that species appear unexpectedly. If you appreciate natural areas as areas where human planning is minimised, you should not create targets but conditions.

20.7   Conditional or operational thinking

20.7.1   Possible and probable nature

The conditional ('means'-directed) approach chooses non-biotic (abiotic) starting points (Jong, 2000) to make vegetations and dependent animal populations *possible* instead of *probable*. It relies less on predictions about spontaneous results developing out of the created conditions. A well-designed house (oikos) does not determine the household; it makes different households possible. Moreover, in a technical sense you usually do not start to build the roof of a house first, but its foundations, its primary conditions (see *Fig. 219*).

20.7.2   Target species

On the other hand, the operational ('target'-directed) approach chooses target species to realise their specific life conditions, based on the expectation that these conditions are sufficient for other coherent species as well. This usual operational approach appeals to administrators and managers, but it is sensible to the accidental popularity of a target species. It often leads to disappointment if the intended species does not appear like Oriolus oriolus (golden oriole) or to surprise if it appears unexpectedly on other locations like Riparia riparia (sand martin) in Zoetermeer (Jong and Vos, 1993-2006).

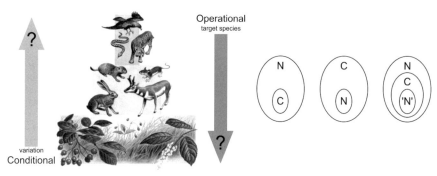

Source:

Fig. 219 Bottom-up or top-down approach   *Fig. 220 Culture (C) and nature*
(Source unknown, arrows added by the author)   *(N)*

20.7.3   Ecocentric and anthropocentric views

Very different opinions underlie the manipulability of nature. The ecocentric view considers culture as part of nature (see the first image of *Fig. 220*). However, the anthropocentric view considers nature as a concept, part of culture (second image). That is the main opinion of administrators and landscape architects. In that view we can talk about more or less 'natural' as added (human) value. Talking about nature forces us to use language. And that is part of our culture. So, the paradox arises in that we cannot exchange imaginations of nature without culture.

On the other hand, from an ecocentric point of view a landscape is more or less 'cultural' (made, planned). Then culture is the added variable of which nature is the starting zero value (not influenced by people, unintended, 'spontaneous'). Within this view nature (N) could contain a culture (C) *talking* about nature ('N') but not covering nature as a whole. Otherwise nothing could be discovered anymore. The Oostvaardersplassen in the Netherlands is a good example of 'unintended' nature.

This controversy can cause confusion of tongues in a planning team as soon as 'nature' is tacitly taken as something to emerge, something to design or as an object of decision.

20.8   Specialists in a planning team

20.8.1   Modalities

Specialists in a planning team not only have different viewpoints on nature, but they also use different (empirical, technical, political) 'language games' (Wittgenstein, 1953), meaning different things by the same statements (Jong, 2002 'Designing in a determined context').

So, in any meeting or poll you should know if any statement is meant as probable, possible or desirable (see *Fig. 221*). Asking people if they agree with a statement has to be distinguished into these modalities. "Do you think it possible, probable or desirable" can produce different answers to the same statement. For example, if the answer is, "It is probable, but not desirable" you have detected a problem. If the answer is, "It is desirable, but not probable", you have detected a target.

	being able	knowing	choosing
modalities	possible	probable	desirable
sectors	technique	research	administration
activities	designing	predicting	organising
reductions of supposed reality into			
character	legends	variables	agendas
place and time	tolerances	relations	appointments

Fig. 221 Three language games

Fig. 222 Different supposed futures, determining fields of problems and aims

20.8.2   Context analysis; fields of problems and aims

However, though every profession has a primary modality, in any profession every modality plays a role. Probable futures as far as they are not desirable consist of a field of problems perceived by the specialist. Desirable futures as far as they are not probable contain a field of aims for acting according to a plan (see *Fig. 222*). The implicit images of a future context within these three modalities determine the judgement of the specialist. However, they can diverge in scale and 'layer' (see *Fig. 212*). So, making these three images of the future context more explicit at the beginning could yield a common field of problems and aims and prevent confusion of tongues.

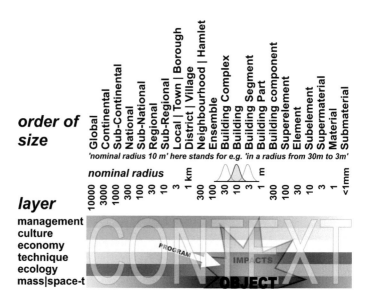

Fig. 223 The future context of the object

What anybody expects or supposes to be possible, probable or desirable, could be different on every different level of scale and layer. The layers of *Fig. 223* give a rough indication of the domain of possible specialists in the planning team.

20.8.3   Including context in any specialism

Specialists in a planning team tend to forget the administrative, cultural, economic, technical, ecological or spatial context within which they participate. That is why making the future context explicit is so important. For example in planning the Dutch newtown of Zoetermeer, the urbanists and traffic engineers involved took three-forked roads as a starting point. They argued three-forked roads deliver much less conflict points than crossings with four access roads. Using three-forked roads would reduce casualties substantially.

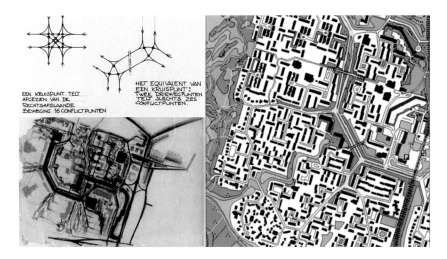

Fig. 224 The traffic concept of three-forked roads design, in Meerzicht, Zoetermeer (Gent, 1999)

However, the number of casualties after realisation did not appear to be less than elsewhere. People drive faster if the situation has clear overall visibility and culture develops into a stage of more indiviual responsibility with less police surveillance. On the other hand, the hexagonal urban lay-out causes more directions of allotment. The inhabitants of Zoetermeer do not complain about that kind of diversity, but services like ambulance, fire brigade, police and delivery services complain they cannot find their way. From an ecological point of view this design principle offers more diversity of expositions to the sun. The observed diversity of plant species in Zoetermeer is relatively high. However that possibility did not play a role in the choice for three-forked roads.

20.9 The role of an urban ecologist

20.9.1 Specialist or generalist

Should an urban ecologist take up a specialist's or a generalist's position? Because human species is one of many I would prefer a generalist. The ecologist can offer arguments others do not think of, while coalitions with nearly all other specialists are obvious. The urban ecologist firstly should go into subjects of urban civil and environmental engineering (Jong, 2006). Specialisms in urbanism, architecture, cultural heritage, building technology, sustainable building, management of public space and greenery are probably primary allies.

For example, if the administrator or planning agency receives conflicting advice from specialists on ecology and cultural heritage separately, this advice could cancel each other out in the decision procedure. Then, the many opportunities of common interest would be lost.

Urban ecology has its own natural contacts with the urban population via nature associations with many local departments present in any town, like the Dutch IVN, KNNV, Vogelwerkgroep.

20.9.2   Local identity

Within the emerging profession of 'area development', (in Dutch: 'gebiedsontwikkeling') a project developer's variant of urban design[a], 'identity' is again a key concept (Geoplan, 2005). Identity is difference from the rest and continuity in itself. The green identity could play a crucial role in urban development. Give any neighbourhood its target species or its unique variant of boundary rich green area. Call street names according to these target species, promote gradual transitions from light to dark, from wet into dry, from accessible to less accessible, from paved into unpaved. Promote the freedom of choice of staying in the public area for different age groups, income groups, life styles. Diversify housing environments that way as a marketing concept. Even urban economy is a potential ally. The price of a dwelling with a view on a park is higher than in a street, in a street with trees higher than in a street without trees. However, that is earlier understood by housing corporations with long-lasting involvement than by project developers.

20.9.3   The task of an urban ecologist

An urban ecologist can be expected to guard (probable), balance (possible) and promote (desirable) the conditions of all life forms in towns in any planning team.

That means attention for monitoring legal and administrative frameworks, but also stimulating imaginative powers, showing possibilities on any level of scale (see *Fig. 225*).

---

[a] www.fellowsgebiedsontwikkeling.nl

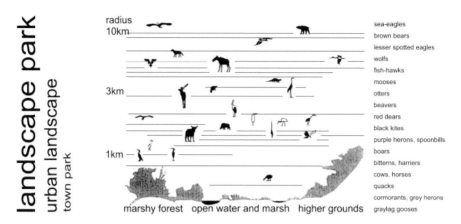

landscape park
urban landscape
town park

radius
10km

3km

1km

sea-eagles
brown bears
lesser spotted eagles
wolfs
fish-hawks
mooses
otters
beavers
red dears
black kites
purple herons, spoonbills
boars
bitterns, harriers
cows, horses
quacks
cormorants, grey herons
graylag gooses

marshy forest   open water and marsh   higher grounds

Fig. 225 Possibilities for nature per level of scale and height position
(after H+N+S, Utrecht)

Because human species plays a crucial role in the town, safety, health, leisure and sports are natural components within that power of imagination (see *Fig. 226*).

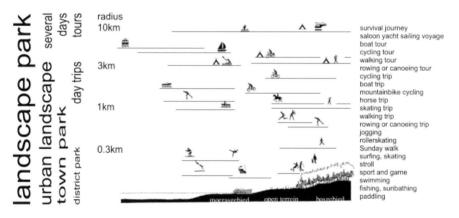

landscape park
urban landscape
town park
district park

several days
tours
day trips

radius
10km

3km

1km

0.3km

survival journey
saloon yacht sailing voyage
boat tour
cycling tour
walking tour
rowing or canoeing tour
cycling trip
boat trip
mountainbike cycling
horse trip
skating trip
walking trip
rowing or canoeing trip
jogging
rollerskating
Sunday walk
surfing, skating
stroll
sport and game
swimming
fishing, sunbathing
paddling

moerasgebied   open terrein   bosgebied

Fig. 226 Possibilities for leisure per level of scale and height position (after H+N+S, Utrecht)

20.9.4   Balancing interests

From these possibilities those increasing freedom of choice for future generations ('sustainability') have to be given priority. Freedom of choice supposes diversity (homogeneity delivers no choice); and diversity is tacitly supposed in any opinion about quality. Too much diversity causes chaos, confusion of choice; too little causes boredom, and enforced choice (see *Fig. 227*). Recognition and surprise, traditional or experimental choice can alternate in between these boundaries. Their balance or alternation determines quality on every level of scale.

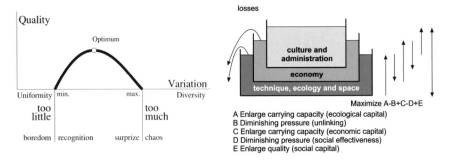

Fig. 227 Quality as a function of variation

Fig. 228 Balancing common interests

The balance of municipal interests can be imagined as vessels filled with liquids of different specific gravity (see *Fig. 228*). The largest vessel is filled with the heaviest liquid: the available space, ecological capital and technology, what the economy can bear. If the pressure of economy is too high it will cause losses of that basic capital, thereby diminishing possibilities for future generations. Economy is represented as a vessel filled with lighter liquid: economic capital capable of carrying an even lighter vessel of social capital.

A municipal policy should therefore:

A.   increase the ecological carrying capacity;
B.   decrease the economic pressure;
C.   increase the economic carrying capacity;
D.   decrease the social pressure on the economy;
E.   increase social capital.

The importance of nature for a town.
The role of an urban ecologist in the balance of *Fig. 228* concerns particularly A and B, not forgetting C, D and E. In the mean-time a number of arguments for and against a 'green' town is superfluous. I will not go into these arguments here. Many publications refer to them (see chapter 15, page 259 and 16.3.4, page 291), supported by interests of economy (the willingness to pay for a representative or for an efficient, useful environment with facilities nearby), well-being and public health (direct and indirect, including social safety and medical care), physical environment (contamination, particulate matter, climate).

The importance of the urban area for nature.
The municipality has to take into account often conflicting nature interests on global, continental, national and regional levels of scale. Especially the actual policy of operational conservation of populations and their habitats

seems to oust a broader conditional approach by its strictness and urgence.

It concerns conservation, protection and development of nature:

1.   the global conservation of diversity (Rio de Janeiro 1992);
2.   the continental passive conservation of species and their habitats (Birds and Habitats Directive) executing 1;
3.   the national passive conservation of species and their habitats (Flora- and Fauna legislation with its protected species, obligation for care, prohibitory clauses, protected habitats and codes of behaviour) executing 2;
4.   the national protection of existing objects ('Natuurbeschermingswet' with its 'Aanwijzingen' and 'Beheersplannen');
5.   the national active policy to protect species (with its approach of habitats, long term programmes like 'Meerjarenprogramma Uitvoering Soortenbeleid 2000-2004', 'Soortenbeschermingsplannen', red lists);
6.   the national active policy for nature development (with its ecological network 'NEN', wet nature and national parks);
7.   the regional, provincial tasks resulting from these policies.

Existing situation
In the Netherlands 2006 the national Flora and Fauna legislation integrating the European Birds and Habitats Directive dominates the scene. Since the Dutch 2006 order of council ('AmvB') the urban policy focuses on protected species of 'list 2' (exemption proviso code of behaviour) and 'list 3' (extended check in case of exemption). So, the protected species of 'list 1' (light check) and the duty of care generally in force ('artikel 2') threaten to lose focus. Codes of behaviour for 'list 2' species are developed fixing the duty of care at the same time. In 2005 the municipality of Leyden developed such a code of behaviour submitted for approval to the Ministry of agriculture and nature preservation ('LNV'). The association of municipalities ('VNG') develops such a code of behaviour preventing the effort of separate municipalities. However, these municipalities themselves should have an overview of protected species and values present in their own territory.

Applying an officially accepted code of behaviour has the advantage that not every several plan for building or demolition needs a separate request for exemption. That prevents delay and inefficient use of financial means.
This procedure has the disadvantage of focusing on a relatively small number of species important on a European level but without much local

public support in the Netherlands[a]. On the other hand nationally rare species ('Red list') are not protected.

More desired situation

National 'Red list' species should be protected more actively. They appeal more to public sentiment and create public support. Apart from legally protected species there are many species, though not rare, with an emotional meaning connected to the identity of the area. Based on that support more possibilities develop for a local municipal policy on biodiversity extending into other species.[b]

Apart from the operational approach forced by legal urgencies, a more conditional approach articulated to scale is needed, focusing on all conditions of urban life.

20.10   Conclusion

The urban ecologist should be prepared for participation in a planning team. Distinguishing other modalities than scientific ones, their language games, distinguishing operational versus conditional thinking, distinguishing levels of scale, will help to survive in a community of specialists, designers and politicians. But as a professional the urban ecologist has also a message to all of them: be careful with ecological tolerances of individuals, communities, systems, do not always obey singular targets, shape conditions for fields of aims and possibilities for future generations, be context-sensitive and modest in what you think to know.

---

[a] For example the Pipistrellus pipistrellus ('pipistrelle bat', 'dwergvleermuis', stricly protected, but general in urban areas) and Anisus vorticulus ('ramshorn snail', 'platte schijfhoorn', a snail species nobody heard of before).

[b] For example the nymphaea alba ('white lotus', 'witte waterlelie'), the passer domesticus ('house sparrow', 'huismus'), the fungus symbionts (in symbiosis with trees and shrubs) for example successful in Zoetermeer.

Infrastructure

INFRASTRUCTURE

Taeke de Jong

Editorial introduction

Half the Netherlands would not exist without dykes, pumping stations and connected waterworks maintaining numerous water levels separated by dykes, weirs and sluices: the 'wet infrastructure'. This wet infrastructure would not exist if there were no economy to sustain that system. And that same economy requires its own network of roads, railroads, cables and pipes, the 'dry infrastructure' crossing the wet one by numerous power pylons, bridges, subways and tunnels. Minor and certainly major water-works like the Zuiderzee- (northern sea intrusion) and Deltawerken (south-west) are prepared by civil engineers in cooperation with economists, agri-cultural engineers, urban designers, landscape architects and many other specialists.

However, the same economy sustaining these systems requires 'beauty' to prevent people escaping a landscape visually dominated by technology, taking refuge to better balanced living conditions elsewhere known from holydays. But that flight undermines economy. Beauty of environment de-termines the real estate value, the biggest part of Dutch national capital. It is a marketing item in the international race to keep globalising companies inside and to attract other ones. Beauty combines recognition and surprise. That combination is the task of integrating Dutch design. Too much recog-nition would produce boredom, too much surprise would produce chaos. It is a kind of human ecological tolerance to be reached by civic art.

Just there, within that completely planned, man-made landscape an in-creasing interest in nature emerges, interest in a refuge for the non-planned time and space. That counterpart for mainly stressful living and working conditions provides a natural balance of recognition and surprise. So, planners, designers and engineers are looking for accessible land-scape ecological knowledge, fitting in their own contexts. They do not have the time for thick books without fast key word entrance, scattered articles on incidental ecological problems buried in the archives of journals. They are looking for simple principles applicable in design like provided by the island theory or the regulation theory. They now embrace one, and then another as soon as they suppose easy application. Landscape ecologists should study these questions of planners, engineers and designers to pro-vide appropriate, applicable knowledge, producing negotiable alternatives in local contexts.

Bruin and Jong (chapter 21, page 405: *Civil engineering of a wet land*) give a short overview of the Dutch 'Waterstaat'. How does this system of controlled rivers, polders, coastal protection, management and policy work?

Bruin (1987) is well-known by his contribution to the 'Plan Ooievaar' combining many specialisms, first regional application of the term 'ecological infrastructure', later translated as 'ecological network' and up to now very influential in Dutch nature policy. This chapter gives some idea of how civil engineers think, how they calculate the risks to cope with, what kind of boundary conditions they have to take into account as not negotiable.

Jong (chapter 22, page 447: *Dry and wet networks*) gives a simple rule of thumb for the mesh-width of these networks, their hierarchy and interference. It tries to give laws and measures for potentially ongoing landscape fragmentation to avoid mistakes in locating 'green' areas.

Bohemen (chapter 23, page 460: *Road crossing connections*) starts again with a description of the impact of roads and traffic on nature, their impacts on habitat fragmentation and potential of de-fragmentation. He describes technical and financial backgrounds of overpasses and underpasses to reduce barrier effects and an overview of (existing and still needed) research on the use of fauna-passages and their effectiveness at the population level. He concludes the need for defragmentation strategies integrating the many sectors involved.

Bohemen (chapter 24, page 491: *Roadside verges*) then firstly describes the flora, vegetation and fauna of roadside verges, subsequently focusing on their opportunities for new nature. The integration of road infrastructure in the landscape has a history of accommodating the users into adapting to the landscape. That produces a need for nature engineering connected with (re)construction, management and maintenance of roads and roadside verges.

Tjallingii and Jonkhof (chapter 25, page 508: *Infranature*) describe the changed meaning of roads in the Dutch landscape, their changing practice of planning and design. To illustrate the increasing role of landscape ecology and other specialisms in planning infrastructure, its consequences of increasing complexity in decision making, they take the Crailo Nature Bridge project as an example. It is mentioned earlier in this book on page 336. They describe the stage of former integration of roads in a landscape without cars developing by growing car ownership from a stage of interest in roadside verges by roadside tourism into a stage of increasing protest, legislation, an increasing number of baffle boards, concern about fragmentation, and measures with an ecological background. The story of the Crailo Nature Bridge is followed step by step and shows how the fragile balance of local interests determines if and how ecological targets on a national level can succeed. The authors conclude that other examples of even bigger infra works will need much political effort to reach enough local acceptation for national ecological goals in between many other targets, especially economic ones.

Bakker (chapter 25, page 508: *Leisure and accessibility*) asks the attention of landscape ecology for the potential of symbiosis with recreating people raising their environmental awareness. The author reports different studies on spare time and outdoor leisure, preferences of people relaxing outdoors and public appreciation of landscape and nature. Based on that knowledge he makes recommendations on planning of leisure networks and nodes for man *and* nature. This chapter is written rather from the viewpoint of a private leisure organisation than as a scientific work. For the landscape ecology we consider it relevant, because it shows the importance of research on leisure behaviour and perhaps it indicates unexpected allies in nature protection and development.

Toorn (chapter 27, page 539: *Ecology and landscape architecture*) describes the recent history and state of the art of a very influential design discipline in Holland. Landscape architecture integrates natural systems and functional requirements on the basis of a program into a new unity that strives for contemporary beauty in the daily environment by design. He states that landscape architecture always works in a natural system, hence the 'natural' relation between ecology and landscape architecture.

Toorn has described and analysed research projects, plans, policy reports ('precedents') by Dutch landscape architects on the relation between ecology and and landscape architecture. He first of all concludes that in landscape architecture the theory of 'Van Leeuwen' — next to other theories — is still used extensively at all levels of intervention. The ideas and theory of 'Van Leeuwen' seems to be almost forgotten by landscape ecologists.

Furthermore he notes recent developments in the increased attention for water and water systems as new approach in landscape planning and design. As such the importance of water for all natural systems and human activities is not new but the world wide shortage of fresh water is reminding all of us. In policy this is reflected in a new approach to water management for the 21st century.

Finally, he mentions the increasing number of projects and plans on urban development and infrastructure. Not surprisingly in our 'culture of mobility' this is a contemporary issue of great importance. The planning and design of these 'spaces of flow' forms a major challenge for Dutch landscape architecture in the future. Again this is a social and cultural dimension that also refers to the changing relation to nature in society at large. From an ecological point of view this also requires a different view on the role of ecology and nature expressed in the search for 'new nature'. This 'new nature' is adding a different dimension to the relation between nature and culture. 'New nature' is appealing to needs in society for new representations of nature. Society is also looking at landscape ecology for new ideas, solutions!

Schrijnen (chapter 27 page 539: *The task of regional design*) shows the full complexity and stagnations of decision making in Dutch urban and regional planning and design. It starts with a sketch of the contemporary context, the four issues he distinguishes as priorities, the barriers to be conquered. Schrijnen was director of public works in the municipality of Rotterdam, director of mobility in the province of Zuid-Holland and now director of the largest contemporary urban extension in the Netherlands, the one nationally assigned to Almere. That municipality faces a future national public housing task of about 45 000 dwellings. In the mean time he occupies the chair of regional and metropolitan design of the University of Technology in Delft. The chapter covers largely his inaugural speech.

This section of the book raises its final question: what is the task of landscape ecology in between other tasks delegated to disciplines determining the quality of Dutch landscape?

21 CIVIL ENGINEERING OF A WET LAND

Dick de Bruin and Taeke de Jong T Delft

21.1   Introduction

21.1.1   History

The colors of *Fig. 229* indicate the area in the Netherlands that would become submerged if there were no flood protection dikes and pumping stations. The flooding area as indicated is supposed to occur during modest river floods (up to 4000 m³/sec at the German/Dutch border) and a normal high tide at sea. However, it was not always like that. In 2000 years that area has increased into the current surface by rising external water levels and falling ground levels (see *Fig. 266*).

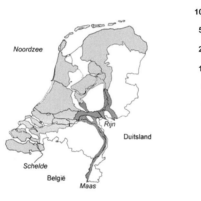

duration	period		issues
1000	100 -	1100	settlement ancestors
500	1100 -	1600 x	Erection dikes, confined contours
250	1600 -	1850 x	Waterlogging control, developing drainage
125	1850 -	1975 x	Riverworks (regulation, normalisation canalisation)
62	1920 -	1982 x	Zuiderzee works
32	1955 -	1987 x	Delta works
16	1975 -	1991 x	Major purification plants
8	1985 -	1993 x	Policy documents tuned (RO, WHH, Trprt, Milieu)
4	1993 -	1997	Pilot schemes, integrated approaches
2	1997 -	1999	Evaluation RWS-200 year
1	1999 -	2000	New water Policy 21st century
	new approaches x		determined by disasters

Fig. 229 Potential threads (RWS, see also Fig. 256)   *Fig. 230 Reverse duration half time of the Dutch water management (author Bruin)*

To cope with regular floods Dutch water management started by erecting terps in the first milennium A.D. and dikes in the next 500 years. At that time the dynamic water surface was confined and the next 250 years the emphasis of water management became waterlogging control and drainage of reclaimed land. Then, in a period of 125 years the Dutch regulated, normalised and canalised their rivers. In a continuing half time of water management policy new priorities developed like Zuiderzee (northern sea intrusion), Delta (south-west) and purification works (see *Fig. 230*). In the last few decades all these continuing efforts were integrated by national policy documents, pilot schemes and evaluation for future safety.

Apart from its threats, water as a medium for trade and transport and as a military barrier for external attacks was also a crucial ally in the development of Dutch independence and perhaps a factor in keeping the nation out of World War I.

Water as military barrier
In the past, the Dutch have created again and again water corridors and water defence systems for the military defence of (parts of) the country. In addition, all major cities developed their own defence system, quite often this is still visible on today's maps of the old cities. In the east and south, huge wild peat areas offered some kind of natural protection against invaders from the east and south east. Where the sub soil contained solid sandy deposits, in other words where realistic chances existed that enemies could penetrate, military fortresses were developed (Nieuwe Schans, Boertange, Coevorden, Grol, Doesburg, Mook, Roermond, etc., see *Fig. 231*) Also along the southern flank of the river area cities developed as military fortresses against invaders from the south (Grave, Den Bosch, Hedel, Willemstad).

Water as primary connection
In parts of the country, through the ages there always have been various options to create water corridors during (threatening) wartime, in particular in north – south direction. These wet corridors were situated in between major military fortresses. To get these systems activated, a well designed (and maintained!) system of sluices, dikes and locks was developed, in combination with natural water systems that could provide sufficient inundation water during critical periods. Today, the remnants of these provisions are cultural elements in the landscape. Quite often money is spent on renovation and restoration, no longer for military reasons but to safeguard a cultural heritage.

Transport
Paved (or railed) roads in the water saturated soft soil areas in the Netherlands gradually started developing from the middle of the 19th century. Around 1800, the best, safest and quickest way to move from the government buildings in The Hague to the navy harbours in Den Helder and Hellevoetsluis was still taking a horse via the beach! That is a major reason why through the ages all the major waterways in the Netherlands were also used for shipping. Until late in the 20th century, most domestic transport of cargo and passengers was done by ship ('trekvaart', beurtvaart). In fact for all important routes and waterways specific (sailing) vessels were developed. The remains of this fleet are now the backbone of the leisure industry. Today, about 35% of all the cargo transport in the Netherlands is still going via waterways; compared to this figure in other countries this is extremely high.

The daily water management of major waterways as shipping routes is still crucial. Shipping developments on the international Rhine also determine the major nautical developments on Dutch domestic waterways. The historic and today's development of cargo transport on the international Rhine (in other words the economic importance of that river), has not been and is not determined by (fluctuations in) the Dutch economy, but first of all by the German economy. The Rhine is the major hinterland connection of the ARA ports (Amsterdam, Rotterdam, Antwerp), and shipping developments have been coordinated and controlled by the International Central Commission for Navigation on the Rhine (CCNR) since the defeat of Napoleon (1813 Waterloo, Vienna Congress 1815). It is the oldest still functioning international body in the world.

International trade
International trade always has been important for the development of the Netherlands. More in particular sea trade on a global scale. It has also determined the intensive navy orientation of society. It is remarkable that for the protection of the capital (Amsterdam, the old trade centre) the so called 'Stelling van Amsterdam' has developed, while for the military protection of the national government centre (The Hague) only a poorly functioning water corridor was available.

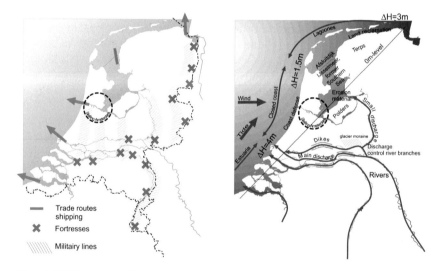

Fig. 231 Water as ally (author Bruin)   *Fig. 232 Water as enemy (author Bruin)*

21.1.2   The distribution of water
The purpose of the Rhine canalisation (3 weirs in the Lower Rhine/Lek branch, plus some bend cuts in the upper reach of the IJssel river) was to gain more control, during low river discharges (of the Rhine at the German Dutch border), of the fresh water distribution via the two bifurcations (Pannerdensche Kop-PK-, IJsselkop -IJK-) to the rest of the country (see *Fig. 233*). Extra fresh water to the north is needed during the dry season, because the IJsselmeer (IJssellake) evaporates about one cm a day during a warm summer day, causing too many shallows in the navigation channels

in the IJsselmeer after some weeks of a dry period. In addition, such a dry period often occurs in the growing season of crops in the adjacent polders around the IJsselmeer , so at that time an extra need exists for fresh water. More fresh water coming down via the IJssel (being the main feeder of the IJsselmeer) can be achieved by closing the weir at Driel.

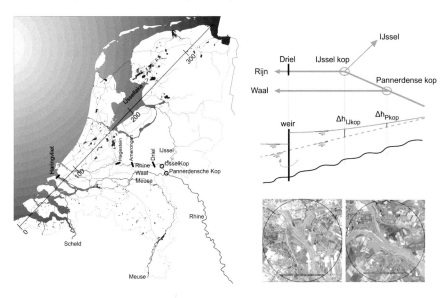

Fig. 233 Weirs directing water
northwards and southwards (author Bruin)

Fig. 234 IJK, PK, Weir of Driel
regulating Dutch water
distribution (Bruin, Google Earth)

The Driel weir is the most important fresh water tap of the country. By lowering (= partly or entirey closing the Lower Rhine) the so called visor gates, a backwater effect is noticeable till upstream Lobith, so also at both bifurcations IJK (more) and PK (less). Because the width of the major channel in the Waal branch is 260 m, and the width of the IJssel major channel only 80 m, the amount of discharge taken from the Lower Rhine will distribute over IJssel and Waal in the order of magnitude 40–60 % / 60-40 %, so as an average 50/50%. However, the lowering of the Driel weir is only possible if first the two other weirs at Hagestein (Lek) and Amerongen (Lower Rhine) are lowered, with the purpose to create sufficient navigable depth in the entire length of the river between IJK and the tidal zone near Rotterdam.

Salt water intrusion
Because the weirs are only closed during dry periods (low discharge of the Rhine at the German-Dutch border), the fresh water discharge coming down the Lek to Rotterdam will be minimised; as a consequence the salt water intrusion from the sea may harm the drinking water inlet east of

Rotterdam along that river. This is not acceptable, so there must be compensation to minimise that salt water effect. It can be done by first closing the Haringvliet sluices, in a way that a backwater effect is created up till at least the Moerdijk zone. Then, all the fresh water coming down both the Meuse and Waal rivers will be sent north to Rotterdam and Hook of Holland. This surplus fresh water is sufficient to stop the salt water intrusion as mentioned.

So one can conclude that a strategic water management of the IJsselmeer is determined by the flush regime of the Haringvliet sluices, via the canalisation of the Lower Rhine.

21.1.3 The threat of floods

The major rivers and the sea always have threatened the Dutch society during severe floods. The tidal characteristics and the regime of the river discharges have determined the development of the flood protection systems in the country. Due to large scale drainage and reclamation over a period of many centuries, major parts of the land where peat deposits at the surface and in the subsoil exist(ed), have subsided. This process is still going on as long as the polders are kept dry with artificial means (pumps, see *Fig. 271*). Due to climate change, expectations are that the sea level will rise and the regime of the major rivers will change (higher peak flows, longer dry periods). As a result, the dense populated areas in the western and centre part of the country will further subside and the river levels and sea level will rise (see *Fig. 252*).

In the past, dike breaches along the rivers have occurred frequently during floods, more in particular during severe winters when ice jams blocked the major streams. There are also well known examples of severe floods by storm surges from the sea, the last major attack was in 1953. During the last 50 years, strong political policy decisions on safety against flooding have determined how flood control measures (coastal defence systems, dike strengthening along estuaries, lakes and rivers) have been designed and implemented. Due to expected climate change, new standards and approaches for adapted policies are considered or already carried out (Room for the Rivers programme). Safety along the major rivers can only be achieved in concert with measures taken by riparian countries in all river basins situated upstream of the Netherlands.

The present map of the Netherlands is fully determined by human intervention with the purpose of flood control and safety. One has to distinguish the rivers and the coastline.

The rivers
Along the rivers, the regulation, normalisation and sometimes canalisation (Meuse, Lower Rhine), in combination with (confined) flood plain management and dike structures (often but not always with a public road

on top) have determined safety; as have the controlled discharge distribution over the various Rhine branches Waal, Lower Rhine and IJssel) during all stages at two bifurcations (Pannerdensche Kop, PK; Ijsselkop, IJK) and the artificial drains at the downstream end of the rivers (Nieuwe Merwede, Bergse Maas, Keteldiep/Kattendiep. Note: the normalised major channels of the river branches are state owned; however the land in the flood plains is mostly owned by private people, including foreign landownership).

The coast
Along the coastline, one has to distinguish at least four major systems of coast development (see *Fig. 232*):

1. estuaries and (clay) island fixation in the south west;
2. a closed sandy coastline in the west (dunes);
3. a fully controlled lagoon in the centre with a primary (Afsluitdijk) and secondary (bunds around reclaimed polders) defence system, and
4. land reclamation in between sandy islands and a clay protection dike in the north (Waddenzee).

There is a litoral drift of the tide along the coast in northerly direction, tidal differences fluctuate between the southwest , the centre and the north east between 5m - 1,5m - 4m (see *Fig. 235*).

Levels and kinds of water
The line on *Fig. 233* between Sluis (Zeeuws Vlaanderen) and Eemshaven (Groningen) is exactly 45 degrees to the north arrow. It is a symbol, representing the 0-line (NAP, normal Amsterdam level, the one and only uniform chart datum in the whole country).

Fig. 235 shows the effort of increasing the elevation of dikes above the sea level along this line after the rare desastrous floods of 1953. They are mainly elevated to 4 metres above regular high tide (different along the coast). It shows also the ground level in Holland, as far as Amsterdam being even lower than the bottom of the IJsselmeer. The blue and red bars left in the drawing show the level of rivers and roads, canals and lakes in the polders. This representation indicates the logic of crossings by tunnels rather than by bridges even if the soil is weak, if dikes have to be crossed and if the densely populated area offers many spatial barriers.

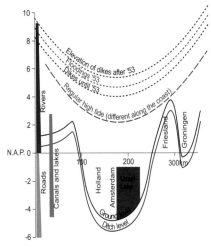

	SURFACE WATER					
	SALT		BRACKISH		FRESH	
	cur.	stag.	cur.	stag.	cur.	stag.
deep	Oosterschelde, Waddenzee	Grevelingen, Veerse Meer	Haringvliet	Biesbosch	Uiterwaarden Maas / Uiterwaarden Rijn	IJsselmeer, Oostvaarders plassen
shallow						
bank						
swamp						
bottom						
	GROUNDWATER					

Fig. 235 Levels on the line of Fig. 233 *(author Bruin)*

Fig. 236 Kinds of water in the Netherlands *(author Bruin)*

The many resulting kinds of surface water (deep, shallow, bank, swamp, bottom, salt, brackish, fresh, current, stagnant) in the Netherlands are an important basis for its ecological diversity (see *Fig. 236*).

Rainfall and seepage
Heavy rainfall and seepage determine also the design criteria of water management measures in the country. In populated and industrialised areas, a severe rainfall with critical intensity must be pumped out completely within a period of 24 to 48 hours. This urges the need for adequate pumping and drainage systems in the flat and low situated areas where due to wind effects, proper drainage by gravity is impossible; in addition proper maintenance of these systems is necessary. This can only be achieved by proper supervision and effective enforcement, so also the institutional aspect of water management (legislation, rules and regulations, set up of management authorities, finances, skill and staff, etc.) is a matter of crucial importance.

21.2   Rivers

21.2.1   Runoff
The Netherlands receives from 700mm in East Brabant to 900mm precipitation in central Veluwe (Fig. 237), but there have been years of 400mm and 1200mm precipitation.

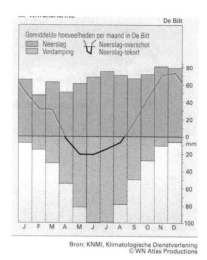

Fig. 237 Distribution of precipitation in the Netherlands
(Huisman, Cramer et al., 1998; page 18)

Fig. 238 Precipitation minus evaporation in the Netherlands
(Wolters-Noordhof, 2001; page 53)

If precipitation exceeds evaporation lakes and subterranean aquifers fill up. As soon as these cannot be filled up in time, water runs off subterranean or along brooks and rivers (*Fig. 239* and *Fig. 240*).

That part of the precipitation that reaches a stream is called runoff. The water during rainfall will gather into rills and streams down the slope. During and after the rain part of the water will soak into the ground. If the soil is saturated with water the remaining water will stream together in small streams and form a river. The groundwater flows also downhill and where the water bearing layer crops the slope a source will come out. The surface water and the subterranean water feed together a river. When the catchment area is large enough a permanent river will be the result. An estimation is made that ⅛ of the annual runoff will reach directly overland the sea while the remainder part will go underground.

the Netherlands receive runoff from catchment areas of the Rhine (entering the Netherlands in Lobith), Meuse and Scheldt rivers.

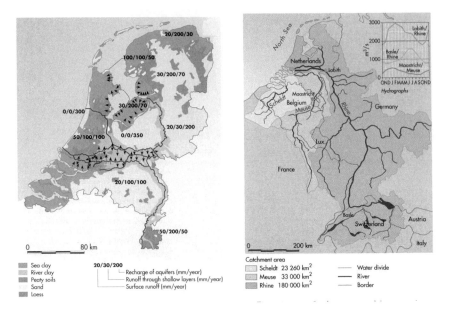

Fig. 239 Major soil types and average
annual runoff in the Netherlands
(Huisman, Cramer et al., 1998; page 21)

Fig. 240 Received runoff in the
Netherlands
(Huisman, Cramer et al., 1998; page 13)

Fig. 241 The river basin of the Rhine
(Paul Maas, Thieme Meulenhoff)

The river Rhine for example

The river Rhine has a catchment area of 160 000km^2 with an annual average of 1 775mm precipitation minus 1 392mm evaporation in the part of that area as far as Lobith. So, approximately 383mm over an area of 160 000km^2 produces 61km^3/year. So, on average 1942m^3/sec of water should run off and enter at Lobith.

Levelling by seasons

Snow and ice in mountains level out seasonal fluctuations of rivers by storing precipitation in winter, releasing it in summer (see *Fig. 241* and *Fig. 243*).

Fig. 242 Source of the Rhine
(*http://www.natuurdichtbij.nl/kennismaken/*)

Fig. 243 Precipitation in the basin
(*http://www.natuurdichtbij.nl/ken nismaken/*)

Discharge related to catchment area

In *Fig. 244* a rough approximation of discharge related to catchment area is shown. A big spot indicates the mentioned values of the river Rhine and a line is drawn for any catchment area producing a discharge in the Rhine circumstances. However, if precipitation is more than the average mentioned the line shifts upward, if evaporation or other reductions are more than mentioned, it shifts downward.

As a rule of thumb the m^3/sec of discharge is 1/100 of the km^2 catchment area, but any river has its own graph, less regular than suggested here.

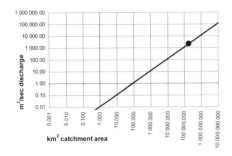

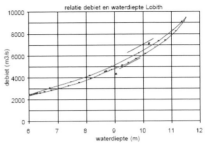

Fig. 244 Discharge Q roughly related to catchment area (author Jong)

Fig. 245 Discharge Q related to water depth H near Lobith (http://www.geog.uu.nl/fg/mkleinhans/teaching/tgrs-hw.pdf#search=%22waterdiepte%20Rijn%22)

Discharge related to depth

The relation of discharge to the water level near Lobith in *Fig. 245* is important for the height of dikes and the draught of ships, but it changes in time because of sedimentation and excavation.

Discharges in time

Because precipitation and evaporation differ much per day, the discharge of the Rhine differs daily (see *Fig. 246*), as unpredictably as the weather forecast.

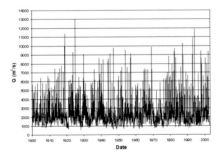

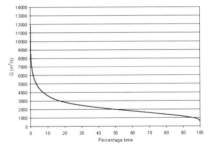

Fig. 246 Daily average discharge of the Rhine at Lobith (Lecture Marc F.P. Bierkens UU Faculty of Geosciences)

Fig. 247 Duration line of Rhine discharge at Lobith (Lecture Marc F.P. Bierkens UU Faculty of Geosciences)

Ranking *Fig. 246* you can derive a 'duration line' as in *Fig. 247*, indicating how often you can expect a given discharge to be exceeded. From that figure you can conclude that 50% of the time the discharge of the Rhine did not exeed 2000m³/sec. The mirrored graph gives the percentages of underspending.

Local impact of rain on discharge
The discharge of a river fed by a catchment area increases some time after the first rainfall (see *Fig. 248*) and after the last rainfall it continues some time, depending on the size of the area.

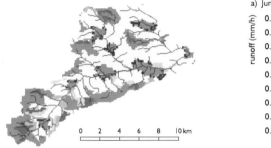
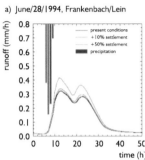

Fig. 248 Local impact of rain in hours R=10km
(*http://www.ncr-web.org/downloads/NCR18nl-2002.pdf*)

Extreme situations
Suppose an unusual system of heavy showers follows the basin around the course of the Rhine and those of its feeding rivers like the Main and Mosel

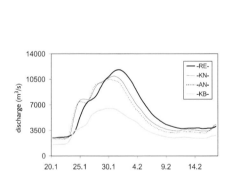
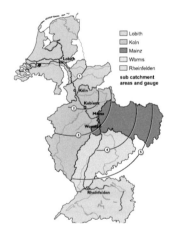

Fig. 249 Flood 1995 (http://www.ncr-web.org/downloads/NCR18nl-2002.pdf)

Fig. 250 Impact of rain in days R=300km (http://www.ncr-web.org/downloads/NCR18nl-2002.pdf)

from Switzerland to Lobith and everywhere in the basin drainage is optimal. A wall of water then nears Lobith. How often will that happen, how long will it last? These are the questions to be answered to calculate risks of flooding.

21.2.2   Risks of flooding

February 1995
At Lobith in February normally a water level of approximately 10m NAP and 3000m^3/sec is measured. But in 1995 it was approximately 17m NAP and 12 000m^3/sec, the second highest discharge of the century (1925: 13 000m^3/sec). Evacuation of 200 000 inhabitants was ordered by the Royal Commissioner of Gelderland Terlouw when floods threatened Betuwe area downstream of Lobith. One million cattle had to be moved. It caused extreme traffic jams on roads the like of which had never been envisaged. The dikes barely held out, becoming wetter and wetter.

Active debate on safety
Afterwards, the real threat of inland floods raised public awareness and the need to make plans to increase safety.[a] If the present state of inland dikes and other hydraulic circumstances is not changed, we apparently have to expect threats of a disaster like 1995 twice a century (a recurrence time of 50 years).

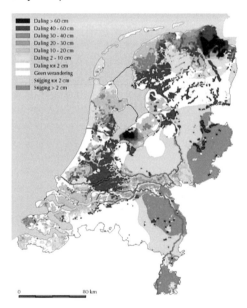

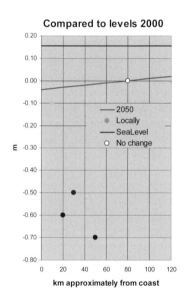

Fig. 251 Subsidence expected by 2050
(RWS)

Fig. 252 Sealevel rise and
subsidence expected by 2050

---

[a] http://www.ruimtevoorderivier.nl/upload/WAAL-MAATREGELENBOEK.pdf

But the hydrological circumstances change. Perhaps we should expect more rain in winter (less in summer) as a result of climate change. Germany and Switzerland have drained their meadows so much, that any rainfall upstream reaches the river Rhine faster than ever. Moreover, the west of the Netherlands faces a general subsidence of at least -3cm until 2050 (locally −70cm, see *Fig. 251*). Increasing the height of dikes along the rivers is necessary, but it does not solve the question how to drain the discharge into the sea while its level rises through climate change (15 cm by 2050?, see *Fig. 252*).

Normal distribution of maximal discharges

Looking at the average yearly maximal discharges[a] of past years (see the 98 years in *Fig. 253*) you can calculate their average maximum discharge (6.6454m³/sec) and their standard deviation (2.1408m³/sec) to draw a 'normal distribution' based solely on these two numbers (see *Fig. 254*).

From that normal probability distribution you can extrapolate the probability per class of 1000m³/sec wide (see *Fig. 255*).

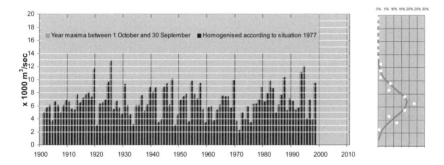

Fig. 253 Extreme discharges of the river Rhine per year   *Fig. 254 Probability*

However, that is only a very first approach, because the formula for an asymmetrical distribution (see *Fig. 247*) or a distribution otherwise different from the normal distribution may fit the data better.

The percentages are represented less precisely and eloquently than their reciprocal value: the number of years you can expect between two occurrences of that class (recurrence time). That measure has political value.

---

[a] http://www.rijkswaterstaat.nl/rws/riza/home/publicaties/rapporten/2002/rr_2002_012.pdf

m³/sec year maximum measured in 98 years		m³/sec class	probability/year		Year/probability (recurrence time)	
average	6 645		⤵			
standard deviation	2 141		⤵			
		>1 000<2 000	0.58%	once in	174	year
smallest observed	2 280	>2 000<3 000	1.77%	once in	57	year
		>3 000<4 000	4.37%	once in	23	year
		>4 000<5 000	8.68%	once in	12	year
		>5 000<6 000	13.87%	once in	7	year
average	6 645	>6 000<7 000	17.81%	once in	6	year
		>7 000<8 000	18.38%	once in	5	year
		>8 000<9 000	15.25%	once in	7	year
		>9 000<10 000	10.18%	once in	10	year
		>10 000<11 000	5.46%	once in	18	year
		>11 000<12 000	2.35%	once in	42	year
largest observed	12 849	>12 000<13 000	0.82%	once in	122	year
		>13 000<14 000	0.23%	once in	439	year
		>14 000<15 000	0.05%	once in	1,961	year
		>15 000<16 000	0.01%	once in	10,881	year
		>16 000<17 000	0.00%	once in	75,115	year
		>17 000<18 000	0.00%	once in	644,950	year
		>18 000<19 000	0.00%	once in	6,887,859	year
		>19 000<20 000	0.00%	once in	91,495,720	year

Fig. 255 Normal probabilities per discharge class of the river Rhine

Risk acceptance
The Parliament of the Netherlands once decided to accept 1 casualty per million inhabitants per year caused by environmental disasters (accepted risk). So, the number of casualties per class of discharge causing floods has to be calculated to plan the measures to meet the accepted risk of that rare discharge. Which area is flooded by which discharge, and how many people live there? Many studies have been executed to get answers on that question. They make clear that 1 casualty per million inhabitants per year would lead to unacceptable measures producing other kinds of risks. So, the Parliament decided in 1960 to accept the higher risk of a disastrous flooding of rivers once in 1250 years. In other areas surrounded by dikes (dijkringen) that risk acceptance is lower or higher according to their economic value (see *Fig. 256*).

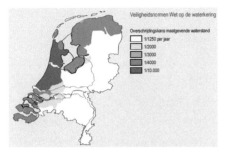

Fig. 256 Current safety standards for floods (MNP, 2004)

Fig. 257[a] Proposed changes of safety standards (MNP, 2004)

However the 'human and economic value' has increased substantially compared to the costs of water safety management. So, these safety standards are in discussion (see *Fig. 257*).

Calculating and extrapolating recurrence time directly from data

If you number the discharges Q from high to low (rank number r), in 98+1 years of experience the first largest maximal discharge has a recurrence time of 99/1 year, the second (including the first!) 99/2 and so on (see Fig. 258).

If you plot them in a graph with a logarithmic x-axis (Gumbel graph, see *Fig. 259*) you can extrapolate the higher discharges to be expected roughly by a straight line.[b] *Fig. 259* shows a discharge of approximately 16 500 m³/sec recurring every 1250 years with a big spot. So, for any river you can indicate every observation y on that graph if you know the last time that level was reached (x years ago)[c]. Nearly any kind of theoretical probability distribution (like the normal one on page 420) will also produce a nearly straight line for the higher levels in the Gumble graph. That method is used for many kinds of natural disasters like earth quakes and eruptions of vulcanoes.

However, the slope 's' and elevation 'e' of the straight line chosen have great effect. In *Fig. 259* a line with formula $Q(r) = s \cdot \ln(r) + b$ m³/sec was chosen, where s = 1.43 and e = 6.36.[d]

---

[a] http://www.rivm.nl/bibliotheek/rapporten/500799002.html
[b] http://www.humboldt.edu/~geodept/geology531/531_handouts/equations_of_graphs.pdf
[c] Download Gumble paper from http://geolab.seweb.uci.edu/graphing.phtml
[d] http://team.bk.tudelft.nl/ > publications 2006 Hydrology.xls

year	m³/sec	rank	recurrence time
	Q	r	99/r
1901	5 058	77	1.3
1902	5 715	68	1.5
1903	6 081	60	1.7
1904	3 731	89	1.1
1905	6 697	44	2.3
1906	6 121	57	1.7
1907	5 058	77	1.3
1908	6 101	58	1.7
...	...	...	...
1925	12 849	1	99.0
...	...	...	...
1992	5 758	65	1.5
1993	11 100	4	24.8
1994	12 060	2	49.5
1995	4 112	84	1.2
1996	7 004	38	2.6
1997	3 912	87	1.1
1998	9 487	11	9.0

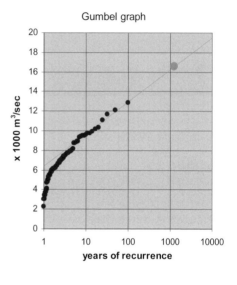

Fig. 258 Ranking maximum discharge per year, calculating recurrence time

Fig. 259 A Gumbel graph of Fig. 258

21.2.3   Measures to avoid floods

Inundation?

One of the proposed measures is, to inundate indicated polders preventively in case of emergency. But a 1m deep polder of 1km² (1 000 000m³) would store 12 000m³/sec water only for 83 seconds at least if it is not sloping. In case of sloping you should half that capacity. If you would like to store 16 000m³/sec during a week to be safe for many centuries because you cannot discharge that amount into the sea because of sea level rising after these centuries, you need 10 000km² (a quarter of the Netherlands). However, you can reduce the needed storage because you still can discharge into the sea, be it at low tide or by huge pumps. But this simple and much too rough calculation shows at least the dimensions of the problem.

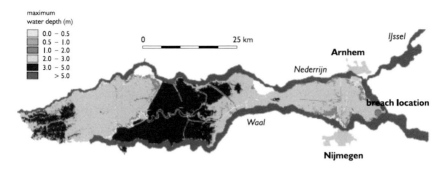

Fig. 260 Maximum water depth during a flooding in Betuwe along the Rhine after a dike breach and a peak discharge of 18.000 m³/s.
(http://www.ncr-web.org/downloads/NCR18nl-2002.pdf)

Other measures

So, construction of retention basins or more general widening of the riverbed in the Netherlands solely cannot be a substantial solution to avoid rare flooding in a river system. Dikes along the rivers have to be heightened, but which height is enough? Deepening the river (filled up quickly with sediment) or making the dikes higher increases the capacity to discharge, but moves the problem to the west where more people live. So, retention in the Rhine basin upstream has to increase to avoid extreme situations downstream. This is discussed by the international Rijncommissie Koblenz.

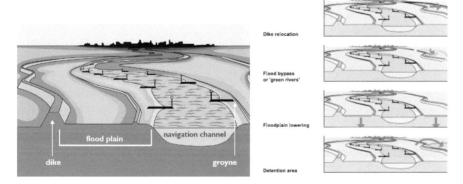

Fig. 261 Schematic representation of a low land river. (http://www.ncr-web.org/downloads/NCR18nl-2002.pdf)

Fig. 262 Measures improving Rhine discharge (http://www.ncr-web.org/downloads/NCR18nl-2002.pdf)

How to design for floods?

To be prepared for floods a landscape will have to be designed mainly as a natural area (see *Fig. 263*).

423

Fig. 263 Anticipated vegetation structure and land use along the Dutch Rhine as a 'green river' (http://www.ncr-web.org/downloads/NCR18nl-2002.pdf)

Room for the river

On 19 December 2006 the Dutch Parliament accepted a Spatial Planning Key Decision (SPKD, in Dutch: Planologische Kernbeslissing PKB) concerning a series of measures along the rivers known as 'Room for the river' (see *Fig. 264*). However, the final set of measures should be determined by commitment of local stakeholders and administrators. To get that commitment Delft Hydraulics has developed a game to determine the effects of any single measure in solving the problem[a].

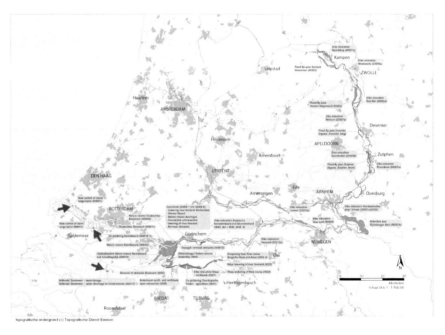

Fig. 264 A series of measures known as 'Room for the river'[b] (RWS)

---

[a] Download from http://www.wldelft.nl/soft/blokkendoos/
[b] http://www.ruimtevoorderivier.nl/

21.3   Polders

21.3.1   Need of drainage and flood control

History
Wetland areas may need drainage to be used for living and agriculture. The draining was started to obtain more space for these activities. The first method of draining was with the help of open ditches and trenches. The water was drained by sluices on lower lying waterways like rivers or at low tide at the sea (see *Fig. 265*). Later when the difference in height of water between the drainage area and the river or sea became too small or even negative, the land was drained by pumps (see *Fig. 266* and *Fig. 271*).

A polder is a piece of land that forms a hydrographical entity. In low lying areas a polder is surrounded by embankments or dikes. Even a lake can be transformed into land (see *Fig. 265*).
This reclamation is also called a polder because the groundwater level is managed in an artificial manner. Such land reclamations are always situated below the surrounding water level.

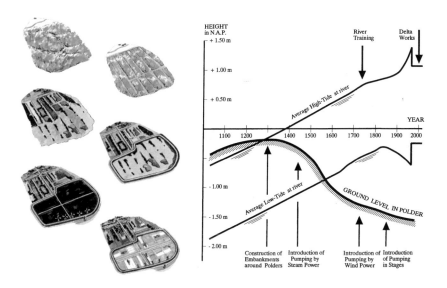

Fig. 265 A short history of   Fig. 266 Rising outside water levels and
polders *(Source unknown)*   dropping ground levels *(Ankum, 2003; page 71)*

Draining an area starts a process of changes in the soil. The ground level will settle and drop depending on the type of soil. Peat soil will actually totally disappear by chemical processes and the ground level will be lowered by the equivalent of the thickness of the peat layer. Also the introduction of better methods and pumps will lower the groundlevel (see *Fig. 266*).

Desired groundwater levels

It is obvious that since the groundwater level is managed artificially , there are several desirable groundwater levels. The depth of the groundwater level depends on the activity that will take place in that area and the type of soil. For grassland a high groundwater level is no problem for growing, but having cattle on that land will be more problematical as the cattle will destroy the grass by walking on it and no food will be left. For crops the depth of the groundwater level is dependent on the type of crop. Grasslands may be wetter, dryland crops should be dryer than 1m below terrain (*Fig. 267*).

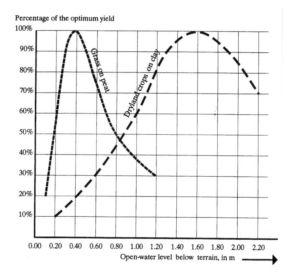

Fig. 267 Crop yields for different open water levels (Ankum, 2003; page 53)

Urban areas

For urban areas the groundwater level is kept at approximately 1m below ground level for different reasons such as foundations and wet crawl spaces. Also the construction of cables and pipes in the streets is easier under dry circumstances.(see *Fig. 268*).

426

Fig. 268 Flooding of a canal in Delft (Paul van Eijk)   *Fig. 269 Deep canal in Utrecht*

Urban areas need dry crawl spaces to keep unhealthy moist out of the buildings but they need wet foundations as long as they are made of wood. Groundwaterlevel is often recognisable from open water in the area. In higher parts of the Netherlands like in Utrecht canals show a level of several metres below ground level (see *Fig. 269*).

The distribution of polders worldwide
Lowlands with drainage and flood control problems cover nearly 1million km^2 all over the world (Fig. 270).

x1000 km2	1 crop	2 crops	3 crops	Total
North America	170	210	30	400
Centra America		20	190	210
South America	60	290	1210	1560
Europe	830	50		880
Africa		300	1620	1920
South Asia	10	460	580	1050
North and Central Asia	1650	520	20	2190
South-East Africa			530	530
Australia		310	120	430
				9170

Fig. 270 *Area of lowlands with drainage and flood control problems*
(Ankum, 2003, page 2)

and nearly half the world population lives there because of water shortages elsewhere (RWS (1998).

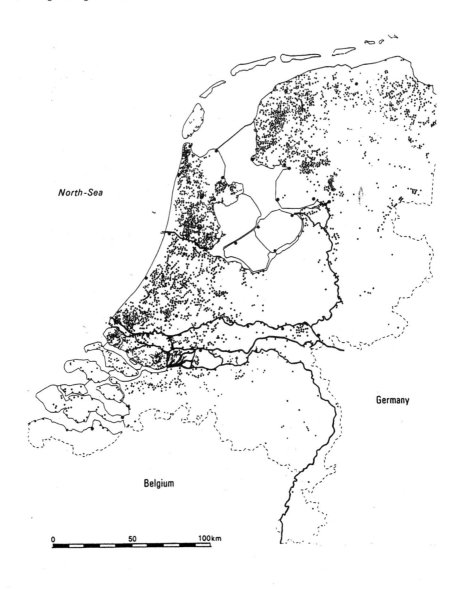

Fig. 271 Pumping stations in the Netherlands (Ankum, 2003, page 78)

21.3.2 Artificial drainage

Inhabited or agricultural areas below high tide river or sea level (polders) have to be drained by one way sluices using sea tides or pumping stations (see Fig. 271, Fig. 275).

428

Fig. 272 is the oldest known example of draining by one way sluices at low tide dating from the 11th century.

Fig. 272 *The oldest one way sluice found in the Netherlands and its modern principle* (Ankum, 2003, page 68 and 38)

One way sluices lose their purpose when average sea and river levels rise and ground level drops mainly because of the subsidence of peat polders (*Fig. 266*). Drying peat oxidates and disappears and so the ground level of the polder will drop below river or sea level.

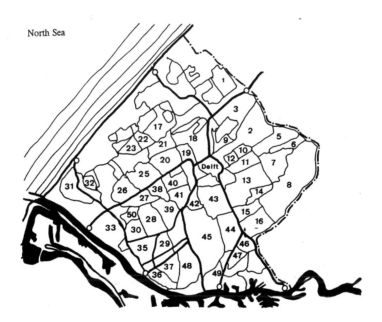

Fig. 273 *The belt ('boezem') system of Delfland* (Ankum, 2003; page 62)

Compartments

The area is divided in smaller entities or compartments that are surrounded by belt canals (boezemkanalen), protected by dikes and internally drained by races (tochten), main ditches (weteringen), ditches (sloten), trenches (greppels), and pipe drains. As the system of outlet canals (boezemkanalen) transports the water from the land to the river or the sea and they are all connected with each other it is also possible to use these waterways for shipping. The area is made accessible for shipping traffic by locks.

Fig. 273 shows the belt system of Delfland and the compartments. Each compartment has its own sluice or pump and outlet canal or 'boezem'.

Methods of impoldering or pumping step by step
The reclamation and drainage of the polders is done by pumps.

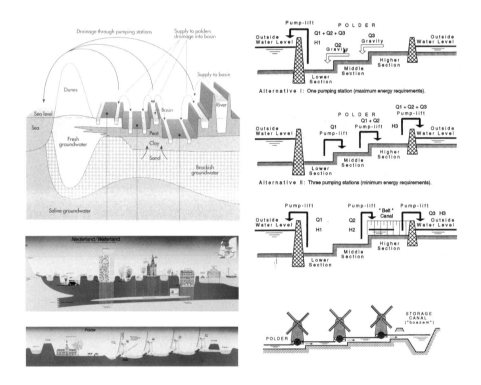

Fig. 274 Lowland system	Fig. 275 Drainage by one to three
(Huisman, Cramer et al., 1998 page 36 ; Veer)	*pumping stations, in earlier times by a 'row of windmills' ('molengang')*
	(Ankum, 2003; page 76 and 55)

The pumps are driven by wind, steam or electricity depending the technical knowledge of the time. The methods used depend on the depth of the polder. Draining marshland is often done by one step of pumping or even by a one way sluice when the land is adjacent to a tidal river or the sea. But after settling of the soil in the course of time it can be necessary to use more steps for pumping. Especially when the only force to drive the pumps was by wind, rows of windmills were used for draining the polder. The most famous row of windmills in the Netherlands are those of Kinderdijk in Zuid Holland.

The methods used for draining polders with different altitudes are pumping at once from the deepest part using gravity by collecting first the water from the deepest level or draining step by step compartments separated by dikes and weirs saving potential energy (Fig. 275).

21.3.3 Configuration and drainage patterns of polders

Polders are optimally drained by a regular pattern of ditches (see *Fig. 276, Fig. 277*).

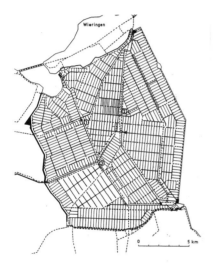

Fig. 276 Wieringermeer polder	*Fig. 277 Hachiro Gata Polder in*
(Kley 1969)	*Japan (Ankum, 2003 page 42 and 82)*

Calculation of distance for drains in a polder
The necessary distance L between smallest ditches (see *Fig. 278*) or drain pipes (see *Fig. 279*) is determined by precipitation q [m/24h], the maximum acceptable height h [m] of ground water above drainage basis between drains and by soil characteristics. Soil is characterised by its permeability k [m/24h] (see *Fig. 280*).

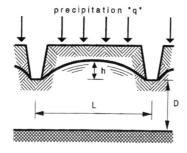

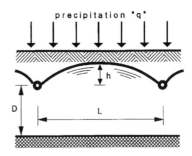

Fig. 278 Variables determining
distance L between trenches
(Ankum, 2003; page 36)

Fig. 279 Variables determining
distance L between drain pipes
(Ankum, 2003; page 36)

$L=2\sqrt{(2kh/q)}$ is a simple formula to calculate L. If we accept h=0.4m and several times per year precipitation is 0.008m/24h, supposing k=25m/24h the distance L between ditches is 100m.

Type of soil	Permeability k in m/24h	
gravel	>1000	
coarse sand with gravel	100	1000
coarse sand, fractured clay in new polders	10	100
middle fine sand	1	10
very fine sand	0.2	1
sandy clay	0.1	
peat, heavy clay	0.01	
un-ripened clay	0.00001	

Fig. 280 Typical permeability k of soil types

However, the permeability k [m/24h] differs per soil layer.
To calculate such differences more precisely we need the Hooghoudt formula described by Ankum (2003) page 35.

21.4   Coastal protection

Disasters stimulating major civil engineering works
As shown in the sketch map of the Netherlands (see *Fig. 232*), there are various major coast forms, differing fundamentally. For the design, strengthening and maintenance of the coastal defence, all these major forms need continuously specific tailor made attention. A universal fact is that disasters are needed to make progress. Also in coastal water management, tragic disasters have determined human intervention in developing the Dutch coast line. One can refer to the big flood in the southern part of the former Zuiderzee in 1916, when severe flooding occurred causing nearly 20 deaths and huge damage; this disaster accelerated the political approval of starting the Zuiderzeewerken

432

(Zuiderzee works) designed by Lely. And of course the storm surge on February 1st, 1953, which initiated the Deltawerken (Deltaworks).

History
In the past, coastal and river works were done by trial and error and on a relatively small scale. If the works that needed to be done were simply too big and complicated, land was given up (again). In those days, coastal engineering was more or less a matter of "If we cannot do what we want, we will do what we can.". Apart from not having proper large tools, current knowledge and practical experience were not enough to justify efforts in coastal development on any sort of large scale. Fundamental coastal research and model investigations were only developed in the Netherlands from the early 1930s. At that time, three major civil engineering works were developed, i.e. the Afsluitdijk (Enclosure dike, whereby the 'Zuiderzee' was renamed the 'IJsselmeer'), the big lock for seafaring vessels at IJmuiden at the end of the Noordzeekanaal (North Sea Canal) and the completion of the Maaswerken (Meuse works; Julianakanaal locks, with the biggest head in the country). Till then, water related research for Dutch clients was often done abroad, for example in Karlsruhe (Rehbock laboratory).

Zuiderzeewerken and Afsluitdijk
The preparations and design for the Zuiderzeewerken in the 1920s urged the need for developing a good mathematical basis for proper tidal computations, to be able to predict with sufficient accuracy changes in water levels along the coast of the Wadden Sea after the closure of the Afsluitdijk. In this respect in particular one name must be mentioned: Lorentz. He developed modern tidal calculations, needed to estimate the impact of the Zuiderzee works (Afsluitdijk) on the tidal regime along the northern Dutch coastline. In fact, one can conclude even after 75 years that the sandy bottom of the Wadden Sea has still not reached a new equilibrium since the closure in 1932, due to the severe changes in the tidal movements as introduced by human intervention at that time.

21.4.1   The Delta project
For all major infrastructure, political approval is necessary by means of a special law being adopted by Parliament. Such a law not only describes the need for the work itself, but also the financing and how institutions are required for design and implementation. The Delta Act was adopted in 1956, three years after the February '53 surge. At the time, repair to the damage and building of new structures was already going full speed ahead. So in fact the financing of those efforts had not yet been approved by Parliament till 1956. The country was in a sense at war, so military means were accepted. For nearly 25 years (in the period 1953 – 1977), the execution of the solid dams in the south west was never a real political question: the need for implementation was simply a political fact because 'safety first' was the guiding motive after the disaster in '53 when about 1850 people were killed. Only in the mid-seventies, when the last episode of the

Deltaworks scheme started with the closure of the Oosterschelde (Eastern Scheldt), socio economic and environmental changes on a national scale prompted the need for a complete revision of the engineering approach to this major work (*Fig. 281*) showing many innovative coastal constructions.

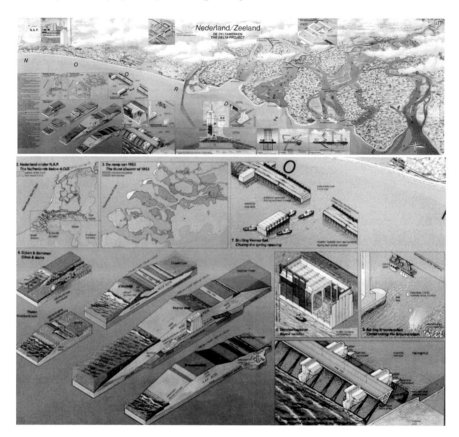

Fig. 281 Delta project (Hettema and Hormeijer, 1986)

A variety of interventions
It is remarkable to notice the huge level of human intervention since 1953, needed to close the estuaries in the south west. As the crow flies over a distance of about 100 km between Hook of Holland and Cadzand/Belgian border, 9 different ways have been used for closing off tidal creeks and estuaries, involving (systems of) primary dams (years as mentioned indicate year of commissioning). From north to south they are: the Nieuwe Waterweg (floating movable barrier, 1998), Brielse Maas (sand supply, 1952), Haringvliet (sluices, dam and by passing lock, 1970), Brouwersdam (caissons and cable, 1968), Oosterschelde (open barrier, 1986), Veerse Gat (caissons, 1961), Westerschelde (open estuary, dike strengthening,

1985), Braakman (sand supply, 1951), and Zwin (gradually closed by natural phenomena).

In addition there are 6 other solutions for the closure of so called secondary dams (some of them located on a former tidal slack) in the Deltaworks scheme, for example the Hollandse IJssel barrier (a main steel gate and a second one just for safety reasons in case the first one has a failure, 1956), the Volkerakdam (caissons plus major locks, and sluices (1969), Grevelingen (cable, minimising the tidal volume in the Brouwershavense Gat before closure (1961), Krammerdam (major locks with a sophisticated salt/fresh water control system, 1982), Markiezaatdam (compartment dam of clay and sand with a lock, to minimise the tidal volume at the Oosterschelde barrier and to control water quality in the Scheldt-Rhine canal, around 1980), Zandkreekdam (sand supply, minimising the tidal volume in the Veerse Gat before closure, 1960). To complete the variety of closure works in this part of the Netherlands, one must also mention the Sloedam and the Kreekrakdam, both needed for the railway connection to Vlissingen (clay and sand dams, 1870).

Funding
Considering all this, in the 20th century the Dutch have reached apparently a point that can now be characterised as 'we can do what we want". Such a huge and costly scheme could only be implemented because the Dutch society was prepared to allocate the necessary funds from its own re-sources, so political support remained consistently positive. On the other hand: if a country in the Third World were to ask a donor organisation (for example the World Bank) to finance a closure scheme in a complicated tidal area with at least ten solutions, this would never been accepted. Such an investment for the safety of only 200,000 inhabitants behind the struc-tures is according to present standards of international donor organisations simply NOT considered as feasible (!).
Note that in 1990, Rijkswaterstaat (part of the Ministry) was awarded the Maaskant Prize for the Deltaworks, in particular for the way the whole pro-ject is flexible in its spatial planning and technical set up, and for the way it has proven to be useful also for new sectors developed after the period of design and execution, for example leisure and environment. For more gen-eral information on these works, see the jury report.

21.4.2   The central coast line
The centre coast line of the Netherlands between Hook of Holland and Den Helder can be characterised by a system of sandy dunes. Because of the lateral drift in northerly direction along the coast, there is some continuous ongoing erosion of the sandy coastline(see *Fig. 282*). The effect over time is visible at the Hondsbosse Zeewering, where the original tow of the revetments at the seaside was constructed (stone construction, 1875) in

line with the low water line on the beach in those days. Today, the low water coastline has moved over about 70 m in easterly direction.

Sand transport

In 1991, Parliament adopted a coastal defence law, giving the green light for regular sand supply (beach nourishment) to maintain the position of the low water line as it was in 1991. Since then, year after year, at some places along the entire coast, nourishment works are carried out outside the tourist season. Like the closure of the IJsselmeer by the 30km Afsluitdijk in 1932 this major project of the fifties caused changes of yearly natural sand transport in the North Sea and Wadden Sea The sand moved mainly from the inland waters as growing islands in front of these works. To stabilise protruding beaches and islands, large amounts of sand from the sea had to be added artificially to these beaches(see *Fig. 283*).

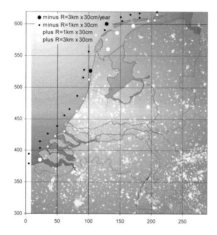

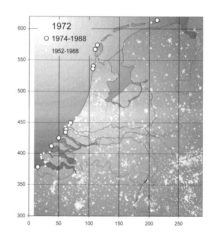

Fig. 282 Natural yearly sand transport (After: Waterman, 1992)

Fig. 283 Artificial incidental sand supply (After: Waterman, 1992)

Fresh water in dunes

Over their entire length, the sandy dunes are important for building up and maintaining a 'fresh water bubble' in the sub soil, floating on the salt groundwater underneath. This fresh water system is an extra (groundwater) protection against salt intrusion in critical areas behind the dunes, for example the Westland. In many cases, the fresh water volume in the dunes is artificially kept above certain levels for drinking water supplies in the west. The inlet water originates from the major rivers in the country, Rhine and Meuse, and is pumped through pipelines.

A special development is de Kerf, west of Schoorl (Noord-Holland). There, in the late nineties, the primary dune ridge was artificially cut to allow the penetration of salt water during rather high tides (about twice a year). The environmental development and habitat have been carefully studied and followed by many institutions since then.

The Afsluitdijk
The Afsluitdijk is presently being renovated, to meet the recent standards for flood protection and safety/reliability. Also the capacity of the sluices may be increased shortly. Sluices, bridges across the locks bypassing the sluices, and dike (alignment) had special design criteria for military reasons. They really have worked: in 1940, Kornwerderzand was the only place in Holland where the invaders could not get through. In the original design of the dam, space was reserved for the construction of a rail track as well. A deep cut for the planned track is still visible on the former island of Wieringen, alongside the motorway to Den Helder. The excavated clay from that deep cut has been used for the creation of the last refuge hill (terp) built in the Netherlands to date; at Wieringerwerf in the Wieringermeer. Indeed it was used by some locals after the German army blew up the surrounding polder dike at the end of WWII. Today, on top of that 'terp' there is a public swimming pool (again the world upside down).

21.4.3   The northern defence system

The sea defence system in the north is rather complicated, because of the sandy islands, the Wadden Sea with all its environmental and morphological extremes, the so called old 'Landaanwinningswerken' and the strengthened long clay sea defence dike between the Afsluitdijk and the Dollard. For the purpose of this chapter, the most interesting aspects are the auxiliaries in the sea coastal defence system, for example the ferry terminals, harbour law outs and terminal structures, the various breakwaters (Harlingen, Delfzijl), navigational aid systems, and the leisure facilities. They all can be used as informative and illustrative examples when designing a specific issue in relation to coastal engineering aspects. Whatever further intervention will be needed in the near future, the fact is that for the 21st century the situation of designing and constructing large scale works can now be described as 'are we still allowed to create what we can?'.

The historical value of the northern islands
Finally, a last aspect when it comes to coastal engineering, the logistics of the execution and implementation of impressive works. It deals with the supply of material in isolated and so far undeveloped areas. This can be illustrated with two examples from the past. For more modern and contemporary equivalents, everyone can use their common sense.

First, when visiting the Wadden islands in the north, many brick houses can be seen that have been built through the ages. This is remarkable, because there have never been brickyards on the islands. Even some lighthouses, like the famous Brandaris (Terschelling), were constructed exclusively with bricks. One may wonder where originally all those bricks came from.

This has everything to do with the flourishing Hanseatic League in the past. Wooden sailing vessels came from the Rhine basin, heading for the Hansa cities in the north and beyond (Baltic Sea). Bricks were transported by ship from brick yards in the river area (flood plain), and handled manually. In those days, where no machinery existed, this was done stone by stone by so called head loading. More astonishingly, each stone of the Brandaris light house must have been handled this way at least six times (or most probably even more), when being moved between the brick yard somewhere in the flood plain to its final place in the structure. En route they were brought on rather small vessels over dangerous and difficult waters.

Second, a similar development can be seen on a larger scale, for distant overseas destinations. The VOC vessels in the 17th-18th century took bricks as ballast on their journey from Holland to the Far East, for example to present-day Jakarta. When visiting the city today, one can still see the typical bricks and tiles of Dutch origin, used in the construction of buildings there.

Design with nature
To stimulate local inland movement of sand and clay from the sea (stopped after these 'hard' defence works) the policy of coastal defence has changed gradually into a 'design with nature' approach.

Fig. 284 Slufter on the isle of Texel (Google Earth)

This involves opening up some 'hard' defences where it is safe (slufters) allowing the sea to come in, bringing sand and clay into these calm inland waters causing the development of beautiful dynamic natural areas calling the original state of the Netherlands to mind.

21.5   Management and policy

21.5.1   Water management tasks in the landscape

Civil engineering offices are involved with many water management tasks (see *Fig. 285*).

01 Water structuring	02 Saving water	03 Water supply and purificatien	04 Waste water management
05 Urban hydrology	06 Sewerage	07 Re-use of water	08 High tide management
09 Water management	10 Biological management	11 Wetlands	12 Water quality management
13 Bottom clearance	14 Law and organisation	15 Groundwater management	16 Natural purification

Fig. 285 Water managemant tasks in lowlands (Das, 1993)

Local water management maps

For a long time now, maps have existed of the Netherlands showing the areas governing their own water management (Waterschappen), and their drainage areas (*Fig. 286* above). Overlays show hydrological measure points (*Fig. 286* below left) and the supply of surface water (*Fig. 286* below right).

Fig. 286 Hydrological maps of Delft and environment: levels, observation points, sluices (RWS, 1984)

On the first map you can find the names of compartments, pumping-stations, windmills, sluices, locks, dams, culverts, water pipes. However, these maps are no longer available in hardcopy anymore by fast development of GIS in the nineties.

21.5.2 Policy

Coordination of different administrative sectors

The storage of water in the lower parts of the Netherlands will put heavy demands on the surface. The 4th National Plan of water management policy V&W (1998, stressing environment), and its most recent successor 'Anders omgaan met water' V&W (2000) (stressing security) marked a change from the accent on a clean to a secure environment, as did the 4th National Plan of environmental policy VROM (2001) compared with its predecessors. Several floods in the Netherlands and elsewhere in Europe have fo-cused the attention on global warming and water management. The future problems and proposed solutions are summarised in the figures below. Storage is a central item in reducing the risks for lowlands.

	RO spatial	WHH water	SVV transp.	NMP environ	?
→'60	1				
→'70	2	1			
→'80	3	2	1		
→'90	4	3	2	1	
→'00	5	4	3	2	1

REVISION   10 YEAR
PLAN HORIZON   25 YEAR
IMPACT   250 YEAR

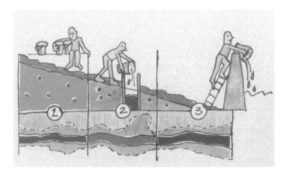

Fig. 287 Dutch Policy documents   *Fig. 288 Strategies: 1 care, 2 store, 3 drain (V&W, 2000)*

Budget
Public sector institutions dealing with infrastructure must spend a lot of money over a time span, always longer than a budget year. Planned expenses must be properly argued (transparency) in annual work plans and need the approval of Parliament (democratic decision making). The approval must be based on a long term policy (political consistency).

Stakeholders
Water related infrastructure facilities are always multi functional; there are always more users and uses, so priorities must be set after political debates (public disclosure) and approval, and the management must integrate the interests that exists in society (integrated water management). The public must be informed on developments and criteria (regular communication with media and NGOs), data must be accessible (preferably for free) reliable and retrievable(web site). All this has to do with good governance.

An acceptable vision first
Integrated water management means that attention must be given to many sectors. Often, first an acceptable vision is needed to start a firm discussion. But usually a vision alone has no legislative status, it is just a recommendation (reference is made to 'Omgaan met Water' –V&W, 1984-and 'Plan Ooievaar' -1986-). More is needed for generating fundamental commitments for the infrastructure sector. In practice, means are always limited so choices must be made based on priorities and criteria. Avoiding

random and un-controlled diffuse discussions, a strong target must be set and made visible to all involved parties in both the public and private sector. Such a well documented target needs political approval in Parliament, its implementation must be feasible in economic terms of course, and also both in technical and socio-economic terms.

Parliamentary approval of long term organisation and finance
Yhis meant that Parliament must not only give its approval to a policy target as such, but also to the finances and the institutional set up needed for implementation over a longer period of years. Such a period is always longer than the ruling period of an elected politician in power. So, there is a need for political consistency to avoid a (sudden) change of major political targets during the implementation period of infrastructure schemes. One may guess how many cabinets with different political colour have ruled the Dutch nation in the period 1953 – 1986, the implementation period of the Deltaworks. During such a long period there always must one ministry as implementing agent and an institution as executing agent that is accountable for the project.

Gradual development of policy documents
The above pleads for a gradual development of one or more Policy document(s) with sufficient legislative status. This cannot be done over night. The way this has been developed in the Netherlands is elaborated hereafter, see also *Fig. 287*.

Rebuilding the nation after World War II
After World War II, in the late forties and fifties the rebuilding of the Dutch nation took shape. In the late fifties it led to a public awareness that at least some coordination was needed on spatial planning; it finally led to a first policy document on spatial planning around 1960. By law it was approved that a revision should take place every 10 to 12 years, and that the planning horizon of a policy document was 25 years. For the implementation, annual workplans of the involved ministries and related public sector organisations needed approval of Parliament (and –of course- still do). Also the way consistent spatial planning had to develop at various levels (national, regional, local) was described. And with additional proper legislation, matters such as disclosure, supervision, enforcement and man-agement (in the public sector) became organised as well.

New public awareness of problems in the sixties
The country developed further, but due to industrialisation and urbanisa-tion, pollution of surface waters became manifest. There was a growing public awareness that a new policy paper was needed on the water man-agement of surface waters. A first version was adopted in Parliament in 1970, a period in which the second version of a revised policy paper on spatial planning was also developed. But because spatial planning and water management were two main responsibilities of different ministries under politicians of different political parties and the public sector organisa-

tions responsible for execution were still working in a top-down approach, there was hardly any coordination between the working floors of the two involved ministries during the preparations of these two policy papers.

Traffic and transport in the seventies

In the late seventies, traffic and transport in the Netherlands became a real problem. In a period where the working culture in the public sector changed from a top-down approach to a bottom-up attitude, and the working floors of separate ministries were allowed to exchange information and views directly with colleagues from other ministries, a first policy document on transport developed. First there were some separate draft versions for different sub-sectors and modes (rail-road-water-pipeline-transmission-telecom).

Integrating policies in the eighties

But Parliament forced the three main ministries involved (Economic Affairs, Public Works, Housing) to prepare a second version in the late eighties on inter modal and integrated transport issues, to be relevant also to water management and spatial planning. In the meantime, a third and fourth version of the policy paper on spatial planning developed, as well as a second and third version of the policy paper on water management (revision compulsory by law, every 10 to 12 years). Also in the late eighties, a first policy paper on nature development and environment got Parliamentary approval, finally leading to a situation at the beginning of the 21st century where four major policy papers on infrastructure sub-sectors were aligned and adopted by parliament: on Spatial Planning, on Water Management, on Transport and on Environment and Nature (respectively the 5th, 4th, 3rd, and 2nd version, see again *Fig. 287*).

Bottom-up and horizontal external contacts on the working floor

An important lesson learned from the development as described is the fact that altogether the time for a more effective alignment of the policy papers could have been shorter from the very beginning if the ministries had accepted an internal working culture, to be characterised as 'bottom-up and horizontal external contacts on the working floor'.

Furthermore it is obvious that when every square inch of land surface has at least a triple function, and every cubic meter of water multi purpose function, adequate planning is only possible when integrated policy plans are adopted by Parliament, and when consistent political support is more or less guaranteed over many years (at least decades).

Public transparency

And it has been experienced during the numerous public disclosure meetings throughout the years, in particular during discussions with well informed NGOs, that the transparency of infrastructure plans and projects is really crucial. Much time (and money!) would have been saved if, as part of the process of public disclosure, relevant files and data had been made public and accessible (web site in recent years) in advance, and if impor-

tant NGOs had been consulted at much earlier stages of planning preparations. We all have noticed the negative image of more recent large scale
projects, such as HSL (High-Speed Line), Betuwelijn (railway), 2nd
Maasvlakte (extension of Port of Rotterdam), dike strengthening, 5th runway
at Schiphol, etc. One may guess why ....

One integrated policy document?
Today, one may ask how the situation will be after a new revision (following
the law) of all these policy documents shortly. It is expected that in the near
future only one integrated policy document will be issued, dealing with the
complete national infrastructure (wet and dry), nature and environment, and
transport, including budget allocations (see last horizontal bar and vertical
column in *Fig. 287*). For an efficient implementation and execution, it
includes that further fundamental reform of public sector institutions is
unavoidable. No doubt more independent Agencies will be separated from
the public sector (as has been done recently with Rijkswaterstaat), and that
as a whole the present number of civil servants in the public sector will
further decrease due to privatisation schemes and the streamlining of
public sector organisations. Legislation, rules and regulations will further
become adapted and aligned to international standards and developments
(EU, global warming, international waters, CO_2 emissions, etc.). Technical
and operational tasks will further shift from the public to the private sector.
EU-directives will further develop and determine the daily management of
infrastructure (water directives, bird habitat directives, etc.).

21.5.3 Spatial plans checked on their impact on water: 'Watertoets'

The text below is derived from official papers[a] concerning the way spatial
plans have to be checked on their impact on water management in the
Netherlands. This could be important for future landscape ecology. From 1
November 2003 onwards the 'watertoets'[b] is legally obligatory in making
regional plans, master plans and zoning plans in the Netherlands.

Scope
The 'watertoets' concerns all waters and all water management aspects like:

1.   guaranteeing the level of safety;
2.   reducing floods, increasing resilience of water systems: care, store,
     drain (see *Fig. 288*);
3.   sewage: care, store, drain; reducing hydraulic load of sewage
     purification installations;

---

[a] http://www.watertoets.net/pdf/aandeslag.pdf
[b] http://www.watertoets.net/pdf/bestuurlijkenotitie.pdf
  http://www.watertoets.net/paginas/helpdesk/handleidingen.html?reload_coolmenus
  http://www.watertoets.net/paginas/contact.html?reload_coolmenus

4.   water supply: right quality and quantity at the right moment; counteract adverse effects of changes in land use on the need for water;
5.   public health: minimising risks of water related diseases and plagues, reducing risks of drowning;
6.   counteracting increasing subsidence and reduction of land use possibilities;
7.   counteracting ground water inconvenience;
8.   surface water quality: achieving and maintaining good water quality for people and nature
9.   preservation / realisation of proper ground water quality for man and nature;
10.   counteracting drying out (verdroging): protecting characteristic ground water depending on ecological values, cultural history and archaeology;
11.   development and protection of a rich, varied and natural wet nature.

Waterparagraph
In any of the plans concerned, a description of the way the consequences of the plan have been taken into account (water paragraph) has to be included.
Beyond safety and water inconvenience the consequences for water quality and drying out have to be mentioned and how the obligatory water advice of the water manager has been taken into account.

Contents of a watertoets
Generally:
1.   elaboration of roles of different participants;
2.   products: appointments, water advice and waterparagraph;
3.   spatially relevant criteria;
4.   the relationship with the obligartory environmental impact assessment;
5.   the environmental impact assessment;
6.   compensation: legislative aspects and examples.

Embedding in procedures:
1.   municipal procedures: master plans, zoning plans, elaborations, changes and exceptions;
2.   regional plans, their elaborations and non-legal provincial plans;
3.   environmental impact assessment procedures for traced out roads;
4.   plans for broadening roads and provincial roads;
5.   reconstruction, land use - and ground clearing plans.

Regional elaborations
The Province of South-Holland published indications of surface claims for water surface in zoning plans[a]: 8,5% times the paved surface and + 1,5% x the unpaved surface.
The Waterboard Rijnland (around Leiden) suggested keeping 6% of the overall urban area to be water surface[b]. The Waterboard Delfland claims volumes of water per specific surface according to *Fig. 289* [c].

	m^3/ha
paved surface (housing, employment, greenhouse areas)	325
unpaved surface (grassland, nature, leisure)	170
arable land	275

Fig. 289 Standards for water reservoirs inside and outside the urban area
(Waterboard Rijnland)

21.6   Conclusion

The typical Dutch landscape is a human creation and will largely return to sea without continuous human intervention. Though it is not always felt that way, civil engineering is its primary condition of existence. However, the solutions are diverse and open to discussion. Keeping the historical heritage, coping with the thread of floods, the need of drainage by different uses of land, configurating drainage patterns, protecting and developing the coastal defence systems, distributing the water, managing the water quality are part of national, regional and local political assessment.

Landscape ecology in the Netherlands has to include civil engineering knowledge and skill if it would like to include that rare landscape below the natural water levels of the sea and the rivers.

---

[a] http://www.watertoets.net/pdf/blauwgekleurd.pdf
[b] http://www.watertoets.net/pdf/rijnland.pdf
[c] http://www.watertoets.net/pdf/delfland.pdf

22   DRY AND WET NETWORKS

Taeke de Jong **T**U Delft

22.1   Introduction

Infrastructure crisscrosses the landscape connecting ever changing human activities along more or less predictable lines of potential. In case of locally increasing economic activity and network density the fragmentation of the landscape will continue along these lines. To anticipate these traces locating quiet areas, this chapter gives an ideal typical categorisation of dry and wet networks as related to their size.

It is ment to give a preliminary overview as long as there is no empirical evidence to generalise a very diverse phenomenon, mainly caused by man (an unpredictable factor, assuming free will and creativity). Like the variety of drainage in a landscape, the opening up of areas occupied by man depends on many local factors, including that of history. Their crossings and bridges as expensive pieces of civic art determine their course for long periods.

However, if you look at the topographical map comparing many fragments of the Netherlands, you can recognise features, a periodicity of roads and drains of different order. Once you establish a range of measures, these are standards to describe deviations making local identity nameable.
Since I proposed an ideal typical range of nominal measures {1, 3, 10, 30 ...m} as a rule of thumb for designers, the diversity of spatial reality determined by networks has become more tangible. A 'nominal measure' is an elastic name for a range of measures. If in this chapter I write '1km', I mean a measure somewhere between 300m and 3km.

22.2   Traffic connections

22.2.1   Demand

Distribution of inhabitants
The distribution of inhabitants (see *Fig. 290*) is the origin of the city, the support for their facilities, work, shopping and leisure. The connections and the technical equipment derive their use from that distribution and therefore they must be traced back to it. Through the increase of car traffic after the Second World War, connections became dominant in each map.

Calculating demands
You can try to model points like origin and destination (by means of locations of work, distribution and leisure), to calculate the demand for roads on the resulting flows like Newtonian forces between mass accumulations. This is the usual way of traffic modelling.

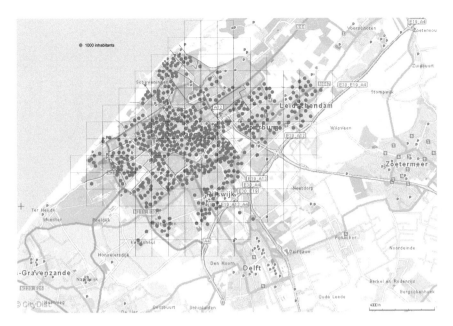

Fig. 290 The Hague, distribution of inhabitants in R=100m spots of 1000 inhabitants (30m² pp) and a grid of 1km (author on CityDisc, 2001)

Intuitively predicting connections

However, the topographical map simply shows that the Netherlands on average have got a district - or country road every kilometre and a motorway every 30km (see also *Fig. 309*). In urban areas, the meshes have been filled in further by neighbourhood roads and residential streets, whereas in larger conurbations every 3km a city motorway seems necessary a and every 10km an orbital motorway. Wherever there are gaps between the 3km city motorway and the 30km regional motorway, traffic jams probably appear because the regional highways are used by local traffic.

22.2.2   Form

From radial into grid

Utrecht was a spider in the radial net (see *Fig. 291*) for centuries. After the Second World War larger ride lengths were demanded, increasingly passing the city by continuing traffic tangents (see *Fig. 292*). The Dutch landscape was transformed from one full of cities to one featuring a grid system with cities added on.

Fig. 291 From radials in 1866 ...
(Source Province Utrecht, 1866)

Fig. 292 ...via tangents to a bigger grid in 2001 (CityDisc, 2001)

Now, cities themselves (like Utrecht in *Fig. 291* as a spider in a thin-lined web) are captured as flies in the peripheral meshes of a larger network. The more distance a city keeps from the central agglomeration (Amsterdam left above in *Fig. 292* on the same scale as *Fig. 291*), the more orthogonal the grid (network city). The question is when that central conurbation in turn will transform from spider to fly in a continental metropolitan net.

The logic of orthogonal grids
The fact that a radial-hexagonal grid is transformed through passing traffic into a tangential-orthogonal grid, can easily be seen through the analogy of a thin sheet of soap bubbles distorted by a long hair (see *Fig. 293*).

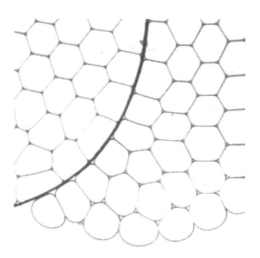

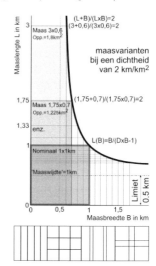

Fig. 293 Right angled by long lines
(Hildebrandt and Tromba, 1989)

Fig. 294 Length and width of meshes with the same network density

450

Although a hexagonal grid has the lowest perimeter/area proportion from the viewpoint of investing in public space, in many respects an orthogonal pattern fits better in urban development (Bach, 2006). An orthogonal grid does not need to contain squares (elongated rectangles are often profitable). You can abstract from the form of a square by expressing the network density (km/km^2) in nominal mesh width M of a rectangle with the same network density (see *Fig. 294*).

22.2.3   Hierarchy

If you leave your home to go on a trip, you start on a 'residential street' (some 20m wide), followed by a larger 'neighbourhood road' (say 30m wide) reaching an even larger 'district road' (say 40m wide) and so on.
The question arises at which mutual absolute distance you have to draw them in urban design. To keep it simple, we take 30m for the smallest residential paths, 100m for residential streets, 300m for neighbourhood roads, and 1000m for district roads (*Fig. 295*).

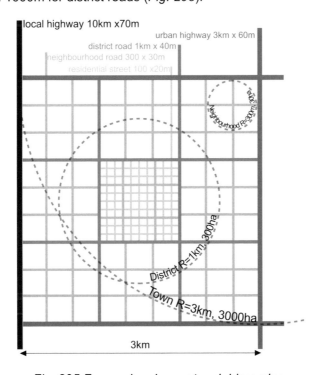

Fig. 295 Four orders in a network hierarchy

This way, every road hierarchy gets a range of different mesh widths. A factor 3 is observed as optimum scale step according to several criteria (Nes and Zijpp, 2000). Do not take that too strictly, it is just a rule of thumb.

In the same time that is the average distance between junctions into the higher order. On average every third road of each level you can make a turn to a road of a higher level (see *Fig. 295*).

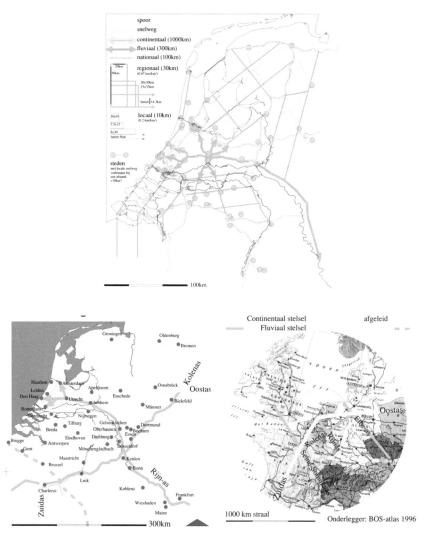

Fig. 296 Potential national, fluvial and continental network systems if nominal densities were applied (Jong and Paasman, 1998)

In the existing networks you might recognise such a hierarchy from the lowest (*Fig. 295*) to the highest (*Fig. 296*) levels of scale.

22.3   Two networks

This hierarchical range applies also to the artificial wet (water) infrastructure where the natural tree-like drainage by rivers and brooks has mainly disappeared, as in the Netherlands. So, the number of crossings as a result of interference between both kinds of networks is easily outlined in different alternatives of mesh form.

22.3.1   Names and scale

Everybody knows many names of wet and dry connections, regardless of their function (*Fig. 297*). They seem to fit nearly logarithmically on a constant difference of scale multiplying the mesh width each time approximately by 3.

NETWORK		BLUE LEGEND		BLACK LEGEND	
density	mesh/ exit interval		NAME	nominal width	NAME
km/km^2	km nominally	width 1%		m	
0.002	1000	≥10000	sea		
0.007	300	3000	lake	120	continental motorway
0.02	100	1000	stream/pond	100	national motorway
0.07	30	300	river/waterway	80	regional motorway
0,2	10	100	brook/canal	70	local motorway
0.7	3	30	race	60	urban motorway
2	1	10	watercourse	40	district road
7	0.3	3	ditch	30	main street
20	0.1	1	small ditch	20	street
70	0.03	0.3	trench	10	path

Fig. 297 Names of networks on the higher levels of scale

However, in reality the interval is sometimes more, seldom less than 3 and often the highest and lowest orders are missing. For example clay grounds do not need trenches and sandy grounds start their drainage by brooks. A wet peat area continues its grid hierarchy into the level of canals, a clay area into the level of ditches and a sandy area or city stops at larger water courses. That makes the soil type globally readable from the topographical map. For its dry opening-up, the rural area is detailed until the road, the industrial area until the main street, the residential area until the street. Similarly, rural areas do not need streets every 300m. In the Netherlands they start with roads every 1km as can be seen on topographical maps

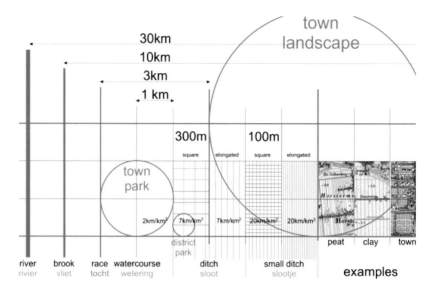

Fig. 298 Styling of wet connections

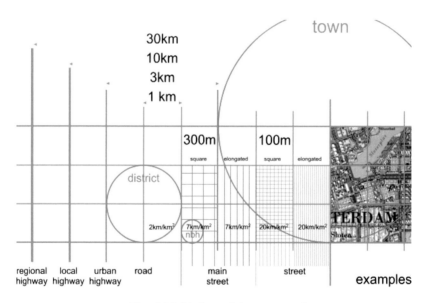

Fig. 299 Styling of dry connections

22.3.2 Functional charge

These functionally neutral names get their time-bound character by changing function. Dry and wet networks get their contemporary meaning by 'functional charge' in *Fig. 300*. Their density implicates the level of investment.

Nominal mesh width	30m	100m	300m	1km	3km	10km	30km	100km
Density (km/km²)	70	20	7	2	0.7	0.2	0.07	0.02

wet connections

	30m	100m	300m	1km	3km	10km	30km	100km
name	trench	small flooded ditch	a flooded ditch	watercourse	race	brook	river	lake
indicative width 1%		1m	3m	10m	30m	100m	300m	1000m
other names			stream	stream	stream	stream		
		urban canal	urban canal	urban canal	urban canal	industrial canal/waterway	canal	canal
functions			draining			drainage pool (from polders)		

Nominal mesh width	30m	100m	300m	1km	3km	10km	30km	100km

dry connections

	30m	100m	300m	1km	3km	10km	30km	100km
name	path	street	main street	road	urban motorway	local motorway	regional motorway	national motorway
an exit every ...km	10m	30m	100m	300m	1km	3km	10km	30km
indicative width	10m	20m	30m	40m	60m	70m	80m	100m
functions	pavement	opening to a hamlet	neighbourhood street	district road, village road, country road	urban motorway, main road	urban motorway	provincial motorway	national motorway
	footpath	residential walk	walking route	cycle route	cycle ride			
'Duurzaam Veilig' (long-term safety)	'Woonpad', free of cars	'Woonstraat', restricted entry for cars	'Erftoegangs-weg', sojourn function	'Gebieds-Onsluitings-Weg', opening to an area	'Stroomweg', throughway			
public					bus	express	fast bus	Interliner

Nominal mesh width	30m	100m	300m	1km	3km	10km	30km	100km
railway line					tram	light rail	regional	national
a supportive base					300m	1km	3km	10km
functions						the underground/metro	local train	intercity train, Argus
					hybrid systems	hybrid systems	hybrid systems	

Fig. 300 The time-bound functional charge of networks

22.4 Superposition and interference

22.4.1 Superposition of levels per network

In connection with the dry and wet legend one can imagine their superposition as shown in *Fig. 301*.

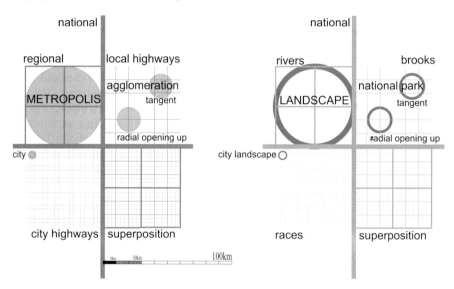

Fig. 301 Superposition of networks

An urban area is radially crossed or tangentially surrounded by infrastructure (see *Fig. 301* above left).

By superposition of the higher order over the lower order, the density of the lower order decreases.

By superposing the wet connections over or under the dry connections, both networks interfere ('interference').

22.4.2 Interference between networks

When one lays different (for example wet and dry) networks over each other, an interference occurs that defines the number of crossings, and, because of this, the level of investment in civil engineering constructions (Fig. 302).This can be done in different ways. Separating instead of bundling them fragments space more. The diversity of interference has important impacts on ecology and cultural identity.

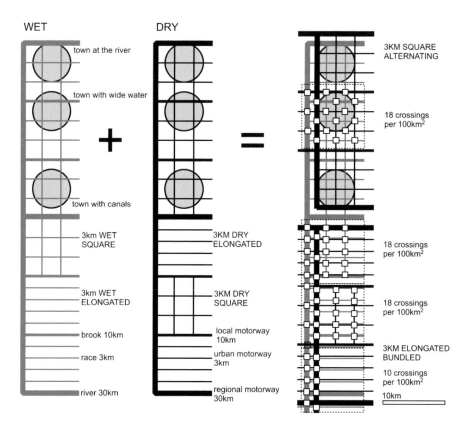

WET   DRY

town at the river

town with wide water

town with canals

3km WET SQUARE

3km WET ELONGATED

brook 10km

race 3km

river 30km

3KM DRY ELONGATED

3KM DRY SQUARE

local motorway 10km

urban motorway 3km

regional motorway 30km

3KM SQUARE ALTERNATING

18 crossings per 100km²

18 crossings per 100km²

18 crossings per 100km²

3KM ELONGATED BUNDLED

10 crossings per 100km²

10km

Fig. 302 Interference between wet and dry networks at equal density.

The position of urban areas with respect to orders of magnitude of water and roads dictates their character to a large extent. The elongation (stretching) of networks reduces the need for engineering constructions when their meshes lie in the same direction. If one bundles them together, this also helps to prevent fragmentation. The aim of the 'Two network strategy' (see Tjallingii, chapter 17.4.1, page 328), on the other hand, is to position water, as a 'green network', as far away as possible from the roads (in an alternating manner). However, this has the effect of increasing fragmentation by roads and watercourses. Moreover, when does the green-blue line meet a next traffic-potential?

22.4.3   Crossings

Mutually crossings of waterways seldom separate their courses vertically (*Fig. 303*) as motorways do (*Fig. 304*).

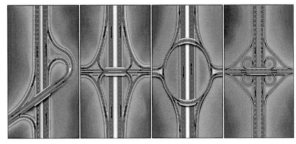

Fig. 303 Crossing of separated waterways (Ankum, 2003; page 160)   *Fig. 304 Crossings of highways (Standaard and Elmar)*

More often their water levels are separated by locks or become inaccessible for ships by weirs or siphons. However, crossings between roads and waterways in full function have to be separated vertically anyhow. This is fairly common.

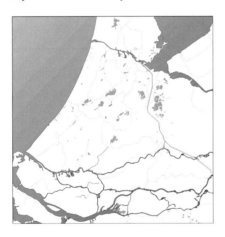

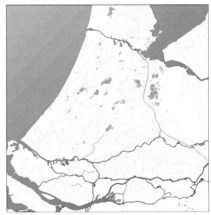

Fig. 305 Rivers, canals and brooks   *Fig. 306 Superposition races*

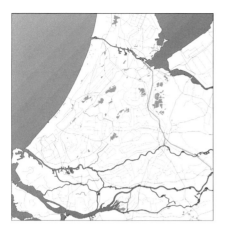 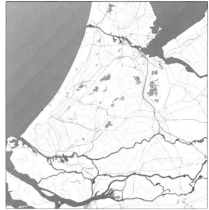

Fig. 307 Interference with highways   *Fig. 308 Interference with highways and railways*

22.5   Conclusion

To protect 'green' areas in a sustainable way, the potential of human connections have to be taken into account. The diversity of networks to be expected can be understood by scale articulation. Every level of scale between an established frame (largest observed measure) and granule (smallest observed measure) shows its unique legend of connections. The 'natural' hierarchy and interference of different kinds of networks determine the fragmentation to be expected. Understanding the regularity of these connections in densifying areas can avoid mistakes in locating the 'green' areas probably once crossed out by an emerging human network. It may help to estimate the realistic possibilities of alternative traces.

23   ROAD CROSSING CONNECTIONS

Hein van Bohemen[a] TU Delft

---

[a] The author acknowledges valuable comments from Ir. Hans Bekker (Road and Hydraulic Engineering Institute Delft).

23.1 Introduction

The construction of road infrastructure and traffic cause negative effects on nature: air and water pollution, altered microclimate, altered hydrological circumstances, traffic mortality, habitat destruction, habitat degradation, barriers for animal movements that can isolate populations.

The extent of the effects vary with the type of impact, the road-effect zone can be from local to global.

Habitat fragmentation, the process in which there is area reduction and bisection of habitats, is one of the main threats to the conservation of biodiversity.

The existing road and rail infrastructure and the expected developments of more roads and railwaysas as well as intensifying landuse will lead to even more barrier effects and habitat fragmentation and isolation.

But by using the following process steps 'prevention, avoidance, minimising, mitigation, physical compensation, financial compensation' it will be possible to reduce negative impacts considerably. This will be illustrated in this chapter.

23.2 The impact of roads and traffic on nature

23.2.1 Four ecological impacts

The construction and use of roads lead in general to adverse effects on the environment and nature.

Ecologically speaking, roads and traffic have four main negative effects:

1. they cause loss of habitat patches
2. they form barriers and isolate habitat patches
3. they create disturbance due to noise, light and air pollution
4. they are responsible for direct mortality (fauna victims due to collision with vehicles).

In the first place, there is the loss of habitat patches: asphalt and concrete replace habitat patches forever.

Secondly, roads form physical barriers: a road can divide a habitat into separate areas, each of which can become too small for the survival of a population. Besides this, behavioural influence can be observed (animals avoiding areas near roads).

Thirdly, habitats are disturbed by roads and traffic (noise, light, pollutants), and for certain species this reduces the quality of habitats adjacent to the road. In the Netherlands, studies of the relationship between breeding birds and motorways have been conducted (Reijnen et al., 1995; Foppen et al., Reijnen and Foppen, 2006),2002), as well as studies about the effects of light on meadow birds and on small mammals.

In addition to the impact of noise on animals, changes can occur in the quality of the air, the soil, the groundwater and the surface water, and this can affect the behaviour and survival of plants and animals (Forman et al., 2003). Road construction also creates microclimatological changes that can affect the quality of the remaining habitat patches (Mader, 1981).

Fourthly, traffic is responsible for the death of many animals. For instance, 10-15% of the Dutch badger (Meles meles) population is killed each year on roads, as are between one and two million birds, and many frogs and toads, along with other small animals are killed. The number of animal-vehicle collisions is steadily growing as infrastructure networks expand and traffic increases (Seiler and Hellding, 2006). In certain cases roadside verges can become a sink for animal species.

Positive effects

However, roads can have positive effects too. Roadside verges and ditches can provide a new type of habitat for species and can provide a corridor function which can be positive, but in certain situations negative. Positive corridor function if they connect separated habitat patches by facilitating for example the daily movement or dispersion; but roadside verges can also have a negative corridor function when they facilitate for example the invasion of non-native species.

23.2.2   The density of a road network

The density of a road network (see *Fig. 309*), the configuration of the road network and the location and design of individual roads in the landscape, its intensity of use and the speed of the cars are important factors concerning the impact of roads on flora and fauna. Many disturbing effects from roads and traffic differ in the extent from the road. The determination of the width of the road-effect zone of the different causes in relation to topography, wind and groundwater flow in combination with road density patterns provides an understanding of the problem in a specific area (Forman et al., 2003).

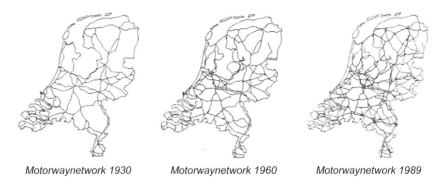

Motorwaynetwork 1930   Motorwaynetwork 1960   Motorwaynetwork 1989

Fig. 309 The development of the road density in the Netherlands

23.3 Habitat fragmentation and defragmentation of roads

23.3.1 The term fragmentation

Fragmentation can in general be described as 'the process in which there is area reduction and the bisection of biotopes and landscapes, and in which the functions of the landscape are changed for people, plants and animals'. A definition more focused on organisms is: 'The dissection and reduction of the habitat area available to a given species caused by habitat loss (for example due to land-take), habitat isolation (for example due to barriers) or increasing distances between neighbouring habitat patches'.

23.3.2 Fragmentation as an environmental theme

Both in the Netherlands and abroad, fragmentation as an environmental theme did not come to the fore until the 1980s. In 1986, the former Research Institute for Nature Management (now Alterra) organised a symposium on the significance of small landscape elements (Opdam et al., 1986), which probably was the first symposium on this subject in the Netherlands. The Road and Hydrolic Institute continued along this line and at its initiative in 1987, a workshop was held on the fragmentation of the Dutch landscape by infrastructure (Watering and Verkaar, 1987) and in 1995 an international symposium about habitat fragmentation (Canters et al. 1997) took place which resulted in a still existing Infra Eco Network Europe (IENE)[a], which initated a European Commission COST programma where 15 countries work together about the topic of defragmentation strategy and mitigation measures[b].

---

[a] www.iene.info
[b] http://const341.instnat.be

23.3.3   Fragmentation in a landscape-ecological sense

For a sustainable way of conservation, restoration and development of plant and animal populations, it is important to take into account the various landscape-ecological levels of scale and the positions and functions of organisms in them (Forman, 1995; Farina, 2001). In this context the population and meta-population concept is essential.

A population is a group of individuals of the same species living, reproducing and interacting in the same area. A metapopulation comprises spatially separated subpopulations that are interconnected via dispersion flows (Opdam, 1987, 1991). In a metapopulation the local populations display their own dynamics and the distance between the local populations still allows for interaction (Hanski and Gilpin, 1991). Practical experience has shown that such a population is more stable than a local, isolated population. Habitat patches can be recolonised after extinction of the local population (Opdam, 1987).

Fragmentation can occur as the result of natural or human processes. Natural processes are, for instance, fire, wind, floods and landslides. Natural landscapes therefore often have a mosaic pattern (Forman, 1995). Also human action has led to mosaic landscapes, which can accommodate additional species as a result of the increase in habitat variation.

The effects of fragmentation can be determined at the level of individuals, populations, living communities and ecosystems. Studies have shown that the number of species in a habitat patch is markedly reduced if more than 80% of the original habitat patch has been lost and the remnants become isolated (Andren, 1994). However, loss of only 40-60% of the habitat patches can reduce the number of species. The actual effect depends on the species' requirements, the mobility of the species as well as the concrete pattern of habitat patches.

For some animal species the degree of interconnection of the landscape elements plays a vital role in the movement of animals in fragmented landscapes. Connectivity is a species-specific property and can play a role both inside and outside the habitat patches. The interconnection of habitat patches can be improved by creating intermediate habitat patches (stepping stones) or corridors that also can fulfil a habitat function (hedges, hedge banks, banks, roadside verges) as well as mitigation measures near or above and under roads, railways and canals (see paragraph 23.4.3).

In addition to movements within metapopulations and the phenomenon of dispersion, also 'source-sink' relations play a role (Dias, 1996). If within a local population the number of births is greater than the number of deaths and all niches have been filled, then the surplus can start to occupy empty habitats or disappear in a "sink".

If there is a negative birth/death ratio, the population can survive if there is continual immigration from a source population. It appears that the source-sink dynamics can have a stabilising effect on the dynamics of metapopulations (Pullian, 1988).

Due to the complexity of systems and the relationships within and between them, it is important for an intervention (for example the planning, construction and use of a road) not only to study the species in the immediate surroundings but also to carry out population-dynamical studies of many species over a relatively large area in order to be able to make a complete picture of the species involved or may become affected in the future. Even the option of (re)introduction of animals in the future can be considered.

23.3.4   Fragmentation by roads

Central to this chapter is the fragmentation of nature as a result of the construction, presence and use of roads and the spatial configuration of roads within the landscape. According to the quoted definitions, the building of roads leads to fragmentation of habitat patches. Furthermore, their use leads to disturbance of habitat patches in the surroundings and death (road-kills) of animals trying to cross the motorway (Bohemen et al., 1994; Bohemen, 1996, 1998, 2004; Bekker et al, 1995; Forman et al., 2003)

In the last three decades, both the number of roads and the traffic intensity have markedly increased in the Netherlands, as elsewhere in the world. In 1996, the mesh size of the motorway network (computed by laying out the road network in an imaginary square raster over the land) was 32 km, that of all governmental roads and provincial roads together 7.5 km, and that of all hard-surface roads 1.2 km. Especially the mesh size of the motorway network decreased during the 25-year period, from 75 to 32 km which can affect many species of large wildlife.

For purposes of the conservation and development of biotopes for plants and animals, not only the density and mesh size of the road network, but also the width of the roads and the traffic intensity are important. Many roads have been widened over the course of time. In the period 1970-1996, the traffic intensity in all road categories in the Netherlands more or less doubled which highly increases the barrier effect of roads and the width of the affected zone.

Despite the fine-mesh network in the Netherlands, widening of existing roads as well as new infrastructure are considered to be still needed . However, there is increasing worry about the negative effects roads and traffic have on the environment and nature. Therefore an increasing number of mitigating measures are being taken. If mitigating measures do not reduce the damage, compensating measures can be carried out (Cuperus, 2005).

In recent years more and more countries have started to pay attention to the phenomenon of habitat fragmentation by roads and traffic. This is also a result of the increasing visibility of the negative effect on nature values due to the increasing number of roads and their traffic-intensity (Trocmé et al, 2003).

23.3.5   Bottlenecks between motorways and nature at various levels of scale

During the preparation of the Dutch Second Transport and Transportation Structure Plan (Ministry of Transport, Public Works and Water Management V&W, 1990) and the Nature Policy Plan (Ministry of Agriculture, Nature and Fisheries LNV, 1990) there was a need for a review of the knowledge about the effects of fragmentation and of the possibilities to limit the fragmentation of populations of animal species due to the existing motorways.

At the request of the Ministry of Agriculture, Nature Management and Fisheries (LNV), the Ministry of Transport, Public Works and Water Management (V&W), the Ministry of Housing, Regional Development and Environment (VROM), the Centre of Environmental Science of the Leiden University (CML) conducted a study of the effects of roads and traffic on mammals and birds (Fluit et al., 1990). The study revealed that the policy objective of conservation, restoration and development of natural and landscape values at certain spots can be contrary to the presence and use of the existing system of motorways. For the first time in the Netherlands, a bottleneck analysis was made in this field.

The following animal species were involved in the analysis, based on a vulnerability determination in which a link was made between the importance of nature conservation and the sensitivity to fragmentation: red deer (Cervus elaphus), roe deer (Capreolus capreolus), wild boar (Sus scrofa), red squirrel (Sciurus vulgaris), otter (Lutra lutra), badger (Meles meles), polecat (Putorius putorius), pine marten (Martes Martes), stoat (Mustela erminea), weasel (Mustela nivalis) and various bird species characteristic for marshlands, meadows, forests, heathlands and small-scale landscapes. The study defined a bottleneck as a location in which a motorway is situated in an actual or a potential habitat or connection zone, in the National Ecological Network (NEN) or at a site where many animals become traffic victims. The results are shown on the accompanying map of *Fig. 310* (from Fluit et al., 1990).

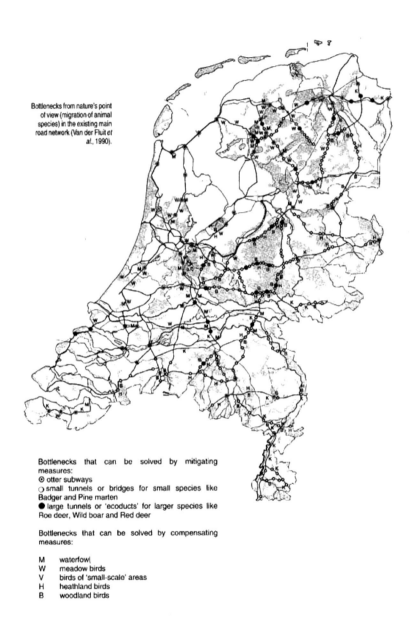

Bottlenecks from nature's point
of view (migration of animal
species) in the existing main
road network (Van der Fluit et
al., 1990).

Bottlenecks that can be solved by mitigating
measures:
⊙ otter subways
◯ small tunnels or bridges for small species like
Badger and Pine marten
● large tunnels or 'ecoducts' for larger species like
Roe deer, Wild boar and Red deer

Bottlenecks that can be solved by compensating
measures:

M	waterfowl
W	meadow birds
V	birds of 'small-scale' areas
H	heathland birds
B	woodland birds

*Fig. 310 Bottlenecks between motorways and areas of ecological
importance in the Netherlands to be solved by mitigation or compensation
measures (Fluit et al., 1990)*

The report also includes a summary of types of measures that can be taken
on roads to eliminate or reduce the effects at an individual, a population

and a species level. All the necessary mitigation measures were included in a proposal for a national action plan. Also the total budget needed to carry out this plan has been included in the report.

In 1992 a survey and descriptive study followed of the effects of fragmentation by the existing and planned network of motorways for the areas in the National Ecological Network (NEN; EHS= Ecologische Hoofdstructuur in Dutch). Map overlays were used to survey the intersections of the NEN by the existing and planned motorways and to include these in a computer data base.

The conclusion of that study is that motorways lead to the highest number of current and future intersections of the NEN, viz. 245 as compared to 182 railroad and 78 national waterway intersections (a total of 505 intersections). After that study, in each province more detailed reports on the fragmentation situation were completed. Herein, the possible solutions formulated have been more focused on the bottleneck locations. For each bottleneck to be eliminated, a detailed survey is conducted on site and a design study will take place.

Provincial and municipal roads form an important part of the road system network and present their own problems. Especially an integrated approach is of importance in which the effects of motorways, provincial and local roads are taken into account. When diverting traffic to a motorway network simultaneously the elimination or limitation of through traffic on the underlying road network should be made possible.

Based on practical experiences (sometimes by trial and error) in the Netherlands as well as abroad and the results in concrete projects and provincial plans a Longterm Defragmentation Program has been formulated in 2004 by three Dutch ministries: the Ministry of Agriculture, Nature and Food Quality, the Ministry of Ministry of Transport and Public Works and the Ministry of Housing, Physical Planning and the Environment.

In 2005 the Longterm Defragmentation Program has been accepted by the Dutch Parlement. In the next 13 years 410 million euros are available for solving botthenecks between existing motorways, railways and waterways and nature as well as an areabased approach has been included. A first evaluation of the Program will take place in 2008. In 2018 thewhole Program as well as the National Ecological Network should be completed.

The next paragraph will show which measures can be taken to counter the threat of habitat fragmentation by roads and traffic.

23.4 Overpasses and underpasses to reduce barrier effects

In order to reduce the effects of roads and traffic on the environment and nature, their planning, design and construction must pay attention to the possibilities to prevent or reduce the negative effects on nature by taking mitigation measures, measures to eliminate or reduce the adverse effects on nature values.

In the planning phase route selection combining human mobility and ecological sustainability and more effective use of the existing infrastructure can reduce the impact of road construction and road usage on the environment. A positive contribution to the reduction of negative effects on nature can be made by formulating regional mobility plans that pay attention not only to the effectiveness of mobility-regulating measures but also to the effects on nature values.

In the design phase, it is possible to apply an integrated design approach. In such an approach, as many functions as possible are harmonised. In the construction and maintenance phase, ecological engineering measures (Bohemen, 2005) provide possibilities to conserve, restore and develop ecological values.

23.4.1   Design sequence of steps to mitigate and compensate adverse effects

In 1978, the US Council on Environmental Quality indicated a desired sequence for assigning priorities to mitigating measures:

1. Prevent adverse effects.
2. Minimise adverse effects by reducing the intervention.
3. Correct the adverse effects by restoring the affected ecosystems.
4. Compensate for the effects.

For the Netherlands, this approach was worked out in the Second Structure Traffic and Transportation Plan (Ministry of Transport, Public Works and Water Management,V&W 1990). In it, the keywords are avoidance, mitigation and compensation for adverse effects.
In this chapter we will deal with the following phases: avoidance and mitigation. For the phase of compensation we refer to the thesis of Cuperus (2005).

23.4.2   Avoidance of negative effects

The first step to counteract negative impact of roads on nature in the planning process is to abandon a proposed road project. If that is not possible an attempt must be made to route and design the road in such a way that the effect on nature is minimal and that any loss of the existing and potential nature values is avoided or prevented.
The prevention of effects is not regarded as part of the mitigation or compensation process. It is a matter of prevention if vulnerable habitat patches are not within the sphere of influence of the proposed intervention.

There can also be avoidance if activities occur outside the critical period for animals. The best return on investment for nature is obtained where the adverse impact on the ecological values is prevented at as early a stage as possible.

23.4.3   Mitigation measures

The damage of an intervention like road construction can be restricted by restoring or improving migration corridors for species, preventing fauna victims and reducing the disturbance. Mitigation means the adoption of measures to reduce adverse effects, for instance, fauna passages over or under roads, fauna exits in canals, fencing, noise-reducing measures and measures for reducing lighting effects (Bekker et al., 1995). The main aim of fauna passages is to conserve local animal populations, provide a corridor for animals regularly moving between resting and foraging areas or migrating between summer and winter areas, and for dispersion (settlement in other areas). A secondary objective is that the passage of animals contribute indirectly the distribution of seeds: seeds can be carried in the fur or in the intestinal track of animals.

The principles of mitigation measures against accidents, barrier effects, isolation and disturbance

Defragmentation measures are based on one or more of the following principles:

1.   restriction (with fencing) of the animal's ability to move between habitat patches or making it safer (fencing combined with safe passages) to reduce the number of accidents;
2.   elimination of the barrier effects of a road (for example modified roadside verges with passageways);
3.   elimination of habitat isolation (for example corridors in roadside verges to guide animals to passageways or as a link to habitat located further away);
4.   elimination/reduction of disturbance by screens
5.   increase of habitat quality of habitat patches in the surroundings.

Source and effect oriented measures

'With all these measures, the location has to be selected first. Source-oriented mitigation measures (reduced traffic intensity or speed, reduced noise level by using low noise road surfaces) not only have a broad scope of operation, they must be implemented over a large area (or, as the case may be, over long stretches of infrastructure). This is different for effect-oriented measures. In this case, the measures have to be tailored to specific locations and closely related to different types of animals. This is usually fairly difficult, because the animals that the measures are designed for live along entire sections of a motorway. In practice, bottlenecks usually occur when infrastructure crosses the national or provincial ecological networks or other areas protected under government policy, but also information from provincial animal conservation plans and data on animal road accident casualties or cases of drowning in case of canals can give indications to locate mitigation measures. The great length, the usually narrow verges and the generally short distances between the entrances that have to be kept open to farms and fields cause practical problems in implementing effect-oriented measures along roads in rural areas. This limits the possibilities for placing continuous fencing and screens.' (Piepers; ed., 2001).

It is interesting to see that for individual species special fauna-passages are developed like cables over roads and motorways for red squirrel and pine marten (Bekker, 2002). The study by Bekker showed all kinds of different constructions (tight ropes, steel wires, rope ladder, adapted motorway portals in use for road signals as well as for fauna passage).

Overview of fauna passages
Fig. 311 and *Fig. 312* give an overview of different construction types of motorway infrastructure in relation to different sized mitigation measures for nature (Bohemen, 2004, Trocmé, 2003).

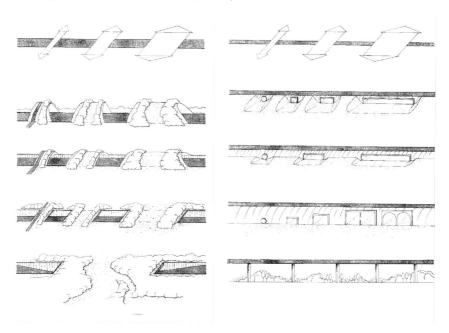

Fig. 311 Examples of different types of overpasses (Bohemen, 2004)   *Fig. 312 Examples of different types of underpasses (Bohemen, 2004)*

It is possible to vary the height of the infrastructure to solve part of the environmental and social effects of the roads and traffic. Beside the situation on surface level the infrastructure can be placed on a height,

472

'half'height embankment or can be constructed on piles (as a long viaduct) or can be brought under the ground (via a bored tunnel, with tunnel elements or cut-and-cover tunnel concept) on different levels of depth. Within the mentioned options three types of constructiontypes for the infrastructure can be distinguished: pipe, U-profile (with or without cover) or sheet. The shown mitigation measures can vary in each situation in size.

23.4.4   Planning of fauna passages

Including all levels of scale and landuses
The effectiveness of measures increases if all roads at various levels are included in the decision process of the measures to be taken. In this context, it is also important that road managers keep in contact with the managers of ecologically valuable sites and areas.

Integrated design
For the sustainability of a solution it is vital that all measures be carried out based on an integrated design approach.. The entire concept must as much as possible be harmonised with the surrounding landscape, including roadside verges and adjacent habitat patches, such as hedges, ditches and habitats for forest and meadow birds. Defragmentation studies usually only pay attention to the effects and measures due to the crossing or intersection of nature areas and ecological connection zones. Apart from (the successfully applied) ecological roadside verge management, more attention must be paid to the ecological connections that run parallel to roads and canals. That is also a big opportunity for increased interest among farmers for what is known as agricultural nature management.

Including future developments
In the planning and execution of measures, it is vital to take into account any future developments, especially if quality impulses for nature development are carried out in the long term. *Veluwe 2010* (Province of Gelderland, 2001) proposes among other measures useful for animals, the reduction of spatial fragmentation and the enlargement of the habitats of the large mammals by eliminating barriers. This can be done by removing fences, reducing traffic intensity on local and regional roads and increasing the number of ecoducts over motorways and railroads in the Veluwe area (currently there are only three) by at least 15, as well as more fauna-passages under roads and railways. The ecoducts are necessary to realise the desired 'sturdy' ecological connection zones between the Veluwe and other actual and potential important areas for large mammals in the Netherlands as in other parts of Europe.

Including other species
In other situations, it is planned that bridges for roe deer will be also made suitable for larger animals like red deer.

23.4.5   Costs of fauna-passages

The costs of mitigating measures play an important role in the decision process of the provisions to be installed. The costs given below are estimates based on rough parameters and estimated lengths and sizes of the adaptations. And there are differences between new and existing situation, which is more expensive. In principle, the following figures can be used (2002 price level) (Bohemen, 2004 based on information from the Directorate-General of Public Works and Watermanagement)):

1   Ecoduct: € 3,5-6,5 million
2   Roe deer tunnel: € 1380 per linear metre (2.75 x 2.5 m; dug-out)
3   Migration tunnel: € 5900 per linear metre (1.5 x 2.5 m; drawn or pressed)
4   Badger pipe: € 68 - € 91 per linear metre (cross section 0.4 m; dug-out under the existing road with concrete without reinforcement)
5   Badger pipe: € 32 - € 45, steel, per linear metre during road construction; reinforced concrete: € 50 - € 68; same; concrete without reinforcement: € 27 - € 45, same.
6   Amphibian tunnel: € 680 per linear metre (1 x 0.75 m)
7   Ecoculvert: € 730 per linear metre (1 x 2 m)
8   Adapting the existing bridge with a stump[a] *) bank: € 2000-8000
9   Purchase of additional land: € 30,000 - € 55,000 per hectare

For comparison:

10   Motorway: per km € 15 - € 20 million
11   Tunnel: per km € 125 - € 225 million

23.5   The use of fauna-passages

One of the ways to check whether the measures to combat fragmentation (all types of fauna passage, including the guiding provisions) are effective, is to establish whether the fauna passages are used by the species for which they were intended. Next, comparative studies can for instance be used to ascertain which type and which dimensions, which material and which maintenance produce the most usage, in order to realise an optimal passage in as far as the design and use are concerned. Beside the knowledge that fauna-passages are used by the organisms for which they have been built it is important to know whether the passages contribute to the long term existence of populations.

---

[a] From a point of view of recycling and sustainable use of material this is a good option as otherwise the owner of the stumps have to pay for dumping or for destruction.

23.5.1 Amphibians and roads: knowledge development and application

First examples
The first fauna provisions were designed for amphibians. At the beginning of the 1960s, Switzerland worked on measures to preserve amphibians on the basis of the first counts of amphibian roadkills (Heusser, 1960). Circa 1975, the Netherlands saw the start of the first actions to protect migrating toads from traffic. As far as mitigating measures are concerned, already various types of tunnels, and the accompanying guiding systems (screens, chutes) are being applied.

Acceptance by amphibians
The acceptance of the provisions by amphibians varies. This depends not only on the type of the provision, but also on its location in the landscape (guidance by landscape elements). Animals should be guided in some way to the tunnel entrance and it is important that the direction of the tunnel does not deviate too much from the migration routes. The diameter of the tube must be at least 60 cm with a length of up to 20 m; for longer tubes, the diameter must be at least 1 m. For tubes longer than 50 m, the diameter should be 150 cm. Furthermore, it is vital that guiding screens should be installed and keep in repair. Their length depends on the willingness of animals to walk along a screen and the width of the migration zone. In order to prevent water from remaining in tunnels, the fall must be approximately 1%. Furthermore, the use of a tunnel depends, besides the occurrence of species and their densities, on the tunnel material, soil substrate, the light entry, the number of and distance between the tunnels, and the characteristics of the screen (material, height, length, exits, vegetation near the screen and the positioning of the screen).

The influence of tunnels on the survival of populations
It is not yet possible to make a statement based on the study of the use of tunnels concerning the influence of tunnels on the survival of populations in the long term. Concerning the effect of tunnels on the exchange between populations at regional level no information exists.

Ecoducts
In addition to tunnels, also ecoducts, culverts with dry banks and viaducts provided with an unpaved zone can be used by amphibians.

Quality of migration routes
In the adoption of measures it is important to keep all migration routes as intact as possible: biotopes should not be separated and the water quality should be protected. The appeal of the water, where mating and reproduction take place, as a factor determining the migration is probably also determined by its scope. The temperature of the water also plays a role (possible on infrared radiation effect) as well as the microclimatological circumstances.

23.5.2   The use of fauna pipes

Beside tunnels for amphibians in the seventies special pipes for badgers have been built. During the last decennia studies have shown the use by badgers of these badger pipes. In order to know if and under which conditions other (small) animal species make use of these provisions 50 fauna pipes (also called: eco-pipes or fauna-tunnels) were studied in the Netherlands in autumn 2001 and spring 2002 (Brandjes et al., 2002).

From the literature-study it became clear that the use of fauna pipes are influenced by: the dimensions of the pipe, the microclimate inside the pipe, the functioning of fences, the structure of the surroundings of the openings, the presence of species in the surrounding, the use of the pipe by predators.

The results of the study of Brandjes et al. (2002) shows the following:

1.   All pipes have been used with an average of 3,8 species/pipe/studied period (3 months).
2.   Amphibians did use the pipes with a diameter of 30-40 cm. The length of the pipe is negatively correlated with the use by marten, weasel and stoat.
3.   Mice probably visit mostly the pipes, but do not pass them completely.
4.   Use of badgers affected negatively the use by hedgehogs, but did not effect other species.
5.   Frequent use by domesticated cats has a significant negative effect on the use by other mammals.
6.   In more humid pipes more passage of amphibians were counted.

In a study by Eekelen and Smit (2002) it was found that differences are caused by the size of the provision as well as the location of the passage within the habitat. When a passage is larger and more humid the use of the number of species is likely to be larger. In an optimal habitat situation (nature reserve) a relatively small fauna pipe will show more frequent use than in suboptimal habitat situations.

23.5.3   The use of ecoducts

Studies of the use of ecoducts by animals have already produced many data to enable better design and realisation. The older French types of ecoducts did not function as predicted. These provisions were small (4.5-8 m wide), not covered with vegetation and poorly situated in the landscape. The newer types which were constructed for example near Mulhouse in 1985 are wider and have a parabolic shape; they provide better results.

The first two ecoducts in the Netherlands over the A50 between Arnhem and Apeldoorn, which were constructed between 1986 and 1989, are used by different animal species (Litjens, 1991; Berris, 1997).1991; Berris, 1997). From almost the first day on the following species made use of the ecoduct of Terlet: red deer, wild boar, fox and rabbit and roe deer.: red deer, wild boar, fox, rabbit and roe deer. Some years later badger crossings were observed too, as well as pine marten and ground beetles. In 2002 and 2003 new counts of animal crossings show again use by mentioned animals (Belle and Frenz, 2005).

In 1993 a study was undertaken of the use of three types of fauna passages (13 pipes, 2 adapted culverts and 1 ecoduct) under and over the A1 near the Boerskotten estate (Nieuwenhuizen and Van Apeldoorn, 1994). The ecoduct was opened in november 1992. The measures were made by counting tracks. This led to the following conclusions:

1. All passages were used.
2. Pipes and culverts were used by five species of mammals (hedgehog, rabbit, fox, pine marter, stoat) and the ecoduct by nine species (hedgehog, red squirrel, rabbit, hare, fox, pine marten, stoat, polecat, roe deer), not including the various species of mouse and vole.
3. Some species seem to prefer a certain type of passage. Foxes and rabbits for instance used the pipes more often than the culverts. The roe deer, hare and red squirrel only used the ecoduct. It was striking that the small species – such as mice and voles – did not use the pipes.

Also outside the Netherlands studies have been made of the use and effectiveness of fauna provisions. In 1991 Germany saw the start of a comparative study of the effectiveness of ecoducts for some groups of organisms. In one part of this study the effect was studied for some five ecoducts before and after the construction of a road, and in the other part a comparison of the use of eleven ecoducts in Germany, France, the Netherlands and Switzerland was made (Keller and Pfister, 1997; Pfister et al., 1997). In the first-mentioned study one of the five fauna passages had a width of 50 m (combination with the local road), two a width of 80 m (one without path, one with agricultural road), one of 29 m (agricultural road with strip of vegetation) and one of 20 m (agricultural road with strips of vegetation for cattle). In the comparative study, three fauna passages were included in Germany (10-35 m wide), four in France (8-12 m wide), two in the Netherlands (Terlet and Woeste Hoeve, 50 m wide) and two in Switzerland (200 m and 140 m wide, respectively). Some supplementary studies were conducted concerning the use by insects, spiders, small mammals and birds and special attention was paid to the badger including a telemetric study of behavior.

The preliminary results show that some forest ground beetles hardly use the passages but remain in the forest. The question is whether the passage of some individuals is not already sufficient to prevent complete genetic isolation. In this case, a real connection can only be realised by full afforestation of the ecoduct.

Furthermore, it appeared important to have a corridor with the same type of vegetation on the ecoduct as the habitat patches to be connected.The use by mammals increases as the bridge becomes wider (50-100 m). Also narrow viaducts are used, but in the study it was discovered that the entire population of red deer eschewed the passage, whereas beforehand they did roam the area. On the other hand, wild boar exhibit learning behaviour, viz. after a time they started to use narrow passages, while at first they had become disoriented.

The use by fauna of the ecoduct over the A1 has been studied (Eekelen and Smit, 2000) and found the use of roe deer and red deer. The wild boar do not only cross the ecoduct but if forms part of their feeding area.

23.5.4   Co-use of viaducts as an underpass

Stump walls
Under Zandheuvel viaduct (A27 near Hilversum), a row of tree stumps was installed underneath the over-dimensioned viaduct in order to investigate to what degree it would function as a faunahabitat and -corridor. Later a viaduct near Zeist (A28) was underneath equipted with a row of tree stumps.

Linden (1997) showed that the construction of a stump wall is an effective mitigating measure to counteract the barrier effect of a motorway for some animal species. The stump wall was used as a corridor by insects, mice, voles and shrews. Also other animal species use the space under viaducts, but their numbers vary depending on the viaduct studied (in the case of Zandheuvel, also mole, red squirrel, roe deer and polecat; underneath a viaduct near Zeist: also rabbit, roe deer, polecat, fox, weasel, stoat, hedgehog and hare).

In 2000 Ottburg and Smit studied the co-use of the Zandheuvel viaduct as well as the viaduct near Zeist which both show use by small mammals and amphibians. The Zeist viaduct could be fully designed as a fauna passage as the planned road had been cancelled. In 2000 beside the mentioned species passage of mice, rat, cat, roe deer and salamander. Hare could not be found probably due to increasing the amount of shrubs near the viaduct. For the first time fish passage has been studied which shows use of the water course underneath the viaduct by different fish species.

23.5.5 Co-use of viaducts as an overpass

Some regional directorates intend to make some existing viaducts over motorways suitable for co-use by animals in order to contribute to the elimination of some habitat fragmentation bottlenecks along motorways. To study the effectiveness of adaptations of viaducts for fauna co-use, measures were made of the current use of these bridges. For this, sand beds and track tubes were used to conduct track studies, and existing observations of mammals, amphibians and reptiles were collected (Dorenbosch and Krekels, 2000; Eekelen and Smit, 2000). Although the current use of the studied, non-adapted viaducts by animals is slight, in the study period seven species were observed on the viaducts: the hedgehog, polecat, fox, wood mouse, brown rat, rabbit and smooth newt as well as insects. A positive correlation was established between the use of the viaduct by animals and the presence of small-scale landscape elements with a well-developed structure nearby the viaducts (see *Fig. 313* and *Fig. 314*).

Fig. 313 A viaduct without a vegetated strip connecting existing roadside verges (Photo Bohemen)   *Fig. 314 A viaduct with a vegetated strip connecting existing roadside verges* (Photo Bohemen)

Based on the distribution study of animal species in the area, for each viaduct a forecast of its use was drawn up. The viaduct and the surrounding landscape must be adapted for co-use by animals.

Eekelen and Smit (2000) found little co-use of viaducts of the A1 on the Veluwe, but by remetal part of the road on the viaduct or by introducing vegetated strips or by placing stumps on the viaduct or increasing the attractiveness by construction of water pools near the entrance could increase co-use by animals.

23.5.6   Co-use of dry edges along waterways under viaducts

Literature study

In order to detect the main gaps in the knowledge of the use of fauna passages in the Netherlands, in 1996 a literature study was conducted on the use of all types of fauna passage in the Netherlands and abroad, including lower passages underneath roads not constructed as fauna passages (Smit, 1996). This study also provided a summary of the methods that were applied or are suitable for establishing the use of fauna passages, along with their advantages and disadvantages. The study shows that for most fauna passages (for example ecoducts, badger pipes and amphibian tunnels) the knowledge has been accumulated about their use by the target or other species. Less is known about viaducts and bridges adapted for the co-use by animals as an underpass.

Exploratory study

Especially about the use of wooden passageways, floating plank bridges, concrete walking strips in culverts and continued banks under viaducts there was a lack of knowledge. Therefore in 1997 an exploratory study of the use of fauna passages along waterways was conducted (Brandjes and Veenbaas, 1998; Veenbaas and Brandjes, 1999), in which also the (partly new) study methods – footprints and tracks on sand beds and on ink bed paper - were tested in practice. Practically all passages appeared to be in some use, but not all were used by the target species. Although the study was focused primarily on mammals, also amphibians appeared to frequently use these passage strips.

Openness and width

An important factor in such use is the openness: the wider and shorter the whole underpass, the more the passageway is used. Also the width of the passage strip is important, especially for mammals: the broader the strip, the more often it is used, and also by more species. In addition to such frequent users as the brown rat and (wood)mouse, those using the passages included the hedgehog, fox, stoat, polecat, stone marten, water vole, muskrat, vole and shrew; most species use only foot planks 40 cm or broader, and the fox and stone marten only travel along continued banks.

Experimental study

The exploratory study was followed in 1998 and 2000 with an experimental study in order to draw up guidelines for design and lay-out. The use and the effect of cover material have been studied. This study showed that (Brandjes et al., 2001):

1.   In 2000, 18 animal species were detected on foot planks (17 in 1998); these animals included the hedgehog, water shrew, weasel, stoat, polecat, stone marten, newt, toad and frog.

2.   In the same year, 30 animal species were detected on continued banks (25 in 1998); these animals included the hedgehog, water shrew, fox, weasel, stoat, polecat, stone marten, badger, roe deer, newt, toad and frog.
3.   Widening the wooden passage way led to significantly more weasel tracks but significantly fewer tracks of all animals combined (this was especially due to the decrease in the number of mouse tracks).
4.   The creation of stump walls on continued banks led to significantly more mouse tracks.
5.   Except for fox and stone marten, the co-use by humans appears to have a negative impact on the use by fauna.
6.   As a result of the significant positive impact that the widening of wooden passageways from 70 cm to 100 cm has on their use by the weasel and of the fact that broader passages generally are used by a wider species range, it is advisable to broaden existing wooden passageways in culverts and under bridges.

23.6   Effectiveness of fauna passages at the population level

The first step in the study of the effectiveness of a fauna passage is to establish whether it is used by the intended species. The next step is research on the viability on population level. For this stage the Road and Hydraulic Institute did 2003 an investigation about the possibilities in this field. started a research project. The study has a long preparation period, since it concerns a new type of study with many complicating factors, such as other developments in an area that influence the survival of a species, and since the data about the dispersion of relevant animal species are only available to a limited degree. In the the proposed research it was adviced to analyse whether and if so where suitable study locations to study the population-effect of fauna-passages are suitable for three animal species – the badger (as indicator to evaluate for badger tunnels), red deer (as an indicator to evaluate ecoducts) and crested newt (as an indicator to evaluate animal walking boards (=foot planks) (=foot planks) under bridges and in culverts). In 2006 an international workshop took place about how to collect the existing knowledge in this field and an international research program is underway (Markensteijn et a., 2006).

23.6.1   Conclusions about the use of fauna passages

Studies of the relationship between infrastructure with traffic and fauna reveals that the effects differ from species to species. Knowledge about the presence, distribution and dispersion of the species that are actually or potentially present in the area where fauna passages should be built, is essential.

An expert-based model for spatial planners, transport infrastructure designers and wildlife conservationists to identity the best locations for defragmentation measures, both from an ecological and a cost-benefit point of view has been developed by the Dutch research institute Alterra and Wageningen Research Center (Grift and Pouwels, 2006). It is called LARCH, Landscape Assessment using Rules for Configuration of Habitat, and it assigns a measure of habitat suitability for a specties to a landscape, based on vegetation, size and the needs of the species. It is a landscape based approach (Pouwels et al., 2002). It should be realised that in a wider context (ecosystemlevel) the problem is more complex (for example the need to include trophic structure).

23.6.2   Economical and social spin-off of fauna measures

The building of provisions for fauna have given different kinds of spin-off in the field of manufacturing, research and in the social field.

Fig. 315 A prefab culvert with a provision for terrestrial fauna passage
(Photo Veenbaas)

Interest of industries

In the past provisions had to be specially designed and manufactured. As the number of concrete provisions are increasing, industries are more and more interested in designing and construction of prefab culverts including passage possibilities for terrestrial fauna, prefab amphibian tunnels and measures like ecoducts (new construction types and other forms).

Combined design

Also in the field of combined design and construct approaches for realising ecoducts have been stimulated, which has been successfully taken up by the private market. As an example of a new tendering approach the ecoduct De Borkeld to be built in Overijssel near the village of Rijssen has been given out and realised in a 'design and build' contract form. A greater variety of disciplines (architects, landscape architects, ecologists, engineers, designers) were involved than normally is the case in such projects and knowledge and ideas of the market will be used for more creative solutions.

A prefab culvert

A prefab culvert with a provision for passage of terrestrial fauna designed and made available by a concrete firm is illustrated in.

23.6.3   Discussion

During the last two decades studies dealing with effects of motorways and traffic on nature have given insight into the intensity and the extent of the effects on the studied groups of animal species.

The research on the use of implemented mitigation measures has given evidence of the usefulness of different types of fauna-passages to certain animal species.

Survival of species on (meta-)population level

Although we know that the different types of fauna passages are in use by different species we are now reaching a phase to learn more about the real effectiveness of mitigation measures for the survival of species on (meta-) population level. Long-term studies are essential as most of the research carried out has a time-span of less than one year till five years. That is often too short. Scott Findlay and Bourgas (1999) studied the response time of wetland biodiversity to road construction on adjacent land and found that in wetland biodiversity loss appeared decades after the intervention.

Other studies needed

Beside there is a need for more studies during the coming years in the field of:

1.   The design of viaducts and ecoducts for combined human and animal use in all possible variants (walking, cycling, horse riding, car traffic).
2.   Adjustment between type and size of the measures and the specific functions and types of species which should use the passages.
3.   The most rational maintenance of the vegetation on ecoducts and other types of passages in relation to the requirements of the different species for which the passage is meant for.
4.   Passages with special provisions, like stump walls, lack detailed information about the optimal size and layout.

5. Looking from the species angle the question arises whether the existing measures will cover all animal groups which show sensitiveness for habitat fragmentation by roads and traffic. Reptiles, certain insects and some other invertebrates groups are more susceptible to habitat fragmentation than other groups.
6. More detailed knowledge is useful about the behavior of animal species using fauna-passages (daily movements, season dependent or use only for dispersion) and the future use (otter, and red deer and other mammals in areas where there are ideas or plans of re-introduction or expectation for 'spontaneous' arrival of such animals) in combination with the minimum number of provisions per km infrastructure. Also the question is relevant what degree of impassability is acceptable or what is the degree of passibility.

Integrated ecosystem approach
A broader approach could be chosen in the future in the form of a more integrated ecosystem approach in order to maintain and upgrade the natural values and functions of roadsides and the surroundings and improving the permeability of roads and motorways.

Cost-benefit analysis
Also cost-benefit analysis is becoming more and more important. The knowledge about the cost-effectiveness of mitigating measures for animal species and ecosystem functioning in relation to integration of sound-barriers, measures against the pollution of air and soil as well as development of ecological, psychological, social and art values in roadsides and surroundings will become relevant.

Optimising the use of ecoducts
In the Netherlands a more habitat approach on the ecoducts is developing, that means to design wider ecoducts(towards 'landscape'bridges) in order to increase the habitat value for animals (part of the daily habitat) instead of only providing a corridor function. In this respect also the size of the animals is relevant: for small animals, which has a small actiradius, the provisions should provide habitat for the species concerned.

23.7   Synthesis: Defragmentation strategies

Coherence
In 1990, an evaluation study of the effects of the existing motorways on nature and the possibilities for defragmentation of nature was conducted (Fluit et al., 1990). Some years later it was examined whether the defragmentation measures at the network of motorways and at that of the railroads contributed to the realisation of the National Ecological Network (NEN) (Ministry of Agriculture, Nature and Fisheries LNV, 1999). The main conclusions were:

1. The current implementation of measures intended to realise the NEN has not yet produced a coherent and sturdy system.
2. Both the knowledge and the understanding of the effects of fragmentation have considerably increased.
3. There is a need for a nation-wide plan, in which choices and priorities for defragmentation are detailed and underpinned.

Strategic defragmentation locations

Based on the spatially coherent larger landscape-ecological units of nature distinguished in the Netherlands, some robust connections (corridors) have been selected, known as strategic defragmentation locations. The selection is a first impulse to make possible the mutual consultation about a more coherent policy context and as well as a coherent defragmentation programme.

Balancing interests

Despite the construction of ecoducts, badger tunnels, ecoculverts and other provisions, there is no equal treatment in the policy processes between traffic, transportation and infrastructure and the 'ecological accessibility' on ecosystemlevel. This applies also and especially to provincial, regional, municipal and local situations. Various approaches to reduce the effect of fragmentation can be distinguished: limitation of the need for new or widened infrastructure, optimisation of the planning process in order to reduce the degree of fragmentation, and improvement of the size and quality of the mitigating works and compensating measures. All measures must be implemented in an optimised context. After the construction of a road, clear quantitative objectives must be formulated in maintenance plans and they should be monitored as well.

Five strategic actions

As reported earlier, in the planning and construction/reconstruction of infrastructure, it is important to adopt the following strategy wherever such is possible: in the first place, prevent, avoid or minimise negative effects; in the second place, implement mitigating measures; and in the third place, compensate for any damage caused. This entails the following five actions:

1. At a strategic level it must first be ascertained how the demand for transport and transportation can be satisfied via means other than the construction of a road at ground level (areal planning coordination of the functions for living, work and leisure, price policy, tunnel construction, public transportation and the use of waterways).
2. Given the decision in favour of a road, its optimum location must be established in a search area that is as large as possible, which give possibilities to minimise negative effects.
3. The road must then be fully integrated (including mitigation measures) into the landscape.

4.   Thereafter, the physical compensation must be realised near the new road and in such a way that the surface area and the quality of habitat patches are not negatively affected.

5.   Finally, financial compensation must not be made until physical compensation of the intervention has turned out to be impossible.

Viewing not only the significance of ecoducts as a passage potential for animals but also the operation of areas as ecosystems, it might become apparent that larger ecoducts (landscape bridges) are necessary or that the road will have to be created under the ground. In order to eliminate a road's barrier effect and to protect both surface water and groundwater, roads on columns might be an answer.

23.8   Conclusions

23.8.1   An internationally recognised problem

All sorts of developments, including the formation of road networks, have made the loss and fragmentation of habitat patches for plants and animals an increasing problem for the conservation of nature. Creating nature reserves will not halt the degradation of biodiversity, for example caused by climate change. All sorts of conventions and EU directives (Ramsar, Bern, the Birds and the Habitats Directive) call for attention to be paid to the importance of nature. In addition, all European countries now recognise fragmentation by roads as a problem. Studies carried out during the COST action 341 learned that even in Norway and Sweden – where the human population density is relatively low – roads and traffic take their toll. In Sweden, over 60% of otter mortalities are caused by traffic; in Norway a local subpopulation of reindeer had been isolated by road construction. In Denmark and Switzerland, fragmentation by infrastructure is a limiting factor for the hare. A study conducted by the World Wildlife Fund on large carnivores showed that the development of infrastructure is limitative for the corridor development for this animal group, unless adequate measures are taken.

23.8.2   Solutions

In the Netherlands, as well as in various countries, studies have been conducted in order to establish where the National Ecological Network leads to bottlenecks with the motorway network (See *Fig. 310*). Ecological techniques or ecological engineering play important roles in the conservation, restoration and development of nature values in relation to the infrastructure. For a large part of the field, the studies discussed provide a scientific underpinning, but future studies must supply data for more encompassing solutions to the usually complex problem of fragmentation and defragmentation. Defragmentation programmes are drawn up.

In addition to defragmentation measures themselves, also more and more value is being attached to their optimum incorporation into the planning process. Also the design and maintenance of roadside verges – not only as habitats for plants and animals, but also as corridors for certain animals – play relevant roles.

23.8.3   Evaluation of mitigating measures

Most mitigating measures have a positive effect on indivual animals. In any case, the measures taken for the Dutch system of motorways have been very important for the badger at population level as well. This rare species in the Netherlands is again gradually increasing in number despite the increasing number of road-kills. In just fifteen years, the Dutch badger population has multiplied by four as a result of the measures taken on roads, the safeguarding of setts and badger routes, and the re-colonisation by young orphaned badgers, which have been accommodated in a centre.

Also at provincial, municipal and local levels more attention is being paid, stimulated among others by the multiyear defragmentation programme to mitigating and compensating measures. The knowledge that is now available could however be used even more effectively if more coherence was created between the national, provincial and municipal networks of dry and wet infrastructures and ecological networks at various scales.

23.8.4   The policy cycle

After the exploration of a problem area and its recognition, the complete inclusion of the policy field into the general policy cycle is necessary for it to play a permanent role in the policy processes. The problem would then receive lasting attention – which has appeared in principle to be very effective based on the study of the implementation of defragmentation in the Netherlands over the last three decades. The policy cycle comprises the following stages: policy memorandums and plans, realisation programmes, management and maintenance programmes, monitoring programmes and the continual adaptation and adjustment of memoranda, plans and programmes based on new knowledge, new insights and practical experiences. In this context it must be said that maintenance of fauna passages is not always optimal and when cutting of budgets have to take place fauna passages as a topic often stay behind. Continuous attention is still needed.

23.8.5   Recommendations

It remains necessary to stimulate the application of the existing knowledge in order to continue a 'no net loss' policy for nature. In addition, studies remain necessary in some subfields. The following are somewhat striking recommendations based on the foregoing:

1. Make widely available a summary of all possible measures related to concrete situations: an up-to-date toolbox for designers is essential (see for instance: www.wildlifecrossings.info as the IENE website www.iene.info, which gives a wildlife crossing toolkit).
2. Better coordination between the design, realisation and future maintenance of fauna passages. A weblog is available to support further discussions about defragmentation measures: http://workshophabitatdefragmentation.blogspot.com.
3. Implement monitoring programmes to observe the use and status of maintenance of the fauna passages in such a way that sound conclusions can be drawn (studying a large number of provisions in the same period with the same method).
4. Harmonise strategic, tactical and operational plans and programmes at the international, national, regional and local scales; this is essential for a more successful application of defragmentation measures.

Design phase
Especially during the design phase of eco-structures and other fauna passages, it is possible to raise the effectiveness, on the condition that the following aspects are taken into account: the surroundings, dimensions (height, width, length) and form (rectangle, convex, road on columns and other forms), material (concrete, steel, wood), set-up (number of eco-structures or ecopipes per km) and form and length of fencing in relation to the function.

Integration of civil engineering
A further integration of civil engineering and the application of ecological engineering can play an important role in the conservation, restoration and development of ecological values along road infrastructure. Though better coordination between the planning of land use and of infrastructure, nature conservation and nature development will produce more possibilities for conserving and developing ecological values. At an early phase of road planning, important nature areas can be avoided. If this is not possible, mitigation and compensation can limit the damage to ecological values. Communication about possibilities to include fauna provisions is important. *Fig. 316* shows the cover of a book which has been published in 1995 and which has stimulated the enthousiams of people involved as well as people to become involved.

Fig. 316 Cover of the book Nature across motorways
(Bekker c.s., 1995)

Combination of large and small provisions
The combination of some large provisions with many smaller provisions, to "perforate" roads as much as possible, is necessary for the sustainable survival of animal populations at various levels of scale. On the one hand, these are provisions specifically for the fauna and on the other hand provisions that allow for the co-use of for instance viaducts and bridges by fauna – not only in rural but also in suburban and urban areas. Bringing nature closer to man, can increase man's valuation of nature. This is certainly important now that it looks like man will increasingly be accommodated in urbanised areas. Of course more research is needed about the differences between one large or several small defragmentation measures. Do we need 1, 2 of 10 ecoducts or 100, 200 or 1000 faunapipes to connect ecologically valuable areas?

Radical vertical separation
If ecological damage cannot be prevented of avoided, large landscape bridges, sinken roads, tunnels and roads on columns (see *Fig. 317*) could help.

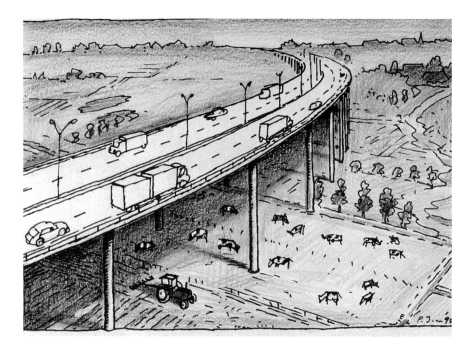

Fig. 317 An art impression of a motorway on columns (RWS/AVV)

23.8.6   Conclusion

The still increasing road and railway network in the Netherlands as well as in the rest of Europe and elsewhere in the world is harming our natural environment. Studies show that by an approach 'to avoid, to prevent, to mitigate and to compensate' negative effects significant improvements between the relationship between human made infrastructure and the ecological infrastructure on all levels of scales can be reached.

Although more research is needed concerning the effectiveness is still needed there is now enough knowledge available for successful implementation of concrete measures in the planning, design as well as during the building and use phase of roads and motorways.

Not only governmental action is important, the effectiveness of local involvement can be of great value which is shown by the development of the ecoduct in 't Gooi in the Netherlands. The longest ecoduct (800 m) has been realised here by local actions (Goois Natuurreservaat, 2006) and has led to higher prioritisation of a governmental ecoduct nearby, to improve connectivity (Markesteijn et al., 2006).

24   ROADSIDE VERGES
integration of road infrastructure in the landscape

Hein van Bohemen[a]  T Delft

---

[a] Valuable comments from Dr. Peter-Jan Keizer (Road and Hydraulic Engineering Institute Delft) on the manuscript are gratefully acknowledged.

24.1   Introduction

In this chapter the function, development and current situation as well as prospects for the ecological value of roadside verges will be presented with an accent on motorway verges. It presents both a general overview of the flora, vegetation and fauna present on Dutch motorway verges. The studies show the ecological importance of motorway verges as well as the maintenance practices to maintain and to increase the ecological values.

Road ecology according to Forman et al. (2003) may help 'to bridge the gap' between 'natural processes and biodiversity'and 'safe and efficient human mobility', and this chapter offers insight into the developments and possibilities of better integration of natural and environmental aspects in the design and construction and maintenance of motorways.

Definition and functions of 'road and motorway verge'
The term 'roadside verge' refers to the space between the edge of a road surface and the edge of the adjacent area with a different landowner, different land use or different native vegetation. It includes grasslands, small woodlands, water (ditches, ponds, small lakes), and the 'artificial' vegetation growing near service stations and road junctions.
Usually the road authority is responsible for the managemant of these areas.

Functions of roadside verges
Roadside verges have six functions:

1.   technical (road engineering);
2.   traffic;
3.   aesthetic; landscape design
4.   ecological;
5.   landscape ecological;
6.   economic

Technical functions
In the days that ecology was not considered, mainly road engineering and traffic functions determined the design, construction and maintenance of roadside verges.Verges and slopes contribute to the sturdiness of the road bed and the road itself, and they accommodate cables and pipes, while also playing a role in the collection of runoff.

Traffic functions
 Roadside verges play a role for placing traffic signs, light poles, crash barriers etc. They also help to increase traffic safety as they can act as refuges for broken-down vehicles or for vehicles that have been involved in an accident.

Aesthetic functions
The visual-landscape or aesthetic functions of roadside vegetation can be subdivided as follows (Schone and Coeterier, 1997):

1. Blending the road into the landscape (In Dutch: 'inpassen').
2. Connecting the road to the landscape (to orient road users) (In Dutch: 'aanpassen'|).
3. Providing a contrast between the road and its surroundings in order to emphasise the road's own character as compared to that of the landscape; the road dominates the landscape (In Dutch: 'veranderen').

Ecological functions
In an ecological sense, a roadside verge can have the following functions for species (Forman *et al.*, 2003; Logeman and Schoorl, 1988):

1. habitat patch (temporary or permanent living and residence space for species and communities).
2. connection or migration route (corridor, connection element between fragmented areas; species movement alongside or "through" a habitat patch).
3. stepping-stone (species can bridge an unsuitable habitat in steps).
4. isolator/barrier (from/against undesirable effects from the surrounding area; there can also be a filter effect).
5. distribution core or source (species find a favourable habitat patch from where the population can expand and spread out over a larger area).
6. sink (a "reception" area for species in areas where they cannot survive independently).
7. regulator (species which can predate pests in agricultural fields from the roadside habitat).
8. refuge (for endangered plant and animal species; plants and animals can also withdraw to it on a temporary basis).

Landscape ecological functions
For the ecological significance of road verges the relations with the hinterland are very important, as are the width of the verge, the degree of 'connectedness' of various habitat patches of verges, and the quality of environmental variables like soil, water and air of the verge and its surroundings. Each of these aspects plays a role at various scales: the verge as such, the verge in the landscape and the network of verges in combination with other ecologically valuable networks at various levels of scale.

Economic functions
The economic effect of road verges on the income of their owner is practically zero, or, in fact, negative. In the past, the grass cuttings and the

wood from the roadside vegetation had some economic value. Given an optimal design, the function of the road verge as a sink for pollutants could result in cost savings, since in many cases (except for groundwater- and drinkwaterprotection areas) no costly sewage measures are necessary (Bohemen and Janssen van der Laak, 2003). An optimal design can prevent or reduce accidents which will result in cost savings.

24.2   Flora, vegetation and fauna

24.2.1   Floristic aspects of road verges
Ecologically laid-out and managed motorway verges can play an important role as habitats for plants and animals, and can form part of the ecological infrastructure at various levels of scale (Sykora *et al.*, 1993; Sykora, 1998; Forman *et al.*, 2003; Zonderwijk, 1978, 1979, 1991).
Road verges host for the greater part common species, but some relatively rare species (for the Netherlands) may be present. They are relatively important for plant species that have disappeared from the intensively used agracultural fields, but can survive in the less intensively used and managed roadside verge environment.

Most plant species are characteristic for grassland communities on dry and moist, more or less nutrient rich (fertile) soils, disturbed or frequently trampled places, shrub areas, nutrient rich forest edges and arable fields. Less frequently, species are found that usually grow in wet or dry relatively nutrient-poor hay fields, dune vegetation, heathland and heath grassland and - very limited - chalk grassland. Examples of roadside verge preferring species are: ox-eye Daisy (*Leucanthemum vulgare*), Oatgrass (*Arrhenatherum elatius*), chicory (*Cichorium intybus*), slender cudweed (*Filago minima*) and others.

Nutrient conditions
In the Netherlands with its intensive agriculture, unfertilised grasslands have become a rare habitat: 19% of plant species of (moderately) fertilised and 35% of species of dry nutrient-poor grasslands have decreased considerably (Plate *et al.*, 1992). Roadside verges, which are not fertilised and are managed with low intensity have become an important refuge area for species and communities of nutrient-poor grasslands.

Inventory of roadside verge plants
A monitoring survey of road verge plants (Tromp and Reitsma, 2001) has shown that the floristic composition of verges can vary markedly. This is because roads cross various landscape types; each has its own characteristic soil and hydrological and geomorphological properties and related flora. The Veluwe and the Utrechtse Heuvelrug areas, for instance, are made up of very nutrient poor, pushed-up preglacial materials along with limeless, coarse sand. Here, heathland plants are found, such as

needle furze (*Genista anglica*), Scotch heather (*Calluna vulgaris*), fine-leaved sheep's fescue (*Festuca ovina* ssp. *tenuifolia*), sheep's-bit (*Jasione montana*), hare's-foot clover (*Trifolium arvense*) and sand sedge (*Carex arenaria*). The sandy soils in the east of the country (Drenthe, the Achterhoek and Twente) are largely situated lower than the Veluwe and the Utrechtse Heuvelrug and are slightly moister and more nutrient-rich. Here grass species of mesic grasslands prevail, like sweet vernal grass (*Anthoxanthum odoratum*), sorrel (*Rumex acetosa*), ribwort plantain (*Plantago lanceolata*), common bent grass (*Agrostis capillaris*).Sweet vernal grass (*Anthoxanthum odoratum*), Sorrel (*Rumex acetosa*), Ribwort plantain (*Plantago lanceolata*), Common bent grass (*Agrostis capillaris*).

Top 10 of characteristic roadside plant species
Using the same data, a 'Top 10' of the plant species that were most specific to road verges was established (Keizer, 2002):

1. Early scurvy grass (*Cochlearia danica*)
2. Milfoil (*Achillea millefolium*)
3. Wild onion (*Allium vineale*)
4. Oat grass (*Arrhenatherum elatius*)
5. Autumnal hawkbit (*Leontodon autumnalis*)
6. Hogweed (*Heracleum sphondylium*)
7. Wild parsnip (*Pastinaca sativa*)
8. Lamb's lettuce (*Valerianella locusta*)
9. Dark mullein (*Verbascum nigrum*)
10. Cow parsley (*Anthriscus sylvestris*).

24.2.2   Grassland and woodland vegetation on roadside verges

Grassland plant communities
A study by Sýkora *et al.* (1993) based on the Braun-Blanquet method (Westhoff and Van der Maarel, 1973) led to 69 different plant communities on road verges. Based on their visual characteristics they were then lumped into 29 road verge vegetation types. The majority of these are grassland vegetation types whose maintenance involves mowing. Mowing is necessary in order to keep the verges free from trees and bushes. The strip immediately adjacent to the road surface is mown most frequently; this zone is usually excluded from vegetation classification. Keizer (2005) has lumped related roadside vegetation types for practical reasons (management), using existing publications. It gives (in the form of a handy guide) a total overview of vegetation types along the Dutch motorways, based on soil types and characteristic plantspecies with which vegetation types can be determined; it gives also information about the most desirable maintenance form for each vegetation type.

Compared to other grassland vegetation (for example those in nature reserves and from historial data) roadside verge grassland vegetation types are often fragmentarily developed (Sykora et al. 1993). Many of the roadside grassland types have been combined into 'trunk' plant communities (communities that are dominated by few species and which cannot be assigned to vegetation types on the level of association). Most vegetation belongs to the unfertilised hayfields (Arrhenatherion), but also pioneer vegetation types are common. Some vegetation types prefer, or are even restricted to roadside verges, such as the Thero-Airion, in shaded places like under rows of planted trees the Melampyro-Holcetea mollis, some type of ruderal vegetation and some shrub vegetation types dominated by bramble (*Rubus sp.*).

Fig. 318 Roadside grassland vegetation which has been mown with removal of cuttings (Photo P.J.M Melman)

Planted woodlands on road verges
The woodlands alongside motorways in the Netherlands are planted primarily for aesthetic reasons, and to reinforce the visual design of the road. Many of them are not older than ca. 30 years – a period much too short for the development of climax woodland plant communities. In addition, the long and narrow shape of many plantations creates strong margin effects, such as a greater input of light and nutrients. The significance of small plantations and shrubberies has been studied only in a restricted area in the northern part of the Netherlands (Stortelder *et al.*,

496

1995). Consequently, their biological value is only partially known. The study showed that species indicative of disturbance and eutrophication are the most common. True forest species are normally absent, and occur only in places where the plantation is located directly adjacent to a long-established woodland.

Comparing other woodland vegetations

The same probably applies to plantations in other parts of the country, as show scattered observations from various localities. Very open, 'transparent' plantations like tree meadows, tree rows, tree lanes have sometimes been created for cultural-historical, visual-spatial (traffic guidance) and aesthetic reasons (Aanen *et al.*, 1990). Leaf fall and shade in combination with mowing usually causes an undergrowth with relatively low biodiversity; on top of that, such plantations provide little hiding and breeding opportunities for birds and insects. They also accommodate tree species that usually are highly cultivated. In the 1970's and the beginning of the 1980's, such plantations were often applied as design tools alongside motorways. They have a tree layer, a dense herb layer (and sometimes a moss layer), a poorly developed bush structure, and no edge or border vegetation. A short study (Anonymus,1989), based on Grime (1979), showed that only a small group of plant species can adapt to the stress caused by shade, limited availability of water, litter accumulation, and mineralisation, and to periodic disturbance caused by mowing. Therefore the ecological value of such tree-meadows is low. Based on these results, only few new tree-meadows have been created since that time along motorways in the Netherlands.

24.2.3   Fauna and roadside verges

Introduction

After growing evidence of the success of ecological management in roadside verges for flora and vegetation, also fauna received gradually more attention in the Netherlands. The significance of roadside verges as insect habitat, as well as in some cases as refuge or as corridor for small fauna species has been found.

However, it became also clear that roads can add negative aspects to the landscape, like disturbance of animal life. For fauna especially the barrier effect and the road killings must be valued negatively for fauna.

Faunistic value

Yet the ecological design and maintenance of verges resulting in relatively species-rich herbaceous vegetation types, where possible with a gradient towards tall herbs and shrubby vegetation types, can add to the local faunistic biodiversity. In several descriptive studies the faunistic value of road verges managed in this way was surprisingly high (Anonymus, 1989; Liebrand *et al.*, 2001; Raemakers et al., 2001). The significance of roadsides concerns mainly small animal species, such as grasshoppers, crickets, (diurnal) butterflies, bees, hoverflies, amphibians (Zuiderwijk, 1989) and reptiles.

Roadside verges can be an important habitat for butterflies. A study of Bink et al. (1996) shows a fair number of butterflies, including species that have decreased on a national scale; 30 of the 57 native butterfly species can be found in Dutch motorway verges.

Also for some mouse species roadside verges can form habitats, or can function as a refuge or temporary refuge (Reest and Bekker, 1990). The corridor function provided by road verges has been shown for ground beetles (Vermeulen, 1995).

Ditches and small wetlands near roadside verges

Waterways are present alongside many roads for water drainage and water discharge, and for separation of fields and/or as cattle barriers. This means that many verges have partly moist to wet environmental conditions. Especially situations with sufficient space for a gradual transition from wet and moist to drier (situated higher) environments provide living space for many different organisms. Recently, in several places ditches with gradients have been designed.

Sometimes when building road junctions or motorway clover-leaf junctions, relatively deep pools have been created. Reservoirs with steep and/or asphalt-covered banks that serve to collect rainwater are of little ecological value. Water sections that have a gentle gradient and natural bank growth provide an environment for many water and bank plants as well as the accompanying fauna.

Effect of ecological maintenance

In 2001, a study was made to establish whether the ecological maintenance of motorway verges was having the intended effect (Kalwij *et al.*, 2001). It was possible to use for comparison data files based on permanent plots from the period of 1986-1988 made for the publication by Sýkora *et al.* (1993). 545 Permanent plots from the period 1986-1988 were resurveyed, spread all over the country. In these plots, a total of 409 higher plant species were found in 1986/1988; in 2001, 410 species were found. Of these species, 333 were found in both periods; in 2001, 76 species had disappeared and 77 new ones had appeared.

A comparison of the absolute distribution over the hour-frequency (UFK) classes showed that especially the number of common species had increased, and the fairly common and the very common species had increased by one and two class units, respectively. The species not found in 2001 were mainly very rare to uncommon species, and the number of endangered species for the Netherlands as a whole decreased by 40% in the second period. This indicates the deterioration in the floristic value of the motorway verges. The number of ruderalised plots had increased by 94%.

Possible adverse effects
Kalwij et al. (2001) summarise the possible adverse effects on the maintenance and development of road verge vegetation:

1.   The incorrect practice of roadside maintenance:

    a.   the incomplete removal of grass cuttings,
    b.   mowing too late in the season,
    c.   leaving the cuttings on the ground for too long (which causes leaching of nutrients to the soil),
    d.   damaging the soil during wet periods, due to the use of too heavy mowing machines),
    e.   spreading mud from the ditch over the verge,
    f.   scraping off the verge surface alongside the traffic lane and spreading this material over the rest of the verge,
    g.   one-time mowing on soil types where mowing twice per year would be ecologically desirable,
    h.   sometimes lack of any mowing for several years in a row.

2.   Insufficient control of the quality of the maintenance measures in the roadsides.
3.   Insufficient finances: the need for sufficient budget must be continually pointed out as a priority. Outcontracting the maintenance of verges assigns the responsibility for the quality of the verge vegetation to persons who may lack the necessary ecological know-how.

Nitrogen deposition
In the Netherlands, there is continuous nitrogen deposition and due to this external eutrophication (the environment becoming richer in nutriens), ruderalisation of the vegetation occurs, i.e. tall, rapidly growing broad-leaved perennial herbs start to dominate the vegetation. Due to this ruderalisation there is a reduction of the percentage of vegetation types of nutrition-low soils with usually less common species.

Deterioration
Assuming that the above mentioned observations are representative of all motorway verges, the management of the roadside vegetation, after an initial level of relatively good vegetation quality did not lead to an improvement of the national floristic composition. In fact a reduction of quality was observed in view of the ruderalisation of the vegetation that had occurred. This was shown by the increase of plant species that are characteristic of ruderalisation and wood formation.
On the other hand, the plots were originally selected to make a survey of roadside vegetation, and not with the aim to make repeatedly comparisons. This could mean that stands richer in species than average have been

selected in the past, with as a consequence a larger chance to find impoverishment.

Potentials
In view of the locally increased botanical values resulting from the creation and maintenance based on nature-technical principles, carefully executed maintenance, monitoring, evaluation and any needed adjustment of the maintenance, it should be possible to increase the botanical value of verges alongside motorways. This is confirmed by the results of the monitoring survey concerning the A58 (Veer *et al.*, 1998).

Maintenance costs and ecological values of motorway verges
Managers and decision makers must have a clear insight in the costs and benefits of the ecological roadside management as applied today. Therefore a study was carried-out which compares the costs of several different management scenarios (Jong *et al.*, 2001). The study was limited to grasslands, tree meadows and ditches, and showed that the change from traditional frequent flailing maintenance to ecological maintenance (mowing once or twice a year and removal of the cuttings) had led to a marked increase in ecological values and to varying effects on the maintenance costs.

The maintenance costs of moderately nutrient poor grasslands and tree meadows decreased as a result of the transition to ecological maintenance. In the moderately nutrient-rich and very nutrient-rich grasslands, a cost increase was noted, except for a cost reduction between 1975-1985. In the case of ditches, maintenance becomes less expensive after a new-design investment.

The current situation of roadside maintenance
In many places, the ecological quality of road verge vegetation is under pressure because traffic has become much more intensive, more traffic related structures like portals, lampposts, cables, etc. are being built and more noise-barriers are constructed in the roadside verges. More frequent disturbance and pollution are important external sources of pressure. Also repeated reconstruction projects do not allow roadside verge ecosystems to reach a mature and stable stage. Moreover, in spite of all the information on the actual and potential ecological value of roadside verges that has been produced, some of those who are responsible for the roadside verge management still lack ecological knowledge. As a consequence roadsides are sometimes still considered and treated as superfluous corners and residual spaces, which do not deserve any special ecological approach.

24.3   Nature engineering and road verges

In the preceding sections the significance of roadsides as habitat and corridor for flora and fauna has been shown and how management can help to develop these values. Based on that knowledge it is now possible to include ecological aspects in all phases of 'road life': routing, design, construction, management plans and actual maintenance measures, reconstruction and demolition. As a kind of summary an overall view is presented about the relationships in the life-cycle of roadside verges.

24.3.1   Roadside management

While applying effective roadside management (and ecological engineering, see Bohemen, 2005) the following aspects should be taken into account:

Not or not easily to influence
1.   Geographical position and position in a certain plant-geographical district.
2.   Type of the hinterland; the potential of a road verge is also determined by the surroundings (type of historical-landscape and whether or not it is a part of an ecological infrastructure).
3.   Exposition towards the sun.

To Influence by design and construction
1.   Relief (both micro and macro); presence of ditches and other waterways.
2.   Ground- and surface water level as well as the quantity and quality of the water; presence of ditches and other waterways.
3.   Soil type, soil texture, humus layer, litter layer; also the presence of gradients between wet and dry, nutrient-rich and nutrient-poor, cold and warm, humus-poor and humus-rich soil conditions.
4.   Presence of soil foreign to the area; compost earth usually has too high a nutrient content, which is unfavourable for species rich communities. Undisturbed soils usually are richer in plant species (including rare and protected species) than disturbed soils. The use of nutrient poor top soils is recommended.
5.   The width, length and surface area of the verge; the wider the verge, the greater the potential as a habitat for diversity of plant and animal species and the larger the significance as dispersal corridor.

To Influence by maintenance measures
1.   Maintenance time and intensity; phases; continuity, type and weight of (heavy) machinery.
2.   Horizontal structural variation: species composition, species diversity, internal structure and degree of rarity.
3.   Vertical structural variation of the vegetation, presence/absence of well-developed margin vegetation types; more variation in the vegetation structure (variation in openness and density from grassland, fringe, mantle till woodland vegetation) means that more insects and small mammals can find a suitable habitat.
4.   Microclimate.
5.   Effects of disturbance from the road as well as from the hinterland.

In road reconstruction projects, sods of the topsoil of botanically interesting road verge sections should be stored in a depot as short as possible and replaced in the new situation.

Buffer zones

Buffer zones between the verge and an area that is in intensive agricultural use can be useful in promoting a local ecological infrastructure in the agricultural landscape, although this aspect was not yet studied in detail. A less intensive zone adjacent to the road side would also be a possibility to increase the ecological value of the roadside. This ecological infrastructure will be more significant if the habitat patches to be connected correspond to each other.

Mowing scheme and dispersion

Spontaneous development of flora, vegetation and fauna should be furthered. In some situations it may be useful to promote the dispersal of species by first mowing species rich grasslands and later less diverse grasslands, or laying out hay from species-rich vegetation on species-poor new-constructed places. The hay should preferably come from nearby fields.

Maintenance of woody vegetation

For woody vegetation spontaneous development or initial planting without intensive maintenance of tree species will give sufficient results. Only in special cases planting and maintenance of tree species are desirable from an ecological viewpoint.

A planned approach

In addition, a planned approach is crucial for achieving predefined objectives. It is essential to adjust the landscape plans, design plans, construction plans, maintenance plans according to the results of regular monitoring and evaluation.

In this process it is essential to communicate the results of the ecological roadside management to the local and central managers, politicians and road users. Only than the funds will be continued which guarantee continuity in the ecological management. Although in the Netherlands there has been established relationship between policymakers, civil engineers, ecologists and road managers and the maintenance people there are still some gaps between the parties concerned. This is partly caused by a lack of knowledge and partly due to ignorance of the ecological values, therefore insufficient recognition of road ecology. It requires more attention of local roadside managers and biologists.

Recommendations for maintenance

Design and construction of road verges are the start, but once established, based on nature technical principles, the basis for ecological road verge maintenance in grassy vegetation is regular, low-intensive mowing and removal of the cuttings. The frequency should be adapted to the current and potential value of the vegetation in relation to the fauna actually present or which can be expected by proper management. The time of mowing should be adapted to the flowering/seed setting period of the most characteristic plant species that are present. Once the target situation has been reached, the maintenance should be continued. There are situations that have low potential ecological values, for example shady places, narrow verges, places with strong influence of agriculture or places which require intense maintenance for safety reasons (for example unobstructed view in bends). Here a large investment in ecological maintenance is perhaps not wise and a balance between cost and output must be found.

Research
Research clearly has shown the positive effects on biodiversity by mowing and removal of the grass cuttings. Two main mechanisms are supposed that cause this biodiversity: changed competition relations (for light and soil nutrients) between plant species and the creation of gaps in the soil, which enable seeds to germinate. So far there is no insight in which circumstances the two might be the most important. In view of optimising mowing a better insight seems highly desirable. An extra complication is that there is still a considerable N-deposition, which causes an increase of soil fertility; high productivity more than 8 ton dry weight/ha per year will lead to low diversity; 3-7 ton dry weight/ha per year can lead to high as well as low diversity due to the influence of low P and K and in some situations addition of P and K may be needed in grassland vegetation. In certain situations (abandoned agricultural lands) sandy soil can have too high phosphate level, 'phosphate mining'using clover and K supply could be an option (Eekeren et al. 2007). For arthropods the situation is easier to predict; most of them need short, open vegetation because of their thermophilous nature.

24.3.2   Opportunities for new nature along roads
Roadside verges form important habitats for plant and animal species and can make an important contribution to connect ecologically valuable areas providing corridors for species. If necessary, species should be prevented from mingling. For example, in areas where the northern vole (Microtus oeconomus arenicola) (an endangered species in the Netherlands) is present, road verges must be laid out and managed in such a way that field mice (Microtus arvalis) cannot reach those areas via road verges.
The design, construction and management of roadside verges should be focused on the local landscape and the plant-geographical district to which the area belongs, and on the animal species that are charactistic of the area concerned.

Where trees, woodlots or plantations are desired from an ecological, landscape ecological or visual-landscape point of view native species should be used, preferably by spontaneous development. Under moist to wet conditions, this means planting or spontaneous development of willow and asn rather than poplar hybirds. In drier areas, the use of oak and beech is recommended. In suitable situations gradual transitions from grassland to forest may be created, with shrubs and tall herbs, although they are relatively difficult to maintain.

Rows of trees can sometimes be desirable not only for guidance of motorists but they can also serve as flying routes for bats and provide suitable habitat for vesicular-arbusculair and ectomycorrhizal fungi.

24.3.3   Instruction guide for management of roadside verges

In 2006 a new Instruction guide for management of roadside verges has become available (Keizer et al. 2006). It gives up-to-date guidelines for the planning as well as the actual execution of concrete maintenance measures.

24.4   Integration of road infrastructure in the landscape

24.4.1   History

Taking into account the period of the last 100 years, one may say that from the beginning of the 20th century until World War II, plantation was central to the integration of roads in the landscape. After that period, aesthetics and safety aspects came to the fore and the landscape in the broad sense was taken into consideration. Through the technical transition from straight roads to a more 'flowing' road line and the increased insight into the composition of the landscape, more integration in the existing landscape took place.

Road furniture

The rapid increase of the intensity of traffic led to motorway expansions, the addition of road furniture (portals, telecommunication masts) and the installation of acoustic baffles at the local level, which in turn resulted in the autonomy of the motorway. This autonomy significantly reduced the visibility from the road of striking, area-specific elements. The route that was carefully designed in the past as well as providing a view on the characteristics of the landscape gradually disappeared. Nevertheless, in the Netherlands a number of sections of roads also may be considered from a landscape integration point of view as 'hotspots' that have great significance from a landscape, historical and visual point of view (Bohemen, 2004).

24.4.2 Planning, design, (re)construction, management and maintenance

In the planning, design, (re)construction, management and maintenance of motorway infrastructure, politics, policy and implementation are faced with challenge of taking into account:

1. Integral spatial quality.
2. The road from the road users'point of view.
3. The road from the point of view of people living in the neighbourhood.
4. The road as ecological and landscape-ecological entity.
5. The road as a work of art.
6. The road as an architectural object.

The challenge is to promote a synthesis between spatial and infrastructure planning, economical and social developments, city planning, architecture, historical-landscape and cultural-historical values and ecological functionality: the question is how to move from an approach in which each discipline is applied separately to an area-oriented, transdisciplinary infrastructure planning. One of the options is to look for possibilities for combining speed, globalised, large scale, macro-economic and social-economic aspects of infrastructure with slowness, regional, small-scale, regional- and local-economic and social-psychological aspects, while taking into account the specific (landscape)ecological patterns and processes.

24.4.3 From accommodation into adaptation

The general observation on integration of roads in the landscape is that the infrastructure is increasingly technological dominant, a type of intervention in the landscape and adaptation of ecosystems to our existence, in which few attention is given to the ecological value, with the exception of local compensatory measures and ecological constructions on the basis of ecological engineering. Ecosystems are complex in nature; although we are gaining more and more insight into its functionality, much knowledge is still lacking. What we get to see is rather the 'surface' of ecosystems (Wood, 1989).

Ecological and technological processes visible in the designs

Because less and less links are made between man and nature, ecological and technological processes should be made more visible in the designs (Thayer, 1994, 1998). The assumption is that what we see and experience may say a lot about the way we deal with the environment and the degree of sustainable activities. Aesthetics play an important role in man's visualisation and sensory perception, but a responsible design approach would also include a 'healing' dimension – a correction process within the landscape based on a normative point of view – (Thayer, 1998; Bohemen, 2002).

24.5   Conclusions

The main conclusion is that roads, as well as other infrastructure, have become a human-made technical component in the landscape, which can also have ecological values. They have become an intrinsic part of the ecosystems and an ecosystem approach can facilitate analyses of the negative as well as the positive effects of road planning and design.

Ecosystem approach

In the Netherlands, construction and maintenance of roadside verges on an ecological basis and the adoption of mitigating measures has in principle become standard practice for both new and existing roads. Although the barrier effect is reduced for some animals, it must be taken into account that such measures sometimes do not combat all negative ecological effects. For instance, soil, geomorphological and hydrological characteristics and location-specific gradient situations with the related plants and animals require an approach that is more focused on ecosystem functions and ecosystem processes, as well as entirely different measures.

Boundaries

An important consequence of the ecosystem approach is the description of boundaries with other systems, so that the boundaries may be considered semi-permeable membranes (Naiman and Decamps, 1989). They suggest that these interfaces (described for riparian zones) 'have resources, control energy and material flux, are potentially sensitive sites for interactions between biological populations and their controlling variables, have relatively high biodiversity, maintain critical habitat for rare and endangered species, and are refuge and source areas for pests and predators'.

Road ecosystem

On the basis of the road ecosystem, it can then be made clear under which boundary conditions and within which boundaries a mobility system is sustainable in the long run. For the ecosystem components of air, soil, ground water, surface water and the effects of noise, the critical amounts of pollution and disturbance can be calculated on the basis of norms (limiting values, target values, intervention values). On this basis it may be decided which space is available for the proposed transportation system based on the maximum permissible emission in relation to the intensity and speed of traffic and the level of sensitivity of the landscape for fragmentation. This is effectively an inversion of the current thinking process, which is based on the mobility system as a given, whereby measures are sought to reduce its impact (Koeleman and Arts, 2005).

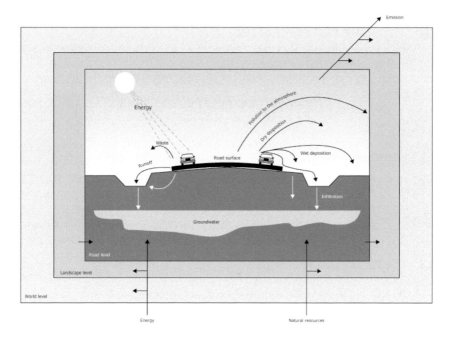

Fig. 319 A road as a system (Janssen van de Laak, Koeleman and Bohemen)

Road ecology, road-systems ecology or infrastructural ecology
Experience has shown that there is a strict division between traffic and transport principles on the one hand, and natural and environmental aspects on the other, despite the fact that these aspects are moving slowly closer to each other and despite the intensification of a more holistic research approach. In practice, however, traffic and transport planning, spatial planning and nature oriented and environmental policy are pointed in several situations in different directions.

Although forecasting models occasionally used data from various sectors, the eventual decisions are made primarily on the basis of financial-political considerations.
In order to achieve full integration of transport sciences, ecology, economy and social sciences, a new discipline, road-systems ecology, 'transportation and infrastructural ecology' or 'infrastructural ecology' might be developed, in which the ecological relationships between roads, traffic and road verges and their 'vicinity' are viewed on different ecosystematic – local, regional, continental and world – levels (Bohemen, 2004).

25   INFRANATURE

Infrastructure works, landscape and nature: experiences and debates in the Netherlands

Sybrand Tjallingii and Jos Jonkhof T U Delft

25.1   Introduction

Changed meaning of roads

A century ago roads were a natural part of the landscape and parkways were designed to make the first motorised travelers enjoy the beauties of natural landscapes. Most traffic moved by the speed of the horse. Today, infrastructure is no more a natural part of green landscapes and dynamic infrastructure arteries have almost become the enemies of quiet rural landscapes and urban parks. Traffic generates speed, noise, air and water pollution, barriers for local traffic and for nice views and all these have become ecological nightmares of environmentalists and local residents.

Sometimes, their concern led to furious and even violent protests and demonstrations that could, of course, not reverse what was seen as the infrastructure of progress. But environmental awareness did lead to a growing number of protective rules and regulations concerning roads and railways.

Baffle boards and other noise protection measures are the most visible results. For the motorist, however, boards and walls make landscapes invisible. Will it ever be possible to restore the natural partnership of landscape and infrastructure in a modern form?

Planning and design

Towards the end of the last century, an increasing number of planners and designers of infrastructure works seemed to struggle with this question. They increasingly started anticipating protective measures by thinking about landscape and ecological issues in the early stages of design. Ecology became a part of their projects, not only in the case of new infrastructure developments but also in restoration projects of existing problematic situations.

The Crailo Nature Bridge project for example

This paper describes such a project: the Crailo Nature Bridge near Hilversum in the Netherlands. The case study illustrates the opportunities and difficulties in the Dutch planning context.

Preceding the case-study section there is a brief introduction to 'planning infrastructure with nature' in the Netherlands. The final section discusses some of the issues in the debate.

25.2 Planning infrastructure with nature in the Netherlands

25.2.1 The 1950s and '60s: road verges and roadside tourism

Motorised traffic gradually became a fact of life in the post-war period. In the 1950s the first motor-ways opened but were still quiet. In the 1960s, as more people became more wealthy, increased motoriscd traffic led to mass tourism. In sunny weekends a huge number of people took the car and a great many of them picnicked along the roadside: roadside tourism (see *Fig. 320*) suddenly came up as a new term. This indicated both the rising popularity of the car and the still relatively quiet road life.

Fig. 320 Roadside tourism, the Dutch way (http://www.gkf-fotografen.nl/album17/bermtoerisme)

Fig. 321 Baffle board (Wikipedia http://nl.wikipedia.org/wiki/Geluidsscherm)

The motor-way as educative cross-section
Twenty years later, young people had never heard of roadside tourism anymore. The attitude of nature loving people in this quiet traffic period of the 1950s and 60s is illustrated by the 'The road to and through nature' symposium, organised in 1970 by the Dutch Association for Field Biology (KNNV 1971). Landscape architect Roorda van Eysinga argued for a role of the motor-way as an illustrative and educative cross-section through Dutch landscapes. His drawing of the A12 in *Fig. 322* demonstrates the idea.

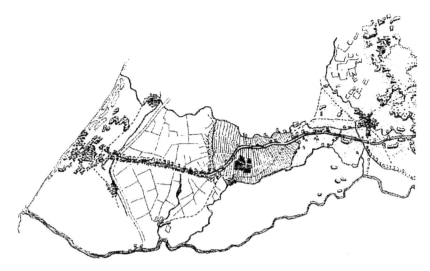

Fig. 322 The A12 as a cross section through Dutch Landscapes by Roorda van Eysinga. 1970 (KNNV, 1971)

Road verges

Ecologists at the symposium criticised the extensive use of pesticides for weed control in road verges and suggested other ways to make roads safe and yet full of attractive wildflowers.

Road professionals, united in the Dutch Road Construction Research Association responded by setting up special working groups for design and maintenance of road verges and roadside planting. The working groups not only discussed experiences in professional practice, they also started experiments to test new ideas. The publication of their conclusions and recommendations (SCW, 1972; SCW, 1975; SCW, 1976) greatly contributed to the ecological basis of road design and road maintenance.

Protection and new plantings of roadside trees is seen as a valuable element in old and new landscapes but at the same time safety on the roads is essential. The working group elaborated detailed technical recommendations for mutual adjustment of roads and trees.

Sandy soils

Road verges got a lot of professional attention too. One of the new ideas was to change current practice of applying nutrient rich soils to roadsides. Instead, the working groups proposed to apply sandy soils. Sandy soils make plants grow slowly. As a result road verges require less maintenance. Mowing two times per year and removing the mown grass would be the recommended regime. Moreover, nutrient poor sandy soils create conditions for a higher diversity of flowering plants and this would make verges more attractive for the users of the roads and more valuable for biodiversity. *Fig. 323* illustrates the design principle for a famous pilot project: the Benelux avenue in Amstelveen, a southern suburb of Amsterdam.

Fig. 323 Ecological principles of roadside design in Amstelveen
(Bohemen, 2005)

25.2.2   The 1970s: protests and legislation

In the 1970s the quiet traffic period was over: road building became a dominating feature of spatial development. Cities started to build ring roads and at the national level the motor-way network took shape. Safety and congestion became more important and the first fly-over was built and also special tunnels or bridges for cyclists and pedestrians became a normal feature.

At the same time a growing world-wide environmental awareness started to question the fruits of technological progress and created a background for many protest actions against new roads. One of the most vehement protest actions unsuccessfully tried to stop the building of a new ring road motor-way that was to cut through Amelisweerd, an old estate park east of Utrecht (Tjallingii, 1996). The cross section in *Fig. 324* illustrates how the growth of traffic changed motor-way design and fueled protests. Mitigating the effects of the infrastructure was not an option. It was yes or no to the road. And in most cases the answer was yes. Road building was unstoppable.

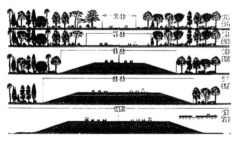

Fig. 324 Successive plans for the A27 section east of Utrecht 1954 - 1970
(Grimbergen et al., 1983)

Fig. 325 Program for tunnels and bridges in the Netherlands
(Fluit, 1990)

Increased environmental consciousness, however, had a major impact on legislation. In the 1970s and 80s a whole system of rules and regulations was adopted that addressed noise and pollution problems. The noise regulation in particular, generated a complete change of the infrastructure land-scapes. Noise barriers in the form of baffle boards (see *Fig. 321*), earthen walls and other constructions created barriers between roads and railways and the built up environment. Ironically, green areas where urbanites seek rest and quiet are unprotected by law and sometimes are among the noisi-est places.

25.2.3   The 1980s: animals, connectivity and ecological networks

In the mean time a far- reaching professional debate among ecologists had started that would have a major impact on the planning of infrastructure. In the 1950s and 60s, botanists and vegetation scientists were leading in the discussions about infrastructure. Naturally, these discussions focussed on road verges and the effects of traffic on habitat qualities. Zoologists only addressed issues like the problems on some roads with spring toad migra-tion.

Island biogeography

This situation changed when *the theory of island biogeography* gained support in the 1980s. According to this theory, not deteriorating habitat quality, but the isolation of populations on small islands, far away from the next population, explains a loss of biodiversity. Whether an individual ani-mal can reach a place *(connectivity)* is considered a more important ques-tion than whether the place is suitable to live *(habitat quality)*.

De-fragmentation

At first, this looked like a merely theoretical debate, naturally both issues are important. Inspired by the theory, however, animal ecologists looked at the Netherlands and saw islands of nature in a sea of agricultural and ur-ban disturbance. In this perspective, populations of certain species became isolated from each other and risked extinction. Infrastructure barriers and *fragmentation* of nature and landscapes, then, were perceived to be major threats to biodiversity that were at least underestimated so far. As a result of these discussions and supported by research findings, *de-fragmentation* became a major objective of spatial conservation policy.

Conservation

Conservation, however, no more adequately described this policy, because restoration and even creating new networks became part of official nature policy. The Nature Policy Plan of 1990 became a landmark for the new approach with the realisation of a National Ecological Network (NEN) as a central focus. Having achieved the status of official policy, conservation of existing and creation of new core habitats is now combined with major ef-forts to create robust ecological corridors. Opdam, one of the founding fa-thers of ecological corridor thinking in the Netherlands, recently has added the climatic change argument in favor of networks at a European scale that enable species to migrate and survive (Opdam and Wascher, 2004; Opdam, 2006).

Tunnels and bridges

At the national scale a comparison of the maps of the NEN and the planned infrastructure network generated a program for building tunnels and bridges for nature in the coming decades. *Fig. 325* illustrates the issue.

Funding

Of course these developments were made possible by a series of pilot projects and experiments that started in the late 1970s. On the other hand official policy for infrastructure and nature creates funds for realising these projects. There is no general and generous national funding scheme, however. Realisation has to be negotiated project-by-project and step-by-step, with national funds as only part of the budget. The Crailo Nature Bridge case will illustrate this point.

Technology

Van Bohemen (2005) made a good survey of technologies and practical issues related to infrastructure and nature.

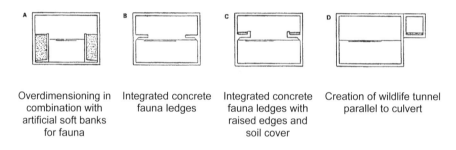

| Overdimensioning in combination with artificial soft banks for fauna | Integrated concrete fauna ledges | Integrated concrete fauna ledges with raised edges and soil cover | Creation of wildlife tunnel parallel to culvert |

Fig. 326 Possibilities for integrating a wildlife passage with large newly installed culverts (Bohemen, 2005)

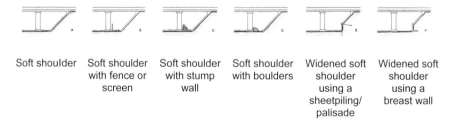

| Soft shoulder | Soft shoulder with fence or screen | Soft shoulder with stump wall | Soft shoulder with boulders | Widened soft shoulder using a sheetpiling/ palisade | Widened soft shoulder using a breast wall |

Fig. 327 Possibilities for creating a wildlife passage under a viaduct (Bohemen, 2005)

The first *ecoduct* was realised in 1986. By 1998 already 551 badger tunnels were realised in various parts of the country. *Fig. 326* and *Fig. 327* illustrate technical details of wildlife passages under a viaduct and in combination with culverts.

25.2.4   The 1990s and the new century: steps towards Integration

Integration

The lines of policy and professional practice that started in the 1970s and 1980s did not, of course, stop after ten years. In the last decade of the old and the beginning of the new century we saw continuity of sector lines while also learning about possibilities for integration. Among the pioneers of an integrated approach to infrastructure and nature it is worth mentioning H+N+S consultants[a]. Their projects in Breda (see *Fig. 328* and *Fig. 329*) may serve as an illustration.

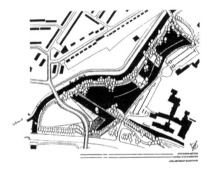

Fig. 328 River-valleys crossing motor-ways 1980s 'ecological design' of crossings between river valleys and motor-ways; Breda, Zaartpark.(H+N+S[a]).

Fig. 329 Breda, Zaartpark
(Waterschap Brabantse delta)

Infrastructure and landscape

In some of the larger projects of urban development both infrastructure and landscape were key elements in planning and design.

Fig. 330 Design for the A2 west of Utrecht (Hoeven, 2001)

---

[a] http://www.hns-land.nl/; http://www.feddes-olthof.nl/home/files/zaartpark_72dpi.pdf

A good example is the new Leidsche Rijn scheme west of Utrecht (see *Fig. 330*). Residential development for 100 000 people combined with a substantial program for commercial and industrial development could be realised only if the double barrier of the canal and the A2 motor-way that separate the old city centre from the area of the new development could be overcome.

25.3   The Crailo Nature Bridge

25.3.1   The bridge project

Problem and plan

On 3 May 2006, the official opening took place of a remarkable project: the Nature Bridge at Crailo near Hilversum (see also chapter 18, page 331), that connects two parts of the Goois Natuurreservaat, a nature park between the towns and villages in the region. The map in *Fig. 331* shows the situation. Amsterdam is just above the top-left corner of the map.

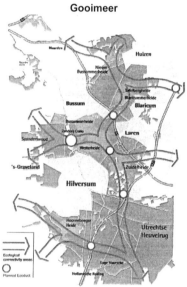
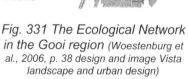
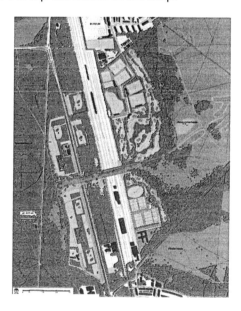

Fig. 331 The Ecological Network in the Gooi region (Woestenburg et al., 2006, p. 38 design and image Vista landscape and urban design)

Fig. 332 Design of the Crailo Nature Bridge (Woestenburg et al. 2006, design and image Vista landscape and urban design)

The bridge

The Nature Bridge is situated between the towns of Hilversum and Bussum and crosses four barriers: a regional road (N 524), the main railway from Amsterdam to the east, a railway yard and a sports field area. *Fig. 332*

(Woestenburg et al., 2006 p. 45) shows the final design of the bridge project. The park's heath lands, woodlands and a ditch are also on the bridge, together with paths for pedestrians and cyclists and a bridal way as can be seen in the cross sections in *Fig. 333* (Woestenburg et al., 2006 p. 46). On the bridge itself the paths for people are shielded from the pathways for animals by shrubs and trees.

Fig. 333 Cross sections of the Crailo Nature Bridge
(Woestenburg et al., 2006, design and image Vista landscape and urban design)

Size and costs
As a nature bridge, the 800m Crailo ecoduct is the longest in the world. The width is 50 - 150 meter. Studies started in 1995 and the actual construction took place between 2003 and 2006. The total costs of 14.75 million euro were funded by 11 public and private partners under the leadership of the Goois Natuurreservaat (GN), a regional organisation founded and led by the six municipalities in the Gooi region, the municipality of Amsterdam and the Province of Noord-Holland (North-Holland). The book published on the occasion of the official opening (Woestenburg, Kuypers and Timmermans, 2006) is the main source of this section. A short summary in English is given by Korten, 2005.

25.3.2   Design with nature for many functions

Natural benefits
Research Institute Alterra of the Wageningen University and Research organisation analysed the significance of a connecting ecological corridor at this location. Among others, roe deer, badgers, squirrels, hedgehogs, pine martens and a number of amphibians and reptile species may have taken advantage of the corridor. Birds, insects and plant seeds also will benefit. The importance of the bridge is that it connects populations and, in doing so, contributes to the survival of many species. *Fig. 331* shows the system of ecological corridors in the Gooi region.

Stepping stones
The first idea about an ecological connection at this location was to plant a series of *stepping stones* between the railway tracks as shown in *Fig. 334* (Woestenburg et al., 2006 p. 39 below). Later, this idea further developed into a full bridge and more.

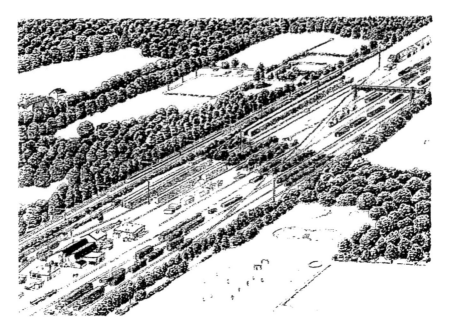

Fig. 334 The first simple stepping-stone proposal
(Woestenburg et al., 2006)

The Gooi landscape
The Gooi landscape is characterised by a wooded sandy ridge that rises about fifteen meters above the surrounding flat and wet polder areas, lying at sea level or below. The contrast between the woods and heaths and the polders make the Gooi region a very attractive area for wealthy Amsterdam people. As a result there has always been a high pressure of residential development. The sand made it easy to build. Near Crailo, the Dutch railways made a large quarry to get the sand they needed for railway construction. A large railway yard was therefore built in the centre of the quarry, parts of it were still in use until 1971. A long time before, the bottom of the quarry west of the railway had changed to farmland. Later, to the east a large sports ground was developed.

Quarry soil
Not only the road and the railway, the quarry as a whole represented a barrier between the wooded heath lands of the nature park to the east and to the west. In the early 1990s, the Goois Natuurreservaat organisation succeeded in buying the farmland and this opened the door for new habitat development on the near groundwater level of the former quarry bottom. *Fig. 332* shows the new sandy wetland habitats realised as a part of the nature bridge project.

Maintenance

Here more and there less digging created a diversity of habitats that was further enhanced by a variety of maintenance practices: doing nothing, grazing and mowing. Some of the participants in the project first raised their eyebrows at the sight of the rectangular forms of the new wetlands, proposed by the designers from Vista landscape architects. They finally convinced all parties that the old quarry forms should remain visible and that the form would not affect the ecological qualities of the new habitats. Why hide the man made character of this landscape?

A coherent program

To the west of the railway, the nature bridge occupies part of the sports fields and this area has been compensated by a new golf course on the flanks of the former quarry between the sports fields and the heath land. The ecological design of the new golf course includes replacing the ecologically poor conifer woodlands by richer native deciduous woodlands and creating dry and wet habitats that are also part of the 'nature development' west of the railway.

Thus, the nature bridge is far more than a bridge that crosses railway and road. In fact the project involves a coherent program of activities that both create habitats and connect them.

25.3.3   The project step by step

The first steps

From the early 1990s to the final result, the idea to make an ecological connection was the driving force of the initiator, the Goois Natuurreservaat organisation (GN). In the earty 1990s, the first steps were relatively easy. The southern farm of the two situated in the area west of the railway was offered for sale and GN bought the land and the farmhouse. The latter was used to house the organisation's offices and technical services. The remaining farmer of the northern farm was unable to further expand his land at this site, so he was not unhappy when GN also bought his land.

The railway

Then, in 1995, the railway company developed the plan to establish a four hundred meter long rail-welding workshop at Crailo. This did not please GN at all because it would considerably increase the barrier effect of the railway yard. The municipality of Hilversum, however, liked the idea of new employment in the area. An official permit for the new business estate required a change of the designated land use plan and the Provincial Regional Plan. The Province, therefore, asked the railway company, the municipality and GN to sit together and negotiate a good solution. At this stage, for the first time, the idea emerged of a real bridge instead of only planted stepping-stones. A bridge would reconcile the demands of the three parties. The problem, of course, was how to finance a nature bridge that would cost millions of euros, compared to the few hundred thousands of euros for the stepping-stones. But the three parties and the Province decided to go ahead and signed a covenant to get the fund-raising process going.

The budget

At first, the railway company was reluctant to contribute a substantial sum to the project. Although, by then, the building of tunnels for roe deer and badgers was already normal practice in the planning of new railways, the case of an existing situation and a complete bridge was different. Yet, the company decided to contribute. On that basis they could go ahead with the plans for the rail-welding estate. GN assigned Vista Landscape architects with the design of the nature bridge. The estimated budget was around 14 million euro.

A new player appeared on the scene as GN manager Henk Korten succeeded in interesting the National Postcode Lottery that donates its profits to charities. GN acquired the status of being entitled to funding by the lottery. A first smaller contribution enabled GN to start with the Vista plan for habitat creation west of the railway.

Animal shelter

In 1996 and 1997 several new obstacles arose. One concerned the animal shelter, situated in the southern part of the area, between the Westerheide and the railway (*Fig. 332*). The animal shelter, supported by the Hilversum municipality, wanted to improve and expand its facilities including a six-hectare training field for dogs. Their preferred location was the northern farm, in the middle of the nature bridge plan and therefore unacceptable to GN. A solution was found by an exchange of land. The shelter offered their land next to the railway yard and this land could be used by a new track for the sports-grounds access road. Once the existing road was removed, GN was able to offer the land east of the existing shelter to be used by new buildings and the training field.

Sports grounds

The next obstacle was related to the sports grounds. The sports park was outdated and two clubs were each planning new golf courses. The Vista plan described above offered a solution, provided the two clubs could agree to merge. A second condition was the required permit from the Ministry of Agriculture, Nature and Food (LNV) for realising a golf course on land that was a designated nature area. GN succeeded in realising both conditions: they facilitated a deal that made one golf club buyout the other one and they convinced the Ministry.

The Ministry

It was less simple, however, to convince the Ministry of the need to make a substantial contribution to the nature bridge project. The experts of the Ministry, in a sectoral line of reasoning, thought that the same money could possibly generate more biodiversity results if spent on other projects in other parts of the country. They asked for proof. Thus, the independent research institute Alterra in Wageningen was asked to make a special study of the expected results of this ecological corridor. The outcome of this study convinced the Ministry that the bridge indeed would significantly contribute to a healthy development of animal populations such as those of the badger, the pine marten and many others. After an official presentation of the research by the Alterra director, the Ministry was prepared to pay.

Five rabbits
In the mean time the local press were also giving much publicity to the plans for the nature bridge. This included critical comments. An opinion-leading columnist put it very clearly: "For the five rabbits that also like to get away for a day, we spend millions to build a viaduct. But what does it mean for the Hilversum people? The Goois Natuurreservaat is far too powerful. While Hilversum is dying for a structural solution to traffic problems, they create a small paradise for nature and dig mosquito ponds!" GN responded by organising a public debate between the columnist and a well-known economist who pointed out that investing in nature is a must to guarantee long-term quality of life in the area. Anyhow, in this case the money was not available for local needs in general. It was not local money at all, but generated from national and regional public and private funds for this specific reason. Over the years of the planning process, GN had been very keen to communicate with all parties concemed, including the local press and this finally created enough public support to go ahead.

Final funding
The financial puzzle was finally solved in 2001, when the Ministry's decision was followed by formal commitments from the Province and the railway company ProRail. The Lottery and some smaller private funds provided the missing part. In millions of euros the parties contributed the following amounts: Province of Noord-Holland: 4.5; National Postcode Lottery: 3.2, Ministry of Agriculture, Nature and Food: 2.4, ProRail: 2.0, EU rural development fund: 0.5, other private funds: 2.1.

Lessons
Finally, in December 2003, construction work for the nature bridge officially started. The main lessons from the planning process, drawn by GN manager Korten are:

1. This was a development process, not the implementation of a preset plan.
2. You need a charismatic person to carry the continuity of the process.
3. The participating parties need to have a real interest in the planning process.
4. A prerequisite for making plans concrete is a starting capital.
5. Creating and sustaining public support by means of communication is essential.

25.4   Infra and nature in debate

Context sensitivity

As long as *nature* refers to the maintenance of road or railway verges, fauna passages or other small mitigating measures, there is only a limited technical debate among professionals about transformations in professional practice. The Crailo case demonstrates that bigger projects also raise bigger issues of public interest and, therefore, public debate. The case study suggests that this debate is local and site specific. Tunnels and bridges are local made-to-measure projects that require local coalitions and local public support. To what extent do more general issues and arguments play a role?

25.4.1   Governance of local nature and national values

Other local values

The *five rabbit story* in the case study reveals that a local majority may not, or perhaps never put nature in first place. In such situations, local politicians and journalists sometimes blame conservationists for being elitists or accuse them of NIMBY (not in my back yard) behaviour if they want to protect nature by rejecting building projects in their backyards. The debate may take a more or less populist turn and the arguments may be more or less justified, but the conflicts between local and higher-level decision-making processes are real and do have some general aspects.

Integrating targets of different levels

At the higher levels of government and within non-governmental organisations people sometimes fail to recognise that planning at the ground level is much more than simply implementing higher-level decisions. The Crailo case clearly illustrates how a sector based national nature policy can only be an input to a local planning process where the sector targets have to be integrated with other targets of other sectors and other interests. Here a new process starts of negotiation and communication, *wheeling and dealing*, finding partners for realisation and seeking public support.

Designing promising combinations

This also involves a new creative design process, looking for promising combinations, for multifunctional synergy if possible and functional segregation if necessary. The precise targets and efficiency criteria of sector departments, in this case illustrated by the nature policy of the Ministry of Agriculture Nature and Food Quality, are often a problem at the local level. Moreover, the sectoral Ministry does not provide the full budget to realise a nature bridge.

Local awareness developing from survival to quality of life

At the local level, however, people sometimes fail to recognise that some values, such as natural and cultural heritage, are of national or even European importance and cannot be left only to local interests. Weaker elements in social and physical planning processes such as nature protection or the interests of deprived groups do need the support of budgets and programs decided upon at higher levels. In our present-day political landscape all civilised political parties agree in general terms with these decisions. This has not always been this way, however. Historically, working-class parties did not care that much about nature and also business interest parties did not put it on top of their list. Nature was neither a socialist nor a capitalist priority. Towards the end of the twentieth century, however, as the economic growth perspective shifted from a survival to a quality of life issue, nature and environmental issues gained public and political support. National and regional governments do have their own right to set priorities, but they need the local initiatives to get things going.

Network and development planning
Related to these issues, current debate in the Netherlands focuses on network planning and development planning, new approaches that seek to combine levels of government and to bridge the gap between passive permit planning and active realisation planning (Priemus, 2005, Teisman, 2006). The Crailo case illustrates both the difficulties and the opportunities of a new approach. In this case the network of project partners developed a shared understanding that created the basis for combining permits and finances and for combining local and higher level interests. No doubt that this practice will actually lead towards a new view on the future of comprehensive planning as well.

25.4.2   Nature and Infra design

The traveller's perspective
In the last 25 years, the debate about nature and infrastructure almost exclusively focussed on the issues of preventing and mitigating the negative impacts of infrastructure on landscape and nature. Dutch architect Francine Houben has drawn the attention to the traveller's perspective: the car as a *room with* a *view* as the point of departure for the aesthetics of mobility (Houben, 2005). As mentioned in the introduction, this was also Roorda's approach in 1970. Between that year and the beginning of the new century the aesthetics of mobility only played a limited role in Dutch professional debate about new infrastructure, as described by Van Bohemen (2004).

Bali-model 'Weidse landschap'
Houben reintroduced the theme in the debate of designers. In the 2002 Rotterdam International Architecture Biennale she highlighted the motorway as a challenge for designers. In the urban centers, she argues, the city is hidden behind noise protecting baffle boards whereas outside city the roads open to a spectacular show of commercial buildings built in ribbons all along the way. Instead, she proposes a typology that may guide designers to create diversity and contrast. Her urban types take Las Vegas,

Paris - La Défense and the Ruhr Area as references. The aesthetic interaction with nature is expressed in the three other types, shown in *Fig. 335*: the eco-viaduct that can be seen as the prototype behind the Crailo nature bridge, the Bali model, a green landscape of villages like islands with no buildings higher than the trees and the panoramic landscape, a type that offers a full view of the open meadow landscape. The latter type served as a guiding model for a design study of the Randstad motor-ways.

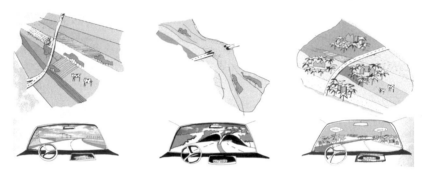

Fig. 335 The aesthetics of mobility: three guiding types for designers
(Houben, 1999[a])

A traveller's connection of landscapes
As shown in *Fig. 336* the larger ecological units link up by means of connecting passages created by raising the roads.

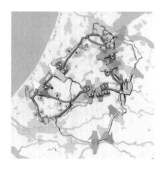 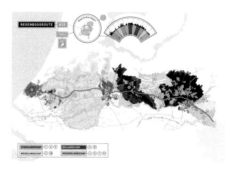

Fig. 336 Panoramas and   *Fig. 337 Reconstruction of the A12 motorway,*
perforations (Houben, 2005)   *a national pilot for spatial quality (RWS[b])*

In this way ecological perforations are combined with panoramas for the traveller.

---

[a] http://www.mecanoo.com/
[b] http://www.rws.nl/rws/dww/home/regenboogroute/regenboogroute/00/index.html

This approach re-introduced a valuable new element in the debate. Further detailed studies should explore more issues such as the solutions to noise problems that are not indicated in the design types so far. A number of further studies are under way.

25.4.3   Big Infra works and nature

High-speed rail connection and Betuwe line
The design of infrastructure works and the way these works fit in the existing landscape is one of the elements in the international debate about big infrastructure works. In the Netherlands this debate concentrates on a number of projects that were planned in the 1990s and are being realised at the moment. Two important projects are the High-Speed rail connection between Amsterdam, Belgium and France and the Betuwe line, a new goods transport railway between the Rotterdam harbor and the German Hinterland. The debate focuses on overspending. In the case of the Betuwe line, the costs were estimated at one billion euros in 1990 when the first plans were discussed. In 2006, as the project nears completion, the actual costs have risen to nearly five billion euros. In their international analysis, Flyvbjerg and his colleagues point at the high ambitions of leading engineers and politicians that make them overestimate the profits and underestimate the risk and the additional costs for making the infrastructure fit in the existing landscape (Flyvbjerg et al., 2003).

Economic criticism goes beyond including fitting-in projects
Yet these fitting-in projects, such as tunnels and bridges, turn out to be essential to get the public support required to go ahead with the project as a whole.
The Betuwe line is a good example. However, in the Dutch debate the critics do not question the need for the extra tunnels (Wee, 2006). Rather, their criticism concentrates on the economic reasons for the project as such and on the manipulative decision-making process. In the case of the High-Speed line to the South there is also hardly any discussion about the extra costs to make the new railway fit in the existing landscape but there is one exception. Near Leiden, a 7-kilometre tunnel has been built to make the new railway pass under a part of the Green Heart, the open meadow landscape in the heart of the Randstad. This plan resulted from a political compromise in the Dutch parliament solving a long lasting impasse that blocked the whole project. Many ecologists and landscape architects consider this tunnel as example of *negotiated nonsense.* At this place the tunnel may have solved a political problem but the importance for landscape and nature is rather limited. For political reasons the more attractive alternatives of combining the new line with existing railways or motorways were rejected.

Lessons learned

Lessons were learned through these experiences. Today the government has appointed 'governmental advisors' for improving the quality of architecture, landscape, built heritage and infrastructure. Their task is to develop ideas and public debate on improving the quality of the design as well as the realisation process, among both principals, politicians and professionals.

One of the pilot projects is the renewal proposal for the A12 motorway, as shown in *Fig. 337.*

Generally speaking, these debates make it clear that there is great public support in the Netherlands for designing infrastructure with nature. The challenge is to make it work in concrete projects at the local level. The Crailo Nature Bridge case demonstrates the design and planning skills that are required to meet this challenge. It finally also demonstrates that a new king of planning practice is about to emerge: planning by networking on the spot!

26   LEISURE AND ACCESSIBILITY

Arjan de Bakker   **ANWB**

26.1   Introduction

Networks for plants, animals and man

Networks for animals and plants and leisure networks have a lot in common. Both are vital to the well being of the species they are intended for. Both have been seriously neglected and have deteriorated over the last decades. Both have been confronted with a lot of blockades, physical or otherwise, with a negative impact on their use and the quality of life. No wonder the call for improvement and reconstruction of both types of network is growing. Unfortunately both support groups speak different languages and, until recently, seemed happier to exclude than to include each other's wishes in their plans. The author of this article is convinced both groups can achieve their goals more easily and more quickly if they are prepared to accept solutions that may not be perfect for nature or leisure on their own, but which serve both goals reasonably well. Landscape ecologists and lobbyists for better leisure networks should therefore be allies, not competitors, for the same hectares or money.

Motives for outdoors activities

Just like mice, badgers, deer and other animals, people roam through their environment. Nowadays seeking food and sexual partners no longer are the primary urges to visit 'the great outdoors'. Still, the drive to familiarise oneself with one's surroundings (on a local scale, cf. 'homing') and to look for favourable climatological conditions to pass some part of the year in (mostly on European scale, cf. 'seasonal migration') is very much alive. A more recent evolutionary trait is the drive to 'see for oneself' and to experience the sights, people, food and other cultural elements of unfamiliar places ('discovery', more and more on a global scale). These positive drives, combined with the rather defensive need for relaxation to balance our rather stressful 21st-century lives, explain a lot about our spare time behaviour.

Demand for outdoor leisure exceeds supply

In the Netherlands, with more than 16 million people on just over 3,4 million hectares, there is a huge demand of possibilities and facilities to relax in a semi-natural environment. This is not (yet) balanced by the supply side: demand exceeds supply, especially in and around urban concentrations. For plants and animals the same holds true. Only 10% of the Dutch land area is managed as nature reserve of some sort. About 60% is agricultural area, since the 1960's mostly stripped of the hedges, ponds, small plots of forest and other linear or spherical microhabitats that once formed a useful migration network for animals and plants.

Reinforcing the leisure network

Therefore both nature and people need networks to be built and rebuilt, and reinforced throughout the country, in particular in agricultural landscapes surrounding and connecting urban areas and nature reserves.

26.2   Behaviour: spare time and outdoor leisure

Fourty hours spare time in a fourty-hour working week

The 20th century has shown a complete popularisation of spare time and leisure. Whereas short breaks and week to several month long holidays (like the Grand Tour) were formerly a prerogative for the elite, shop-floor workers were also given holidays after 1900 and even free weekends from around 1960. Nowadays the average Dutchman aged 12 years and older has about 40 hours of spare time in a normal week outside holiday periods.

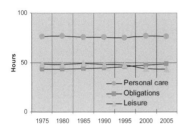

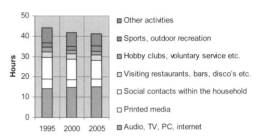

Fig. 338 The number of hours of spare time per week for Dutchmen aged 12 years and older over the last 30 years (holiday periods excluded)
(Breedveld, 2006).

Fig. 339 Number of hours per week the average Dutchman spends on spare time activities in the years 1995, 2000 and 2005 (time for mobility and holiday periods excluded)
(Breedveld, 2006)

Personal care (sleeping, eating, drinking and body care) and obligations (paid work, education, care for others) cost us respectively over 76 and nearly 49 hours per week (See *Fig. 338*). For the average Dutchman aged 20 to 64 'obligations' nowadays take up more time than 'leisure', in contrast to the situation before 1995.

Less than three hours per week for outdoor leisure

Of these 45 hours of free time we use only two to three hours per week for outdoor leisure (see *Fig. 339*). Every week about fifteen hours are spent (wasted?) behind computers and in front of television screens. Of course in holiday periods we spend a greater percentage of our time outdoors, so the study does not give insight in our time budget on a yearly basis. Still it adds some realism to the discussion on the role and importance of outdoor leisure in our present day life.

The number of trips

Another way of getting a grip on outdoor leisure is looking at the number of the ample 2.5 billion day trips per person and per type.

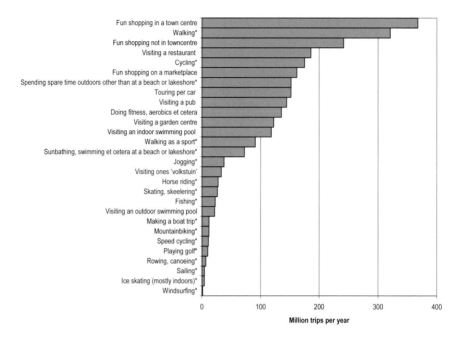

Fig. 340 The number of day trips with a minimum duration of 1 hour the average Dutchman makes on a yearly base, holiday periods excluded. Only the 12 most popular types of day trips, supplemented by the most relevant outdoor activities* (CVTO, 2005)

Since 2004 we have reliable data on all daytrips with a minimum duration of 1 hour (travelling time excluded). Unfortunately activities that take less than one hour like 'walking the dog' or 'short strolls to get some fresh air' are not available, but still these data provide us with some insight in what people think important in their lives and the relevance of outdoor leisure (see *Fig. 340*).

Trips for 'Quiet' outdoor activities

Holidays apart, the Dutch spend only 37% of their daytrips on outdoor activities. Within this segment relatively 'quiet' activities like walking and cycling are by far the most popular; together they account for more than 26% of all outdoor day trips.

Walking s.l. (walking, walking as a sport, jogging)	449	17%
Cycling s.l. (cycling, speed cycling, mountainbiking, skating, skeelering)	225	8%
Water sports s.l. (making a boat trip, sailing, windsurfing, rowing, canoeing)	25	1%
	699	26%

Fig. 341 Segment 'quiet' outdoor trips (millions) of Fig. 340

26.3   Preferences of people relaxing outdoors

A 'profile' of preferences per activity outdoors

Environmental factors that are very important to practitioners of one form of outdoor leisure activity can be totally unimportant tot practitioners of another form. Each form has its own 'profile'. For cyclists silence is most important, for walkers three other aspects are more important: accessibility of the area, type and quality of the landscape and safety (see *Fig. 342*).

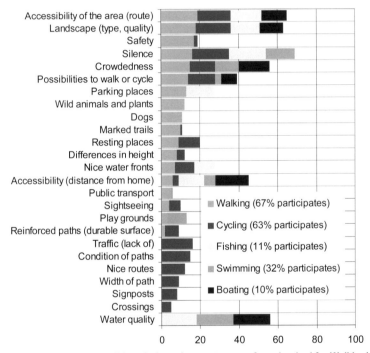

Importance (20 - priority nr.) per category of use (ranked for Walking)

Fig. 342 Ranking of aspects of the surroundings for different leisure activities (19 = most important; for swimming, fishing and boating not mentioned aspects also important). (Goossen, Langers et al., 1997)

Preferences in experience for walkers/hikers

Even within the activity 'walking' one can discern different groups that are looking for different types of experience and (therefore) rank the characteristics of their surroundings in a different order.

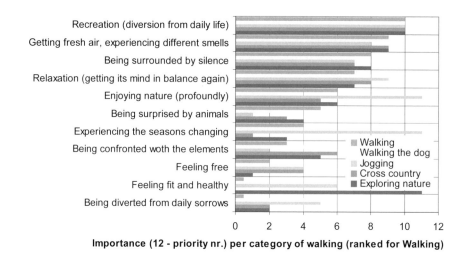

Fig. 343 Importance of characteristics of nature for different groups of people practising forms of walking/hiking. (Reneman, Visser et al., 1999)

Preferences of affordances of the landscape

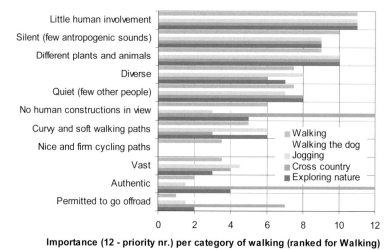

Fig. 344 Importance of characteristics of the surroundings for different groups of people practicing forms of walking/hiking.
(Reneman, Visser et al., 1999)

Different experiences through different levels of walking?
Perhaps dividing the activity 'walking' in different categories according to the energy level that is involved and the degree to which people notice the landscape they are in, may be helpful in the future when networks for both people and nature are being designed

	(Very) low interest in landscape	Moderate interest in landscape	(Very) high interest in landscape
High energy level	Track and field running	Race Survival run	Mountain running Orientering
Moderate energy level	Fitness	Jogging Brisk walking Power walking Nordic walking Walking a fixed number of K's as for sport / achievement	(Half) day hiking Voluntary service (trimming hedges etc)
Low energy level	Meditating Social walking Walking the dog	Geocaching Walking with children	Bird watching Hunting Fruit picking (blue-berries, mushrooms etc)

Fig. 345 Distinguishing different kinds of walking according to the level of energy invested

Moderate interest in nature
Most people enjoying outdoor leisure probably only take a moderate interest in nature and landscape. The landscape for most people forms a relaxing green or blue background, not a source of inspiration in itself. This would explain people spending time on locations that are not particularly beautiful or rich in nature. That is not to say landscape is unimportant. A lot of people often appreciate the landscape, but for most of them is not crucial to their outdoor experience.

Importance of accessibility
If asked what leisure facilities they find most important, people stress the importance of parks, cycle paths and foothpaths. These facilities are deemed less important than shops, but more so than restaurants, swimming pools, pubs, playing grounds, cinemas et cetera.

Missing facilities in the residential area
However if you ask people what they would like to see in their residential area (defined as a circle five kilometers in diameter) the list changes dramatically. Shops, restaurants and pubs go down, playgrounds and cinemas go up. What is interesting is that foothpaths in this list also appear at the top. So here we have an important amenity that is perceived as being in too short supply. The percentages differ slightly per age group.

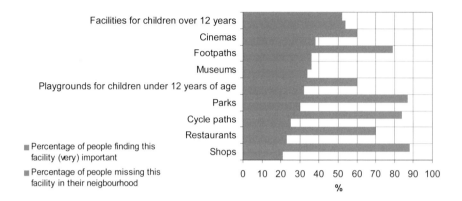

Fig. 346 Percentage of people who believe certain facilities (very) important versus percentage of people who experience a shortage in the supply of that facility. (ANWB, 2004)

Supply and demand for walking and cycling possibilities

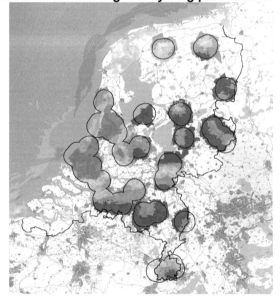

Fig. 347 Balance of supply of and demand for possibilities for walking and cycling (Vries and Bulens, 2001)

26.4   Public appreciation of landscape and nature

As we have seen most people spend only a few hours per week in a green environment.

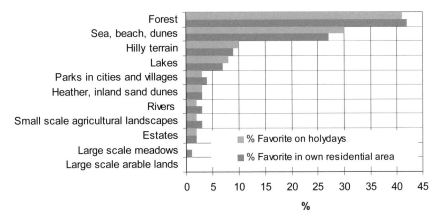

Fig. 348 Appreciation of different landscapes, measured as a percentage of people who prefer a certain type of landscape to live in or to go to during their holidays (Reneman, Visser et al., 1999)

	Score
Natural landscape (forest, heather et cetera)	8.4
Landscapes with lakes, swamps et cetera	8.0
Rivers	7.7
Small scale agricultural landscapes	7.6
Open areas with meadows	7.6
Estates	7.3
Areas with small villages	7.0
Large scale arable lands	6.8
Polders	6.3
Modern (large scale) agricultural landscapes	6.0
Intensive agriculture landscapes	5.6
Urban areas	4.5

Fig. 349 Appreciation of different landscapes, measured as indicated by people in a questionnaire (1=very low, 10=very high: NIPO, 2004).

This behaviour however is not a good parameter for the attitude towards nature conservation and protection of 'unspoilt' landscapes. Although we

535

make little use of it for ourselves, we still think it extremely important tot protect nature worldwide and give nature in the Netherlands more space.

26.5 Networks and nodes for man and nature

People, other animals and plants need networks with the right dimensions to feel good. These dimensions differ per species, but also per type of leisure activity.

Paths and routes

Before we get into the optimal networks for several types of outdoor leisure, let us first stress the distinction between paths and routes. A path is a strip of lnd intended for for example walking, cycling or riding a horse. A route is just a description of which paths to take to get from A to B or via B back to A. The paths the route follows may or may not be marked or signposted. In the Netherlands there are many routes, but many more paths. A path can be part of one or more routes, but it need not be. Routes are described to help people find their way in an area they are not familiar with. Routes therefore give you the opportunity to divert visitors from spots that can be easily damaged or disturbed.

Small is beautiful (if we do not hae much time)

Apart from holidays most people only have an hour at most to spend on outdoor leisure. There seems to be a strong relationship between the amount of time we have available and the landscape we think acceptable. City parks like the Amsterdamse Bos or the Kralingse Plas are crowded places with a lot of noise, but on a day when you have a lot of other obligations it might be the best substitute for the real thing further away. And of course they have their own qualities, like a variety of food and drink shops.

50%Travelling time

In general people seem to have a maximum for their travelling time of 50% of the time they intend to spend on a certain location. So for a four hour activity we seem to find one hour to the location and one hour back the maximum. For a one hour activity only fifteen minutes to and from the spot seems acceptable.

Differences in walking and cycling trips

Walking and cycling trips can both originate from the same starting point, but apart from that there are huge differences. In fact these differences in attitude and speed can be so great that conflicts arise when walkers and cyclists are forced to use the same path. Of course there are people that alternate walking and cycling, but sometimes the population seems to be divided into walkers and cyclists.

	Walking	Cycling
Starting point	Home Campground, bungalow resort Parking lot Train station	Home Campground, bungalow resort Parking lot (Twinny Load) Train station
Average duration	30 minutes – 1 hour	2-3 hours
Average pace	4 km/u	12 km/u
Average distance	2-4 km	25-40 km
Preferred path	Soft: grassy, mulch, sand	Hard: asphalt, bitumen
Dense grid	250 x 250 m	2000 x 2000 m
Distance from outer limit of dense grid to starting point	1000 m	7500 m
Interest in landscape	Moderate to high	Moderate

Fig. 350 Differences between walking and cycling trips

Strolls of more and less than an hour
According to the CVTO, the average walking tour day trip takes about 2,2 hours (CVTO, 2005). On average people travel 12,8 kilometers from their home to the starting point of their walk. Only 41% leaves home on foot; the others use a car (28%), a bike (8%), or public transport (23%). These results are based on day trips with a minimum duration of 1 hour, which means that short strolls, with or without a dog, are omitted. The average walk presumably takes about one hour. Ideally all concentrations of houses (villages, towns but also camp sites and bungalow resorts) should have a dense grid of footpaths both within the built up perimeter and some 1000 meters around. The paths should not be more than about 200-300 meters apart to give people several options for their daily walk. For a half hour walk you can combine two different cells, for an hour, four cells, et cetera.

Curved footpaths
Of course these networks should not be constructed at right angles. Paths with lots of bends, such as those that follow small streams or slopes, are generally more popular. But in landscapes that are developed in straight lines, like the 'slagenlandschap', straight footpaths fit perfectly. Whatever the case, there are many places where the density of the grid of footpaths should be much greater. Planting hedges and making footpaths can often be combined.

Grid density
The density of the grid also should 'follow' (be adapted to) the landscape. In forests paths can be as close as 50 meters apart without people on them hindering each other. In open landscapes on the other hand spotting another hiker or group of hikers at 500 meters can be quite a turn off.

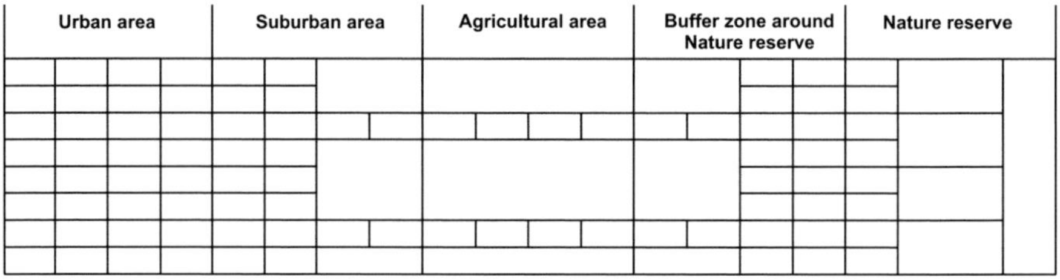

Urban area	Suburban area	Agricultural area	Buffer zone around Nature reserve	Nature reserve

Fig. 351 Proposed network of footpaths in and around urban areas
(author, inspired by Bruin and Burger, 2004)

Attractions
A view, an open space, a remarkable tree, a bench and a building: attractions are very diverse. Some of them are part of the cultural history of an area. You cannot alter them at will. Others are fairly recent additions. There is no need to invent a fixed formula for where to place new attractions. At most you can say there is a big need for benches on remarkable spots, like the edge of villages and towns, water fronts et cetera. With the aging of the populations this need will further increase. For people on a hiking tour a pub or restaurant every two hours (5-10 km) is very welcome, but eight hours without them can also be a gratifying experience.

Distributing visitors
In nature areas of 10 000 hectares and more, concentrating visitors is quite simple: offer one or several 'gateways' to the park with all kinds of facilities like parking places, a restaurant, a visitor centre and a dense grid of foot-paths within a 1km radius. More than 90% of your visitors won't go outside this radius and still leave content with 'nature experience'. In a country like the Netherlands, where most nature reserves are much smaller than 10 000 hectares, visitor management asks for improvising. In and around the most vulnerable spots simply less or no paths should be made. Else-where much 'steering' can subtly be done by offering marked routes, differ-entiating path surface and 'creepiness' of the vegetation.

26.6 Conclusion

Hiking, biking, horse riding and boating all require networks of linear structures very similar or even identical to the networks landscape ecologists advocate for (other) animals and plants. Assuming the use of these networks by man doesn't deter animals from using them as well (at night), teaming up by landscape ecologists and lobbyists for better leisure networks could benefit boths groups in reaching their goals more quickly or more completely.

27   ECOLOGY AND LANDSCAPE ARCHITECTURE

Nature and culture in the Dutch landscape

Martin van den Toorn Ṫ Delft

27.1   Introduction

In this overview we want to research what the role of ecology is in landscape architecture. We mostly focus on the Dutch situation. Dutch landscape architecture has built up a relation with ecology over a long period of time and from different backgrounds like vegetation science, soil science, hydrology, nature conservation, forestry, geology, geography. We have chosen the second part of the last century as the period of study.

Motive and reasons for the research

Landscape architects work with natural materials in natural systems; the relation with ecology and landscape ecology is evident and universally used. This is also one of the key differences between architecture and landscape architecture. Main reason for the research is to give an overview of how this relation between ecology and landscape architecture has been worked out in plans and projects and what the backgrounds have been for different approaches. Two different aspects contribute to the reasons of the research as well.

It is remarkable there is no book dealing exclusively with the relation between ecology and landscape architecture in the Dutch context. Even more striking is that in the Dutch journal of landscape ecology 'Landschap', I found only one article that deals with the subject specifically (Streefkerk, 1980). In this article an overview is given of the different approaches at that time. The important relation between ecology and landscape architecture was strongly emphasised.

Scope of the research

In this chapter examples have been selected of research projects, planning and design projects and texts that have contributed in one or another way to the development over time. A second main line of thought in this chapter is that the concept of 'nature' has evolved during the period of study. This is also visible in these projects and texts; they make clear how the idea of existing nature (nature reserves) evolved to the 'making of nature' to the 'design of nature development'. Finally the research pays special attention to the aspect of 'flows', in the natural environment worked out in the water systems approach and in the attention for 'infralandscapes', landscapes that are directly or indirectly related to infrastructure.

What is landscape architecture?

There are a great many definitions of what landscape architecture is.
The European Council of Landscape Architecture Schools (ECLAS) defined 'landscape architecture' as:

Landscape Architecture is both a professional activity and an academic discipline. It encompasses the fields of landscape planning, landscape management and landscape design in both urban and rural areas at the local and regional level. It is concerned with the conservation and en-

hancement of the landscape and its associated values for the benefit of current and future generations[a].

It is a very general definition that covers the work field of landscape architects and the scientific part as a discipline in academia.

Vroom (2006) is more extensive in his description and mentions two aspects that are interesting in the context of this chapter. First of all he describes the scope of the profession within what he calls 'the triad of garden, landscape and nature'. In this part he quotes Hunt (2000) who distinguishes three 'types of nature:

1.   'First nature' being nature untouched by man, wilderness.
2.   The 'second nature' is the agrarian landscape, the cultivated land by man.
3.   The 'third nature' is the explicitly designed landscape like gardens, parks and landscapes.

Vroom remarks that this is a more or less historical development that still is valuable as a distinction nowadays. It does not say how landscape architects have dealt with the planning and design tasks though. The 'three natures' are not completely separated entities they are more conceptual distinctions that can clarify the present attitude towards nature in planning and design.

Jong (1993) brings in another point that is interesting in this context. In his book he describes and analyses the making of gardens in the 17th century in the Netherlands. He notes that in that period there was a passion for gardens, plants and garden architecture, for the 'cultivating of nature'. The design of so many gardens had definitely also to do with this attitude of being actively engaged with the growing of plants, enjoying the garden, the plants and the layout of the garden as an expression of natural beauty. So it was not so much nature itself but the cultivating of nature that really was part of the culture of that time period.

Vroom (2006) makes a second distinction between 'landscape planning, landscape design and urban design'. He views that the distinctions are becoming 'blurred' as he mentions it. The scope of the profession is developing so rapidly that these distinctions do have a general headline of approach, methodology and practice but are more difficult to precisely limit the boundaries between the three. The yearbooks of Dutch design projects in 'landscape architecture and town planning' (Harsema et al., 1996; 1998; 2000; 2003) but also other sources like (Landschapsarchitectuur, 1985;

---

[a] http://eclas.org/content/landscape-architecture/landscape-architecture_main.php, 2004

Bolhuis, 2004; Laurie, 1976; Lynch, 1974; Motloch, 2001; Simonds, 1961; 1997) are demonstrating the same.

History
Ecology and landscape architecture have a different tradition and background from a historical point of view. Ecology as a discipline developed from the natural sciences and has a long scientific tradition from the natural sciences and from biology. From the 19th century on the cross-disciplinary developments have marked the scientific development of the biological sciences in general; ecology being one of them. Landscape ecology as a special attention for the relation between biology and physical geography, came later. Ecology developed at the end of the 19th and the beginning of the 20th century, landscape ecology developed as a specialty of ecology. A more recent example is for instance biotechnology.

Landscape architecture is — like all design disciplines — a practice-based discipline with no tradition in the natural sciences. Glacken, 1990) cites Alberti, who in his 'Ten Books' already considers 'an understanding of the environment fundamental to the practice of architecture' (landscape architecture did not exist at that time as a profession or discipline). In the course of history this practice is very much influenced by scientific developments like the invention of perspective, the study of plants, the developments in geography, hydrology and hydraulics and geodesy. From the 19th century on the discipline has been directly and indirectly influenced by engineering and technology. Equally important but totally different has been the influence from the (visual) arts, like landscape painting, impressionism, cubism and more recently landart in the development of different modes of representation.

Nowadays, modern landscape architecture finds its roots in three knowledge domains: science, art and technology. Note that for this context, nature is treated very differently in various domains; in science nature is reduced, in art nature is sublimated whereas in landscape architecture nature is transformed.

The Dutch situation; the Netherlands as a natural environment
the Netherlands being located in a temperate climate zone under influence of the sea and the geological material of a relatively young age, define the global ecological conditions in general (Lambert, 1985; Ven, 2004). This Dutch situation in the domain of ecology is in many ways also very special from different perspectives. First of all because of the density of population and the pollution that the country still faces. Secondly because ecology and specially landscape ecology has developed in the Netherlands in a very specific way as discipline that at the same time has been very influential on different levels of planning and design. The scientific basis for the NEN, an ecological network at the national level, is a remarkable example of this. Landscape ecology, nature conservation and the awareness of environ-

mental problems are highly developed in the Netherlands. The history of nature conservation is certainly part of the background for this remarkable attitude to nature, ecology and environmental problems (jaar, 1956; Gorter, 1986).

Ecology, nature, landscape ecology

In this chapter we follow the more or less general descriptions for 'ecology', 'landscape ecology' and 'nature'. We consider 'ecology' as a form of natural science, 'landscape ecology' as a subset of 'ecology'. We consider 'nature' as both abstraction from 'natural environment', so what our view is, and the physical aspects of the natural environment as such. For all of them there are probably hundreds of definitions. The limitation of boundaries and the precise description of these terms is not a primary goal in this chapter. We distinguish on the basis of the general descriptions; also here the goal is more to give an overview of relations that are important and to give insight than to focus on limiting boundaries for terms.

Problem definition

The general research problem in this chapter is: what is the relation between landscape ecology and landscape architecture?
First of all some of the typical ecological problems in the Netherlands.

Foremost problem — and for that matter typical Dutch — is the population density and the intensity of land use. The term 'carrying capacity' used to be very popular in the eighties of the last century but did not bring solutions for the problem that since that time even has gotten more serious. Note that the Dutch population is still growing at this moment. The whole problem of pollution, over use etc. is partly related to this high population density.

Global ecological problems nowadays are first of all the problem of water, more precisely of water management for the future. The shortage of fresh water, the water quality is probably the foremost problem in the Netherlands that also has far reaching ecological aspects.
Ecological problems at a local scale are the wind erosion in the dune area and limited areas of wind erosion in the southern part of the country in the province of Limburg.

Relation to nature; culture and nature in contemporary society. This is much more a social problem in Dutch society at large. The view of what nature is in contemporary society has very much evolved towards an 'artificial image', a representation that is totally different from what ecologists see as 'nature' (Feddes et al., 1998; Boelens, 1998; Mensvoort et al., 2005).

Research questions:

1.   What is the state of the art on the relation between ecology and landscape architecture?
2.   What are specific approaches in the use of landscape ecology in design projects?
3.   What is the theoretical background of these approaches?

27.2   Relation between ecology and landscape architecture

27.2.1   Some key texts and issues

The relations between landscape ecology and landscape architecture in the Netherlands are first of all reflected in the large number of plans and projects. Before analysing the plans and projects we start with some key texts that can be considered as crucial for understanding the development of the relation between ecology and landscape ecology in the second part of the last century. What are publications that have influenced this relation?

Vegetation science as a basis for the native plant discussion

The 'native plant discussion' in landscape architecture (Bijhouwer, 1951; Wolschke - Bulmahn, 2004) started a long time ago and still continues.

The issue of native plants in plantation design can be seen as both technically ('how to?') but also as a normative point of view ('you should use native plants and keep out exotics'). The underlying idea was that the best way to grow plants for plantation, is to understand their natural habitat and use indigenous species. It resulted in the view of optimal conditions for plant growing, in this case mostly trees and shrubs, was to make use of native plants. Vegetation science or at that time also called 'plant sociology' gave the scientific basis for this viewpoint. Meltzer & Westhoff (1942) described the view of plant communities in their natural habitat as 'plant sociology'; the relations between plants, soil, hydrology and land use formed the basis for the distinction of 'plant communities'. For this reason it drew much attention of landscape architects at that period of time.

Bijhouwer, being landscape architect but also very knowledgeable in the field of use of plant materials, vegetation science and planting techniques in landscape architecture, contributed very much to establish an ecological sensitivity and practical working experience for his students in landscape architecture and practitioners alike. Already in the years before WWII, the discussion about the choice of plant material, especially in the rural landscape, came up. When in the 1930s Bijhouwer was asked to make plantation plans for the Wieringermeer polder, he was clear about the choice of plantation. He choose only indigenous plantation both for roadside plantation and for the plantation plans of the farmsteads (Bijhouwer, 1933).

After WWII, while the discussion was still going on, he published an article in which he emphasised the 'scientific background' of the 'plant sociology' or vegetation science, to legitimise this choice (Bijhouwer, 1951). Especially from the side of the tree nurseries there was a growing pressure to also use exotics in the rural landscape. They saw the growing tendency of making landscape plans as an interesting market for their products.

In 1959, Van Leeuwen & Doing Kraft published an overview on the relation between natural landscape and its natural vegetation. On the basis of soil conditions and drainage conditions, lists of potential vegetation types and trees were indicated. They not only based their guidelines on vegetation science but also on soil science and hydrology. It has been — and still is — used by many landscape architects as a guideline for planting design.
The 'native plant discussion' referred mostly to trees and shrubs in the rural landscape although later on Broese in Amstelveen, applied the use of native plants in the urban landscape (Lörzing, 1992).

Ecological conditions for planning and design; an ecological basis for landuse planning
In 1948, Bijhouwer published an important article that went a step further than only the choice of native species for plantation. Bijhouwer pleaded for the use of soil maps as a basis for new urban development (Bijhouwer, 1948). Actually this can also be seen as a plea for an ecological basis for urban planning; urban development should take place at places that were most fit for that. In this article he used the case of the extension of the village of Kethel, close to Rotterdam.

In 1970 a conference was organised on soil science and the relation to planning and design. It can be seen as the continuing search for ecological conditions; soil as a basis for land use planning (S&V, 1970). In this publication an overview has been given of the possibilities of use of soil maps and the suitability for different types of land use. Similar to the native plant discussion but applied to a broader scale of land use in general, the conditions of soil and water formed the basis for the optimal land use. This land use was first of all agricultural or forestry but also urban development. Don't forget that at that time the production of food was of prime importance and yields and mechanisation of production were not as optimised as nowadays. So food production got priority over almost all other types of land use; ecological conditions were used as a basis for land use planning. This approach was also widely used in all IJsselmeerpolders for the planning of the different types of landuse (Toorn, 2006). Only in the last polder, 'Zuidelijk Flevoland' forest was not planned on land unfit for agriculture. What Bijhouwer had pleaded for in 1948, now was put in a more general context of landscape and urban planning on the basis of soil maps. This was not only referring to urban development but also to different forms of land use, especially agricultural land use like horticulture, fruit growing, diary farming etc. In very similar ways Steiner (1991) or Glickson & Mumford (1971) do treat this issue where they refer to landscape planning and land use planning.

27.2.2   Theories in ecology and landscape ecology

The relation theory of Van Leeuwen
The theory of Van Leeuwen (1966) drew much attention among landscape architects because Van Leeuwen related space and time (or: pattern and process). In landscape architecture this is also one of the key issues. Moreover Van Leeuwen was capable of relating theory and practice; during his time in the Forest Service as an ecologist, he advised many landscape architects on the use of plant material (see chapter 12, page 208).

Island theory, ecological networks and NEN
The island theory (McArthur and Wilson, 1967)drew attention at the time the discussion on an ecological infrastructure at the national scale was introduced by the Ministry of Agriculture in the Netherlands. Up till now the discussion on the theoretical basis of this NEN is still going on but the island theory certainly played a role.

The systems approach in ecology
Odum (1971) introduced the systems approach in ecology for a large audience. In the Netherlands this book was widely used at that time. It was also popular because Odum included both the qualitative and the quantitative aspects of the ecosystem in his approach. The relation with the theory of Van Leeuwen was apparent although in publications so far it is difficult to derive whether Van Leeuwen himself also saw this relation between Odum's book and his own theory. McHarg (1971) — being educated as an architect in his early education — refers to Odum's book that made him realise for the first time how intricate and important ecological relations were for planning and design in landscape architecture.

The environmental crisis; Silent spring
The publication of 'Silent Spring' (Carson, 1962) marked the start of the environmental movement; pollution, quality of the environment became serious issues for landscape architects, non-specialists and for society in general. Especially in the Netherlands with its dense population and land use, this was very apparent. Maybe even more serious than in the U.S. where the book was firstly published.

'Design with nature'
The first and most important book from the viewpoint of landscape architecture where the relation with landscape ecology is key issue, certainly is 'Design with nature' of McHarg (1971). It introduced a systematic survey of the landscape as a natural system as a basis for planning and design. It can also be seen as a reaction to the environmental crisis as described by Carson (1962) but also in general how to deal with the environmental crisis

from a design point of view. Besides emphasising the relation between natural system and design in landscape architecture it was also important from a methodological point of view. It was firstly published in 1969. McHarg departs from a personal experience of growing up close to Glasgow and serving in the military service during WWII in Italy. In his practice, office and in his teaching at the University of Pennsylvania, McHarg developed a special approach in which ecology and ecological conditions were the prime basis for landscape development. He focused very much on the regional scale.

His book became very well known in the Netherlands, also outside the school of landscape architecture in Wageningen. Dutch landscape architecture students in Wageningen in the period of Vroom got a special introduction to the ideas of McHarg and his book because Vroom himself had been one of McHarg's students during his studies at the Dept. of Landscape architecture at the University of Pennsylvania. You can say that this book developed a sensitivity and an awareness for the environmental crisis and what to do in the context of planning and design. The book is built up of texts and plans; almost every project is preceded by a chapter referring to backgrounds of the approach that has been used in the plans. In the texts it becomes clear how comprehensive this approach is. Despite the title, it is not only about ecology but also about history, the characteristics of the land, the urban landscape and the role of society in relation to the landscape. The book is cleverly written and expresses a vast and comprehensive view on land, people and landscape development. Special mention needs to be made to the section on health and environmental conditions in the metropolitan areas.

A new look at the role of water in planning and design
The study named 'Multiplex' (Sijmons et al., 1990; Tjallingii, 1996; 1998)was one of the first studies to draw attention to the importance of water, water systems and water sheds as a basis for landscape planning and design. The title is metaphorical in the sense that 'multiplex' refers to adding the 'layer of water' to the existing model of 'triplex'. 'Triplex' was a model for landscape planning based on the three essential layers in the landscape:

1.   the abiotic layer of the geological underground
2.   the biotic layer of flora and fauna
3.   the layer of human occupation as an interaction between the two other layers and human use of the land

Initiated by the Ministry of Housing and spatial planning, a team headed by Dirk Sijmons made a research on the role of water in spatial planning. Key issue was the water system approach; from 'triplex to multiplex'. Actually it gave a formal and in some respects also scientific basis of a landscape planning based on water systems; the watershed as a starting point for all spatial interventions.

Before 'Multiplex' water was already acknowledged as an important factor in the landscape as a natural system, but its crucial role in the processes that take place in the natural system was emphasised. So in the dynamics of landscape form and design, water forms a key factor in understanding the system and being able to intervene at different levels.

Again ecological conditions were a basis for land use but two different aspects were added:

1. Water was considered much more important as a basis for the eco-system. This was partly learned from the environmental pollution of the seventies were pollution of groundwater became more and more apparent (Carson, 1962).
2. The insight that the conservation of nature reserves was becoming more and more dependent on control of the water and the water system.

The study 'Multiplex' introduced the water systems approach in the circles of ministries and officials. After the NEN was accepted and the planning was started, it became clear that the role of water became more and more important. Sijmons (1992; 1998), being a landscape architect himself, approached this problem much more from the viewpoint of landscape architecture; integrating planning and design of water with technical and ecological aspects of water as part of the natural system.
Tjallingii, as an ecologist, played a major role with his publications (Tjallingii, 1996; 1998) to draw attention for this viewpoint among landscape architects.

The most recent development: the importance of urban ecology
the Netherlands is more and more getting urbanised, some people even state that the Netherlands as a whole is an urban landscape. The work and publications of Van Zoest (2006) play a key role in the awareness that 'nature is also part of the urban landscape'. His research is quite remarkable; there are examples of a greater diversity of plants and animals in the urban landscape than in the traditionally rural landscape. For landscape architects the role of urban green space gets a new dimension, both at the structural level and at the level of elements and artefacts (see chapters 16, page 267 and 20, page 372).

27.3   Dutch landscape architecture; a choice of relevant precedents

The study of 'precedents' is a way of learning from previous experiences, from 'learning by doing' and trying to make these experiences explicit and

accessible for day-to-day design. This will result in well-represented knowledge so that designers can use it when necessary.

Contrary to the natural sciences, where object of research is to understand
nature's general principles in order to be able to predict, design disciplines
deal with the creation of objects, artifacts and environments that do not
exist yet. Jong & Voordt (2002) go one step further in abstraction and view
design as 'the search for what is possible in the context of what is desirable'. They see empirical science as the 'search for the probable' while
design searches for 'less probable possibilities'.

The 'scientific approach' in design disciplines is based on the making explicit of earlier design experiences, development of design methods and the
making explicit of viewpoints that are basic to all design processes. In design disciplines the study of case-studies, plans, examples or more generally what is called 'precedents', plays a key role in building up design
knowledge.

In this chapter we have analysed a number of such cases, precedents next
to a selection of texts that have influenced the relation between landscape
ecology and landscape architecture.
We have made a limited choice of different types of precedents; research
projects; plans and competitions. The list is far from complete but does give
an idea of the relation between ecology and landscape architecture in research, design projects and texts. Common to all is that they are related to
planning and design where ecology played a role. In all cases the research
of the landscape as a natural system was used as a basis for planning and
design. The projects are placed in a chronological order to show the development over time.

27.3.1   Research projects

Early Warning
We have chosen this study (Patri et al., 1970) from abroad because of its
indirect influence, together with the book of McHarg 'Design with nature'.
The study was done in the same period that McHarg published his book
'Design with Nature' that drew a lot of attention. The Early Warning System
did not draw that much of attention but it was based on a different methodology and planning approach. It is similar to most of the projects McHarg
dealt with in his book in one respect; it deals with the planning and design
of the landscape at the regional level.

The study area of the EWS study was the area south of San Francisco,
part of the Santa Cruz mountains and was done in the Dept. of Landscape
Architecture of the University of California in Berkeley. Note that this is also

the area where the San Andreas Fault is passing; an area highly susceptible for earth quakes.

The major goal of the study was to develop a model that would enable to
incorporate natural land resource information into the planning process.
The model provides a predictive tool for locating potential development and
land dynamics conflicts. Five forms of development are distinguished; selected residential, logging, tree farming, grazing and specialty crops. The
mapping results in a series of critical areas for the distinguished forms of
development, usable both for developers and users, local groups alike.
The study is a thorough and systematic analysis of the study area based on
existing data at that time, giving extensive information on the natural qualities, possibilities and limitations of the land.

The research methodology is based on the research of vertical relations,
making use of the technique of overlays. At that time still done by hand. It
results in an overview of areas fit for different types of land use, but also
providing information on the impacts of other types of land use. So it can be
seen as a study of impacts, not as usual after interventions are proposed
but before.

Especially the susceptibility for earth quakes ('hazardous areas') are of
course very clear. Don't forget that large areas of San Francisco and the
rest of the Bay Area are built close to the San Andreas Fault! Besides the
earth quake areas there also sites on the coast that are very susceptible for
erosion and landslides. These areas come also out very clearly as well.
The title 'Early Warning System' is partly related to this environmental risks
and hazards that are specific for this area.

The EWS study does provide information before interventions were to be
done, so that in case interventions are initiated this information can be used
as a basis for taking decisions and for further research in the context of that
specific intervention. In the projects that McHarg describes, he more or less
defined himself, on the basis of the analysis, where and how new interventions should take place. Ingmire & Patri (1971) criticised McHarg's approach on two grounds. First that McHarg treated the environmental assessment of a site and its environmental qualities as a 'technical problem'
('how to'), whereas the choice of what type of analysis, how to analyse and
what criteria to use are at the heart of the problem that McHarg did not
discuss at all.

Secondly they were critical about the one-dimensional approach of
McHarg; the ecological analysis seemed to be the only ground for planning
and design of new interventions in the landscape. Natural determinism was
in fact the basis for his planning. That's why they asked in this article for a
new book, complementary to 'Design with Nature' in which the human ap-

proach was also worked out; 'Design with man'. Indeed, in all landscape architectural projects the integration of these two approaches — the natural system and the human use of the land — are basic for the planning and design approach and finally also the plan proposals. This second argument is not quite the case as you can read in McHarg's book; he definitely takes into account also the non-ecological factors of the program and tries to seek a balance.

In Wageningen, Vroom (1982) always emphasised these methodological differences that were quite fundamentally and far reaching in their consequences. He did not choose for the one or the other but stressed the importance of being explicit about these type of choices that were very influential for the outcome of the analyses independent of the methodology. A viewpoint that up till now is shared by most landscape architects.

Kring Midden Utrecht; Van Heuvelrug to Kromme Rijn
Boekhorst and Boekhorst-van Maren (1971) made one of the first studies in landscape architecture where the ecological input as a basis for planning and design was systematically researched. Goal of the study was to research the potential for spatial development in the long run. The case study area was located around the city of Utrecht in the province of Utrecht. This landscape was very much influenced in its development by the northern course of the river Rhine. This river was in the Roman times the northern border of the Roman Empire. In the west it crosses the peat district. The landscape analysis was based on soil maps, vegetation maps and hydrology of the area. The vertical relations were mapped in order to show the genesis of the natural landscape but also to find the potentially interesting areas for ecological development and other types of land use. One of the outcomes of the study was a map showing potential land use for agriculture, horticulture, urban development, nature development. This map and the insights in the working of the landscape as a natural system formed the basis for a strategy for the landscape development in the long run. The basic approach was very much influenced by the ideas of Van Leeuwen. The search for ecologically interesting boundaries and transition zones — according to Van Leeuwen — formed the basis for these type of ecological studies (Leeuwen, 1966).

Volthe De Lutte
The project of 'Volthe De Lutte' (Bijkerk et al., 1971) was a regional study where the problem of transformation in a small scale and sensitive rural landscape of Twente, was researched. The approach was based on the making of different 'models' as concepts of development in the long run. The different models each had a special focus in the form of an optimalisation of a certain type of land use relative to the others like (modern) agriculture, urbanisation, nature conservation and leisure.
This 'model study' was inspired on the study of the 'Early Warning System' (Patri et al., 1970). The EWS was a reaction and a critique on the approach

of McHarg (1967; 1971; 1972) proposed in his book 'Design with nature'. A book that still is considered as a milestone in the development of landscape architecture. The approach of McHarg was — also by others — considered as too much top-down and too implicit for decision makers and users.

The EWS was an analysis of the landscape as a natural system that gave insight into natural qualities, problems and hazards for users and decision makers so that they themselves could get insight into the problem and base their decisions on that. Even though the intention was clear and the possibilities were there, it did not have the intended effect because the number of choices was so large that it proved to be difficult to come to a choice. In the project Volthe De Lutte these choices were limited by the 'model approach'; there were four models to choose from for landscape development in the long run. In that sense it can be seen as a reaction and further elaboration of the EWS that offered a 'complete choice' but in practice too many to make a choice.

Mergelland

The study of Mergelland (Vrijlandt & Kerkstra, 1976) was an assignment by the Marl Company ENCI in the southern part of the Netherlands to the Dept. of Landscape architecture of Wageningen University. Goal of the study was to research potential sites for mining marl, from a landscape point of view. How could these type of huge interventions take place in a landscape of high visual quality? The problem of size and scale of the interventions was core of the problem; how to deal with these type of interventions at this scale. The potential sites were the outcome of a geological research done by the ENCI. The site is extremely interesting for its landscape quality; it is also 'un-Dutch' in the sense that it is not flat but has — for Dutch circumstances — remarkable differences in elevation. It is a huge plateau, that is incised by river valleys.

The ecological part of the study was based on study of the geomorphology, hydrology, soils and vegetation done by a team of ecologists. It resulted in a mapping of the natural landscape as a system. The spatial interventions that were proposed in the study were also based on this ecological survey. The dominant topographical form of rivervalleys and plains that were uplifted, formed also the basis for the design proposals of how to deal with the major intervention of marl mining in such a landscape.

The ecological theory that was used in this study was for a large part based on the theory of Van Leeuwen (1966; 1975) and Odum (1971). In this study, the ecological aspects were distinguished into ecological functions; to regulate the landscape as a natural system. Implicit of this approach is that diversity of the ecosystem contributes to this ability to regulate. So, already in this period biodiversity was seen as basic.

552

Schoonebeek

The 'Schoonebeek project' (Boekhorst et al., 1977)was an assignment of the NAM (the Dutch national oil and gas company at that time) to the Landscape section of the research institute 'De Dorschkamp' in Wageningen.

Goal of the study was to research impacts of oil drilling to the landscape. As a case study area Schoonebeek in the eastern part of the country was chosen; a site where oil drilling was omnipresent that time. Key problem in this research was how the activity and artifacts of oildrilling could be related to the existing landscape structure and perception, since the oil has no visual references to the present landscape. It is related to the landscape of former geological periods with different climates and different landscapes.

The study of the landscape as a natural system was extremely thorough and has been beautifully mapped and presented. It was done by a large number of ecologists with different backgrounds. Don't forget that in this period the implicit idea in planning and design was 'the more you know, the better the plan!' (Boekhorst, 2006). Actually, Te Boekhorst who also worked on the 'Kring Midden Utrecht' extended his research in this project further. He elaborated further on the relation between human settlement and occupation and how they were determined by ecological factors and land use.

The study has two major components; the research of the landscape as a natural system on the one hand and the landscape as a 'psychological space', what he calls in Dutch the 'betekenisruimte'. The analysis of the local landscape resulted in a concept of the site as an 'asymetrical landscape'; the front side being mostly seen and used and the backside as a kind of 'wilderness', not well known, not belonging to the daily experience of local inhabitants.

Knowledge of the landscape as natural system was used to assess and react to interventions related to drilling and the associated activities of pollution, warming up locally, spill etc. The landscape as 'psychological space' was a basis for making clear what the intervention of the oildrilling did for the local landscape and its inhabitants, the location of these activities.

Planning and design of high tension lines in the the landscape; KEMA

The KEMA - study (Vrijlandt et al., 1980) was commissioned by the Dutch Electricity Board, responsible for the planning and design of the high tension electricity transport at the national level. At that time the high tension lines and the pylons were regarded as a type of 'visual pollution' of the horizon. The research question was what could be done in the planning and design of this infrastructure to improve this situation.

The study was mainly focused on the methodological and theoretical aspects; the type of approach from a landscape architectural point of view. Ecological aspects were included but were part of other aspects like land use, relation to other infrastructure and visual and spatial aspects.

One of the outcomes of the study was the planning of high tension lines in relation to the regional landscape structure like river valleys, glacial ridges.

553

Different ways of working out this relation were distinguished like the creation of contrast by crossing such phenomena perpendicular or plan them parallel to those features. Especially river crossings of the lines were an important issue in the Netherlands. So the distinction of different levels of intervention and relating them to the ecological structure at that level, was an important issue. There were also specific ecological issues regarding the high tension lines; especially the hindrance these lines gave to (migrating) birds. At specific places where large flows of migrating birds passed, sometimes special measures were taken like the lowering of the lines next to Muiden.

Design methodology and landscape plans; Groesbeek

The 'Groesbeek' project (Nieuwenhuijze et al., 1986) was primarily focused on methodology. The goal of the study was to develop a methodology that could serve as a basis for making landscape plans for reallotment schemes all over the Netherlands. After WWII, the Dutch rural landscape was modernised to improve agricultural production and the living and working conditions of farmers (Groeneveld, 1985; Andela, 2000). All plans for reallotment schemes were based on four different studies and plans (Groeneveld, 1985). First of all there was the plan for roads and waterways, secondly there was the study of the natural qualities and the impacts of the reallotment scheme on those qualities, thirdly a study of leisure facilities and potential for leisure development and finally there was the landscape plan. The landscape plan has proven in the course of time a basis for integration of all three; it was most comprehensive and integrated. On top of that the spatial aspects of future developments were made visible, which gave it also easy access for different groups of interested people (Staatsbosbeheer, 1985; Vroom, 1992; Lörzing, 1992).

For the making of landscape plans a methodology was needed first of all for landscape architects as an aid during the planning and design process but also to make explicit to stakeholders, decision makers how the plan was developed.
Key issues in this methodology were:

1. the distinction between different levels of intervention, each with their own design means
2. the conceptual approach, at each level a new concept was developed on the basis of the earlier one. A concept was used as guiding principle for working out that level
3. the relation between research and design; in this study a distinction was made between rational thinking/knowledge and intuitive thinking/knowledge. The conceptual approach and the role of concepts in landscape architecture was earlier developed in the KEMA - project (Vrijlandt et al., 1980).

Ecology was approached in three different ways in this study:

1. As a basis for the planning and design of nature reserves in the area; the Bruuk nature reserve was worked out specifically.
2. As basis for the framework of the landscape structure that organised the use and functioning of the landscape as a natural system
3. As a basis for natural qualities in the daily environment as a whole, like the road side plantations

The casco-concept in the sandy areas of the east; Zandgebieden

The study (Kerkstra and Vrijlandt, 1988) was initiated by the Dept. of Landscape architecture of the Forestry and Landscape Planning Division of the Ministry of Agriculture. The 'Zandgebieden' project was focused on the making of landscape plans at a regional scale in the sandy areas of the eastern part of the country (mostly Achterhoek and Brabant). One of the motives of the study was the problem of dispersed small nature reserves amidst agricultural land in the context of the many reallocation plans that were executed as part of the restructuring and modernisation of the rural landscape. Those small nature reserves were especially vulnerable for diminishing in size or in many cases complete disappearance also partly due to reallotment schemes that favoured upscaling of farms and modernisation of the drainage system. As planning approach — that was developed earlier (Bruin et al., 1987) — the casco concept was further elaborated in this areas.

In this project, the casco concept was used to upscale farm sizes and on the other hand create a framework that also formed the basis for the ecological structure since this would be more or less stable over a longer period. Backbone of this framework was the water system, very important in this area. On top of that existing small nature reserves were 'put together' at sites that had a high ecological potential for those nature reserves. So, you could say that both agriculture and nature reserves were upscaled. At the same time a network of roads and waterways was created to optimise agricultural use (road system and water system) and natural qualities (water system and road system).

The search for integration of methods, approaches in regional planning; Limburg

The 'Limburg' project (Houwen and Farber, 2003) was an assignment of the Province of Limburg to advice and make plans for the landscape development in the southern part of the Province. The study was done in the Dept. of Landscape Architecture of Wageningen University. The Limburg project is in many ways an elaboration of earlier developed concepts, approaches but at the same time it is a major step forward in its integrated and comprehensive approach of landscape planning at the regional level.

Special mention has to be made of the application of GIS as a design tool in regional planning by Peter Vrijlandt and some of his students.

The research of the landscape as a natural system is fully integrated in the study at all levels of intervention and digitally available at all scales due to the GIS - based information. This GIS - basis besides offering a tool for designers, also offers other resources:

1. It is interactive at all levels of intervention. Available for different groups of stakeholders and professionals for the whole region both for the Province and for users. For the Municipalities this information offers immediate information at the practical level of day-to-day planning.
2. Availability and detailed information in digital form at different levels can also be used as a way of 'visualising' potentials, possibilities and new concepts.
3. Patterns and processes in the existing landscape can be made visible in the existing landscape any time and anywhere in the region.
4. Vertical relations in the landscape can now easily be researched by means of overlays; both the number of overlays is larger but also the time you need for doing that is much shorter.

27.3.2   Plans & competitions

Design with nature
'Design with nature' (McHarg, 1971) was published in 1969. The book is divided in chapters on projects and chapters that deal with background information in relation to the projects. Altogether five projects are described; 'Criteria for highway route selection'; Philadelphia Metropolis; 'Plan for the valleys' (Baltimore region); Staten Island; Washington and the Potomac River Basin. All deal with problems at the regional scale. All projects have in common that they take the ecological conditions as a starting point of the project. In all cases the methodological part gets most attention. For this chapter it is interesting to see how the ecological background information is developed and how it is used in planning and design.

The approach of ecological research of a study area is very much based on mapping various aspects like geology, topography, water system, vegetation cover, land use etc. and secondly to overlay maps in order to gather new knowledge. The overlay technique is used to research the vertical relations; horizontal relations and patterns are reflected in the map images as such. Vertical relations do refer sometimes to forces behind processes, for instance if you overlay water system and topography you can derive the direction of the water flow how the whole system works. Another type of knowledge that McHarg frequently uses is to overlay maps of the same area in different time periods. From that analysis you can derive the historical development based on the physical properties of the land.

The technique of overlays is not new, on the contrary. What is remarkable in McHarg's projects is the systematic way of landscape analysis by map analysis and the regional scale. The explicit attention for the research of the ecosystem was at that time also new; it is largely inspired on Odum (1971). McHarg's approach is very much focused on processes and dynamics of landscapes. Again, this is not new as such but had never been done in such a systematic way. In his book he has a special chapter on 'Process and Form', here again with a reference to Odum (1971). So apart from the attention for the background of the environmental crisis and how to deal with that in planning and design at a regional scale, McHarg's projects are very important for the methodology and systematic approach.

Ooievaar

'Ooievaar' (Bruin et al., 1987)is the winning entry (first prize) for the first 'EoW'-competition. 'EoW' stands for EoWijers, a planner who, already at an early time, drew attention for the importance of planning and design at the regional level. The study area was the river landscape of the Betuwe in the centre of the Netherlands. The assignment was to develop proposals for the future of the river landscape in the Netherlands.

The overall strategy of the plan first of all proposes a strict division in the area of the river and its forelands & the land behind the (winter)dike. The river forelands used to be largely in agricultural use. In the plan this area was proposed to 'give back to the river' for nature development. Thus creating a 'river framework' at the national scale. The river forelands which comprised a large area should become completely available for natural development. So at the level of strategy a division was made in two zones; the river and its forelands for ecological development and the land behind the (winter)dike for agriculture and other types of land use. Then at the structural level the land behind the (winter)dike was proposed to be restructured so that parts would be made partly fit for modern agriculture by upscaling the sizes of the farms and the parcels. Other parts were to be developed residentially, recreationally and also for ecological development mainly in the newly to be created frameworks. The special importance of this study was the comprehensive approach that was worked out for the river area as a whole both from a land used point of view and from and ecological point of view.

Successors of 'Ooievaar'

Bijlsma (1995) describes two projects that can be seen as a result of 'Ooievaar', a working out of the ideas in headlines in the competition. In both projects the division between the river system with its river-related nature development and the land system inside the dikes where nature development is related to agricultural production, has been worked out.
In the project for the 'Millingerwaard' the river-related nature development is a result of the dynamics of the river system. It can be seen as comple-

mentary to the land-based system inside the dike where nature development is related to the agricultural production system. Both forms of nature development contribute to the identity of the area as part of the river landscape, by their design in form and site.

The second project is the nature development inside the dikes at the 'Land van Maas en Waal'. In this project, 'wet nature' relies also on the seepage water of the river. The 'dry nature' development is related to differences in soils and land use. The coherence between the land-based nature and the river-related nature contributes to the identity of the area both for local inhabitants and tourists.

Haarlemmermeer; Uit de klei getrokken
In the 80-ies a special regional planning board for the Randstad asked the office of Vista to develop ideas for nature development for the polder the Haarlemmermeer (Vista, 1988). Already at that time the polder was almost completely urbanised and under high pressure of spatial development due to its proximity to the national airport Schiphol. In the strategy the parcelling and landscape structure of the polder were taken as a basis for creation of different ecological conditions for nature development. In this case human use and even interaction were explicitly part of the plan. One of the goals of the plan was to offer local people an opportunity to experience different types of nature, a type of nature that represented an aspect of wilderness, of wild nature 'untouched' by man. It is typical for the approach of the office of VISTA (see also chapter 18, page 331 of Wardenaar & Veen on the work of VISTA).

Crailoo; nature development at the site of a former sandpit
Crailoo (see chapter 25.3, page 516) is a former sandpit for the Dutch Railways and a large working area related to railway construction. The pit was closed and activities were for a great deal replaced to other sites. The site being located between two nature reserves gave an excellent opportunity for making an ecological connection between the two. That is also the most spectacular and best known part of the reconstruction of the site; the construction of the largest ecoduct in Europe (Feddes et al., 1998). The office of VISTA was asked to make a landscape plan for the area as a whole, not only the area of the ecoduct. In the plan the difference between east and west sides is worked out to strengthen the identity of the place mainly for orientation. The design of a new nature development on the west side is extremely interesting in the sense that this was done in rational and linear forms. The principle of creation ecologically interesting differences to stimulate nature development is the same as in other nature development projects. This project is one of the rare examples where design and ecology are closely related and enhance each other; they create new forms and types of use.

Nature development
The making of 'new' nature is an important aspect of the relation between ecology and landscape architecture. Over the years many projects have

been realised in very different situations. They all have in common that ecological conditions were created by human intervention in a very short time. In some cases technical interventions, sometimes changes of land use.

A series of examples can be shown in this respect; they show the variety of design solutions and like in Crailoo, they show how design and ecology can enhance each other and lead to results that are more interesting than the two apart.

In Zeeland the project of the 'overhoeken bij de Vlakebruggen — Field remnants at Vlake Bridges' (Harsema et al., 1996, p 150-152) is an example how nature development and infrastructure — in this case the motorway A58 — can be integrated.

At Enkhuizen, a totally different situation but the same approach resulted in a spectacular project in the Markermeer. Like in the 'overhoeken bij de Vlakebruggen, new conditions for nature development were created by the special design of the barrier. The 'Baggerdepot bij Enkhuizen' (Harsema et al., 1998, p 54-57) was designed by the office of Baljon in Amsterdam.

The most extreme example is without doubt the design of a new park landscape on a site that is considered as one of the most polluted spots in the Netherlands; the Volgermeer. The most polluted part is completely isolated and 'covered up'. On top of that a new landscape is created for nature development and leisure. The office of VISTA from Amsterdam, designed this new landscape, taking into account the limitations of such a site and at the same time creating a new future for landscape development and human use. (Harsema et al., 2002, p 190-193). See also the contribution of Wardenaar & Veen in this book.

Water systems approach

Characteristic for the water systems approach is that the water system is taken as a point of departure for landscape development in the long run. Sometimes this takes place at the level of a park, but also at other levels.

In the case of the city of Breda in the south of the country, an urban park was designed by the office of H+N+S on the basis of the regional water system at that site; 'Het Zaartpark in Breda (Harsema et al., 1996, p 156-160; Oosterhoff, 1991; Roggema et al., 1994). An example how on the basis of ecologically interesting sites in an urban context new qualities can be added to the urban landscape. Special attention is being paid in the design to the different ways of experiencing the water.

At the level of a city, in the new urban extension of Leidsche Rijn, a new urban water system was designed by the office of H+N+S (Harsema et al., 2000, p 106-107). It is an interesting example how from the very beginning of the planning and layout of a settlement such an approach can be planned and show that the water systems approach also in an urban landscape can lead to interesting solutions. Unfortunately the office did not have the possibility to integrate this appraoch in the masterplan as a whole.

They did so — together with the office of Palmboom & Van de Bout — in another site: in the plan for IJburg, a new urban extension east of Amsterdam.

The water systems approach was also applied at a regional level, the northern part of the Randstad (Harsema et al., 2000, p 130-133)) between Leiden, Haarlem and Amsterdam, designed by the office of H+N+S.

Use of plant material in relation to soil and landscape character

In projects where the choice of plant material is based on ecological conditions like soil and groundwater levels, a 'natural' ecological setting is used as a basis for the design of the program. That is not always the case; in urban landscapes and gardens exotics are used frequently. Still, also exotics do influence the micro climate, work as wind screen or bring shade in case of sunshine. So they also influence the ecological conditions of a site. This means in case ecological conditions determine the choice of plantmaterials, ecological conditions are important. In the other case, these plantations do influence and change the ecological conditions of the site.

In Tilburg the office of H+N+S designed a new urban forest outside the city at a site of a former treatment fields for purification of polluted urban water (Harsema et al., 2000, p 166-169). The rectangular pattern of the former treatment fields is used as underground for the plan. The choice of trees was also determined by the polluted situation of the former treatment fields that need to be treated with chalk every twenty years. Oak is the main species, it is planted in different densities and different typologies, for instance there is also coppice plantation included. The access and maintenance routes are also planted with linear plantations of oak. There is a curvilinear plantation of Red Oak included to show the influence of the seasons; in autumn they will turn into their characteristic red colours. The plan shows an interesting combination of forestry, design and urban use on polluted land.

A totally different approach of planting design is worked out by the office of MTD for the University Campus at Tilburg (Harsema et al., 2004, p 176-179). Interesting for the context of this chapter is the choice of Scots Pine for the entire campus. The sandy soils around Tilburg were mostly planted with Scots Pine. MTD has used the same species but in different planting typologies and planting systems. It contributes to a certain coherence in the atmosphere and visual ambiance of the Campus.

Integrated approach at the regional level

In an integrated approach not only ecological conditions and guidelines are taken into account but they are integrated in the planning and design as a whole. In most cases this is only possible at the regional level.

As a first project we have chosen 'Duurswold' in the province of Groningen (Harsema et al., 1998, p 64-70). Designers were asked to develop a re-

560

gional strategy for the landscape between Groningen and Delfzijl. This area is the site of a huge natural gasfield. Due to this gas production, the level of the land has sunken considerably. The concept developed by West 8, landscape architects and H+N+S, first of all proposes to further lower the land above the gasfield by creating a lake named 'the great gas lake'. Further on the different landscape types have been worked out quite differently as part of the NEN. Moreover both offices have suggested to start in the area also leisure development, forests and (sub)urban development for instance in the form of new 'rural villa's', similar to the old mansions that were established a long time ago. This project is an interesting example where the historical landscape typology has been the basis for the development of nature development in the NEN.

In the new urban extension east of Amsterdam, IJburg, a series of six islands in the IJmeer (Harsema et al., 1998, p 90-94). The plan will comprise 18.000 dwellings, shopping centra, leisure facilities and new nature developments. The plan is remarkable from a design point of view. The landscape plan forms the basis for the urban design and forms an autonomous design principle since it integrates land form and water movement. Existing channels and streams determine location and form of the islands. At the same time it creates fantastic possibilities for nature development in the future. Contrary to the organic forms of the islands, the urban design is based on a simple urban structure connected by a public transport line that connects all islands.
Contrary to Leidsche Rijn, the plan for IJburg is a cleverly designed urban extension where design and ecology are beautifully integrated, also on the level of the Masterplan. In IJburg the integration of the characteristics of the site, the new urban design in relation to the water and the possibilities for new nature development are extraordinary rich and promising.

In Deventer, the office of H+N+S has developed a regional strategy based on the water systems approach (Harsema et al., 2002, p 178-181). Because of the problems in agriculture in that area, the need for leisure facilities for people from the city of Deventer, there was a need for an integrated approach of the urban and rural landscapes. Different zones around Deventer are distinguished on the basis of their location in the water system. The strategy for the landscape development defines types of land use like residential, leisure or agricultural use on the basis of the relation to the water. The underlying goal is water conservation and water management.

In Twente, exactly the same approach is applied at the scale of the region of Twente by the office of Zandvoort Ordening & Advies (1994). In this case the situation was more complicated from a hydrological point of view than in Deventer but the principles are the same. It always starts with the defining of the watersheds, then analysing how the system works and how different types of land use are related to the different water qualities. In planning and design the first thing is defining the location of the high quality water and making sure this quality does not get mixed up with lower qualities. Organising land use types according to water quality is a general rule in the water systems approach. Finally it is important to define how you organise and design the water storage.

27.3.3 Texts

In this part a series of texts have been selected that can be considered as important steps in the idea development on the relations between landscape ecology and landscape architecture. Many of these texts have been used as background in the projects that have been discussed earlier.

Vroom, 1992
Vroom (1992) gives an overview of recent projects of Dutch landscape architects. Vroom was not only heading the group of editors, he also wrote an introduction giving an overview of backgrounds and approaches in landscape architecture of that time in the Netherlands. One of the aspects of landscape architectonic design he describes in his introduction, is what he calls 'symbolic nature'. Referring to the relation between man and nature that is a key issue in any design project in landscape architecture. In this book he makes clear that not only the ecological basis of plans but also different forms of representations of nature in gardens, parks and the landscape as public space are part of the work of landscape architects.

Lörzing, 1992
In the same year as Vroom, Lörzing (1992) published a book that also gave an overview of landscape architectural projects and approaches. He described a time span between the first project — 'the Amsterdamse Bos', that was started just before WWII — till the beginning of the 90-ies with the last project of the Floriade park. His choice of precedents was different from Vroom (1992). Lörzing emphasised very much the chronological development whereas Vroom made a main division on the basis plan types; 'garden', 'park' and 'landscape'.
In the chronological order he distinguishes certain periods and approaches. Among them he also distinguished the 'ecological movement in planning' in the 60-ies and 70-ies with several projects (Lörzing, 1992, pp. 58-72)

The design approach of this 'ecological movement', Lörzing describes as a design approach where, like in all plans natural materials are used but also where these natural materials are not getting an explicit architectonic form but are left in their natural form and context. So it is not the cultivation of nature in architectonic forms but the showing to the visitor 'how nature really is if man would not interfere'. He mentions the Thijssepark in Amstelveen as an early example of this approach; a kind of 'botanical' garden where choice of plants is based on their natural habitat. He mentions the name of Louis Le Roy, who designed many parks at that period, the best known at Heereveen. Le Roy's approach was slightly different in the sense that he also wanted the people to take part in construction, maintenance and support for their park. In this context Lörzing also mentions the 'Oostvaardersplassen', by now world famous as wetland but relatively young. In the design of landscape plans for the rural landscape, he mentions the project 'De Weelen' close to Enkhuizen as a nature development project in an agricultural reallotment scheme.

From nature conservation to design of nature development

Bijlsma (1995) gives an overview of approaches and projects of nature development in the Netherlands in the second part of the last century. He describes the historical development from nature conservation of existing nature reserves to the making of new nature and design of nature development. The NEN plays a key role in contemporary nature policies (Leeuwen, 1998). He makes clear that both in policy and in practice nowadays nature policy and landscape policy are totally separated.

His book is built up in three parts:

1.   Views of nature
2.   Landscape architecture and design of nature
3.   Examples of nature development

In the design of nature development the key issues are on the one hand separation and integration at different levels and on the other hand the different views of nature, 'ecocentric' where man is considered part of nature and 'anthropocentric' where man is opposing nature.

In the relation between nature and agriculture in the Netherlands, Bijlsma describes the historical development from totally integrated in early times to the contemporary situation with different stages of separating and integration at different levels (see also the chapter 20, page 372 of De Jong: 'Connecting is easy, separating is difficult').

The relation between nature and leisure is different. After the Industrial Revolution the experience and appreciation of nature became accessible for large parts of the population. From that time on man was no longer directly dependent on the production of his own daily food. The formalisation of leisure time, the sunday off for work in the beginning of the last century, marked the beginning of leisure in many forms. Nature leisure is one of the many forms of leisure. There is also a large part of leisure activities that people experience as 'being in nature' but have not so much to do with the nature seen by ecologists such as   sea sailing, bronzing on the beach, fishing, delta flying etc.

Relating nature development and landscape planning

Visie Stadslandschappen: 'Ecologie–inclusieve planning' (Croonen et al., 1995) was a government white paper (LNV, 1995) to stimulate discussion, it was not an agreed upon policy that was to be implemented. It is one of the exceptions of government papers that is well written, with good references and well illustrated. The paper deals with 'urban landscapes' that do imply the immediate surroundings of cities. There are 8 parts, here we focus on number 1 that deals with 'ecology-inclusive planning'. A somewhat strange term that refers to a planning that complies with ecological guidelines and restrictions and that takes into account the ecological potentials. It is also an integrated approach in the sense that also other than ecological factors are taken into account so it is not 'ecological planning' that focuses on ecology as such. The paper was brought out by the Ministry of Agriculture that also deals with nature conservation, nature development.

The outline of the paper is set up in four parts:

1.   Problem
2.   A new approach; management of flows
3.   A new planning strategy; planning on the basis of water systems approach and on networks
4.   Implementation

In the first part the problem is described. In short, the environmental problems of lowering of the groundwater table, the 'over-manuring' of the land and the environment as a whole getting more acid due to acid rain, are mentioned as the most important to tackle. Note that all these problems are directly or indirectly related to water. To tackle these environmental problems and to improve the ecological quality of the environment, the paper proposes a way of sustainable urban development. This approach is based on three principles:

Management of flows
Flows do refer to people, material, energy etc. It is a broad term. The water system is the core of these flows for the natural system, in having influence on the quality of the natural environment at a regional level. Thinking in cycles, understanding the system as a whole is considered basic for the approach of planning and design.

Focus on networks
A distinction is made into three types of networks; grey, green and bio networks. Grey networks stand for infrastructural networks, green networks for those related to green structures and bio networks are those related to flora and fauna.

Focus on planning at the regional level
The regional level is considered the prime level of intervention for achieving ecological qualities to improve the natural environment.
The water systems approach is the core of this newly proposed strategy. It was for the first time that this approach was put forward by a government source, as being an important basis for a planning strategy. Important aspect of this water systems approach is that any new planning problem, be it agricultural, urban, or infrastructural, should be regarded in the context of

its water system, that is the water shed. In the western part of the Netherlands the water sheds are actually man made, the polders. In the eastern part they are natural, mostly changed by man.

This is an important step forward especially to make planners think differently. Landscape architects are quite used to the importance of water and water systems but many architects and sometimes urban planners not. Remarkable is also the plea for a focus on the regional level, this is mainly due to the water systems approach.
In the planning and design of civil engineering works, Bohemen (2004) gives an excellent overview of limitations and possibilities, a reference work for all who work in planning and design of infrastructure, in fact also 'spaces of flow'.

The distinction between urban and rural landscapes becomes increasingly blurred especially in the Netherlands. On the other hand the role of infra-landscapes — landscapes that are directly or indirectly influenced by infrastructure — is getting more and more important. In the Netherlands you could state that the planning of future infrastructure, including the development of the two mainports, is actually determining the spatial development of the western part of the country. Right now, looking back, it is easy to conclude that the 'Visie Stadslandschappen' (LNV, 1995) with its explicit attention for 'landscapes of flow' was far ahead of its time.

An 'outsider' looks at the relation between ecology and landscape architecture

In 18 — what he calls 'vignettes' — headings, Van der Staay (1998) describes his experiences as relative outsider in the selection committee for the 'Yearbook of Landscape architecture and Town planning in the Netherlands'. He is critical in several respects, especially on Dutch culture, he is not always enthusiastic.

In the context of this chapter it is interesting to see what he remarks on the relation between ecology and landscape architecture.
First of all he makes a distinction between between 'environment', 'nature as culture' and 'water as the most important aspect of Dutch nature'. 'Environment' is mainly referring to conditions for life like clean air, enough fresh water and the struggle against pollution. Dutch nature and nature policy, he considers as culture. This is a very similar viewpoint also expressed by Bijlsma (1995) and others. Apparently from the discussions in the selection committee, he got the impression that water was considered the most important component and aspect of Dutch nature. He is missing the large scale, since nature functions at a world scale, he remarks that there are no examples of projects on the European scale. He notes a great preference in the Netherlands for man-made or designed nature, as already earlier mentioned 'nature as culture' (Bijlsma, 1995; Vroom, 2006). Even though water is omnipresent in all Dutch projects, he misses the cultural dimension

of water like for instance in Chinese garden architecture (Motloch, 2001; Simonds, 1961; Simonds, 1997; Passmore, 1974).

Finally it is interesting to read his remarks on gardens, and garden art. For him the garden is the ultimate expression of the relation between nature and culture, a viewpoint that also can be found with other authors like for instance Vroom (2006). Van der Staay notes that the professional level of garden design— as experienced in this selection committee — was not visible in the projects. He sees more developments, aspirations in the non-professional practice of gardening, garden making and experiences.

Another interesting point is that he misses the integration of the dynamics of natural processes in contemporary garden design. As an example he mentions the changing of the seasons in the Dutch climate that is very prominent but lacking in Dutch garden design.

All together a remarkable contribution from a non-designer that seems to be very much engaged with the relation between nature and culture at large.

27.4 Results and conclusions

27.4.1 General remarks on ecology and landscape architecture

In headlines we conclude how landscape architects use ecology in their planning and design projects could be distinguished in two distinct ways of thinking.

Ecological conditions for technical interventions, planting design and landuse

Ecology is first of all used to describe the ecological conditions of the land; especially how the different ecosystems at different levels are related to each other and to human use. It can help to describe the landscape as a natural system; how it works. It forms a basis for land use planning at different levels to find the 'best fit' between (ecological) conditions of the land and type of land use. It seems that the systems approach, the landscape as a natural system, is still used as a basis for landscape planning.

Furthermore ecology provides a 'technical basis' ('how to') and a 'normative basis' (you should use native plants') in the form of design means at different levels. In landscape architecture the natural materials like water, ground and plants are crucial in the realisation of plans. Ecology also gives information how these design materials can be used in a technical way; for example where you can grow what trees. The technical aspects of how to grow plants in the most optimal conditions and at a larger scale how to find the best type of land use for ecological conditions and the other way round still is important. The same goes for the conditions for land use planning.

Nature and landscape development
On the basis of the studies and projects described before, a distinction has been made between three approaches regarding the relation between nature and landscape development.

Nature reserves first
Attention for 'top-rated nature' from an ecological point of view; nature in nature-reserves. It is highly diverse (biodiversity!), it is stable as long as the maintenance regime stays the same over a long time. On the other hand it is not specifically meant for daily use for people. In most cases these are areas where the more people stay out, the better it is. In many cases the 'islands' in the NEN are part of this category.
Once in a while Dutch landscape architects are also engaged in the design of nature reserves; a recent example is the landscape plan for Crailoo as a site for nature development, designed by the office of VISTA (Feddes et al., 1998).

Frameworks and ecological structure as backbone for the landscape as a natural system
A second approach is the design of frameworks, in Dutch very well known as the 'casco-approach' (Sijmons, 1992). Basic to this approach is the focus on the design of a framework ('casco') that should also include the ecological structure. How such an ecological structure should be designed depends very much on the local conditions and the existing situation. This framework is supposed to be more or less permanent over a certain time. That means that the rest of the landscape can be more flexible. For instance in the change of land use or intensity of use.

For landscape planning and design at the structural level, the challenge is the search for location and the design of those frameworks. We distinguish three types of relations between new interventions and the existing landscape structure; insertion, adaptation and total change.

Nature is everywhere
A third approach is the one that considers the natural system as part of the everyday living environment. So, nature is everywhere and not only nature reserves are important. In fact you could say this is the ecosystems approach; the landscape as a natural system. This approach is very common among landscape architects. Nature being omnipresent and part of the system creates possibilities for nature reserves but also for direct use and experience of the natural elements and processes. Not being rare, or in any other way ecologically valuable, is not really a problem. For instance a tree for children for climbing can be a great resource for play, for experience of the environment. The same can be done with the other design materials like ground and water.

567

We have seen examples in the series of this type of projects, like for instance the Groesbeek project (Nieuwenhuijze et al., 1986), discussed earlier although landscape architects are also involved in the first and second approach.

The three can be distinguished typologically but in practice are mostly overlapping, like we have seen in all projects and case studies in this chapter.

27.4.2   Theory in ecology and landscape architecture

Ecological theory

It is not so easy to tell precisely on what ecological theories these studies and texts were based. In the early studies, the theory of Van Leeuwen — which was only written down partially in some articles (Leeuwen, 1966) before he started teaching in Delft — was in most cases the basis for the research of the landscape as a natural system. In Van Leeuwen's theory the ecological potential of borders and transition zones between different ecosystems played a key role (see chapter 12, page 208). Moreover the management of the landscape in the long run was very important in his theory. Van Leeuwen was the first ecologist to relate space and time in an ecological context, that made his theory so interesting for landscape architecture. The 2nd Policy document for National Planning was from the viewpoint of natural quality, based on the ecological theory of Van Leeuwen.

Up till now there was only the study of M.Th.M. de Jong (2002) that gave a short overview of the work of Van Leeuwen. One of the reasons there seems to be a certain controversy about the work of Van Leeuwen is the lacking of a general overview and analysis of his work, both in theory and practice. After 'Multiplex' the first studies appeared where the role of water was explicitly emphasised. This has been a major shift in Dutch Landscape Planning. The attention for water and the importance of the role of water came from society; it is also a global problem of increased shortage of fresh water worldwide. It is essentially based on hydrology, the hydrological cycle. Here again the systems theory seems to be an underlying theoretical concept. Still, in all later studies we have seen, you could say that Van Leeuwen's theory was used, the role of water was added to that.

A new ecological theory that was developed later was the 'island theory' (McArthur & Wilson, 1967). It can be seen as the theoretical basis for the NEN. The NEN is partly based on the island theory in the sense that the connections between the islands are very much emphasised. The islands had to be connected to enable exchange of species. This would lay the basis for the primary goal of the NEN; sustaining and enlarging biodiversity. In ecology, the role of a theory is that it can be used to predict; 'what if' - questions can be answered. It seems that the three theories mentioned here can do so, but that reality seems to be more complex than these theories can handle. We do have a lot of insight in and knowledge of the natural system but there is still a vast terrain to be explored.

Theory in landscape architecture

Theory in landscape architecture has a different goal and consequently also content. Design is not about prediction but the search for possible futures (Jong & Voort, 2002). Design theory is not based on natural science but on 'designerly ways of knowing' as Cross (2006) puts it. This 'designerly ways of knowing' are based on the experience from plans, projects and designers. Because the tradition of making gardens is already very old, this is a rich source of information for developing and making explicit this design knowledge. Describing, analysing and comparing plans, design projects — or nowadays called 'precedent analysis' — is an important way of developing such knowledge.

Another aspect of theory in landscape architecture is design methodology. Methods can be generalised, compared and can be made explicit. The choice of a method does not mean a predictable outcome; as you can read in the projects before. Design methods are used as a basis for a systematic approach that is explicit and also is needed to collaborate with other professionals.

Time and space in planning and design; levels of intervention and levels of scale

Another theoretical viewpoint that can be derived from the case studies and texts, is the relation between time and space, between process and pattern. In landscape architecture — contrary to architecture — 'time is dominant over space' or 'process is dominant over pattern'. This means that in landscape architecture, besides the levels of space (national, regional, local), the time scales are even more important. These time scales are called 'levels of intervention'; they consist of 'element', 'structure' and 'process' (Toorn, 2005). 'Element' stands for the types of short term interventions, whereas 'process' stands for the long term.

If we make this distinction in the projects we have discussed, they can be summarised as follows.

First is the level of process where the strategy for the landscape development in the long run plays a key role. At this level, from an ecological viewpoint, water is the prime design material, mostly at the level of national/regional water systems.

At the level of structure, frameworks are important. They become apparent in and form the basis for the landscape structure. Infrastructure, both the road system and the water system, is always part of the landscape structure. Such a framework can form also the basis for the ecological structure of the landscape and at the same time enables flexibility to landuse, as we have seen in several projects. This seems to be a very Dutch approach; the 'casco approach' in landscape architecture, especially at the regional level.

The third level of intervention is the materialisation of form of the daily environment. Here the three 'classic' design materials in landscape architecture come into picture; water, ground and plantation. This is the most direct and also most visible part of ecological input into landscape architectural projects; working with natural materials.

27.4.3 Approaches, viewpoints and the representation of nature

'Approaches' and 'viewpoints' are related but can also be distinguished separately. 'Approaches' are based on viewpoints but do have a methodological focus. 'Viewpoints represent a philosophical component, a matter of perception of the problem.

Approaches

We first want to put forward 'Ecological design' — a frequently used term — in the context of this chapter.

'Ecological design' has first of all to do with a point of view towards ecology, towards nature maybe. It is a normative viewpoint that views ecology of prime importance beyond other needs. This point of view has of course also consequences for the design approach and methodology. In landscape architecture even the approach of McHarg (1971) would be considered more comprehensive in the sense that it takes into account more aspects of planning and design than only ecological. On the other hand no landscape architectural project could be done without any ecological input. Depending on the problem the input of ecology can be different but is always there. Vroom (2006) is clear about 'ecological design'; in his view 'ecological design' is the design for ecological sustainability. The earlier mentioned 'deterministic' approach is comparable, 'ecological design can be seen as a deterministic approach.

Ecology and landscape architecture

A general conclusion that everybody would agree upon is that ecology is an indispensable foundation of any landscape architectural project. This statement will definitely not give rise to much discussion (Laurie, 1976; Lynch, 1974; Motloch, 2001; Simonds, 1961; 1997). The question is how ecology is applied and even more; from what point of view? In the projects presented here, we have seen a number of approaches; they can all be categorised in different attitudes towards the relation between nature and culture. A major distinction can be made between deterministic vs. designerly approaches. 'Deterministic' refers to the ecological conditions determining form, use and meaning of the plan. In the examples 'Crailoo' is an interesting example. Also in the case of 'Mergelland', you could say that the natural system is determining the design proposals in headlines. 'Designerly' refers to different ways of making use of ecological conditions. Also the creation of new ecological conditions like for instance at the 'Vlakebruggen' in Zeeland can be viewed as belonging to this category.

Different approaches in landscape architecture from an ecological point of view

Weddle (1973) makes a distinction between two different approaches:

- the 'territorial approach'
In the territorial approach, the future development of a certain geographically determined area is researched. This includes the search for distinct properties and resources that determine the landscape characteristic of that area. Ecology and ecological conditions are part of this approach and integrated with other issues like cultural, historical and economical aspects. The study 'Volthe - de Lutte' (Bijkerk et al., 1971) and the 'Early Warning System' (Patri et al., 1970) are examples of this category from this chapter.

- the 'problem solving approach'
In the problem solving approach, the starting point is the (mostly) technical intervention like the construction of a new road or a reallotment scheme for a rural landscape or an erosion problem.
The study 'Mergelland' (Vrijlandt and Kerkstra, 1976) or 'KEMA' (Vrijlandt et al., 1980) are examples in this category.
In both cases the ecological research as a basis for these type of studies, will be quite different. In the territorial approach, the research will be broad in order to understand the landscape as a natural system at large. In the case of the problem solving approach, the research will first of all be focussed on the specific type of intervention.

- Interactive approach
On the basis of the case-studies in this chapter, we could add a third category; the interactive approach which was the original goal of the Early Warning System and more recently comes back in the regional study in 'Limburg' (Houwen and Farber, 2003). Interaction based on the idea of giving people insight into design proposals and its consequences for the future landscape development.

The role of viewpoints

Landscape architecture — and for that matter all design disciplines — is always normative. Glacken (1990) gives a magnificent overview or the historical development of the relation between nature and culture from a cultural point of view. Ecology seems to be split in two major streams, the scientific and the normative approach. What is more confusing is that quite often scientific approaches have an implicit normative viewpoint. For planning and design these normative viewpoints are extremely interesting because in the end all planning and design projects in urban design and landscape architecture imply a viewpoint on the relation between culture and nature. The different attitudes towards nature play an important role in the normative aspect of the relation between nature and culture.

Amstel et al. (1988) distinguished five basic viewpoints and approaches to nature conservation and nature development.

First is the viewpoint of 'classical nature'. Key issue is the protection, maintenance and where needed the reconstruction of natural and landscape values. These values are mostly related to old cultural landscapes. In this vision, human activities can play a positive and essential role in the conservation of natural values that are related to these old cultural landscapes.

Second is the viewpoint of 'nature development'. Key issue is the self regulation, authenticity of processes and the completeness of natural communities. In this vision natural values are independent of landscape values but are related to an evolutionary frame of reference. Minimalising human intervention is considered as condition for maximalising of natural values.

Third is the viewpoint of 'functional nature'. Key issue is the functionality of use of space, flexibility in land use and technical design. Harmony with human activities and accepting the dynamics of use of space are additional components of this viewpoint. Conservation and development of both natural and landscape values are a result functional use and maintenance of the land and optimal use of space. Technology adapted to scale, forms an integral part in this viewpoint

Fourth viewpoint is the 'ecosofical nature'. Key issue is harmony with nature, a society based on human scale with worldwide responsability, democratic decision process and spirituality also in relation with nature. Technology should be applied functionally in harmonic relation with nature.

Fifth viewpoint is the sustainable technology viewpoint. Key issue is the optimal use of innovative and sustainable technology. 'High Tech' is considered as condition for survival of man and environment. Optimalisation and spatial separation of strong and weak functions, makes it possible to concentrate strong production functions. In this way space is created for the weak functions that include nature conservation and development.

For that period this distinction worked as clarifying the different viewpoints of that time. Ten years later, in 'Oorden van onthouding' (Feddes et al., 1998), these viewpoints seemed to be outdated and this issue comes back in a larger context of planning and design from an ecological point of view. In this publication already the first contours of the changing view of nature in society in general comes to the surface. It has and will have immediate and far reaching consequences for the future of nature conservation, nature development but also for landscape planning and urban planning not only in the Netherlands (Donadieu and Rumelhart, 1991). In a recent publication of Vroom (2006), he placed this contemporary development in a historical perspective in which the role of gardening and garden design is important. The creation of illusions has always been an important aspect of garden design.

Nature and visual culture; in search of new representations?

As mentioned before, in all landscape architectural projects the relation between nature and culture plays a role. Glacken (1990) describes the historical development extensively. We see similarities between his descriptions and the ones we found in the case studies in this chapter. From the projects, texts discussed in this chapter, we have distinguished three key aspects in the relation between nature and culture.

Historical development of nature in land and landscape

Human occupation and settlement of the land starts with the making of clearings in the forest to create land for agricultural production. These open spaces increased because of the growth of the population and the ability to grow enough food for this increasing population. In the Netherlands right now, we are at the point where agricultural production still increases at an ever smaller surface. It means that for the first time that former agricultural land becomes available for other uses, among them nature development. This process can be seen as inversion of the historical situation.

Change of views and viewpoints regarding nature

In early times till far in the Middle Ages, nature was considered not only as dangerous and threatening but also something to be 'cultivated'. This meant, making it fit for human use in some or another way.

The situation at this moment is in many ways the opposite. We need less land for agriculture and 'nature' is by many seen as an alternative. Sometimes this 'nature' is interpreted as 'untouched' by man, 'nature as wilderness'. Modern man seems to seek the challenges of 'wild nature' although this wilderness is very well controlled by high tech and carefully selected circumstances for instance in the case of 'wild river canoeing' of mountain climbing.

The design of 'new nature' or the design of virtual environments like 'wilderness' as representation of nature but not in the traditional sense. In the traditional sense, ecological knowledge forms the basis of planning and design of nature development. In the design of 'new nature' this is not necessary anymore, although still possible. This responds to the desire of people for 'images that represent nature' but are not the same as the ones that ecologists are looking for. From the viewpoint of landscape ecology this is 'fake nature' for other people it is real in the sense that it offers them the experiences they see as 'nature', 'natural'. In 'Oorden van Onthouding' (Feddes et al., 1998), several contributions do refer to this development for instance Boelens (1998).

Vera (1998) takes a slightly different viewpoint; he wants 'nature untouched by man', like it was before the Romans reached the low countries. For him the choice of what is 'real' or not lies in the criterion whether it is untouched by man.

Jackson (1994) points out another development that is related; the perception of the landscape in the U.S. is no longer related to real contact with the landscape. He states that more and more people experience the landscape from the motorway, very few people actually still experience and know the landscape from close by. The landscape of the motorway is increasingly the 'living world' of most people with their own restaurants, shops, meeting places etc. Even though Jackson is referring to the American situation, this can also be said for large parts of Europe nowadays.

Mensvoort et al. (2005) go a step further; the title of their book is 'Next Nature'. This is referring to the total opposite of what an ecologist would see as 'real nature'. Even though their approach is anecdotical, it gives hints to this complete change of viewpoint of 'nature' in contemporary society. In this context planning and design then come very close to what used to be done in the theatre and is done in Hollywood nowadays; the creation of virtual environments to provide new experiences.

28   THE TASK OF REGIONAL DESIGN

Joost Schrijnen 🔥 U Delft

28.1   Introduction

This chapter[a] tries to describe a confusing multitude of competing interests coming together in a contemporary landscape at the scale of a region. From an ecological and evolutionary point of view you could have confidence in a natural developement by survival of the fittest producing the best mutual adaptation at last. However, that could take a very long time of disturbing developments, unacceptible from a human point of view. The contemporary technical possibilities of separate human projects and activities have produced an unprecedented and often irreversible impact on other activities and species, decreasing the freedom of choice and possibilities of future generations. So, the question arises if these interests could be brought into a fruitful symbiosis imagining alternatives by design and planning beforehand. This chapter looks for a contemporary assignment for that kind of design and planning. How could we describe the field of problems and aims we are facing and what means we have got to reach a desirable future for the present and future generations?

We cannot rely on the methods of design and planning we developed in the previous century. The context has changed dramatically causing a 'culture shock', loss of identity and reliable expectations. Our image of the surrounding landscape has changed, our historical awareness is decreased. People are forced to immediate action in a context of temporary opportunities lost if not catched. Public institutions are busy with short-range objectives because they cannot rely on expectations of a longer term. Their small-scale administration did not grow in the same pace as private enterprises forced by globalisation. Any scenario is overtaken by unexpected projects emerging with sudden economic or political urgency resulting in an opportunistic management of our landscape. The increasing complexity of research questions forces to split up knowledge in numerous specialisms with decreasing context sensitivity giving opposite advice and canceling each other out in decision making.

To direct these interests and expectations we need images of possible and desirable futures by design, perspectives bringing them into symbiosis before these possibilities disappear forever.

28.2   The contemporary context

The changing image of surrounding landscape
Within thirty years, the Randstad (the cities of Amsterdam, The Hague, Rotterdam and Utrecht around a 'Green Heart') surrounded by an urban

---

[a] This chapter is based on the inaugural speech accepting the chair of Design Town and Region in 2005

park landscape called the Netherlands, can expand to become a more or less cohesive metropolis rich in green space and water based on a sound economic perspective. But how to imagine that? In the previous thirty years the image of the surrounding landscape has changed by its focus on Randstad: the urban area many of us live in (see *Fig. 356, Fig. 352*). Firstly we lived in separate cities, then in the Noordvleugel (North Wing) or the Zuidvleugel (South Wing) and perhaps, if we do not want to lose the global competition, in a Delta Metropolis.

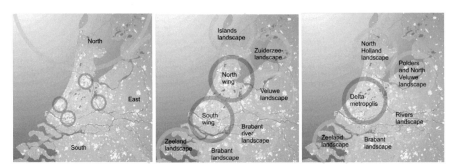

Fig. 352 Landscapes around Randstad in the public image developing into landscape parks. (Jong and Voordt, 2002)

That means a changing public perception of the originally Green Heart between four 'large' cities at the boundary of three very different landscapes into two conurbations (North and South Wing) separated by the Old Rhine between Leiden and Utrecht, the Dutch equivalent of Oxford and Cambridge. The sprawl of population fragmented surrounding landscapes into landscape parks. Originally the Old Rhine was flanked by peat areas from where the rivers Rotte and Amstel flowed south and north until their dam: Rotterdam and Amsterdam. Technical and economic development, the pressure of local population, their administrative autonomy and fading resistance of agriculture transformed the villages of the drained peat once called Green Heart into small towns with modern extensions in deep polders. Now globalisation forces to think about a smart division of tasks between nodes of a well connected urban network, a more or less cohesive metropolis rich in green space and water to stay in the race of global competition: Delta Metropolis.

A culture shock

The spatial changes occurring in the Netherlands over the last thirty years are resulting in a culture shock that we have as yet to resolve and which is at least equal to the changes resulting from the globalisation of the economy and changes in the composition of the population in this country.

1965   1995

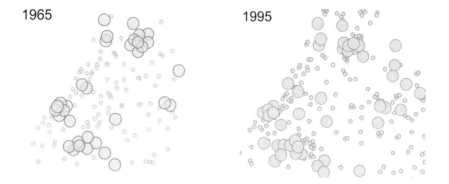

Fig. 353 Change of Randstad occupation in 30 years represented in real scale spots of 10 000 and 100 000 inhabitants (Jong and Voort, 2002)

Decreasing historical awareness

Because of this, the *Dutch* identity crisis is a very specific one, and it is intensified by the fact that the recent immigrant groups living here lack a historic perspective of the Netherlands' history of creating its own landscape.

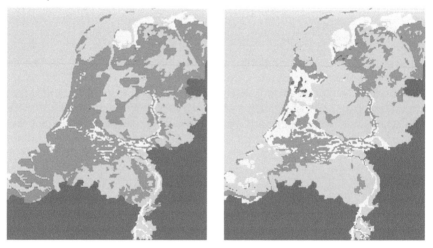

Fig. 354 Dramatic Dutch change in occupation of the landscape by drainage between AD 1000 and 1100
(Universiteit Utrecht ca. 1987 after Lambert, 1985)

Planning for short-range objectives

After an initial start in the mid-1980s, spatial planning authorities in the Netherlands have done little to address the necessity of dealing with the spatial consequences of climate change and the globalisation or Europeanisation of our society and economy (see *Fig. 355*).

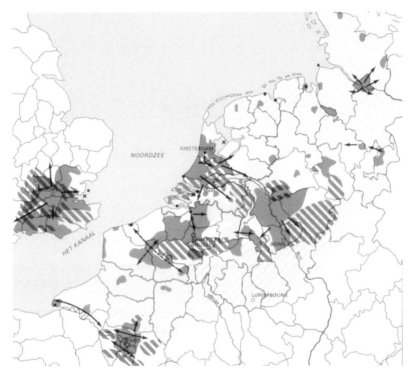

Fig. 355 Power of urbanising forces (Bobbert, 1984)

The recent major policy documents issued by the Ministry of Housing, Spatial Planning and the Environment (VROM); the Ministry of Agriculture, Nature and Food Quality (LNV); and the Ministry of Transport, Public Works and Water Management (V&W) about the spatial layout of the Netherlands focus only on short-range objectives.

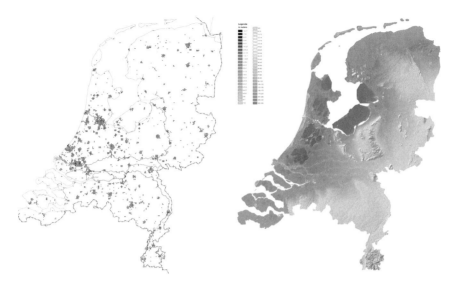

Fig. 356 Invested capital in spots of €500 000 000 (Jong and Paasman, 1998)   *Fig. 357 Hights compared to average sea level (V&W, RMD: AHN)*

Small-scale administration

The general scaling up of spatial issues has led to the growth of unofficial administrative networks and control as a supplement to the official administrative column of Thorbecke (Cammen and Klerk, 1993, 1986) holding responsibility for the issue (see *Fig. 358*). Although the government is not withdrawing from this issue, the responsibility for it is becoming divided. When it comes to dealing with spatial strategy, over the last thirty years there has been a shift from the municipal level via the level of urban regions, to the level of regions within the country. In regard to agreements for implementing strategy in the Randstad, this has shifted to the wing level.

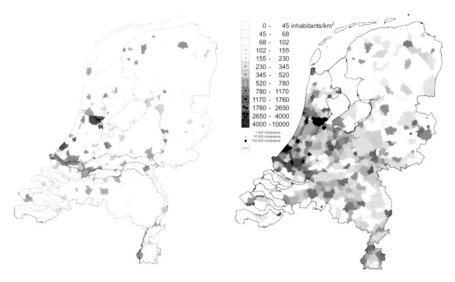

Fig. 358 Municipalities and their densities A.D. 1880 and 2000
(*http://avn.geog.uu.nl/* *CBS, data elaborated by RIVM (Henriette Giesbers), 2001*)

Opportunistic management of landscape

The future of large parts of our cultural landscape in the more urbanised areas is clouded by their current managements as well as the opinions that exist about them.

Fig. 359 Internationally important landscapes (LNV, 2002)

Fig. 360 Transformation into a utilitarian landscape (Photograph: Jong)

The original Dutch culture is associated with the large-scaled transformation of landscape into a utilitarian landscape, the function of which is constantly changing.

Conquering specialists' views in Universities by design

There is a number of metropolitan control and design issues that the national government and the regions have to address by means of research. The universities will have to start playing a designing role in this process instead of a sectoral role. The hardware of a metropolis in a delta is as yet a scarcely defined but the potential area of knowledge could be a relevant export product. The curriculum for the urban designer is extremely broad, the final purpose of this training is simple: a legible, beautiful and functional environment as a basis for identity and social cohesion.

28.2.1   Consuming space without concept

Consuming the past
We are not yet ready for the future in regard to the spatial domain in the Netherlands. Dutch people are now consuming our communal spatial past while we do not have a concept for a new spatial future.

Globalisation and climate change
We came in for a rough landing in the 21st century. At first, the next hundred years looked like an era of limitless communication and territorial dissipation. Nothing could have been less true. In the spatial domain alone, we are being confronted with the economic, social and land-use consequences of globalisation of an open society. We are also confronted by the implications of climate change. Each of these issues has spatial repercussions.

The Netherlands as a nation, a landscape and an urbanised conglomerate
These major issues form the context for the spatial issues confronting the Netherlands: the Netherlands as a national entity, the Netherlands as landscape, and the Netherlands as an urbanised conglomerate. Town and country, thus, in all their meanings and stratification. And the context is even more complicated because we don't know as yet what's happening to our territorial identity.

Randstad
I would now like to elaborate on four aspects relating to the Netherlands as space and as a spatial issue. These could be described as the designing of future scenarios and having a vital physical environment as the foundation for these scenarios. Involved here are the landscape, the transportation network and the city, the administrators, and the designers. The essential question is this: how can we better meet the probable, possible and desirable future? In this discourse, the focus is on the Randstad, the wet west of the country, the city in the Dutch Delta.

28.2.2   Randstad Holland within a European perspective

A distribution country with key projects
The regional design applied to the Netherlands can only be placed properly within an international perspective. This has been done since the Fourth Policy Document on Physical Planning (VROM, 1988). In those years, this international perspective was still being interpreted mainly as the idea of the Netherlands as a distribution country, a concept that was elaborated into the national main port policy. Besides this, there was the first generation of national urban key projects in the Fourth Policy Document on Physical Planning-Plus (VROM, 1992) issued in 1990, examples of which are Rotterdam's Kop van Zuid, the railway station and vicinity project in The Hague, and the Amsterdam IJ Waterfront Project (the South Axis in Amsterdam didn't even exist ten years ago).

These projects were intended to improve the conditions for establishing businesses in the Netherlands in order to benefit our internationally oriented service economy. Most of them were devised on the basis of standpoints developed by individual cities, although the four municipal real estate departments presented the projects as a group at the international real estate fair in Cannes.

Delta metropolis, Randstad Holland
In particular, the Delta Metropolis Association (Deltametropool, 2000) has been promoting a metropolitan concept for the Randstad by the name of 'Deltametropool' since the late 1990s. This name doesn't refer to any city in particular because most metropolitan regions and metropolises are situated in deltas. The concept made it to the Fifth Policy Document on Physical Planning (VROM, 2001) but disappeared in the next Policy Document on Space (VROM, LNV, V&W, 2004). At this time, Randstad Holland is the name used by the national government and by administrative organs working together in the Joint Scheme for Randstad Collaboration (VROM, 2001).

Geographic position as an economic advantage
The international orientation of the Netherlands is a condition for its spatial-economic position, although this orientation has to be acquired time and again on government agendas and within the social and socio-cultural context of the Netherlands. The individual cities that have grown together in this trading delta are the physical and cultural expression of this centuries-old geographic position, which forms a part of this nation's heritage and for this reason has shaped our country. This choice in favour of an international economic approach in policy development and among the business community seems obvious.

From competing independent cities into a cohesive green delta metropolis

This is not automatically considered in the choice before us now to – in the words of the predecessor to my chair in Delft, Dirk Frieling – "forge the individual swarm of cities in the Randstad into a cohesive green delta metropolis" (Vereniging Deltametropool, 1998) for the sake of a better internal and international performance. The lack of cohesion is due to various factors. In the first place, the Netherlands has traditionally been more anti-urban than pro-urban, although there was a certain revival of urban culture in the 1990s. The tradition of independent cities, each with its own identity and administration, is very strong in the Netherlands – and is also accompanied by a usually healthy cultural rivalry.

Global city

However, internationally oriented economic development, as Saskia Sassen (2001) clearly states in *The Global City*, demands a relevant concentration of services, which does not exist in a single centre in the Netherlands but which is distributed over various centres.

The concept favouring a Dutch Metropolis, a realistic prospect for the Netherlands between now and 2030 was "invented" in the 1990s. Randstad Holland already accommodates certain metropolitan functions, which, if given coherence, could form the foundation for the metropolis. For the sake of brevity, I will not elaborate on this here.

The spatial layout and organisation of Randstad Holland still has to win an essential battle, however, before it can function as a metropolis.

28.2.3　Four issues

Landscape

Our landscape is the created landscape of the past. Like a scrap of old cloth, it just barely hangs together. If no special measures are taken, this landscape will be incapable of absorbing the effects of rising sea levels, subsidence, increasing water discharges in its rivers, and more extreme rainfall and drought caused by climatic change. And still we are pumping out water (see *Fig. 271*), oxidising and shrinking the peat soils so typical in these areas. Our landscape is not automatically suited for increasing urbanisation or for taking advantage of new demands from industrial agricultural production – and even less for the needs of the urban dwellers of the Netherlands.

Physical networks

Our cities and their road networks continue to develop as based on their origins and are clashing with other cities, without the development of a unified system within which this is being defined as a new spatial order. This is urban planning without acknowledging the true scale of urban planning.

Administrative networks
From their territorial roots and in a variety of functional connections, administrators in our country continue endless discourse within the intermediary administrative networks of our consensual democracy. As democratically elected representatives of the people, politicians feel a very strong territorial responsibility. Only a handful of them ever look beyond their own borders.

Integration by design
Designers gnaw on their coloured pencils, powerless due to an exorbitantly increasing corset of rules and the excessive stratification of the issues, the desired connection between the past and the future, and the addition of process managers who are added to the process not for their knowledge of the situation but who will (wish) to be held accountable for short-term results in any case.
Anyone who has been following spatial planning in our country over the last decades knows how quickly these developments have been occurring and how many claims for space have been tumbling over one another. Apart from content-related objectives, there is in any event a considerable schedule of functional requirements.
The discipline is striving for cohesion in the spatial integration of these requirements. This striving for cohesion is reflected in the fact that since 1990 we have had four government architects (in Dutch: Rijksbouwmeesters)[a] to represent the discipline in addressing the spatial issue at the level of national government. Integrated design is one of the options.

28.3 Discovering the regional issue

28.3.1 Sectoral policy, science and design

Sectoral ministries involved
The fact that these disciplines are embedded in the three executive ministries - the Ministry of Housing, Spatial Planning and the Environment (VROM); the Ministry of Agriculture, Nature and Food Quality (LNV); and the Ministry of Transport, Public Works and Water Management (V&W) - is also cause of the segregation as it exists today.

Segregated disciplines
Each discipline has its own course of study, its own flow of funds and its own pressure groups – and they used to have their own territory, too. Meanwhile, actual spatial development shows that all these territories display an enormous amount of overlap.

---

[a] Rijksbouwmeesters Rijnboutt, Patijn, Coenen, Crouwel

The city is everywhere, the landscape is everywhere, and the infrastructure is everywhere.

Increasing technical means
The Netherlands continues on its road to higher and even higher density. In this world, design engineers are inventors. Over the centuries, this artificially created land has been a demonstration of what the inhabitants have made possible. Their role has changed, however, because they have to create their issue in collaboration with other sectors as well as devise possible solutions and find a party to implement them.

28.3.2 Scale

The regional scale
The discovery – or rediscovery – of the regional scale and applying it to the design of the Randstad occurred simultaneously with the issuing of a long series of national policy documents and concepts about the development of the Netherlands since the mid-1980s.

Emerging from lower scale
And this discovery has many parents, such as the EO Weijers Foundation[a] and the discovery it made together with Riek Bakker, Frits Palmboom and Frank de Reede of the end of the district concept in 1983 with the regional positioning of the Prinsenland district (Bakker, Palmboom and Reede, 1983; Palmboom, 1990; Goossens, Guinée and Oosterhoff, 1995). Fruitful ideas were exchanged during the discussions about the South Wing of the Randstad along with 400 colleagues in the auction halls of Bleiswijk in 1994 initiated by Elsbeth Hylckama Vlieg, Hein Struben and myself (1994), all three of us then directors in The Hague, Dordrecht and Rotterdam.

Emerging from national scale
Steef Buijs, as a designer with the National Spatial Planning Agency, worked on "The Randstad on its way to 2015" (Buijs, 1990) and "The Randstad and the Green Heart: the Green Metropolis" (Buijs, 1996). Also influential was the ARCAM Map[b], The New Map of the Netherlands (Metz and Pflug, 1997), the Netherlands Scientific Council for Government Policy, WRR, with its 1998 report entitled "Spatial Development Politics" (WRR, 1996), its authors including Maarten Hajer, and The Inter-provincial Design Agency for the Randstad with Luuk Boelens and Axel Hartman (1998).

Delta Metropolis Association
And, of course, initiated by Dirk Frieling, the founding in 1998 of the Delta Metropolis Association by the city councillors for spatial planning in the four largest Dutch cities (Frieling and Venema, 1998). Or Margreet de Boer who, as the Minister of Housing, Spatial Planning and the Environment and

---

[a] http://www.eowijers.nl/
[b] ARCAM http://www.arcam.nl/

as based on her experience as a member of the Provincial Executive of North Holland, created the Randstad Executive Committee in which the national government and the region could communicate with one another. It would be valuable if a scholarly study and description could be made of the history of this process, the starting point and renewed awareness of the regional design, followed by the strengthening of its position during the period from 1985 to 2005. All those involved are still to be consulted.

Park city between court and port
Special attention is reserved for the plan known as "A park city between court and port" by Loudi Stolker-Nanninga, a member of the Provincial Executive of Zuid Holland and head designer San Verschuuren from the Province of Zuid Holland (Stolker-Nanninga and Verschuren, 1989). This was a plan for the area between The Hague and Rotterdam and a courageous attempt to go against the tide to formulate an issue of developing urban networks and larger city networks. It perished in the violence of the negotiations about the Fourth Policy Document on Physical Planning-Plus (known as VINEX; VROM,1992) where the concept of "first in, then on the city" was the central maxim, reinforced by the emergence of the urban provinces that opposed a provincial opinion.

Legislation
Hopefully, this chapter will now be concluded by the new spatial planning legislation that is presently being submitted to the Lower House. This will provide the regional level with the same legislative power as the local level always had: the crowning glory to this period and a fresh start for the new period.

28.3.3   Landscape and function

The decreasing carrying capacity of agriculture, new claims
Our landscape was created for an agricultural function and used that way for centuries, which is why it looks the way it does today. In certain parts of the Netherlands, this agricultural objective is coming to an end. Certain forms of agricultural production are moving elsewhere, the succession within these farming operations is a problem, and there are other issues. For example, the demand for space to be used for water management, the demand for land for advancing urbanisation, and the question what use the landscape is to the urban dwellers. It is thus unavoidable that our national identity, the design of new landscape coupled to our traditional landscape, will once again have to be reformulated. Maintaining, of course, the large structural elements such as rivers, lakes, canals, polders, dunes, functional dikes and the sea.

National identity
Historian Simon Schama (1995) recognises this relationship in his intriguing book entitled *Landschap and Herinnering (Landscape and Memory)*. He notes "landscapes are intentionally designed to express the

virtues of a particular political or social community. National identity would lose much of its ferocious enchantment without the mystique of a particular landscape tradition: its topography mapped, elaborated, and enriched as a homeland." What, then, is the true meaning of our 'traditional landscape' "?

Continuously man made changes on different levels of scale

The essence of our culture is that we have created new landscapes over and over again. The value assessment of our existing landscape is getting in the way of starting to exercise that true skill: to restructure the land in order to survive. The aesthetics that are currently supplanting this dynamism are paralysing us. At the same time, this poses no dilemma in our daily reality. We haven't suddenly stopped our interventions in the landscape – many changes are taking place on a massive scale as we speak. An important difference, however, is that much of the landscape revision in the past was done to aid economic-agricultural or culture-technical pursuits, and to manage water levels. These were large-scaled projects whilst our current urbanisation is a comparatively small-scaled process. Their management is located in cities and lacks a large-scaled landscape-oriented vision. In many places, too, the transformation from agricultural use to nature conservation is being implemented on a small scale within the original large-scaled system.

Fragmentation by small-scale projects not solved by nostalgy

All these mini-interventions are leading to a landscape that is being uncontrollably fragmented by outside forces and impoverished from within. The nostalgic perspective offered by Geert Mak (1998), for example, does not address the problems involved in the issue but reinforces this understandable defensive attitude. The question is: when and how will we become aware of the urgency of the new issue? Not now, evidently.

Who is financing the large scale?

The funding needed to address the urgency involving water management has not yet been calculated. The financing of green urban space and the National Ecological Network (LNV, 1990) display inconceivable shortages. Neither of these, evidently, are urgent in political programmes. The balance between urban development and green space, denoted as red versus green on maps, is very much favouring the red. The Dutch polder landscape is simply being bought up by city dwellers and no thoughts are spent on how long such a situation can continue. Money from the EU won't provide the means to save the Dutch landscape either. Yet another factor in this is that during the twentieth century, our focus on two separate worlds continued and grew stronger: on the one hand through urban and economic developments and on the other through the rural landscape being optimised by farmers to increase their agricultural yields.

No resistance to urbanisation

In many places, the functional landscape of the past offers little if any resistance to urbanisation; apparently, it's no longer valuable enough for this purpose. Even so, there are major differences in the scenic value of landscapes in the Netherlands.

The need for larger administrative regions without sectoral boundaries
In the end, there is but one road to tomorrow: to re-establish the relationship between town and country. Create a single Ministry of Spatial Planning. Decentralise the organisation of the entire urban and rural area once and for all into regions and make them responsible for the issues, the resources and the institutions so that they can no longer hide behind the Ministry of Agriculture, Nature and Food Quality. And give them a powerful set of tools to control the price of land.

28.3.4   Two models for the metropolitan network

Interlocking regions
By the end of the twentieth century, the main features of the map of the Randstad had changed from individual cities into an interlocking system of regions. This had everything to do with an accelerated rate of urbanisation. An urban region developed that gradually changed from a monocentric to a polycentric metropolis.

Cities transforming from traffic poles into nodes of a network
The development of the North and South Wings and such regions as Arnhem-Nijmegen or Groningen-Assen are fundamentally the beginnings of urban networks that may continue to grow into even larger network cities. In many cases, however, the public transport lines and road networks of the cities themselves are actually still aligned with the former main centres of these cities.

The network as a driving force
The designing of city maps is also still done from the monocentric viewpoint of reality. The first breakthroughs, however, have already been realised or will be soon: Randstad Rail between Rotterdam, Zoetermeer and The Hague, the N470, the road linking these cities, the Zuidtangent in the Amsterdam region, and the light rail between Nijmegen and Arnhem. The content-related misunderstanding involving the designing of new spatial developments in the city or in growth centres that focused on a single spatial orientation existed for a long time.

Locally isolated housing projects for a market covering the region
Surprisingly enough, however, the same mistakes are still made to this day. Housing projects in the area between The Hague and Rotterdam are earmarked for conurbation whilst the housing market is not. People living throughout the entire area can freely choose their house, no matter what region they come from, but the infrastructure is still not geared to the kind of universal accessibility this free choice demands. In a similar vein, the demand for housing in the North Wing is directed not just to Almere or the

region north of the North Sea Canal but just as much to Leiden and the Green Heart.

It's time to start learning from this.

Conflicting network principles

We now come to main question number two: the design of the network. The network of the Randstad has always had two dominant levels: a monocentrically oriented network for the city and a national motorway and railway network. These motorways were drawn for the first time as an independent national network in the Second Policy Document on Physical Planning (RPD, 1966). In the Randstad, the national network was placed in a superior position above that of the local network. No longer were cities located at the ends of motorways but rather at their exit ramps. It's still that way today.

Differentiation and integration of network levels

An intermediate network – the regional or metropolitan network designed for the city – does not exist. Meanwhile, in many places, urbanisation has once again pushed its way up against both sides of the national motorways. Infrastructure as defined in the Second Policy Document can no longer be interpreted as independent infrastructure located outside the cities but it has to be reinterpreted as a part of the city. This applies to both the function of the road and to its detailing and incorporation. In a few cases, this has already been accomplished. A good example is the A10 west around Amsterdam. This urban motorway avant la lettre has been grafted onto an urban structure plan carrying the same weight as that of the national motorway network.

The model of a Deltanet

So there is a choice. Over the upcoming years, we'll have to elaborate two models both at Delft University and in practical situations. In the first model, we'll continue with that superior position allotted to the infrastructure and use it to create a metropolis. Similar to what Siemens proposed as a "floating tour of Randstad" which was then freely interpreted by Luigi Snozzi in his submission to the government architect's "Metropolis Design Studio" (Frieling and Schubert, 2003). In this proposal, the major hubs – the new centres – actually have prime access to the main infrastructure and are located centrally within the metropolis but eccentrically in relation to the former urban centres. This would become a free interpretation of the Deltanet proposal presented during the weeklong design project organised by the Delta Metropolis Association in 2004 known as "Werkweek Deltanet" (Zandbelt and Berg, 2003).

The model of an intermediary network

The second choice is the model that produces an intermediary network. In this model, the wing city will explode into a new urban dimension and intensity. We'll see a single city within the wing, which will ensure and necessitate a more finely meshed connecting system for public transport and motorways. The existing transport system will become a part of this new city itself, in which process it will obtain another connecting characteristic. A number of the existing motorways will once again have to be considered components of the city. The city map would be the map of the Randstad, beginning with its wings.

Connecting stakeholders

I prefer the second model, the interwoven network, even though I realise it would involve having to overcome many problems. It would be an extremely complex project. For one thing, all the institutions that now have an interest in the existing infrastructure, such as owners, concession granters, managers or licensees, would have to start working together on this new project and do so in an apparently open market. It would require real system innovation. Not just in designing or defining the project but, even more importantly, in regard to the organisational and administrative component. This would go further than the local and regional area development to which Hugo Priemus (2003) referred in his inauguration as Dean of the Technology, Policy and Management Faculty.

Balancing urbanisation and network design

What's more, according to Henk van Zuylen, the newly appointed Scientific Director of TRAIL (the Netherlands Research School on Transport, Infrastructure and Logistics, civil planning and traffic engineering) discontinued its methodological research into improving the relationship between urbanisation and accessibility ten years ago. Since the models always have to be provided with the desired urbanisation first, this makes it impossible to objectify the transport product as input. In the case of our network city, the infrastructure explorations of the Ministry of Transport, Public Works and Water Management (V&W) will in the near future turn out to be based on quicksand. Science is losing its lustre in the light of this particular social issue.

Restoring urban connection by separation in the third dimension

Back to the two models. Is reconciliation possible? Yes, with the important and very recent example being the South Axis of Amsterdam. Here, because a bold decision was made to sink the national infrastructure below ground, the regional city map is becoming a fusion between Berlage's Amsterdam Zuid and Van Eesteren's Amstelveen, connected as it is by tram, metro, motorway and high-speed train. This modification should serve as an example for such sections of the national infrastructure as the A20 through Rotterdam, the A2 in Utrecht and the concentration of national motorways running through the middle of The Hague conglomerate. And this concept applies not only to the Randstad but just as much to Maastricht, Arnhem-Nijmegen, Eindhoven, and so on.

Tasks for research and design study

The assignment for the Delft University of Technology thus relates to researching and designing other concepts in regard to the network; the development of its interchanges; the integration of civil engineering and urban planning at various scales; the institutional system innovation to serve the interests of the owners, managers and licensees of the infrastructure; and urban development.

28.3.5 A networking administration

The owners of the task
The planning done at larger scales is often communicated in verbal form only, face-to-face. According to Andreas Faludi (2002), "Planning is not about design, but about policy." And there are two more opinions that provide food for thought: many believe that the era of creatability is over and that government intervention is waning. The simultaneous changes involving the increase in scale in many spatial issues, the stronger position of the private sector, and the change in the roles of the various levels of government are making clear observation a complicated matter. In my opinion, the government is still very much involved but it is no longer the only important player in the field. What's more, today's government is no longer the highly hierarchical organisation of the past with its clear territorial and sectoral executive power. It also requires official and administrative commissioning parties that see themselves as genuinely responsible owners of the task to be undertaken.

Problems going beyond small territorial democracy
In the private industry, a specific organisational unit is established to handle every new issue. The government, however, is bound to its territorial democracy for its formal decisions. This dilemma is bridged by means of specific administrative alliances, with or without shared regulations. An ordinary municipal councillor can no longer keep track of this, but administrators are resigned to this way of conquering the inadequacies of the existing administrative system. This requires a lengthy administrative agenda.

Separated administrative mandates based of territory or sector
In the Netherlands, there are three parallel organisational and administrative structures and control mechanisms.

Firstly, there are the official territorial and integrated administrative organisations with their directly representational, democratically elected administrations: the Dutch house of Thorbecke with three stories: municipality, province and nation.

Secondly, there are the functional territorial administrative bodies such as the water boards (Waterschappen) and the legally authorised transport districts.

Thirdly, the Netherlands has formal and informal administrative structures such as the Joint Scheme for Randstad Collaboration, the platforms in the North and South wings, and many other inter-municipal, inter-provincial and inter-governmental alliances. Within these alliances, the relationships between the administrators are non-hierarchical. It is within this informal column that agreements on strategy, programmes and projects are made. If these are successful, priorities are negotiated.

Because the projects become the "property" of their corresponding formal column, these administrative organs will have to approve of the informally made agreements. In addition to these three control mechanisms, there are also independent administrative organisations: agencies that provide the administrative column with services and that have inter-governmental ownership. There is a whole range of these organisations in which a complex structure of owners impedes the demand for control and accountability.

A transport authority for example
The Randstad should have established a transport authority long ago but very few transport authorities in the Randstad show responsibility for this urgent assignment, let alone willingness to restructure authorities for this necessary operation. But for that matter, no rearrangement of the administrative scale of any kind can put an end to this continuing development.

In another ten or twenty years, other issues and levels of scale will apply. An example might be an approach to transport at the megapolis level: the Randstad, the Ruhr District, the Flemish Diamond. We could define it thus: a German-Dutch-Flemish management of freight mobility using a common pricing policy and a common environmental approach to freight transport over the road in favour of rail and water transport. Or an administrative reaction to a merger of the independent port businesses of Rotterdam and Antwerp into a European Delta Harbours Authority by a Flemish-Dutch Delta Council.

Administrative networks connecting levels and sectors
Geert Teisman (1995) already warned us in his dissertation in 1992: "there's no end in sight – it just keeps coming". Or L. Geelhoed (Ministry of Economic Affairs EZ) who talks about "the army barracks of public administration". Or in my words: the Dutch house of Thorbecke still has three stories but it now has an infinite number of rooms with many mezzanines and staircases and an infinite number of connecting passageways that do not necessarily intersect. Work in progress goes on indefinitely, but the place is still rented. This is administrative renewal in our network society. Importantly, the latest Spatial Planning Act of 2005 indirectly achieves the same objective. Every administrative layer can be awarded its own extra authorities if its existing possibilities offer no solution.

The role of private parties on an increasing scale

In this environment, what is the role of private parties? Obviously, they, too, are increasingly taking the initiative when it comes to the spatial layout of the Netherlands. But the real estate sector is primarily a player at the site level and until now, financial equalisation has been restricted within the zoning plan.

Meanwhile, the scale of the task is increasing and voluntary regional equalisation is being tried on for size. The private parties are also experiencing an increase in scale. This can be seen, for example in the mergers taking place between corporations. What's more, there is added competition from parties such as the National Forest Service, Staatsbosbeheer in the Netherlands that can play an outstanding regional development and control role. New for these parties is the linking of urban development with green space and space reserved for water management, or vice versa. However, in their book, *Rood voor Groen(Red for Green)*, Frans Evers and Pieter Winsemius (2003) state that legislation opposes a more operational collaboration.

Public-private partnerships (PPP) and roles

The role of creative party and the engaging of designing disciplines by private parties is of great importance in the frequently snarled performance of the administration. It is with the assistance of private parties in particular that the creative intervention between the final realisation and the design can result in a substantial improvement of the final product. The final steps taken during the design process involving the South Axis in Amsterdam demonstrated this. Nevertheless, the government must always accurately define its own role in public-private relationships so that it can properly pursue its discussion with the private parties. In the field of transport, there are as of yet no powerful parties that could place an integrated supply of transport on the market.

The network city as an ecosystem

Frans Oswald, Professor of Architecture and Design, and Peter Baccini, Professor of Resource and Waste Management, both at the University of Zurich, have an interesting definition of the network city (Oswald and Baccini, 2003). They see it as a non-hierarchical urban system and use the metaphor of an ecosystem to describe it in their book, *Netzstadt: Designing the Urban*. The networking administrator is a central concept in this city and in our network society. The success of the collaboration, the commissionship, and the regional development also depends more and more on the personal efforts of key figures. After all, the pecking order supplied by Thorbecke is no longer relevant. A role is also reserved for designers in this interaction, on the condition that they can base their creativity on a concerned consideration of the roles and concepts of each party sitting around the negotiations table.

28.3.6 The design issue

Landscape, town and infrastructure

At this time, there are three entirely different issues: urban issues, new landscapes and networks. In the first place, there are three kinds of issues at the regional scale in the Netherlands in which the city plays a central role: the independent city and its surrounding region, the regional urban network and the network city, and the metropolis.

Key projects

Many Dutch cities have been rediscovering their urban potentials over the last ten to twenty years. The development of railway station sites or the restructuring of former industrial areas near city centres is injecting new vitality into these cities. Perhaps the success of the Fourth Policy Document on Physical Planning is even more evident in its key projects in these medium-sized cities in the Netherlands than in the VINEX sites located on the fringes of these cities. In all three of these matters, not only is the position of the cities different; so is that of the landscapes. The medium-sized cities have developed new relationships with their surrounding landscapes in recent years. Consider the examples of Groningen Meerstad, Den Bosch Haverleij, and so on.

New landscapes in the National Ecological Network (NEN)

A second new series of issues involves the creation of new landscapes in conjunction with the development of the National Ecological Network (NEN). The power of regional development is being tested in the coupling of landscape development with urbanisation and water management issues. New modes of living and leisure are being developed, for which co-finance can be arranged thanks to the fact that more people are living in rural areas. Development planning is also becoming a regional issue and is being realised in public-public and public-private alliances. Examples are the Wieringerrandmeer, the restructuring project known as "The Heart of the Heuvelrug Area" in Utrecht, the Ronde Venen in the Green Heart, and landscape transformations developed by regional Space for Space schemes.

Connections

A third issue is the redefining and redesigning of the existing infrastructure networks in connection with the urbanisation projects into the three functional scales for the purpose of improving the connections between the networks used by both public transport and passenger cars. Good examples are the Randstad Rail, Rijn Gouw Lijn, Light Rail Arnhem-Nijmegen, Brabantstad and Rail and the concept of linking the large cities in the Randstad with a provincial metro system – the meanings and magnitudes of which I have already explained.

So what will the Randstad Holland Project as a budding metropolis look like?

Shifting focus in the Randstad landscape

Looking down from high above, one can truly admire the strength of the Dutch landscape. Zooming in to the superregional level, the Randstad is completely unlike the rest of the Netherlands and the issues involved within the Randstad are also highly diverse.

North Wing

In the North Wing of the Randstad, there is a central city with a finger-like structure at the regional level: urban lobes with green wedges between them. All these elements have to be viewed within a larger regional context. Once upon a time, the central train station in Amsterdam replaced the IJ as the main transport hub. Nowadays, Schiphol is the portal to the world and the South Axis has become the new Central Station for the Amsterdam region. In the next round, the IJ and the IJmeer could become the new link between the twin city of Amsterdam-Almere. In that case, this water should be understood as central metropolitan space. In fact, this is happening at Teun Koolhaas' Atelier DUTCH thanks to a bold initiative by Hans Bijl, an alderman for the Municipality of Almere. Designers are creating a series of plans and projects that gives the IJmeer and Markermeer new ecological, leisure and urban meaning. Orienting Utrecht to the polders is the next step.

South Wing

The South Wing of the Randstad displays an entirely different constellation. Here, there are two large central cities, whilst parts of the coast and the underlying landscape have little scenic value. And the competition for land by horticulture under glass, the port and the city is much fiercer here. The South Wing's rapid development into a network city stretching from Leiden to Dordrecht with a population of 3.5 million entails that all the meanings of the original cities, centres and green spaces are changing. All functions are becoming part of a larger urban body and will have to be analysed and programmed anew (PZH, 2005).

New centres of higher scale in between the older ones

At this point, the South Wing as a design issue can only be seen as a budding metropolis, while it is itself part of the Randstad as a whole. In short, a challenge in which the Delft University of Technology and the Province of Zuid Holland should take the lead to elaborate concepts with many other fellow professionals in the region within the framework of a research-oriented studio.

The conceptual framework for this phenomenon, however, is still underdeveloped. The concepts involving the network city and the urban network may have been introduced in the Fifth Policy Document on Physical Planning, but very little has been done in the way of operational research and they have as yet to receive any qualified content. The proper vocabulary hasn't been established yet either. In *Cities without Cities,*

Thomas Sievert (2003) uses the term "Zwischenstadt" to indicate a regional *and* urban character.

The Green Heart
The Green Heart involves an entirely different set of problems and issues. A policy of preservation will eat away the region to its core because it will decay from the inside out. If no economic support can be found for farmers that can guarantee the continuity of the existing landscape, the region will become cluttered, fill up with urban sprawl and degenerate. Instead, the Green Heart should become a huge outdoor area offering Randstad inhabitants scenic landscape and water leisure. The proposals made in 2003 by Dirk Sijmons in the studios of the government architect and the studies such as "Waterrijk", a water management proposal, made by the Delta Metropolis Association in 2000 (Deltametropool, 2001) are waiting to be implemented in one way or another.

New owners of the issue
Designing, calculating and legislation are now the next steps on the way to finding new owners for the issue. More and more often, the new owners of these new issues are regional governments and the alliances between cities and regional administrative bodies. Their organisations and their accompanying instruments and flows of funds are now being readied for this. Therefore, universities should be training their students to handle these complicated issues. What lies ahead is a new complexity involving regional planning, regional land-use policy, regional programming, regional design, and the management of regional spatial processes and transformation.

Strengthening remaining and new tasks of national government
The national government's role will involve decentralisation, and it will have to persevere in its resolution to stay on this course. Otherwise, regional ownership of the issue will be nipped in the bud. In addition, recognition of this change of scale among the profession within the national architecture policy could help in the content-related design of the coordinating ministry now being employed in working out the Policy Document on Space. The issue is so new to both region and national government that stimulating its integrated quality will be extremely important in the studios that will be working on these issues.

28.4 The designer

Integrating landscape, town and infrastructure
Glancing through the five editions of the biennially published books entitled *Landscape Architecture and Town Planning in the Netherlands*, the first of which appeared in 1996 (Hartman and Hellinga, 1996), anyone will notice an enormous planning and implementation production as well as an abrupt increase in both the quality of design and the variation in production – a

tacit compliment to the groups of professionals and their commissioning parties.

Navigating in a context of flow

Everything's flowing, said Luuk Boelens (2005) at his inauguration as Professor of Urban Planning, in which he redubbed the name of his profession as fluviology. To the urban planner and the fluviologist, this urban planner and landscape architect says, "Having heard and seen everything, we want to make sensible proposal for somewhere where we can cast concrete, have trees planted, or let water flow." After all, when it comes to urban development, it's not just about strategy or meaning related to content but also about the design of actual physical interventions and how they impact society. This requires not only technical insight but also socio-cultural insight into the meaning of a location and its context. Designers are responsible for these actions even though they cannot predict all the consequences. Because once realised, users provide the finished product with their own interpretations and use it in their way.

Creative interventions in a context of specialisations and regulations

When engaged in actual issues, the urban planner has to master an unprecedented number of disciplines and regulations that make no use of creative intervention. This is what's expected of him. But it is only one side of the coin. The other side is that he in turn has to possess the skill to inspire his commissioning party or parties with his design. What's more, in his role as designer, he applies research methods to construct the assignment and the commissionship.

A 'universal' approach of separate sub-systems

Here lies a task for the University. Urban planners and landscape architects are not general specialists in the operation of systems that are determining factors in regions such as traffic systems, water systems, administrative systems and their related economic clusters. If a designer wants to make decisions about the layout of space, he will have to become familiar with the operation of one or more of these sub-systems. The University has extensive knowledge of all of these sub-systems, not necessarily connected to the creativity of the designing discipline. And then there's the time factor.

Study by design instead of final images by design

These questions are missing pieces of information that often pose a problem for an architect, considering his training. The designer's specialty has always been based on the Beaux-Arts tradition of sketching the final image of a development, despite the fact that there are post-modern architects such as Rem Koolhaas (Sigler, 1995) who denounce this final image doctrine by idealising research as the way to arrive at designs.

Integrating professions by design

In the second place, the modern concept and CIAM dogmas have resulted not only in a division of functions in space but also in a division of professions, each one preaching to his own parish. Many engineering issues are approached in terms of sectors instead of in terms of a cohesive spatial system.

Integrating technology, policy and management
In the third place, the problem of overlapping urban regions is relatively new and still fairly unexplored as an element in design research. The new design projects, or reinterpretations of existing projects resulting from growing mobility, have only now started to make it through to university and administrators. Or, translated into University terms, urbanism and urban management; civil engineering with water, network development and infrastructure design; and technology, policy and management form identical triplets and an opportunity card for a broad professional practice.

Integrating different futures
In Ways to Study: Design in Strategy published in 2002, Dirk Frieling (2002) distinguishes four kinds of futures: potential, possible, probable and unavoidable – or, better said – actual futures. As part of Delft's modest research portfolio and its associated courses, we have students engage in exercises involving the design of strategic perspectives within a region. Using decision-making simulation, we show how different politically defined strategies can still result in feasible projects when they fit into various strategies.

Design connecting levels into realisation
The regional designer thus creates a spatial perspective based on a strategy. In doing so, he creates certain conditions for possible developments that will be realised by other designers and parties. A strategy developed for the system or the territory is accompanied with the definition of certain key projects. Based on these projects, the designer can show what a regional intervention means at the local level. And within this framework, crucial details such as a sightline, a detail, how a route through an area is experienced, and a programme can arise that will unite the local with the regional upon realisation. This returns him again to the role of traditional designer. To Frieling's dialect, I would thus expressly add this last design step as an essential part of the skills for a regional designer.

New Charter of Athens
In its recent version of the New Charter of Athens launched at its conference in Lisbon in 2003, the European Council of Town Planners (ECTP, 2003) formulated four core competencies for spatial planners and designers: the scholar and researcher, the designer and visionary, the policy consultant and intermediary, and the manager of spatial transformations. An outstanding and inconceivably broad curriculum for the urban planner. After all, just as a designer, his insights as a consultant and manager are of equal importance in generating spatial proposals. Stimulating spatial proposals creates the issue, the assignment and the

commissioning party. This is what gives our designing discipline such exceptional potential.

28.5   Conclusion

Selecting crucial projects, organisation of finance and large scale strategy
Broad international research that The Greater London Council commissioned to be conducted by such institutions as the London School of Economics show how successful designing at larger scales and structure planning are. The success of cities and regions is also determined, of course, by their strategic capacity, by their potential to select projects, and by their organisation of the required financing. But no success can be achieved without a spatial strategy. Where the interrelationships of these aspects is strong, the region will be strong as well. Dutch examples are countless. Amsterdam has a famous structure plan tradition (Amsterdam, D.P.W.G., 1934) and continues this tradition by constantly working on a strategic project portfolio. The same goes for Rotterdam as can be seen from its successful waterfront and city centre development. The Hague previously lacked an urban and regional force and usually worked with a disconnected project portfolio. Now, however, (after Berlage in 1908 and Dudok in 1950 see Meyer and Burg, 2005) it's making up for lost time thanks to its new structure plan that is placing it within the context of the South Wing and allowing The Hague to claim its position.

Symbiosis of regional practice and University study
As an extension of our professorial duties, we conduct research into the formal plans of certain large metropolises or metropolitan regions and their proposals for strategic interventions and projects resulting from such decision-making processes. Comparing these plans and projects clarifies the selection issues in Dutch regional planning. At the same time, the first cautious attempts are being made to clarify the meaning of the spatial configuration of metropolises for their economic development. The design research into the new landscapes, the models for the network city, is usually done by students and in collaboration with people in professional practice. We can also make grateful use of the background information about the Randstad as supplied in *Over Holland*, published in 2005 and written by the Department of Architecture under the leadership of Henk Engel (2005) in which the cities and infrastructure are explained in relevant time scales for purposes of analysis and design.

Top-down and the reverse
I see two different ways of viewing things here at the faculty. The architects definitely consider the object-related and urban side, whilst we consider the regional and metropolitan side. And this within a perspective that has yet to be designed. For me, this approach with all its ramifications, inevitably leads to the question: can I educate designers in Delft to keep their footing

at this scale, in this whirlwind of continuous and rapid change? Students who choose to study architecture know that they are entering a well-defined area in which the context is much clearer and manageable than that of the urban planner. It used to be said that those who were not good enough to design should switch to urban planning. Such a way of thinking slights the importance of the profession as it is manifested in the Netherlands. And taking more courses in urban planning is also necessary for the professional development of future designing architects, just as the reverse is true for the urban planner and for a productive relationship between urban planning and landscape architecture. My professional field demands a considerable understanding for and insight into complex social forces and changes and an unimaginably creative and investigative capacity from the designer. I see it as my task to stimulate this and to promote this at this educational institution. It is my contribution to Town and Country and to the creation of the issue.

Landscape Ecology

29 THE TASK OF LANDSCAPE ECOLOGY

**Aat Barendregt, Rob Jongman, Jacques de Smidt,
Martin Wassen**

29.1 Introduction

This is a personal reflection of the authors of this chapter on the book. To find an answer to the question what the task is of landscape ecology, we will split the question in two parts. The first part of the question is about science for society: what is the task of landscape ecology in a changing society and more specifically what is the role of landscape ecology between tasks delegated to single issue disciplines and how does this add to improving the quality of the Dutch landscape. The second part concerns science for science: which scientific challenges do we see for landscape ecology? The order of both questions is characteristic for Dutch landscape ecology. Often social problems define landscape ecological research.

It appears helpful to go back to the statement of Carl Troll, the godfather of landscape ecology who used the word 'Anschauungsweise' to emphasize that landscape ecology is a way of looking at the landscape. The way of looking at an object is the outcome of the chosen concept, the methodology and implicit or explicit values.

Landscape ecology covers a wide field of research and application. The origin of landscape ecology is in biology and physical geography and in the 1970s also landscape architects and urban and rural planners joined. Since landscape ecology has chosen for the holistic concept and methodologically focuses on the higher integration level (ecosystem, landscape) it has developed as an umbrella with the aim to understand the complexity of ecosystems and landscapes by integration of patterns and processes at different scale levels. Although landscape ecology roots in natural sciences the cultural aspect of the landscape has gradually gained importance in landscape ecological studies. In 1998 IALE defined landscape ecology as the scientific basis for the analysis, planning and management of the landscapes of the world, implicitly acknowledging the social aspect.

The Dutch perspective
This book shows that the working sphere of landscape ecology in the Netherlands is a densely populated man-made country where space is limited and technology dominates developments in land use. This is the reason why Dutch landscape ecologists make sustainable development the core of their work. Since sustainable development has an ecological, social and economic dimension, all of these aspects appear to be part of landscape ecology. This also reflects the Dutch attitude towards landscape. National policy in the Netherlands aims for preservation of biodiversity, restoration of ecosystems and creation of new nature. Economic development, however, is the prime interest. The social dimension of the landscape receives recently more and more attention. This is because land use shifts are envisaged for a variety of reasons. European and global developments lead to a diminishing relative economic importance of land bound agricul-

tural production in the Netherlands. Because the population is still growing the need for housing increases which is even fastened since families tend to get smaller and one- or two person households increase. Higher income people prefer living on the countryside, while working in the city. In an urbanised country such as the Netherlands this is possible. This also causes a growing need for infrastructure and outdoor recreation, both requiring space. These trends enhance the urgency of answering questions about 'the Future of our Landscape' such as: How do people perceive landscape? What makes them feel happy? How is this perception related to the history of the individual or the group? How can we develop an innovative transport system that prevents large-scale fragmentation? How can we prevent developments in the rural areas that lead to the loss of the identity of the landscape? All in all the future of the Dutch landscape is an issue nowadays. Recently, 'De Volkskrant', one of our high-quality daily newspapers, initiated a project called 'Spatial Agenda'. In this project experts as well as laymen express their worries, discuss, search for solutions and set priorities for this agenda. This project exemplifies the growing concern about undesirable trends in the Dutch landscape that are even facilitated by the national government through step-back policy in land use planning during the last decade.

29.2 The task of landscape ecology in a changing society

Most societies wish to change their surroundings, to develop as a society or to improve the conditions for their benefit. This often causes unwanted negative effects. Solving these problems requests an interface of co-production of knowledge and solutions between science and policy. In this interface communication between scientists, planners, developers, politicians and practitioners such as resource or nature managers is difficult. They have different inputs and different languages, although they might aim the same: sustainable development.

In this problem field landscape ecology has obviously a task as a problem solving science. It is therefore logic that it had a great boost around 1970. Society in the industrialised world suffered heavily from the pollution of air and water. Apart from human health problems, also the rapid decrease of animal and plant species became a source of concern. Landscape ecological studies helped to analyse and to make the inventory of what was left of natural resources. It carried out cause-effect relationship studies and tried to model society in an ecological way. The 'Global Ecological Model' initiated by the Ministry of Housing and Environment of the Netherlands in the 1970s was one of the early predecessors of the present models and ecosystem service studies and this knowledge was a prerequisite for designing solutions. The terms environmental studies and human ecology were introduced. Both terms have an anthropocentric character, but they also express the concern of humans with other species, which makes clear that all those species are part of the human habitat.

The concept of landscape layers as a tool to explain its complexity
Landscape ecology deals with a continuous growth of the complexity of its scientific object because of the growing availability of knowledge produced by other disciplines. Because of its holistic approach it integrates this growing body of knowledge in its own discipline. This produces growing understanding of the system and predictability of responds on changing conditions. As in other sciences that deal with complex systems and accept this complexity as a characteristic of its object, the scientific output is not always easy to communicate to those who want to use it for management, design or decision-making. Landscape ecology integrates aspects of ecology (which by definition already integrates abiotic and biotic components), history and land use, merging these in analyses, descriptions and models of processes and patterns of a landscape. This allows also for predictions of the effects of alternative processes or rates of change. The recent growing concern with the effects of climate change on nature and the human society are a great new challenge for landscape ecology to design mitigation, adaptation and prevention measures, for instance against floods or restyling the National Ecological Network in anticipating on climate change effects on species distribution and migration in landscapes. Landscape models are available that after new parameterization can predict the effects of climate change on the interaction between vegetation, soil and water. Ecosystem models of landscapes are already in use to predict possible effects of new roads and new towns on the populations that live there, including the effects on the human population. Early assessments are supporting careful decision-making.

It is useful to make a distinction between the different components and aspects of the landscape and nevertheless approach the landscape in a holistic way for understanding its complexity. In this respect the concept of landscape-layers proved to be very useful for communication with policy makers and developers about the complex interrelations and feed backs in the landscape:

1. The physical processes such as changes in climate, soil, chemistry and hydrology
2. Species populations and inter-specific interactions
3. The human occupation, with changes in land use including urbanisation
4. The spatial relations in the landscape such as fluxes of water, energy and matter, wet or dry spatial linkages and infrastructure
5. The cultural aspects, with the presence of archaeological, historical and architectonic aspects

Knowledge, alternatives, choice and public involvement

The task to prevent problems can be accomplished by means of predicting models only in case of sufficient availability of quantitative data on the relation between the organism and the environmental factors. The ability of a society to prevent problems depends on many factors. What are the needs, problems and aims; who are the stakeholders; what are the alternatives; what is the available knowledge; which of them is the choice. The whole process starts with the definition of needs and problems to be translated into objectives. From the perspectives of stakeholders and available knowledge and technologies alternatives can be elaborated. Then choices can be made. In all steps in this policy making process landscape ecology can contribute, as stakeholders can do. Landscape ecology can help to articulate needs and problems in the agenda-setting phase. It can help to find alternatives. A key task is to show opportunities and options on the one hand and to assess the impact of each of the alternatives on the other hand. Such cooperation or co-production is completely depending on the available knowledge of cause and effect in the landscape where the project is planned. In such cases landscape ecological knowledge has two levels of integration. One level is the general ecological knowledge on pattern and process in landscapes, laid down in theories and models. The other level is the region specific interactions that are decisive for the structure and the function of the landscape on the spot where the project is planned. It is important to realise that because of the complexity of the landscape the available knowledge of cause - effect relationships in practice is barely enough to understand and predict all details. That is why such analyses or assessments need close cooperation between scientists, stakeholders and decision makers. Nevertheless at the end of the process the decision makers will have to make the choice. Making explicit what are the uncertainties, explaining with the available knowledge what the possible risks are as a consequence of the uncertainties, provide the decision makers with a tool to gain public support.

Relations of humans with their landscape

There is a strong understanding that landscape contributes highly to the enjoyment and relaxation in human life. However, the preconditions for mental health and happiness are not expressed in quantitative landscape values. Here is a task for landscape ecology to extend the available models with quantitative parameters. Problems with mental health and happiness may be prevented when people know what is important for feeling comfortable. Currently, interview techniques are available to gather information on such issues. Landscape ecology can communicate on this issue by explaining the relation between the structure of the landscape and the conditional processes. It is good to realize that here landscape ecology integrates techniques from social sciences into its own sphere of beta science. The next step - appreciation of landscape - is a social process in which landscape ecologists can help. The human population has to express what properties of their landscape are considered as important for feeling good.

The (beta) landscape ecological knowledge that was produced in the first step feeds the people to understand their landscape and provides them with concepts and words to express their priorities. These priorities are called: values. On the other hand, such values can open (or close) eyes for characteristics of the landscape. For example, in the current debate about landscape in the Netherlands the historical-cultural aspects are often emphasized whereas before there was a focus on naturalness / wilderness. Van der Zande in his chapter advocates a landscape ecological approach in which historical-cultural values are more explicitly considered. We agree. However, we stress that historical geographers in terms of age, completeness, uniqueness, rarity, setting in the landscape, et cetera describe the historical-cultural aspects of landscape in the objective way. The next step - determining the value - is however subjective. This subjectiveness should be acknowledged explicitly and determining the value should be separated from the description.

Knowing such values is a great help to avoid problems when new land use plans are made. This means again a task for landscape ecology: to find out if there is a way to design a plan that integrates the values considered as important by inhabitants. If this is not possible and the population nevertheless wants to agree with the plan, the landscape ecologist again is in charge to design possible ways to compensate (to a certain extent) the loss of value. The outcome can be discussed with the stakeholders. The combination of landscape ecological knowledge and the values that are denominated by the people, provide the decision makers with information to make the final step. The process described above is already practiced in the Netherlands in certain complex planning tasks. In our view better facilitation of this process by landscape ecologists is an innovative new challenge in the development of landscape ecology.

29.3   The task of landscape ecology as a science

Landscape ecology evolved a lot since Troll introduced it in 1939. This evolution is very much connected to the available knowledge about the processes in the landscape. At the starting point of WLO in 1972 or IALE in 1981, landscape ecology still principally offered a holistic view on the relations in the landscape. It was a way to deal with complexity of interrelations in the landscape, which were ignored or regarded as too difficult by reductionistic scientists in geography and ecology. Nowadays, according to the IALE-website (2007) landscape ecology is defined as:

The study of spatial variation in landscapes at a variety of scales. It includes the biophysical and societal causes and consequences of landscape heterogeneity. Above all, it is broadly interdisciplinary and transdisciplinary. The conceptual and theoretical core of landscape ecology links natural sciences with related human disciplines. Landscape Ecology can be portrayed by several of its core themes:

1.   The spatial pattern or structure of landscapes, ranging from wilderness to cities;
2.   The relationship between pattern and process in landscapes;
3.   The relationship of human activity to landscape pattern, process and change;
4.   The effect of scale and disturbance on the landscape.

The core of landscape ecology
If we mirror this description of what landscape ecology is with the content of the present book we note a number of points. In the 28 chapters of this book contributions are grouped around three themes: nature, town, and infrastructure. The book gives a specific cross-section of Dutch landscape ecology. Most contributions are rather personal reflections; very diverse, some quite scientific others very applied, but all placing the content in a historical context. Here we will restrict ourselves to some main conclusions.

The first conclusion is that the core of landscape ecology (linking natural sciences with related human disciplines) is almost present in every contribution. The three sections of the book deal with application of science in the society; of ecological theory applied in the National Ecological Network, of advances in ecohydrology and its contribution to preservation and restoration; of ecological and social sciences in urban ecology; of technological and ecological theory in infrastructure planning and of water in urban areas, although the latter subject is slightly under-represented in this book. Most chapters pay explicit attention to one or more of the above-mentioned IALE core-themes. Many chapters are implicitly interdisciplinary. Moreover, almost all chapters start from the point of human activities to be performed, to be evaluated or to be solved. Therefore, our overall conclusion is that this book is an excellent example of what modern landscape ecology should be.

Diversity
We note that there is not one ruling concept prescribing how to investigate or to apply landscape ecology. Questions are too diverse and in most cases in answering a question only one or a limited number of landscape characteristics are considered. Sometimes, the approach is more inclusive leading to ready-for-practice suggestions for managing or designing landscapes in a sustainable way. In other cases the reductionistic approach seems to be the best way of answering a question. So, we refrain ourselves from judging what is the best landscape ecology. The diversity we noted is in our view the biggest selling point of landscape ecology. An ecohydrologist, a spatial ecologist, a landscape architect, a social scientist … they all are landscape ecologists as long as they realize that their part con-

tributes to the complete landscape. It is the 'Anschauungsweise' that counts.

Just as there is no minimum or maximum spatial dimension for an ecosystem, there is no fixed dimension for a landscape. During the first years of landscape ecology, the perception of a human view of a landscape (region with the same conditions and visual appearance) was the fundamental scale of investigation. Nowadays we observe that local processes in a landscape operating on a short time scale (e.g. turnover of organic matter by micro-biological processes) might explain the distribution of species, but at the same time research on the above-regional relations in spatial constellation (e.g. regional groundwater flow, migration) are prominent. So, landscape ecology is the ecology of scale. Also, as emphasized above, the same is true for the object of investigation, as also in landscape ecology frequently very specialized research is performed on species or ecosystems, which is just ecology. The developments of specialized WLO working groups with research in ecohydrology, historical landscape ecology or urban ecology might be seen as an illustration of this specialising trend in landscape ecology. They focus on one aspect of the landscape; however, we think results of such studies should always be placed in the integrated multidisciplinary context of relations in landscape ecology.

The task of landscape ecology in future science
Although we do not see a strict difference between scientific and applied landscape ecology we observe three important 'scientific' challenges for landscape ecology. Firstly, new results from disciplinary research have always been incorporated in the landscape context, and the need for this will increase in the future. Landscape ecologists should be aware of this since they are the ones who have to pick up the pieces of fundamental insight delivered by the basic disciplines and incorporate them into their landscape models that are used to communicate with society. Secondly, climate change/global change will affect many relations in the landscape leading to complex changes. Landscape ecologists have to provide essential knowledge needed for the design of prevention, adaptation and mitigation measures. Thirdly, communication with the society, with stakeholders, in participatory settings will be more and more important. This will be a challenge for the communication skills of landscape ecologists. Communication should become an essential part of the academic education in this respect. The future of the world is about communication and science is no exception in this. The future of science is in communication with other disciplines.

29.4   Final conclusions

This book illustrates the present state-of-the-art of landscape ecology in three sectors. It is not an encompassing review of all that is performed in the Netherlands, but it illustrates very well the impressive progress made the last decades. Due to the high population pressure in this country, the international relevance is that the Netherlands may function as a laboratory, an experiment, a case to learn from, exemplifying a region where urbanisation and human land use dominate and where even nature is culture. The difficult tasks set for the realisation of the National Ecological Network prove that nature is regarded as important in national policy but at the same time economy rules and inhabitants put their personal landscape perception including their values on the agenda. It is remarkable to note that sometimes coincidental events such as floods or heavy storms triggered new landscape policies. We also note that urban ecology is important as a science base to develop a sustainable built-up area, relevant for many locations in this country. One might even wonder where urban ecology is not applicable in the Netherlands. All the aspects of transport and the 'freedom to go wherever you want' put a heavy pressure on the landscape. However, ex-ante impact assessment and innovative solutions might lead to sustainable planning of infrastructure. Dutch landscape ecology might keep serving as an example of understanding a constructed landscape, as a case to learn from for areas where conflicting claims are competing for the same sparse space.

REFERENCES

Aanen, P.; Alberts, W.; Bekker, G.J; Bohemen, H.D. van; Melman, P.J.M.; Sluijs, J. van de;
Veenbaas, G.; Verkaar, H.J.; Watering , C.F. van de (1991) *Nature engineering and
civil engineering works* (Delft/Wageningen) Ministerie van V&W/RWS/DWW en Pudoc

Ahern, J. (2004) *Greenways in the USA: theory, trends and prospects* In: Jongman, R.H.G.;
Pungetti, G. [eds.] *Ecological networks and greenways, concept, design and implementation* (Cambridge) Cambridge University Press; 34-55

Aitchison, J. (1995) *Cultural landscapes in Europe: a geographical perspective* In: Droste, B.
von; Plachter, H.; Rössler, M. [eds.] *Cultural landscapes of universal value.*
(Jena/Stuttgart/New York) Fischer; 272-288

Alaimo, M.G.; et al. (2000) *The mapping of stress in the predominant plants in the city of
Palermo by lead dosage* (j) Aerobiologia 16[1]: 47-54

Alerstam, T.; Lindström, Å. (1990) *Optimal bird migration: the relative importance of time,
energy and safety* In: Gwinner, E. [ed.] *Bird Migration. Physiology and Ecophysiol-
ogy* (Berlin) Springer Verlag; 331-351

Algemene Rekenkamer (2006) *Ecologische Hoofdstructuur* (Den Haag) Sdu Tweede Kamer,
vergaderjaar 2006-2007, 30 825, nrs. 1-2.

Amstel, A.R. van; Herngreen, G.F.W.; Meyer, C.S.; Schoorl-Groen, E.F.; Veen, H.E. van de
(1988) *Vijf visies op natuurbehoud en natuurontwikkeling - knelpunten en perspec-
tieven van deze visies in het licht van de huidige maatschappelijke ontwikkelingen*
(Rijswijk) R.M.N.O-publicatie nr. 30

Amsterdam (1934) *Grondslagen voor de Stedebouwkundige Ontwikkeling van Amsterdam
Algemeen Uitbreidingsplan* (Amsterdam) Stadsdrukkerij; Dienst Publieke Werken
Gemeente Amsterdam

Amsterdam (2006) *Amsterdam in cijfers 2005* (Amsterdam) Gemeente Amsterdam;
www.amsterdam.nl/pdf/2005_jaarboek_hoofdstuk_01.pdf

Amsterdam.nl (2006) *De ecologische voetafdruk van de Amsterdammer*;
www.amsterdam.nl/wonen/afval_milieu/de_ecologische

Andel, J. van ; Aronson, J. [eds.] (2005) *Restoration Ecology The New Frontier* (Oxford)
Blackwell

Andela, G. (2000) *Kneedbaar landschap, kneedbaar volk - De heroïsche jaren van de ruilverkavelingen in Nederland* (Bussum) THOTH

Andelman, SJ; Fagan, WF (2000) *Umbrellas and flagships: efficient conservation surrogates
or expensive mistakes?* (j) Proceedings of the National Academy of Sciences of the
United States of America 97: 5954-5959

Andeweg, R. (2002) *Muurplanten in Rotterdam: de stand van zaken* (Rotterdam) Natuurhistorisch Museum; SER 1- Muurplanten in Rotterdam; Natuurlijk Rotterdam 2002-1

Andren, H. (1994) *Effects of habitat fragmentation on birds and mammals in landscapes with
different proportions of suitable habitat: a review* (j) Oikos 71: 355-366

Andrewartha, H.G. (1961) *Introduction to the Study of Animal Populations* (Chicago) University
of Chicago Press

Ankum, P. v. (2003) *Polders en Hoogwaterbeheer Polders, Drainage and Flood Control* (Delft)
Technische Universiteit Delft, Fac. Civiele Techniek en Geowetenschappen, Sectie
Land- en Waterbeheer: 310

Anonymus (1978) *De Werkgemeenschap Landschapsecologisch Onderzoek; een plaatsbepa-
ling tussen verleden en toekomst* (j) WLO-Mededelingen 5(3): 4-6

Anonymus (1982) *Beleidsplan van de WLO voor de jaren 1982-1987* (j) WLO-Mededelingen
9(2): 67-82

Anonymus (1989) *Ecologische infrastructuren rond rijkswegen en -kanalen op Zuid-Beveland
en Walcheren* (Wageningen) i.o.v. RWS/Directie Zeeland, De Groene Ruimte

Anonymus (1989) *De ecologische betekenis van 'boomweiden' langs rijkswegen* (Delft) DWW
Note MIOL 89-09

Antrop, M. (2001) *The language of landscape ecologists and planners* (j) Landscape and Urban Planning 55: 163-173

ANWB (2004) *Groot ANWB Vrijetijdsonderzoek 2003* (Den Haag) ANWB

Arcadis Heidemij Advies (1999) *Evaluatie-onderzoek ontsnippering van infrastructuur* (Arnhem) Arcadis Heidemij Advies

ARCAM *http://wwwarcamnl/*

Arends, G. J. (1994) *Sluizen en stuwen* In: Arends, G. J. [ed.] *De ontwikkeling van de sluis- en stuwbouw in Nederland tot 1940* (Delft) Delftse Universitaire Pers / Rijksdienst voor de Monumentenzorg

Atkinson-Willes, G.L. (1972) *The international Wildfowl censuses as a basis for wetlands evaluation and hunting rationalization* Proceedings *Int. Conf. on Conservation of Wetlands and Waterfowl* (Ramsar 1971); p. 87-110

Atkinson-Willes, G.L. (1976) *The numerical distribution of ducks, swans and coots as a guide in assessing the importance of wetlands in midwinter* Proceedings *Int. Conf. on Conservation of Wetlands and Waterfowl* (Heiligenhafen 1974); p. 199-271

Aykac, R.; Daalhuizen, F.; Piek, M.; Pieterse, N.; Wagt, M. van der (2005) *Het gedeelde land van de Randstad Ontwikkeling en toekomst van het Groene Hart [The divided land of the Randstad, Development and future of the Green Heart]* (Rotterdam/Den Haag) NAi/RPB

Baaijens, G.J.; Everts, F.H.; Grootjans, A.P. (2000) *Bevloeien van grasland in Nederland* (Groningen) OBN/RU

Baaijens, G.J.; Everts, F.H.; Vries, N.P.J. de (2004) *Vloeiweidesysteem Klein Bieler; leven op kwelkraters* (Groningen) RUG/EGG consult Everts and De Vries

Bach, B.; Jong, T.M. de (2006) *Traffic and design of the urban fabric* (Ede) CROW

Baird, A.J.; Wilby, R.L. (1999) *Eco-hydrology: plants and water in terrestrial and aquatic environments* (London) Routledge

Baker, L.A.; ; (2002) *Urbanization and warming of Phoenix [Arizona, USA]: Impacts, feedbacks and mitigation* (j) Urban Ecosystems 6[3]: 183-203

Bakker, A. de (2007) *Leisure and accessibility* In: Jong, T.M. de; Dekker, J.N.M.; Posthoorn, R. [eds.] *Landscape ecology in the Dutch context: nature, town and infrastructure* (Zeist) KNNV-uitgeverij

Bakker, J.P.; Poschlod, P.; Strykstra, R.J.; Bekker, R.M.; Thompson, K. (1996) *Seed banks and seed dispersal: important topics in restoration ecology* (j) Acta Botanica Neerlandica 45: 461-490

Bakker, J.P.; Berendse, F. (1999) *Constraints in the restoration of ecological diversity in grassland and heathland communities* (j) Trends in Ecology and Evolution 14: 63-68

Bakker, Riek; Palmboom, Frits; Reede, Frank de (1983) *Bestemmingsplan Prinsenland* (Rotterdam) Gemeente Rotterdam

Bal, D.; Beije, H.M.; Hoogeveen, Y.R.; Jansen, S.R.J.; Reest, P.J. van der (1995) *Handboek natuurdoeltypen in Nederland* (Wageningen) IKC Natuurbeheer

Bal, D.; Beije, H.M.; Fellinger, M.; Haveman, R.; Opstal, A.J.F.M. van; Zadelhoff, F.J. (2001) *Handboek Natuurdoeltypen* (Wageningen) EC-LNV; Rapport Expertisecentrum LNV

Barendregt, A.; Stam, S.M.E.; Wassen, M.J. (1992) *Restoration of fen ecosystems in the Vecht River plain: cost-benefit analysis of hydrological alternatives* (j) Hydrobiologia 233: 247-258

Barendregt, A. (1993) *Hydro-Ecology of the Dutch Polder Landscape* (Utrecht) PhD Thesis University of Utrecht

Barendregt, A.; Wassen, M.J.; Smidt, J.T. de (1993) *Eco-hydrological modelling in a polder landscape: a tool for wetland management* In: Vos, C.C.; Opdam, P. [eds.] *IALE Studies in Landscape Ecology 1 Landscape ecology of a stressed environment* (London) Chapman & Hall; 79-99

Barendregt, A.; Wassen, M.J.; Schot, P.P. (1995) *Hydrological systems beyond a nature reserve, the major problem in wetland conservation of Naardermeer (the Netherlands)* (j) Biological Conservation 72: 393-405

Barendregt, A.; Klijn, J.A. (2006) *Het landschap bestudeer je van groot naar klein, niet andersom* (j) Landschap 23 (4): 161-169

Barendregt, A.; Dekker, J.N.M. (2007) *Landscape ecology and the national ecological network in the Netherlands* In: Jong, T.M. de; Dekker, J.N.M.; Posthoorn, R. [eds.] *Land-*

scape ecology in the Dutch context: nature, town and infrastructure (Zeist) KNNV-uitgeverij

Barendregt, A.; Jongman, R.H.G.; Smidt, J.T. de; Wassen, M.J. (2007) *The task of landscape ecology* **In:** Jong, T.M. de; Dekker, J.N.M.; Posthoorn, R. [eds.] *Landscape ecology in the Dutch context: nature, town and infrastructure* (Zeist) KNNV-uitgeverij

Barends, S.; Baas, H.G.; Harde, M.J. de; Renes, J.; Stol, T.; Triest, J.C. van; Vries, R.J. de; [red.], F.J. van Woudenberg (2000) *Het Nederlandse landschap; een historisch-geografische benadering* (Utrecht) Matrijs

Batenburg-Eddes, T. van; Berg Jeths, A. van den; Veen, A.A. van der; Verheij, R.A.; Neeling, A.J. de (2002) *Slikken in Nederland; Regionale variaties in geneesmiddelengebruik [Regional variations in use of pharmaceuticals]* (Bilthoven) RIVM ; http://www.rivm.nl/bibliotheek/rapporten/270556005.html

Bateson, G. (1980) *Mind and nature - A Necessary Unity* (Toronto) Bantam Books

Bateson, G. (1983, 1980, 1973) *Steps to an Ecology of Mind* (New York) Chandler Publishing; Balantine Books

Batten, D. (1995) *Network cities - creative urban agglomerations for the 21th century* (j) Urban Studies 32[2]: 313-327

Baudry, J.; Merriam, HG. (1988) *Connectivity and connectedness: functional versus structural patterns in landscapes* **In:** Schreiber, K.F. [ed.] *Connectivity in Landscape Ecology. 2nd International Seminar of the International Association for Landscape Ecology* (Münster) Münsterische Geographische Arbeiten 29; 23-28

Baumeister, R.F. (1991) *Escaping the Self: alcoholism, spirituality, masochism and other flights from the burden of selfhood* (New York) Basic Books

Baumeister, R.F.; Heatherton, T.F.; Tice, D.M. (1994) *Losing Control: How and Why People Fail at Self-Regulation* (San Diego) Academic Press

Beckett, K.P.; Freer-Smith, P.H.; Taylor, G. (1998) *Urban woodlands: their role in reducing the effects of particulate pollution* (j) Environmental Pollution 99(3): 347-360

Beckett, K.P.; Freer-Smith, P.; Taylor, G. (2000) *Effective Tree Species for Local Air-Quality Management* (j) Journal of Aboriculture 26(1): 12-19

Begon, M.; Harper, J.L.; Townsend, C.R. (1996) *Ecology* (Oxford) Blackwell Science

Beier, P.; Noss, R.F. (1998) *Do Habitat Corridors provide Connectivity?* (j) Conservation Biology 12 (6): 1241-1252

Bekke, S. ter; Dalen, H. van; Henkens, K. (2005) *Emigratie van Nederlanders: geprikkeld door bevolkingsdruk* (j) Demos 21: 25-27

Bekker, G.J.; Hengel, B.H. van den; Bohemen, H.D. van; Sluijs, H. van der (1995) *Natuur over Wegen/Nature across Motorways* (Delft) RWS/DWW

Bekker, G.J. (2002) *Lopen op hoogte; hoe steken in bomen levende zoogdieren wegen over?* (j) Zoogdier 13(4)

Belle, F. van; Frenz, W. (2005) *Wildviaduct Terlet werkt* (j) Groen 61(10): 22/28

Belle, J. van; Barendregt, A.; Schot, P.P.; Wassen, M.J. (2006) *The effects of groundwater discharge, mowing and eutrophication on fen vegetation evaluated over half a century* (j) Applied Vegetation Science 9: 195-204

Bellrose, F.C.; Trudeau, N.M. (1988) *Wetlands and their relationship to migrating and winter populations of waterfowl* **In:** Hook, D.D. [ed.] *The ecology and management of wetlands* (Oregon) Timber Press; Vol. 1 Ecology of wetlands; 183-194

Beltman, B.; Broek, T. van den; Barendregt, A.; Bootsma, M.C.; Grootjans, A.P. (2001) *Rehabilitation of acidified and eutrophied fens in the Netherlands: Effects of hydrologic manipulation and liming* (j) Ecological Engineering 17: 21-31

Bengston, D. N.; Y.C.Youn (2006) *Urban Containment Policies and the Protection of Natural Areas: The Case of Seoul's Greenbelt* (j) Ecology and Society 11(1): art 3; http://www.ecologyandsociety.org/volll/issl/art3

Bennett, G.; Wit, P. (2001) *The development and application of ecological networks, a review of proposals, plans and programmes* IUCN/AIDEnvironment

Bennett, G. (2004) *Integrating biodiversity conservation and sustainable use, lessons learnt from ecological networks* (Gland) IUCN

Berg, A.E. van den ; Berg, M.M.H.E. van den (2001) *Van buiten word je beter; een essay over de relatie tussen natuur en gezondheid [The outside heals; an essay on the relationship between nature and health]* (Wageningen) Alterra

616

Berg, A.E. van den (2004) *De verborgen angst voor natuur: Essay en verslag debat [The hidden fear of nature]* (Den Haag) InnovatieNetwerk Groene Ruimte en Agrocluster

Berg, A.E. van den (2005) *Health impacts of healing environments: A review of the benefits of nature, daylight, fresh air and quiet in healthcare settings* (Groningen) Foundation 200 years University Hospital Groningen

Berg, A.E. van den; Ter Heijne, M. (2005) *Fear versus fascination: Emotional responses to natural threats* (j) Journal of Environmental Psychology 25 (3): 261-272

Berg, A. van den (2007) *Public Health* **In:** Jong, T.M. de; Dekker, J.N.M.; Posthoorn, R. [eds.] *Landscape ecology in the Dutch context: nature, town and infrastructure* (Zeist) KNNV-uitgeverij

Bergh, J. van den ; Barendregt, A.; Gilbert, A.; Herwijnen, M. van; Horssen, P. van; Kandelaars, P.; Lorenz, C. (2001) *Spatial economic-hydroecological modelling and evaluation of land use impacts in the Vecht wetlands area* (j) Environmental Modeling and Assessment 6: 87-100

Berris, L. (1997) *The importance of the ecoduct at Terlet for migrating mammals* **In:** Canters, K.J. [ed.] *International Conference Habitat Fragmentation, Infrastructure and the role of Ecological Engineering 17-21 September 1995.* (Delft) Ministerie van V&W/DWW; 418-420

Bervaes, J. (2004) *Respect voor het landschap van het verleden* (j) Landschap 21(1): 47-52

Beunen, R. (2006) *European nature conservation legislation and spatial planning: For the better or for the worse?* (j) Journal of Environmental Planning and Management 49: 605-619

Biesmeijer, J.C.; Roberts, S.P.M.; Reemer, M.; Ohlemüller, R.; Edwards, M.; Peeters, T.; Schaffers, A.P.; Potts, S.G.; Kleukers, R.; Thomas, C.D.; Settele, J.; Kunin (2006) *Parallel declines in pollinators and insect-pollinated plants in Britain and the Netherlands* (j) Science 313: 351-354

Bijhouwer, J.T.P. (1933) *Beplanting in den Wieringermeerpolder* (j) Tijdschr. voor Volksh. en Stedebouw 14: 158-160

Bijhouwer, J.T.P. (1948) *Een bodemkartering ten behoeve van de stedebouw* (j) Boor: 97-102

Bijhouwer, J.T.P. (1951) *Plantensociologie en Tuinarchitectuur* (j) De Boomkwekerij 6(8): 67

Bijkerk, C.; Mörzer Bruijns, M.F.; Vroom, M.J. (1971) *De landinrichting van het gebied Volthe - De Lutte - Verkenning, analyse en modellen* (Wageningen) Landbouwhogeschool

Bijlsma, M.P. (1995) *Natuurontwikkeling en vormgeving; de relatie tussen cultuur en natuur en de betekenis daarvan voor het ontwerpen en vormgeven van natuur* (Wageningen) IKC-Natuurbeheer; Studiereeks Bouwen aan een Levend Landschap 30.

Bink, F.A.; Maaskamp, F.I.M.; H.Siepel; Hengel., m.m.v. L.C. van den (1996) *Betekenis van wegbermen voor dagvlinders* (Wageningen en Delft) IBN-DLO en DWW/RWS

BirdLife International (2004) *Birds in the European Union: a status assessment* (Wageningen) BirdLife International

Biris, I.A.; Veen, P.H. (2005) *Inventory and Strategy for sustainable management and protection of virgin forests in Romania* (Bucharest/Utrecht)

Biris, I.A.; Marin, G.; Stoiculescu, C.; Maxim, I.; Verghelet, M.; Apostol, J. (2006) *Country Report Romania Protected Forest Areas in Europe - Analysis and Harmonisation* (Bucharest) PROFOR, Cost Action E27

Biró, E.; Bouwma, I.; Grobelnik, V. (2006) *Indicative Map of the Pan European Ecological Network in Southeastern Europe* (Tilburg) ECNC-European Centre for Nature Conservation ECNC Technical Report series

Bischoff, N.T.; Jongman, R.H.G (1993) *Development of rural areas in Europe: The claim for nature* (Den Haag) WRR Netherlands Scientific Council for Government Policy; Preliminary and background studies V79

Bixler, R.D.; Carlisle, C.L.; Hammitt, W.E.; Floyd, M.F. (1994) *Observed fears and discomforts among urban students on school field trips to wildland areas* (j) Journal of Environmental Education 26: 24-33

Black, J.M.; Prop, J.; Larsson, K. (2007) *Wild goose dilemmas. Population consequences of individual decisions in Barnacle Geese* (Groningen) Branta Press

Blaeu, Joan (1652) *Toneel der Steden* (Amsterdam);
http://www.nikhef.nl/~louk/LKW/PICS/blau-utrecht.jpg

617

Bloemmen, M.H.I.; Sluis, T. van der; Baveco, H.; Bouwma, I.M.; Greft, J. van der; Groot Bruinderink, G.; Higler, B.; Kuipers, H.; Lammertsma, D.; Ottburg, F.G.W.A.; Smits, N.; Swaay, C.; Wingerden, W.K.R.E. van (2004) *European corridors: example studies for the Pan-European Ecological Network: background document* (Wageningen) Alterra

Blumstein, D.T.; ; (2003) *Testing a key assumption of wildlife buffer zones: is flight initiation distance a species-specific trait?* (j) Biological Conservation 110[1]: 97-100

Bobbert, E. (1984) *Atlas van Nederland Deel 18 Ruimtelijke Ordening* (Den Haag) SDU; Stichting Wetenschappelijke Atlas van Nederland; http://avn.geog.uu.nl/

Bobbink, R.; Hornung, M.; Roelofs, J.G.M. (1998) *Essay review: The effects of air-borne nitrogen pollutants on species diversity in natural and semi-natural European vegetation* (j) Journal of Ecology 86: 717-738

Bobbink, R.; Ashmore, M.; Braun, S.; Flückiger, W.; Wyngaert, I.J.J. van den (2002) *Empirical Nitrogen Critical Loads for Natural and Semi-natural Ecosystems: 2002 Update* (Berne, Switzerland) The expert workshop on empirical critical loads for nitrogen on [semi-]natural ecosystems

Boekhorst, J.K.M. te; Boekhorst-van Maren, E.N. te (1971) *Tussen Heuvelrug en Kromme Rijn* (Wageningen) Kring Midden Utrecht, Deelrapport 3

Boekhorst, J.K.M.; Coeterier, J.F.; Harms, W.B.; Vos, W. [eds.] (1983) *Omgaan met het landschap - Vriendenboek voor Pieter Tideman bij zijn afscheid* (Wageningen) Rijksinstituut voor onderzoek in de Bos- en Landschapsbouw 'De Dorschkamp'

Boekhorst, J. te (2006) *Landschapsarchitectuur en onderzoek - een korte geschiedenis van de landschapsarchitectuur binnen DLO* (Wageningen) Alterra

Boekhorst, J.K.M. te; Farjon, H.; Harms, W.B.; Leeuwen, R.J. van; Nooren, M.J.; Schouten, M.G.C.; Vos, W.; Zeeuw, P.H. de (1977) *landschapsbouwkundig onderzoek voor het RW2 - oliewinningsproject te Schoonebeek* (Wageningen) RBL De Dorschkamp

Boelens, L. (1998) *Nature à diffuse vitesse. De verschillende werkelijkheidsbereiken van de natuur* In: Feddes, F.; Herngreen, R.; Jansen, Sj.; Leeuwen, R. van; Sijmons, D. [ed.] *Oorden van onthouding - Nieuwe natuur in verstedelijkend Nederland* (Rotterdam) NAI; 118-122

Boelens, L.; Hartman, A. (1998) *Ruimtelijke ontwikkelingspolitiek* (Den Haag) WRR

Boelens, L. (2005) *Fluviologie, een nieuwe benadering van ruimtelijke ordening* (Utrecht) Universiteit Utrecht

Boersema, J.J.,Copius Peereboom, J.W. et al. (1991) *Basisboek Milieukunde* (Meppel) Boom

Boersema, J.; Kwakernaak, C. (1993) *Integratie: nuttig mits helder en groen Een conceptuele analyse* (j) Landschap 10(3): 3-20

Boerwinkel, H. W. J. (1994) *Identiteit in de landschapsplanning* (j) Landschap 11(4): 33-42

Boeye, D. (1992) *Hydrology, hydrochemistry and ecology of a groundwater fed mire* (Antwerpen) PhD Thesis University of Antwerp

Bohemen, H.D. van; Padmos, C.; Vries, J.G. de (1994) *Versnippering-ontsnippering; beleid en onderzoek bij Verkeer en Waterstaat* (j) Landschap 11(3): 15-26

Bohemen, H.D. van (1995) *Stadsecologie* (j) Natura 1: 14-15

Bohemen, H.D. van (1996) *Mitigation and compensation of habitat fragmentation caused by roads; strategy, objectives, and practical measures* (Washington) Transportation Research Record 1475

Bohemen, H.D. van (1998) *Habitat fragmentation, infrastructure and ecological engineering* (j) Ecological Engineering 11: 199-207

Bohemen, H.D. van (2001) *Sustainable road infrastructure; ecological design, construction and management of the road infrastructure* In: Hendriks, Ch.F. [ed.] *Sustainable Construction* (Boxtel) Aeneas Technical Publishers

Bohemen, H.D. van (2002) *Infrastructure, ecology and art* (j) Landscape and Urban Planning 59: 187-201

Bohemen, H.D. van; Janssen van der Laak, W. (2003) *Water, soil and air pollution and motorways* (j) Environmental Management 31(1): 50-68

Bohemen, H.D. van (2004) *Ecological Engineering and Civil Engineering Works - A Practical Set of Ecological Engineering Principles for Road Infrastructure and Coastal Management* (Delft) PhD Thesis TUD

Bohemen, H.D. van (2005) *Ecological Engineering; Bridging between ecology and civil engineering* (Boxtel) Aeneas Technical Publicers

Bohemen, H.D. van (2007) *Road crossing connections* In: Jong, T.M. de; Dekker, J.N.M.; Posthoorn, R. [eds.] *Landscape ecology in the Dutch context: nature, town and infrastructure* (Zeist) KNNV-uitgeverij

Bohemen, H.D. van (2007) *Roadside verges; integration of road infrastructure in the landscape* In: Jong, T.M. de; Dekker, J.N.M.; Posthoorn, R. [eds.] *Landscape ecology in the Dutch context: nature, town and infrastructure* (Zeist) KNNV-uitgeverij

Bolck, M.; Togni, G. de; Sluis, Th. van der; Jongman, R.H.G. (2004) *From models to reality: design and implementation process* In: Jongman, R.H.G.; Pungetti, G. [eds.] *Ecological Networks and Greenways, Concept, design and implementation* (Cambridge) Cambridge Studies in Landscape Ecology

Bolger, D.T.; Scott, T.A.; Rotenberry, J.T. (1997) *Breeding bird abundance in an urbanizing landscape in coastal southern California* (j) Conservation Biology 11[2]: 406-421

Bolhuis, P. van (1995) *Natuur en de toekomst van het verleden* (j) Landschap 12 (5): 51-55

Bolhuis, P. van (2004) *Het verzonnen land - The invented land* (Wageningen) Blauwdruk

Bolman, J. (1976) *Wilde planten in en bij Amsterdam* (Zutphen) Thieme

Bond, B. (2003) *Hydrology and ecology meet - and the meeting is good* (j) Hydrological Processes 17: 2087-2089

Bonnes, M.; Carrus, G.; Bonaiuto, M.; Fornara, F.; Passafaro, P (2004) *Inhabitants' environmental perceptions in the city of Rome within the framework for urban biosphere reserves of the UNESCO programme on man and biosphere* (j) Annals of the New York Academy of Sciences 1023: 175-186

Bootsma, M.C.; Wassen, M.J. (1996) *Water quality of fen vegetation types in three European lowland mires* (j) Vegetatio 127: 173-189.

Bouwma, I.M.; Jongman, R.H.G.; Butovsky, R.O. (2002) *Indicative map of the Pan-European Ecological Network for Central and Eastern Europe* Technical background document (Tilburg/Budapest) ECNC; Technical Report Series

Bouwmeester, H. [ed.] (2003) *Met groen meer stad Nieuwe impulsen voor stedelijk groen* (Den Haag) Ministerie van Volkshuisvesting, Ruimtelijke Ordening en Milieubeheer VROM; Claudia Bouwens, SEV Realisatie, Karien van Dullemen

Brackenbury, J. (1984) *Physiological responses of birds to flight and running* (j) Biol. Rev. 59: 559-575

Brandjes, G.J.; Veenbaas, G. (1998) *Het gebruik van faunapassages langs watergangen onder rijkswegen in Nederland Een oriënterend onderzoek* (Delft) Dienst Weg- en Waterbouwkunde; DWW-ontsnipperingsreeks deel 36

Brandjes, G.J.; Veenbaas, G.; Tulp, I.; Poot, M.J.M. (2001) *Het gebruik van faunapassages langs watergangen onder rijkswegen in Nederland Resultaten van een experimenteel onderzoek* (Delft) DWW; DWW-ontsnipperingsreeks, deel 40

Brandjes, G.J.; Eekelen, R. van; Krijgsveld, K.; Smit, G.F.J. (2002) *Het gebruik van faunabuizen onder rijkswegen; resultaten literatuur- en veldonderzoek* (Delft) Bureau Waardenburg/DWW; DWW/Ontsnipperingsreeks deel 43

Brandt, J.; Vejre, H. (2004) *Multifunctional Landscape - Volume I - Theory, Values and History* (Southampton) Witpress

Braun-Blanquet, J. (1964) *Pflanzensoziologie* (Wien) Springer-verlag

Breedveld, K.; (2006) *De tijd als spiegel - Hoe Nederlanders hun tijd besteden* (Den Haag) Sociaal Cultureel Planbureau

Brenninkmeijer, B.; Dekker, J. (2001) *Verbinden verbindingszones ook natuuronderzoek en natuurbeleid?* (j) DLN 102: 265-269

Breuste, J.; Feldmann, H.; Uhlmann, O. [eds.] (1998) *Urban ecology* (Berlin) Springer

Brockerhoff, M.P. (2000) *An Urbanizing world* (j) Population Bulletin of the Population Reference Bureau 55[3]: 3-44

Bruin, D. de; Hamhuis, D.; Nieuwenhuize, L. van; Overmars, W.; Sijmons, D.; Vera, F. (1987) *Ooivaar De toekomst van het rivierengebied* (Arnhem) Stichting Gelderse Milieufederatie

Bruin, D. de (2001) *WatErStaat* (Den Haag) RWS, Directie Kennis

Bruin, T. H. M. d.; Burger, J. E. (2004) *Ruimte voor de ommetjesmaker en zijn habitat* (Amsterdam) Uitgeverij Op Lemen Voeten

619

Bruin, D. de; Jong, T.M. de (2007) *Civil engineering of a wet land* In: Jong, T.M. de; Dekker,
    J.N.M.; Posthoorn, R. [eds.] *Landscape ecology in the Dutch context: nature, town
    and infrastructure* (Zeist) KNNV-uitgeverij
Brumm, H.; Todt, D. (2002) *Noise-dependent song amplitude regulation in a territorial song-
    bird* (j) Journal of Animal Behaviour 63: 891-897
Brussaard, L.; Weijden, W. Van der (1980) *Biogeografie van eilanden* (j) Intermediair 16: 9-17
Bryant, D.M.; Westerterp, K.R. (1980) *Energetics of foraging and free existence in birds* In:
    Nöhring, R. von [ed.] *Acta XVII Congressus Internationalis Ornithologici* (Berlin) Ver-
    lag der Deutschen Ornithologen Gesellschaft; 292-299
Buchwald, K.; Engelhardt (1980) *Handbuch für Planung und Schutz der Umwelt Die Bewer-
    tung und Planung der Umwelt* (München/Wien/Zürich) BLV Verlagsgesellschaft
Buijs, A.E. (2000) *Natuurbeelden van de Nederlandse bevolking* (Wageningen) Alterra
Buijs, A.; Pedroli, B.; Luginbühl, Y. (2006) *From Hiking Through Farmland to Farming in a
    Leisure Landscape: Changing Social Perceptions of the European Landscape* (j)
    Landscape Ecology 21(3): 375-389
Buijs, J. (1995) *Algemene natuur Een realistische referentie* (j) Landschap 12 (2): 51-55
Buijs, S. (1990) *De Randstad op weg naar 2015 Studienota verstedelijking Randstad* (Den
    Haag) Ministerie van VROM
Buijs, S. (1996) *Randstad en Groene Hart De Groene Wereldstad* (Den Haag) SDU Ministerie
    van VROM
Buijs, G.; Kaars, S.; Trommelen, J. (2005) *Gifpolder Volgermeer, van veen tot veen* (Wormer-
    veer) Stichting Uitgeverij Noord-Holland
Bureau Stroming bv (2006) *Natuurlijke klimaatbuffers; Adaptatie aan klimaatverandering;
    Wetlands als waarborg* (Nijmegen) Vereniging Natuurmonumenten; Vogelbescher-
    ming Nederland; Staatsbosbeheer; ARK Natuurontwikkeling; Waddenvereniging
    http://www.vogelbescherming.nl/documents/pdf-files/hier_klimaatbuffers.pdf
Burger, J. (2001) *The behavioral response of basking Northern water [Nerodia sipedon] and
    Eastern garter [Thamnophis sirtalis] snakes to pedestrians in a New Jersey park* (j)
    Urban Ecosystems 5[2]: 119-129
Burgi, M.; Hersperger, A.M.; Schneeberger, N. (2004) *Driving forces of landscape change -
    current and new directions* (j) Landscape Ecology 19: 857-868
Burkhardt, R.; Jäger, E.; Mirbach, E.; Rothenberger, A.; Schwab, G. (1995) *Planung Vernetz-
    ter Biotopsysteme: design of the habitat network of Rheinland-Pfalz, Germany* (j)
    Landschap 12(3): 99-110
Burm, P.; Haartsen, A. (2003) *Boerenland als natuur; verhalen over historisch beheer van
    kleine landschapselementen* (Utrecht) Matrijs
Burny, J. (1999) *Bijdrage tot de historische ecologie van de Limburgse Kempen [1910-1950];
    tweehonderd gesprekken samengevat* (j) Publicaties van het Natuurhistorisch Ge-
    nootschap in Limburg 42(1)
Buuren, M. van; Dorp, D. van; Ertsen, D.; Hoogveld, J.; Molenaar, M.J.; Wassen, M.J.; [eds.]
    (1999) *Eutrofiëring van grondwaterafhankelijke ecosystemen via vervuild grondwa-
    ter* (j) Landschap Special Issue 16
Cammen, H. v. d.; Klerk, L. A. de (1993, 1986) *Ruimtelijke ordening* (Utrecht) Spectrum Aula
Canters, K.J.; Piepers, A.A.G.; Hendriks-Heerma, D. (eds.) (1997) *Habitat fragmentation and
    Infrastructure* (Delft) V&W/DWW-Delft; International Conference on Habitat frag-
    mentation, infrastructure and the role of ecological engineering 17-21 September
    1995
Canters, K.J.; Vos, J.G.; Schimmel, H.L.; Bruyn, G.J. de; Schouten, M.G.C. (2000) *Kansen
    voor natuur in de stad* (j) De Levende Natuur 101: 227-228
Carey, P.D.; Preston, C.D.; Hill, M.O.; Usher, M.B.; Wright, S.M. (1995) *An environmentally
    defined biogeographical zonation of Scotland designed to reflect species distribu-
    tions* (j) Journal of Ecology 83: 833-845
Carey, C.W. [ed.] (1996) *Avian Energetics and Nutritional Ecology* (New York) Chapman &
    Hall
Carson, R. (1962) *Silent spring* (Greenwich) Fawcett
Castree, N. (2005) *Nature* (London/New York) Routledge; Key Concepts in Geography
Castro, G.; Stoyan, N.; Myers, J.P. (1989) *Assimilation efficiency in birds: a function of taxon
    or food type?* (j) Comp. Biochem. Physiol. 92A: 271-278

620

CBS (2006) *Stedelijke omgeving blijft zich uitbreiden* (j) CBS Webmagazine

Chapin III, S.F.; Zavaleta, E.S.; Eviner, V.T.; Naylor, R.L.; Vitousek, P.M.; Reynolds, H.L.; Hooper, D.U.; Lavorel, S.; Sala, O.E.; Hobbie, S.E.; Mack, M.C.; Diaz, S. (2000) *Consequences of changing biodiversity* (j) Nature 405: 234-242

Charnov, E.L. (1976) *Optimal foraging: the marginal value theorem* (j) Theor. Popul. Biol. 9: 129-136

Ciais, Ph.; et al. [33 authors] (2005) *Europe-wide reduction in primary productivity caused by the heat and drought in 2003* (j) Nature 437: 529-533

CityDisc (2001) *Street guide CDRom* (Den Haag) CityDisk - Topografische Dienst

Coesèl, M. (2001) *Laten wij Amsterdammers het voorbeeld geven* In: Halm, H. van; Timmermans, G.; Koningen, H.; Bouman, R.; Melchers, M.; Kazus, J. [eds.] *De wilde stad, 100 jaar natuur voor Amsterdam, een eeuw Koninklijke Nederlandse Natuurhistorische Vereniging, afdeling Amsterdam 1901-2001* (Utrecht) KNNV; 12-18

Cohen, S.; Herbert, T.B. (1996) *Health psychology: psychological factors and physical disease from the perspective of human psychoneuroimmunology* (j) Annu. Rev. Psychol. 47: 113-142

Collinge, S.K.; Prudic, K.L.; Oliver, J.C. (2003) *Effects of local habitat characteristics and landscape context on grassland butterfly diversity* (j) Conservation Biology 17[1]: 178-187

Collins, J. P.; Kinzig, A.; Grimm, N. B.; Fagan, W. F.; D. Hope, J. Wu; Borer, E. T. (2000) *A new urban ecology* (j) American Scientist 88: 416-425

European Commission (2000) *Managing Natura 2000 Sites The provisions of Article 6 of the 'Habitats' Directive 92/43/EEC* (Luxemburg) European Communities

Cormont, A.; Radix, J.; Segers, M. (2004) *The spatial distribution of biodiversity over the Netherlands - the role of landscape age* (Utrecht) MSc Thesis Utrecht University NRM

Cosgrove, D.E. (1984) *Social formation and symbolic landscape* (London) Croom Helm

Cramer, J.; Kuiper, M.; ., C. Vos (1984) *Landschapsecologie : een nieuwe onderzoeksrichting ?* (j) Landschap 1(3): 176-184

Cramer, J.; Wulp, W. van der (1989) *De ontwikkeling van de Nederlandse landschapsecologie als interdisciplinair wetenschapsgebied in internationaal verband* (j) Landschap 6(1): 47-65

CRM; VRO (1981) *Structuurschema Natuur- en Landschapsbehoud* (Den Haag) Ministerie van Cultuur, Recreatie en Maatschappelijk werk; Ministerie van Volkshuisvesting en Ruimtelijke Ordening

Cronon, W. (1995) *The trouble with wilderness; or, getting back to the wrong nature* In: Cronon, W. [ed.] *Uncommon ground; toward reinventing nature.* (New York/London) Norton; 69-90

Croonen, R.J.; Hazendonk, N.; Wiel, K. van der (1995) *Ecologie-inclusieve planning* In: LNV [ed.] *Visie stadslandschappen - Discussienota* (Den Haag) LNV

Cuperus, R. (2005) *Ecological compensation of highway impacts Negotiated trade-off or no-net-loss?* (Leiden) PhD Thesis Universiteit Leiden

CVTO (2005) *Continu VrijeTijdsOnderzoek 2004-2005* (Amsterdam) NIPO/CVTO

Dahlström, A.; Cousins, S.A.O.; Eriksson, O. (2006) *The history [1620-2003] of land use, people and livestock, and the relationship to present plant species diversity in a rural landscape in Sweden* (j) Environment and History 12(2): 191-212

Dam, A. van (2002) *De Blauwe Kamer - nieuwe natuur langs de Nederrijn* (Abcoude) Uniepers

Dantzer, R. (2001) *Can we understand the brain and coping without considering the immune system?* In: Broom, D.M. [ed.] *Coping with challenge: Welfare in animals including humans* (Berlijn) Dahlem University Press; 7; 102-110

Dauvellier, P.L.; Littel, A. (1977) *Toepassing van het Globaal Ecologisch Model en de Landelijke Milieukartering in de ruimtelijke planning* (j) Mededelingen van de WLO 4 (2): 3-9

Davenport, John; Davenport, Julia L. [eds.] (2006) *The ecology of transportation: Managing mobility for the environment* (New York/Berlin) Springer; Environmental Pollution , Vol. 10

Davies, P.; Robb, J.G.; Ladbrook, D. (2005) *Woodland clearance in the Mesolithic: the social aspects* (j) Antiquity 79: 280-288

621

References

Deci, E.L.; Ryan, R.M. (1985) *Intrinsic motivation and self-determination in human behaviour* (New York.) Plenum

Deelstra, T. (1998) *Towards ecological sustainable cities* In: Breuste, J.; Feldmann, H.; Uhlmann, O. [eds.] *Urban ecology* (Berlijn) Springer; 17-22

Dekker, J. (1993) *De ontdekking van het kultuurlandschap; de bijdrage van de Werkgroep voor de Cultuurlandschappen van de Contactcommissie voor Natuur- en Landschapsbescherming 1944-1950; een voorstudie* (Utrecht) Rijksuniversiteit Utrecht, Vakgroep Natuurwetenschap en Samenleving

Dekker, J. (2002) *Dynamiek in de Nederlandse natuurbescherming* (Utrecht) PhD Thesis University of Utrecht

Dekker, J.N.M. (2007) *Urban ecology, an exploration* In: Jong, T.M. de; Dekker, J.N.M.; Posthoorn, R. [eds.] *Landscape ecology in the Dutch context: nature, town and infrastructure* (Zeist) KNNV-uitgeverij

Dekker, J.N.M. (2007) *Nature, editorial introduction* In: Jong, T.M. de; Dekker, J.N.M.; Posthoorn, R. [eds.] *Landscape ecology in the Dutch context: nature, town and infrastructure* (Zeist) KNNV-uitgeverij

Dekker, S.C.; Rietkerk, M.; Bierkens, M.F.P. (2007) *Coupling microscale vegetation - soil water and macroscale vegetation - precipitation feedbacks in semiarid ecosystems* (j) Global Change Biology 13: 671-678

Delft, B. van; Kemmers, R.; Waal, R. de (2002) *Ecologische typering van bodems onder korte vegetaties - Het humusprofiel als graadmeter voor standplaatsontwikkeling* (j) Landschap 19(3): 153-164

Deltametropool (1998) *Verklaring* (Rotterdam) Vereniging Deltametropool

Deltametropool (2000) *Actieplan 2000* (Delft) Voorlopig Werkbestuur Vereniging Deltametropool

Deltametropool (2001) *Waterrijk* (Delft) Vereniging Deltametropool

Denevan, W.M. (1992) *The pristine myth: the landscape of the Americas in 1492* (j) Annals of the Association of American Geographers 82: 369-385

Denters, T. (1994) *Het urbaan district; een eigen district voor de stadsflora* (j) Natura 91: 65-66

Denters, T.; Ruesink, R.; Vreeken, B. (1994) *Van muurbloem tot straatmadelief* (Utrecht) KNNV-uitgeverij

Denters, T.; Vreeken, B. (1998) *Flora-atlas van de regio Amsterdam* (Haarlem) Provincie Noord-Holland

Denters, T. (1999) *De flora van het urbaan district* (j) Gorteria 254: 65-76

Denters, T. (2004) *Stadsplanten Veldgids voor de stad* ('s-Graveland) Fontaine

Denters, T. (2005) *Beschermingsplan flora; beschermde en kwetsbare, bijzondere plantensoorten in Amsterdam* (Amsterdam) Adviesbureau EcoStad

Denters, T. (2006) *De ecologische identiteit van de stad* (j) Landschap 23(3): 127-132

Denters, T. (2007) *The Urban District, a biogeographical acquisition* In: Jong, T.M. de; Dekker, J.N.M.; Posthoorn, R. [eds.] *Landscape ecology in the Dutch context: nature, town and infrastructure* (Zeist) KNNV-uitgeverij

Derckx, H. (1995) *Wit gebied of bleek beleid?* (j) Landschap 12 (1): 35-40

Deyn, G.B. de; Raaijmakers, C.E.; Zoomer, H.R.; Berg, M.P.; Ruiter, P.C. de; Verhoef, H.A.; Bezemer, T.M.; Putten, W.H. Van der (2003) *Soil invertebrate fauna enhances grassland succession and diversity* (j) Nature 422: 711-713

Dias, P.C. (1996) *Sources and sinks in population biology* (j) Ecology and Evolution 11(8): 326-331

Diggelen, R. van (1998) *Moving Gradients; Assessing Restoration Prospects of Degraded Brook Valleys* (Groningen) PhD Thesis University of Groningen

Diggelen, R. van; Grootjans, A.P.; Harris, J.A. (2001) *Ecological Restoration: State of the Art or State of the Science?* (j) Restoration Ecology 9: 115-118

Dijk, A. van (2004) *Ecohydrology: it's all in the game?* (j) Hydrological Processes 18: 3683-3686

Dirkmaat, J. (2005) *Nederland weer mooi; op weg naar een natuurrijk en idyllisch landschap* (Den Haag) ANWB i.s.m. Stichting Nederlands Cultuurlandschap

Dirkx, G.H.P.; Hommel, P.W.F.M.; Vervloet, J.J. (1992) *Historische ecologie Een overzicht van achtergronden en mogelijke toepassingen in Nederland* (j) Landschap 9(1): 39-51

Donadieu, P.; Rumelhart, M. (1991) *Ecologie et paysage: la nature comme projet* (j) Paysage et Aménagement 17: 20-24

Dooren, N. van (2001) *De buigzaamheid van het landschap (Sanering Volgermeerpolder)* (j) Blauwe Kamer, tijdschrift voor landschapsarchitectuur en stedenbouw 2001(3)

Dorenbosch, M.; Krekels, R. (2000) *Medegebruik door fauna van tien viadukten over auto-snelwegen; uitgangssituatie 2000* (Delft/Nijmegen) RWS/DWW/Bureau Natuurba-lans

Dorp, D. van; Canters, K.; J., Kalkhoven; Laan, P. (1999) *Landschapsecologie: Natuur en landschap in een veranderende samenleving* (Amsterdam) Boom

Drent, R.H.; Daan., S. (1980) *The prudent parent Energetic adjustments in avian breeding* (j) Ardea 68: 225-252

Dûfrene, M.; Legendre, P. (1997) *Species assemblages and indicator species: the need for a flexible asymetrical approach* (j) Ecological Monographs 67: 345-366

Duijvestein, C.A.J.; Jong, T.M. de; et al. (1993) *Begrippen rond bouwen en milieu* (Rotterdam) Stichting Bouwresearch

Duinen, G.A. van; Brock, A.M.T.; Kuper, J.T.; Peeters, T.M.J.; Smits, M.J.A.; Verberk, W.C.E.P.; Esselink, H. (2002) *Important keys to successful restoration of character-istic aquatic macroinvertebrate fauna of raised bogs* Proceedings *Proceedings of the International Peat Symposium, 3-6 september: pg 292-302* (Pärnu, Estonia)

During, R.; Specken, B. (1995) *Schoonwaterplan, helder water voor de stad Utrecht en de Kromme Rijn* (j) Landschap 12(2): 33-45

Duuren, L. van et al. (2003) *Natuurcompendium 2003 - Natuur in cijfers* (Voorburg) CBS

Eagleson, P.S. (2002) *Ecohydrology: Darwinian expression of vegetation form and function* (Cambridge) Cambridge University Press

ECLAS (2004) *A critical light on landscape architecture* Proceedings (Ås, Norway) ECLAS Dept. of Landscape architecture and Spatial Planning

ECNC (2004) *The Pan-European Ecological Network and People Background paper as well as "Conclusions and Recommendations"* (The Hague) ECNC/LNV, The 2004 semi-nar on "People and PEEN"

ECTP (2003) *The New Charter of Athens 2003* (Lisbon) European Council of Town Planners; Conseil Européen des Urbanistes

Eekelen, R. van; Smit, G.F.J. (2000) *Het gebruik door dieren van kunstwerken in de A1 op de Veluwe* (Culemburg) Bureau Waardenburg

Eekelen, R. van; Smit, G.F.J. (2002) *Gebruik van fauna passages onder rijkswegen in Zuid-Holland* (Culemburg) Bureau Waardenburg/RWS-Zuid-Holland

Eekeren, N. van; Iepema, G.; Smeding, F.W. (2007) *Natuurherstel in grasland door klaver en kalibemesting* (j) De Levende Natuur 108(1): 29-32

Eerden, M.R. van (1997) *Patchwork. Patch use, habitat exploitation and carrying capacity for waterbirds in Dutch freshwater wetlands* (Lelystad) PhD Thesis Groningen Univer-sity; Rijkswaterstaat Van Zee tot Land 65

Eerden, M.R. van (2007) *Habitat Scale And Quality Of Foraging Area determining carrying capacity of Dutch wetlands: lessons learned from a changing delta* **In:** Jong, T.M. de; Dekker, J.N.M.; Posthoorn, R. [eds.] *Landscape ecology in the Dutch context: nature, town and infrastructure* (Zeist) KNNV-uitgeverij

Eggink, G.; Aarts, H.; Wiertz, J.; Hesselink, J. (2002) *Nationale Natuurverkenning 2 - 2000-2030* (Bilthoven) Rijksinsituut voor Volksgezondheid en Milieuhygiëne RIVM

Egmond, P.M. van; Koeijer, T.J. de (2005) *Van aankoop naar beheer - Verkenning kansrijk-heid omslag natuurbeleid I* (Bilthoven) Milieu- en Natuurplanbureau

Eisenbeis, G.; Hanel, A. (2003) *Artificial night lighting and insects with remarks to increasing light pollution in GelDlany* (Melbourne) Presentation congres Comparative Ecology of Cities and Towns, Melbourne, Australia, 19-22 July 2003

Elliott, R. (1997) *Faking nature; the ethics of environmental restoration* (London) Routledge; Environmental Philosophies

Emlen, J.T. (1974) *An urban bird community in Tucson, Arizona: derivation, structure, regula-tion* (j) The Condor 76: 184-197

Engel, H. (2005) *Randstad Holland in kaart, OverHolland 2* **In:** Engel, H.; Pané, I.; Bogt, O. van der [eds.] *Atlas Randstad Holland* (Delft) Technische Universiteit Delft, Faculteit Bouwkunde; 23-73

Engelen, G.B. (1981) *A system approach to groundwater quality* **In:** Van Duijvenbooden, R.W.; et al. [eds.] *Studies in Environmental Sciences* (Amsterdam) Elsevier; 17; 1-25

Engelen, G.B.; Jones, G.P. (1986) *Developments in the analysis of groundwater flow systems* (Wallingford) IAHS Publications 163

Ertsen, A.C.D.; Frens, J.W.; Nieuwenhuis, W.; Wassen, M.J. (1995) *An approach to modelling the relationship between plant species and site conditions in terrestrial ecosystems* (j) Landscape and Urban Planning 31: 143-151

Etienne, R.S.; Heesterbeek, J.A.P. (2000) *On optimal size and number of reserves for meta-population persistence* (j) Journal for Theoretical Biology 203: 33-50

Etteger, R. van (1999) *Het Kunstwerk van het millennium: het Nederlandse landschap* (j) Landschap 16(2): 123-125

European Commission (2002) *Interpretation Manual of European Union Habitats - EUR25* (Brussel) adopted by the Habitats Committee on 4 October 1999 and consolidated with the new and amended habitat types for the 10 accession countries as adopted by the Habitats Committee on 14 March 2002

European Commission (2002) *Atlantic Region Conclusions on representativity within pSCI of habitat types and species* (Den Haag) Seminar June 2002, Doc. Atl./C/ rev.2, July 2002

European Commission (2004) *Commission Decision 2004/813/EC of 7 December 2004 adopting, pursuant to Directive 92/43/EEC, the list of sites of Community importance for the Atlantic biogeographical region* Oj EC L 387

European Environment Agency (2005) *The European environment - State and outlook 2005* (Luxemburg) Office for Official Publications of the European Communities

Europees Milieuagentschap (2003) *Het milieu in Europa: de derde balans [Samenvatting]* (Luxemburg) Bureau voor Officiële Publicaties van de Europese Gemeenschappen

Europees Milieuagentschap (2004) *EMA-signalen 2004* (Luxemburg) Bureau voor Officiële Publicaties van de Europese Gemeenschappen

Evers, F.; Winsemius, P. (2003) *Rood voor Groen, van filosofie naar resultaat* (Nieuwegein/Tilburg) Amstelland MDC

Everts, F.H.; Vries, N.P.J. de (1991) *De vegetatieontwikkeling van beekdalen; een landschapsecologische studie van enkele Drentse beekdalen* (Groningen) PhD Thesis University of Groningen

Eyck, A.E. van; Parin, P.; Morgenthaler, F. (1968) *Ecology in Design / Kaleidoscope of the mind / Miracle of Moderation / Image of Ourselves* (j) Via 1: 129

Faludi, A. (2002) *European Spatial Planning* (Cambridge Massachusetts) Lincoln Institute of Land Policy

Farina, A. (2001) *Landscape Ecology in Action* (Dordrecht/Boston/London) Kluwer Academic Publishers

Farjon, J.M.J. (2005) *Pakt de Nota Ruimte de verrommeling aan?* (j) Landschap 22(2): 83-91

Feddes, F.; Herngreen, R.; Jansen, Sj.; Leeuwen, R. van; Sijmons, D. [eds.] (1998) *Oorden van onthouding - Nieuwe natuur in verstedelijkend Nederland* (Rotterdam) NAI Uitgeverij

Feddes, F. [ed.] (1999) *Nota Belvedère Beleidsnota over de relatie cultuurhistorie en ruimtelijke inrichting* (Den Haag) VNG Uitgeverij; Ministeries van OCW, LNV, VROM, V&W; Belvedere

Fernandez-Juricic, E. (2000) *Bird community composition patterns in urban parks of Madrid: the role of age, size and isolation* (j) Ecological Research 15[4]: 373-383

Fernandez-Juricic, E. (2000) *Local and regional effects of human disturbance on forest birds in a fragmented landscape* (j) Condor 102: 247-255

Fernandez-Juricic, E.; Telleria, J.L. (2000) *Effects of human disturbance on spatial and temporal feeding patterns of Blackbird Turdus merula in urban parks in Madrid, Spain* (j) Bird Study 47: 13-21

Fernandez-Juricic, E. (2001) *Avian spatial segregation at edges and interiors of urban parks in Madrid, Spain* (j) Biodiversity and Conservation 10: 1303-1316

Fernandez-Juricic, E.; Jimenez, M.D.; Lucas, E. (2001) *Bird tolerance to human disturbance in urban parks of Madrid* **In:** Marzluff, J.M.; Bowman, R.; Donnelly, R. [eds.] *Avian*

Ecology and Conservation in an Urbanizing World. (Dordrecht) Kluwer Academic Publishers

Fernandez-Juricic, E.; Jokimaki, J. (2001) *A habitat island approach to conserving birds in urban landscapes: case studies from southern and northern Europe* (j) Biodiversity and Conservation 10: 2023-2043

Fernandez-Juricic, E. (2002) *Can human disturbance promote nestedness? A case study with breeding birds in urban habitat fragments* (j) Oecologia 131: 269-278

Finlayson, M.; Moser, M. [eds.] (1991) *Wetlands* (Slimbridge) International Waterfowl and Wetlands Research Bureau

Florida Greenways Commission (1994) *Creating a Statewide Greenways system, for people-for wildlifefor Florida* (Tallahassee)

Fluit, N. van der; Cuperus, R.; Canters, K.J. (1990) *Mitigerende en compenserende maatregelen aan het hoofdwegennet voor het bevorderen van natuurwaarden, inclusief een nadere uitwerking voor drie regio's in Nederland* (Leiden) CML

Flyvbjerg, B.; Bruzelius, N.; Rothengatter, W. (2003) *Megaprojects and Risk, an anatomy of ambition* (Cambridge, UK) Cambridge University Press

Foppen, R.; J.Graveland; Jong, M. de; A.Beintema (1998) *Naar levensvatbare populaties moerasvogels* (Wageningen) IBN-DLO; IBN-rapport 393

Foppen, R.; Kleunen, A van; Loos, W.B.; Nienhuis, J.; Sierdsema, H. (2002) *Broedvogels en de invloed van hoofdwegen, een nationaal perspectief; een analyse van de gevolgen van wegverkeer voor broedvogels aan de hand van landelijke aantals- en verspreidingsgegevens* (Delft) SOVON i.o.v. RWS/DWW

Forman, R.T.T. (1995) *Landmosaics: The ecology of landscapes and regions* (Cambridge) Cambridge University Press

Forman, R.T.T.; Reineking, B.; Hersperger, A.M. (2002) *Road traffic and nearby grassland bird patterns in a suburbanizing landscape* (j) Environmental Management 29[6]: 782-800

Forman, R.T.T.; Sperling, D.; Bisonnette, J.A.; Clevenger, A.P.; Cuttshall, C.D.; Dale, V.H.; Fahring, L.; France, R.; Goldman, C.R.; Heanue, K.; Jones, J.A.; Swanson, F.J.; Turrentine, T.; Winter, T.C. (2003) *Road Ecology* (Washington, D.C.) Island Press

Fox, R.J. (1999) *Enhancing spiritual experience in adventure programs* In: Miles, J.E.; Priest, S. [eds.] *Adventure Programming* (Pennsylvania) Venture, State College; 455-461

Franken, E. (1955) *Der Beginn der Forsytienblute in Hamburg 1955* (j) Met.Rdsch. 8: 113-114

Frederickson, L.M.; Anderson, D.H. (1999) *A qualitative exploration of the wilderness experience as a source of spiritual inspiration* (j) Journal of Environmental Psychology 19: 21-39

Frerichs, R. (2004) *Gezondheid en natuur; Een onderzoek naar de relatie tussen gezondheid en natuur [Health and nature; a research into the relation between health and nature]* ('s Gravelande) Vereniging Natuurmonumenten

Friedmann, J.; Miller, J. (1965) *The urban field* (j) Journal of the American institute of planners 31[4]: 3u-320

Frieling, D. H.; Venema, H. [eds.] (1998) *Delta Metropool Verklaring van de wethouders Ruimtelijke Ordening van Amsterdam, Rotterdam, Den Haag en Utrecht over de toekomstige verstedelijking in Nederland* (Delft) Deltametropool

Frieling, D. H. (2002) *Design in strategy* In: Jong, T. M. de; Voordt, D. J. M. van der [eds.] *Ways to study and research urban, architectural and technical design* (Delft) DUP Science; 491-500

Frieling, D.; Schubert, L.L. [eds.] (2003) *Atelier Deltametropool - identiteit en imago* (Delft) TUD Bouwkunde

Ganchev, I.; Bondev, I.; Ganchev, S. (1964) *Vegetation of meadow and pastures in Bulgaria* (Sofia) BAS

Ganzhorn, J.U.; Eisenbeift, B. (2001) *The concept of nested spedes assemblages and its utility for understanding effects of habitat fragmentation* (j) Basic and Applied Ecology 2[1]: 87-95

Gavareski, C.A. (1976) *Relation of park size and vegetation to urban bird populations in Seatde, Washington* (j) The Condor 78: 375-382

Gelderland (2000) *Veluwe 2010; een kwaliteits impuls!* (Arnhem) Provincie Gelderland

Gent, B. v. (1999) *Zoetermeer, ontwikkeling van een nieuwe stad* (Zoetermeer) Gemeente Zoetermeer

Geoplan (2005) *Gebiedsontwikkeling* (Rotterdam)

Gilbert, O.L. (1989) *The Ecology of Urban Habitats* (London) Chapman & Hall

Gill, J.A.; Sutherland, W.J.; Watkinson, A.R. (1996) *A method to quantify the effects of human disturbance on animal populations* (j) Journal of Applied Ecology 33: 786-792

Giurgiu, V.; Donita, N.; Radu, S.; Cenusa, R.; Dissescu, R.; Stoiculescu, C.; Biris, I.A. (2001) *Les forêts vierges de Roumanie* (Louvain la Neuve) ASBL Foret Wallone

Glacken, C.J. (1990) *Traces on the Rhodian shore - Nature and culture in Western Thought from Ancient times to the end of the eighteenth century* (Berkeley) University of California

Glastra, M.J. (2000) *Venster op de Vallei* (De Bilt) Het Utrechts Landschap

Glastra, M. (2007) *The National Ecological Network in practice; experience in land management* In: Jong, T.M. de; Dekker, J.N.M.; Posthoorn, R. [eds.] *Landscape ecology in the Dutch context: nature, town and infrastructure* (Zeist) KNNV-uitgeverij

Glaudemans, M. (2000) *Amsterdams Arcadia: de ontdekking van het achterland* (Nijmegen) SUN

Glikson, A.; Mumford, L. (1971) *The ecological basis of planning* (The Hague) Martinus Nijhoff

Gobster, P.H. (2001) *Visions of nature: conflict and compatibility in urban park restoration* (j) Landscape and Urban Planning 56: 35-51

Goldstein, D.L. (1989) *Estimates of daily energy expenditure in birds: the time-energy budget as an integrator of laboratory and field studies* (j) Am. Zool. 28: 829-844

Gomez, F.; Tamarit, N.; Jabaloyes, J. (2001) *Green zones, bioclimatics studies and human comfort in the future development of urban planning* (j) Landscape and Urban Planning 55[3]: 151-16i

Goois Natuurreservaat (2006) *Over de brug; Hoe Zanderij Crailoo een natuurbrug kreeg* (Hilversum) Goois Natuurreservaat

Goossen, C. M.; Langers, F.; et al. (1997) *Indicatoren voor recreatieve kwaliteiten in het landelijk gebied* (Wageningen) SC-DLO

Goossens, J.; Guinée, A.; Oosterhoff, W. ([1995]) *Buitenruimte - Ontwerp, aanleg en beheer van de openbare ruimte in Rotterdam* (Rotterdam) 010

Gorter, H.P. (1986) *Ruimte voor natuur - 80 jaar bezig voor de natuur van de toekomst* ('s Graveland) Vereniging tot Behoud van Natuurmonumenten in Nederland

Goss-Custard, J.D. (1980) *Competition for food and interference among waders* (j) Ardea 68: 31-52

Goss-Custard, J.D.; Caldow, R.W.G.; Clarke, R.T.; Durell, S.E.A. le V. dit; Sutherland, W.J. (1995) *Deriving population parameters from individual variations in foraging behaviour I Empirical game theory distribution model of Oystercatchers Haematopus ostralegus feeding on mussels Mytilus edulis* (j) J. Anim. Ecol. 64: 265-276

Goss-Custard, J.D.; Durell, S.E.A. Le V. dit; Clarke, R.T.; Beintema, A.J.; Caldow, R.W.G.; Meininger, P.L.; Smit, C.J. (1996) *Population dynamics: predicting the consequences of habitat change at the continental scale* In: Goss-Custard, J.D. [ed.] *The Oystercatcher: from individuals to populations* (Oxford) Oxford University Press; 352-383

Griffioen, M. (2005) *Openbare ruimte met allure* (j) Stad en Groen 2/3

Grift, E. van der; Pouwels, R. (2006) *Restoring habitat connectivity across transport corridors identifying high priority locations for defragmentation with the use of an expert-based model* In: Davenport, J; Davenport, J.L. [eds.] *The ecology of transportation: Managing mobility for the environment. Springer.*; 205-232

Grimbergen, C.; Huibers, R.; Peijl, D. van der (1983) *Amelisweerd: de weg van de meeste weerstand [Amelisweerd,the road of most resistance]* (Rotterdam) Ordeman

Grime, J.P. (1979) *Plant Strategies and Vegetation Processes* (Chichester) Wiley

Groeneveld, J. (1985) *Veranderend Nederland - Een halve eeuw ontwikkelingen op het platteland* (Maastricht / Brussel) Natuur & Techniek

Groot, W.T. de; Udo de Haes, H.A. (1984) *Landschapsecologie en milieukunde: enkele paradigmatische danspassen* (j) Landschap 1(2): 85-91

Groot, J.C.J.; Rossing, W.A.H.; Jellema, A.; Stobbelaar, D.J.; Renting, H.; Ittersum, M.K. Van (2007) *Exploring multi-scale trade-offs between nature conservator, agricultural profits and landscape quality - A methodology to support discussions on land-use perspectives* (j) Agriculture, Ecosystems and Environment 120: 58-69

Grootjans, A.P. (1979) *Effecten van grondwaterstanddaling op een beekdalreservaat van de Drentse Aa* (j) WLO-Mededelingen 6(3): 11-16

Grootjans, A.P. (1980) *Distribution of plant communities along rivulets in relation to hydrology and management* In: Willmanns, O.; Tüxen, R. Epharmoni [eds.] *Berichte der Int. Symp. der IVFV* (Vaduz) Cramer Verlag; 143-165

Grootjans, A.P. (1985) *Changes of groundwater regimes in wet meadows* (Groningen) PhD Thesis University of Groningen

Grootjans, A.P.; Schipper, P.C.; Windt, H.J. van der (1985) *Influence of drainage on N-mineralization and vegetation response in wet meadows I: Calthion palustris stands* (j) Oecologia Plantarum 6: 403-417

Grootjans, A.P.; Diggelen, R. van; Wassen, M.J.; Wiersinga, W.A. (1988) *The effects of drainage on groundwater quality and plant species distribution in stream valley meadows* (j) Vegetatio 75: 37-48

Grootjans, A.P.; Engelmoer, M.; Hendriksma, P.; Westhoff, V. (1988) *Vegetation dynamics in a wet dune slack I: rare species decline on the Wadden island of Schiermonnikoog in the Netherlands* (j) Acta Botanica Neerlandica 37: 265-278

Grootjans, A.P.; Sival, F.P.; Stuyfzand, P.J. (1996) *Hydro-geochemical analysis of a degraded dune slack* (j) Vegetatio 126: 27-38

Grootjans, A.P.; Wirdum, G. van; Kemmers, R.H.; Diggelen, R. van (1996) *Ecohydrology in the Netherlands: Principles of an Application-Driven Interdiscipline* (j) Acta Botanica Neerlandica 45(4): 491-516

Gutzwiller, K.J.; et al. (1998) *Bird tolerance to human intrusion in Wyoming Montane Forests* (j) Condor 100: 519-527

H+N+S (1996) *Over scherven en geluk Report* (The Hague) LNV/V&W/VROM

Haaren, C. von; Reich., M. (2006) *The German way to greenways and habitat networks* (j) Landscape and Urban Planning 76: 7-22

Haartsen, A.J.; Klerk, A.P. de; Vervloet, J.A.J. (1989) *Levend verleden; een verkenning van de cultuurhistorische betekenis van het Nederlandse landschap* ('s-Gravenhage) SDU; Achtergrondreeks Natuurbeleidsplan 3

Haartsen, A. (1995) *Natur versus cultuur? Natuurontwikkeling, landschapsbehoud en de identiteit van het cultuurlandschap* (j) Landschap 12(1): 31-35

Haas, W. de (2006) *Planning als gesprek Grondslagen voor ruimtelijke planning en beleid in de eenentwintigste eeuw* (Utrecht) Uitgeverij de Graaff

Hall, P. (1996) *Cities of Tomorrow, updated edition* (Oxford) BlackwellA.C. et al. [eds.] *Green structure and urban planning. Final Report of COST action CII* (Luxembourg) Office for Official Publications of the European Communities; 267-274

Halm, H. van; Timmermans, G.; Koningen, H.; Bouman, R.; Melchers, M.; Kazus, J. [eds.] (2001) *De wilde stad, 100 jaar natuur van Amsterdam, een eeuw Koninklijke Nederlandse Natuurhistorische Vereniging, afdeling Amsterdam 1901-2001* (Utrecht) KNNV

Ham, W. van der (2002) *De Historie, Een wijd perspectief; een historische verkenning van het Nederlandse landschap in relatie tot het waterbeheer* In: RIZA [ed.] *WaterLandschappen de cultuurhistorie van de toekomst als opgave voor het waterbeheer* (Lelystad) RIZA; work document; 15-56

Hammond, K. A.; Diamond, J. (1997) *Maximal sustained energy budgets in humans and animals* (j) Nature 386: 457-462

Hannah, D.M.; Wood, P.J.; Sadler, J.P. (2004) *Ecohydrology and hydroecology: 'A new paradigm'?* (j) Hydrological Processes 18: 3439-3445

Hanski, I; Gilpin, M.E. (1991) *Metapopulation dynamics : brief history and conceptual domain* (j) Biological Journal of Linnean Society 42: 3-16

Hanski, I. (1999) *Metapopulation Ecology* (Oxford) Oxford University Press

Harcourt, A.H. (2000) *Coincidence and mismatch of biodiversity hotspots: a global survey for the order primates* (j) Biological Conservation 93: 163-175

627

Harrison, P.; Pearce, F. (2001) *AAAS Atlas of population and environment* (Berkeley and Los Angeles) University of California Press

Harsema, H.; Bijhouwer, R.; Cusveller, Sj.; Keulen, N. van; Luiten, E. [eds.] (1996) *Landschapsarchitectuur en stedebouw in Nederland 93-95 - Landscape architecture and town planning in the Netherlands 93 - 95* (Bussum) Thoth

Harsema, H.; Cusveller, Sj.; Bijhouwer, R.; Bolhuis, P. van; Keulen, N. van; Meyer, F. [eds.] (1998) *Landschapsarchitectuur en stedebouw in Nederland 95-97 - Landscape architecture and town planning in the Netherlands 95 - 97* (Bussum) Thoth

Harsema, H.; Cusveller, Sj.; Bijhouwer, R.; Bolhuis, P. van; Keulen, N. van; Meyer, F. [eds.] (2000) *Landschapsarchitectuur en stedebouw in Nederland 97-99 - Landscape architecture and town planning in the Netherlands 97 - 99* (Bussum) Thoth

Harsema, H.; Cusveller, Sj.; Bijhouwer, R.; Bolhuis, P. van; Keulen, N. van; Meyer, F. [eds.] (2002) *Landschapsarchitectuur en stedebouw in Nederland 99-01 - Landscape architecture and town planning in the Netherlands 99 - 01* (Bussum) Thoth

Harsema, H.; Cusveller, Sj.; Bijhouwer, R.; Bolhuis, P. van; Keulen, N. van; Meyer, F. (2002) *Volgermeerpolder - Waterland Noord Holland: een sawa op een gifpolder. Volgermeerpolder - Waterland Noord Holland: paddy fields in the poison polder* (Bussum) Thoth p 190-194

Harsema, H.; Cusveller, Sj.; Bijhouwer, R.; Bolhuis, P. van; Keulen, N. van; Meyer, F. [eds.] (2004) *Landschapsarchitectuur en stedebouw in Nederland 01-03 - Landscape architecture and town planning in the Netherlands 01 - 03* (Bussum) Thoth

Hartig, T. (1993) *Nature experience in transactional perspective* (j) Landscape and Urban Planning 25: 17-36

Hartman, W.; Hellinga, H. [eds.] (1985) *Algemeen Uitbreidingsplan Amsterdam 50 jaar* (Amsterdam) Amsterdamse Raad voor de Stedebouw

Hazelworth, M.S.; Wilson, B.E. (1990) *The effects of an outdoor adventure camp experience on self-concept* (j) Journal of Environmental Education 21[4]: 33-37

Health Council of the Netherlands; RMNO (2004) *Nature and health The influence of nature on social, psychological and physical well-being* (The Hague) Health Council of the Netherlands; Dutch Advisory Council for Research on Spatial Planning RMNO

Heeling, J.; Jong, M. de; Meyer, H.; Westrik, J. (1998) *De Kern van de Stedebouw - in het perspectief van de 21e eeuw Deel 3, Het ontwerp van de stadsplattegrond* (j) Blauwe Kamer, tijdschrift voor landschapsontwikkeling en stedebouw 5

Helmer, W. (2000) *Proefproject Meers: Begin van een levende Grensmaas* (Maastricht) Stroming, Bureau voor Landschapsontwikkeling

Hendriks, Ch.F.; Duijvestein, C.A.J. [eds.] (2002) *The ecological city - impressions* (Boxtel) Aeneas

Hennekens, S.M.; Schaminee, J.H.J. (2002) *Synbiosys: kennissysteem vegetatie voor natuurbehoud, natuurbeleid en natuurontwikkeling* (Wageningen) Alterra, CD-rom

Hermy, H (2005) *Groenbeheer een verhaal met toekomst; de stad als ecosysteem* (Berchem) Velt

Het Utrechts Landschap (2001) *Hart van de Heuvelrug* (De Bilt) Het Utrechts Landschap

Hettema, T.; Hormeijer, P. (1986) *Nederland/Zeeland* (Amsterdam) Euro-Book Productions

Heusser, H. (1960) *Ueber die Beziehungen der Erdkroete zu ihrem Laichplatz II* (j) Behaviour 16: 93-109

Heuvel, C. van den (2005) *'De Huysbou' A reconstruction of an unfinished treatise on architecture, town planning and civil engineering by Simon Stevin* (Amsterdam) Koninklijke Nederlandse Akademie van Wetenschappen KNAW

Heyd, T. (2005) *Nature, culture, and natural heritage: toward a culture of nature* (j) Environmental Ethics 27: 339-354

Hidding, M.; Kolen, J.; Spek, T. (2001) *De biografie van het landschap; ontwerp voor een inter- en multidisciplinaire benadering van de landschapsgeschiedenis en het cultuurhistorisch erfgoed* In: Bloemers, J.H.F.; Wijnen, M.H. [eds.] *Bodemarchief in behoud en ontwikkeling; de conceptuele grondslagen.* (Assen) Van Gorcum; 7-109

Hidding, M. (2006) *Planning voor Stad en Land* (Bussum) Coutinho

Hildebrandt, S.; Tromba, A. (1989) *Architectuur in de natuur, de weg naar de optimale vorm* (Maastricht/Brussel) Wetenschappelijke Bibliotheek Natuur en Techniek

628

Hoeven, C. v. d. ; Louwe, J. (1985) *Amsterdam als stedelijk bouwwerk Een morfologische analyse* (Nijmegen) SUN

Hoeven, F. van der (2001) *RingRing: ondergronds bouwen voor meervoudig ruimtegebruik boven en langs de ring in Rotterdam en in Amsterdam [RingRing: subsurface building for multifunctional use of space on top of and alongside the ring roads of Rotterdam and Amsterdam]* (Delft) PhD Thesis Technische Universiteit

Honnay, O.; et al. (2002) *Satellite based land use and landscape complexity indices as predictors for regional plant species diversity* (j) Landscape and Urban Planning 969: 1-10

Hooimeijer, F.L.; Kamphuis, M.I. (2001) *The Water Project, a nineteenth century walk through Rotterdam* (Rotterdam) 010 Publisher

Hooimeijer, F. (2007) *Water city design in waterland* In: Jong, T.M. de; Dekker, J.N.M.; Posthoorn, R. [eds.] *Landscape ecology in the Dutch context: nature, town and infrastructure* (Zeist) KNNV-uitgeverij

Hosper, S.H.; Meijer, M.-L. (1986) *Control of phosphorus loading and flushing as restoration methods for Lake Veluwe, the Netherlands* (j) Hydrobiol. Bull. 20: 183-194

Houben, F. (1999) *Mobiliteitsesthetiek en ingenieurskunst. Architectuur en de openbare ruimte* In: *De dynamische delta 2* (Den Haag) Ministerie van Verkeer en Waterstaat; Mecanoo architecten; 20-41

Houben, F. (2005) *From centre to periphery: the aesthetics of mobility* In: Charlesworth, E. [ed.] *Cityedge: case studies in contemporary urbanism* (Amsterdam) Architectural Press; 102-117

House, J.S.; Landis, K.R.; Umberson, D. (1988) *Social relationship and health* (j) Science 241: 540- 544

Houwen, J.; Farber, K. [eds.] (2003) *Ontwerpen aan de zandgebieden van Noord- en Midden-Limburg 2030 - Drie over dertig* (Maastricht) Provincie Limburg

Howard, E. (1965) *Garden Cities of To-morrow* (Massachusetts) MIT Press, Cambridge

Huisman, P.; Cramer, W.; et al. [eds.] (1998) *Water in the Netherlands* (Delft) Netherlands Hydrological Society; NHV-special

Huizinga, J. (1974) *Homo ludens-proeve eener bepaling van het spelelement der cultuur* (Haarlem) H.D. Tjeenk Willink

Huizinga, J. (1980) *Homo ludens, a study of the play-element in culture* (London) Routledge and Kegan Paul; International library of sociology

Hunt, J.D. (2000) *Greater perfections - The practice of garden theory* (London) Thames & Hudson

Hylckama Vlieg, Elsbeth; Struben, Hein; Schrijnen, Joost (1994) *Nakaarten over de Zuidvleugel van de Randstad* (Rotterdam) Gemeente Rotterdam e.a.

Ingmire, T.H.; Patri, T. (1971) *An early warning system for regional planning* (j) J. American Institute of Planners XXXVII(6): 403-410

Iso-Ahola, S.E.; Park, C.J. (1996) *Leisure-related social support and selfdetermination as buffers of stress-illness relationship* (j) Journal of Leisure Research 28: 169-187

IUCN; UNEP; WWF (1985) *World conservation strategy* (Gland) International Union for the Conservation of Nature

Jackson, J.B. (1994) *A sense of place, a sense of time* (New Haven/London) Yale University Press

Jalink, M.H.; Jansen, A.J.M. (1989) *Indicatorsoorten voor verzuring, verdroging en eutrofiëring van grondwaterafhankelijke beekdalgemeenschappen* (Nieuwegein) KIWA report SWE 89.029

Jalink, M.H. (1991) *Indicatorsoorten voor verzuring, verdroging en eutrofiëring in laagveenmoerassen* (Nieuwegein) KIWA report SWE 90.037

James, P. (Undated) *Ecological networks: creating landscapes for people and wildlife*

Jans, L. [ed.] (2001) *Monitoring nevengeulen; Integrale jaarrapportage 1999/2000* (Lelystad) RIZA Werkdocument 2001.062X

Jansen, S.R.J.; et al. (1993) *Ontwerpnota Ecosysteemvisies; werkdocument IKC-NBLF 48* (Den Haag) IKC-NBLF; Ministerie van Landbouw, Natuurbeheer en Visserij

Jansen, S.; et al. (1997) *Vleermuiskelder en faunapassage in geluidswal* (j) Zoogdier 8(4): 24

Jenerette, G.D.; Wu, J. (2001) *Analysis and simulation ofland-use change in the central Arizona - Phoenix region, USA* (j) Landscape Ecology 16[7]: 611-626

Jokimaki, J. (1999) *Occurrence of breeding bird species in urban parks: effects of park struc-ture and broad-scale variables* (j) Urban Ecosystems 3: 21-34

Jong, E. de (1993) *Natuur en kunst - Nederlandse tuin- en landschapsarchitectuur 1650-1740* (Amsterdam) Thoth

Jong, H. de (1996) *Handboek Civiele Kunstwerken* (Den Haag) TenHagen Stam; losbladig 3 mappen

Jong, J.J. de; Spijker, J.H.; Wolf, R.J.A.M.; Koster, A.; Schaafsma, A.H. (2001) *Beheerskosten en Natuurwaarden van groenvoorzieningen langs rijkswegen Een vergelijking tus-sen traditioneel en ecologisch beheer van grazige bermen, boomweiden en berm-sloten*

Jong, M.D.T.M. de (2002) *Scheidslijnen in het denken over Natuurbeheer in Nederland - Een genealogie van vier ecologische theorieën* (Delft) PhD Thesis Technische Universi-teit; DUP Science

Jong, T.M. de (1978) *Milieudifferentiatie, een fundamenteel onderzoek* (Den Haag, Delft) PhD Thesis Technische Hogeschool; Rijksplanologische Dienst.

Jong, T.M. de; Vos, J. (1993-2006) *Verspreide artikelen over de oeverzwaluw en de wielewaal in Zoetermeer* (j) Kwartaalbericht KNNV Zoetermeer:over de oeverzwaluw in Zoe-termeer nr. 2p13, 16; 3p5; 9p12; 12p20,21; 15p29; 16p28,29,30; 24p9 en over de wielewaal 9p14; 10p19; 14p27; 20p15; 26p26; 33p14; 35p27; 37p28

Jong, T.M. de; Achterberg, J. (1996) *25 plannen voor de Randstad* (Zoetermeer) Stichting MESO

Jong, T.M. de; Paasman, M. (1998) *Het Metropolitane Debat Een vocabulaire voor besluit-vorming over de kaart van Nederland* (Zoetermeer) MESO, Stichting Milieu en ste-delijke ontwikkeling; HMD, Het Metropolitane Debat

Jong, T.M. de; Paasman, M. (1998) *Een vocabulaire voor besluitvorming over de kaart van Nederland* (Zoetermeer) MESO

Jong, T.M. de (2000) *De abiotische uitgangssituatie in de stad* (j) De Levende Natuur 101(6) (Wageningen) Alterra en DWW

Jong, T.M. de (2001) *Standaardverkaveling 11exe* (Delft) Technische Universiteit, Faculteit Bouwkunde; software; http://team.bk.tudelft.nl_ > Publications 2001

Jong, T.M. de (2001) *Ecologische toetsing van drie visies op Almere Pampus* (Zoetermeer) Stichting MESO

Jong, T.M. de (2002) *Designing in a determined context* In: Jong, T. M. d.; Voordt, D. J. H. v. d. [eds.] *Ways to research and study urban, architectural and technical design.* (Delft) DUP Science

Jong, T.M. de (2002) *Het begrenzen van hiërarchisch geordende gebieden in de stadsecolo-gie* (Utrecht) WLO lezing

Jong, T.M. de; Voordt, D.J.M. van der [eds.] (2002) *Ways to study and research urban, archi-tectural and technical design* (Delft) DUP Science

Jong, T.M. de (2005) *The Ecology of Health in regional perspective* Proceedings 2nd WHO International Housing and Health Symposium (Vilnius) World Health Organisation

Jong, T.M. de (2006) *Stadsecologie, schaal en structuur* (Amsterdam) Lezing WLO-symposium Stadsecologie 6 april 2006

Jong, T.M. de; Moens, R.; Akker, C. van den; Steenbergen, C.M. (2006) *Sun wind water - Earth life and living - Legends for design* (Delft) TUD

Jong, T.M. de (2007) *Connecting is easy, separating is difficult* In: Jong, T.M. de; Dekker, J.N.M.; Posthoorn, R. [eds.] *Landscape ecology in the Dutch context: nature, town and infrastructure* (Zeist) KNNV-uitgeverij

Jong, T. M. de (2007) *Urban ecology, scale and structure* In: Jong, T.M. de; Dekker, J.N.M.; Posthoorn, R. [eds.] *Landscape ecology in the Dutch context: nature, town and in-frastructure* (Zeist) KNNV-uitgeverij

Jong, T.M. de (2007) *Dry And Wet Networks* In: Jong, T.M. de; Dekker, J.N.M.; Posthoorn, R. [eds.] *Landscape ecology in the Dutch context: nature, town and infrastructure* (Zeist) KNNV-uitgeverij

Jong, T.M. de (2007) *Town, editorial introduction* In: Jong, T.M. de; Dekker, J.N.M.; Post-hoorn, R. [eds.] *Landscape ecology in the Dutch context: nature, town and infra-structure* (Zeist) KNNV-uitgeverij

Jong, T.M. de (2007) *Infrastructure, editorial introduction* **In:** Jong, T.M. de; Dekker, J.N.M.; Posthoorn, R. [eds.] *Landscape ecology in the Dutch context: nature, town and infrastructure* (Zeist) KNNV-uitgeverij

Jong, T.M. de; Dekker, J.N.M.; Posthoorn, R. [eds.] (2007) *Landscape ecology in the Dutch context: nature, town and infrastructure* (Zeist) KNNV-uitgeverij

Jongman, R.H.G.; Troumbis, A.Y. (1995) *The wider Landscape for Nature Conservation: ecological corridors and buffer zones* (Tilburg) European Centre for Nature Conservation; Project-report 1995, submitted to the European Topic Centre for Nature Conservation in Fulfilment of the 1995 Work Programme

Jongman, R.H.G. (2002) *Homogenisation and fragmentation of the European landscape: ecological consequences and solutions* (j) Landscape and urban planning 58: 211-221

Jongman, R.H.G. (2004) *The context and concept of ecological networks* **In:** Jongman, R.H.G.; Pungetti, G.P. [eds.] *Ecological networks and greenways, concept, design and implementation.* (Cambridge) Cambridge University Press; 7-33

Jongman, R.H.G.; Külvik, M; Kristiansen.I. (2004) *European ecological networks and greenways* (j) Landscape and Urban Planning 68: 305-319

Jongman, R.H.G.; Pungetti, G.P. (2004) *Ecological networks and greenways, concept, design and implementation* Cambridge University Press

Jongman, R.H.G.; R.G.H.Bunce; M.J.Metzger; C.A.Mücher; D.C.Howard; V.L.Mateus (2006) *Objectives and Applications of a Statistical Environmental Stratification of Europe* (j) Landscape Ecology 21: 409-419

Jongman, R.H.G.; Veen, P. (2007) *Ecological Networks across Europe* **In:** Jong, T.M. de; Dekker, J.N.M.; Posthoorn, R. [eds.] *Landscape ecology in the Dutch context: nature, town and infrastructure* (Zeist) KNNV-uitgeverij

Joosten, J. H. J.; Noorden, B. P. M. (1992) *De Groote Peel: leren waarderen Een oefening in het waarderen van natuurelementen ten behoeve van het natuurbehoud* (j) Natuurhistorisch maandblad 81(12): 203 e.v

Joustra, D.J.; Vries, C.A.d. (2004) *Het brilletje van Van Leeuwen* (Leeuwarden) NIDO; Duurzame stedelijke vernieuwing; www.nido.nu

Justice, B. (1994) *Critical life events and the onset of illness* (j) Comprehensive Therapy 20: 232-238

Kalwij, J.M.; Sykóra, K.V.; Keizer, P.J. (2001) *Worden rijkswegbermen goed beheerd? Herinventarisatie van Nederlandse rijkswegbermen; een vegetatiekundige en floristische vergelijking van 1986-1988 met 2001* (Wageningen/Delft) WUR en DWW

Kam, J. van de; Ens, B.; Piersma, Th; Zwarts, L. (2004) *Shorebirds. An illustrated behavioural ecology* (Utrecht) KNNV Publishers

Kamphuis, M.I.; Hooimeijer, F.L. (1997) *De stad en de dood* (Rotterdam) De Hef Uitgeverij

Kaplan, G.A.; Salonen, J.T.; Cohen, R.D.; Brand, R.J.; Syme, S.L.; Puska, P. (1988) *Social connections and mortality from all causes and from cardiovascular disease: Prospective evidence from Eastern Finland* (j) American Journal of Epidemiology 128[2]: 370-380

Kaplan, R.; Kaplan, S. (1989) *The Experience of Nature: a psychological perspective* (London) Cambridge University Press

Karasov, W. H. (1996) *Digestive plasticity in avian energetics and feeding ecology* **In:** Carey, C. [ed.] *Avian Energetics and Nutritional Ecology* (New York) Chapman and Hall; 61-84

Kati, V.; Devillers, P.; Dufrêne, M.; Legakis, A.; Vokou, D.; Lebrun, P. (2004) *Testing the value of six taxonomic groups as biodiversity indicators at a local scale* (j) Conservation Biology 18: 667-675

Katoele, H. (1987) *Vleermuizenhotel Beilen* (j) Noorderbreedte 1987(5)

Keessen, J. (1997) *Een raar landschap in Amsterdam* (j) Landschap 14(3): 123-130

Keizer, P.J. (2002) *Planten langs de weg* (j) Via Natura 14: 3-7

Keizer, P.J. (2005) *Overzicht van vegetatietypen langs de rijkswegen* (Delft) DWW

Keizer, P.J.; Hengel, L.C. van den; Groshart, C. (2006) *Leidraad Beheer Groenvoorzieningen* (Delft) DWW

Keller, S.E. et al. (1994) *Human Stress and Immunity* (San Diego) Academic Press

Keller, V.; Pfister, H.P. (1997) *Wildlife passages as means of mitigating effects of habitat fragmentation by roads and railway lines* **In:** Canters, K.J.; Piepers, A.A.G.; Hendriks-Heerma, D. [eds.] *Habitat fragmentation and Infrastructure* (Delft) International Conference on Habitat fragmentation, infrastructure and the role of ecological engineering 17-21 September 1995 V&W/DWW-Delft

Kemmers, R.H.; Jansen, P.C. (1988) *Hydrochemistry of rich fen and water management* (j) Agric. Water Management 14: 399-412

Kemmers, R.; Waal, R. de; Delft, B. van; Mekkink, P. (2002) *Ecologische typering van bodems - Actuele informatie over bodemkundige geschiktheid voor natuurontwikkeling* (j) Landschap 19(2): 89-103

Kemmers, R.H.; Delft, S.P.J. van; Jansen, P.C. (2003) *Iron and sulphate as possible key factors in the restoration ecology of rich fens in discharge areas* (j) Wetlands ecology and management 11: 367-381

Kerkstra, K.; Vrijlandt, P. (1988) *Het landschap van de zandgebieden - Probleemverkenning en oplossingsrichting* (Utrecht / Wageningen)

Keulartz, J.; Swart, S.; Windt, H. van der (2000) *Natuurbeelden en natuurbeleid; theoretische en empirische verkenningen* (Den Haag) NWO Ethiek & Beleid

King, J.R. (1974) *Seasonal allocation of time and energy resources in birds* **In:** Paynter, R.A. [ed.] *Avian Energetics* 15 (Cambridge, Massachusetts) Nutttall Ornithological Club; 4-70

Kingdon, J.W. (1984) *Bridging Research and Policy Agenda's, Alternatives and Public Policies* (New York) Harper Collins

Kirkwood, J.K. (1983) *A limit to metabolisable energy intake in mammals and birds* (j) Comp. Biochem. Physiol. 74: 1-3

KIWA; EGG (2005/2006) *Challenges and opportunities analysis Natura 2000 areas ['Knelpunten- en kansenanalyse Natura 2000 gebieden]* (Den Haag) Ministerie van Landbouw, Natuur en Voedselkwaliteit

Kleijn, D.; Berendse, F.; Smit, R.; Gilissen, N.; Smit, J.; Brak, B.; Groeneveld, R. (2004) *Ecological effectiveness of agri-environment schemes in different agricultural landscapes in the Netherlands* (j) Conservation Biology 18: 775-786

Kleijn, D.; Baquero, R. A.; Clough, Y.; Díaz, M.; Esteban, J. De; Fernández, F.; Gabriel, D.; Herzog, F.; Holzschuh, A.; Jöhl, R.; Knop, E.; Kruess, A.; Marshall, E. J. P.; Steffan-Dewenter, I.; Tscharntke, T.; Verhulst, J.; West, T. M.; Yela, J. L. (2006) *Mixed biodiversity benefits of agri-environment schemes in five European Countries* (j) Ecology Letters 9(3): 243-254

Klerken, B.; Bruggencate, P. ten; Janssen, H. (1997) *[Muizen en vleermuizen in] Een Groene geluidswal in Groningen* (j) Natura 94: 78-[82]-83

Klijn, J.A. (1983) *Biogeografische inzichten en het ruimtelijk beleid* (j) WLO-Mededelingen 10(4): 155-159

Klijn, J.A.; Harms, W.B. (1990) *Natuurbeleidsplan en onderzoek - Visie, onderbouwing en perspectief* (j) Landschap 7(2): 121-128

Klijn, F.; Witte, J.P.M. (1999) *Eco-hydrology; groundwater flow and site factors in plant ecology* (j) Hydrogeology Journal 7: 65-77

Klijn, J.; Vos, W. [eds.] (2000) *From Landscape Ecology to Landscape Science* (London) Kluwer Academic Publishers

Klundert, A.F. van der; Saris, F.J.A. [eds.] (1983) *Special issue "Ecologische infrastructuur"* (j) Mededelingen van de Werkgemeenschap Landschapsecologisch Onderzoek 10(4): 154-205

KNAG (2000) Retrieved 16 March 2007; http://ww2.knag.nl/pagesuk/geography/engels/news95engelstekst.html

Knight, R.K.; Grout, D.J.; Temple, S.A. (1987) *Nest-defense behavior of the American crow in urban and rural areas* (j) The Condor 89: 175-177

KNNV (1971) *De weg naar en door de natuur [The road to and through nature]* (j) Wetenschappelijke mededelingen KNNV 87

Knopf, R.E. (1987) *Human behavior, cognition, and affect in the natural environment* **In:** Stokols, D.; Altman, I. [eds.] *Handbook of environmental psychology, Vol. I.* (New York) Wiley

Koekebakker, O. (2005) *Cultuurpark Westergasfabriek. Transformatie van een industrieterrein* (Rotterdam) NAi Uitgevers i.s.m. het Projectbureau Westerpark

Koeleman, M.; Arts, J. (2005) *Environmental infrastructure; Towards a new perspective in impact assessment* **In:** Bohemen, H.D. van [ed.] *Ecological Engineering; Bridging between ecology and civil engineering* (Boxtel) Aeneas Technical Publicers; 383-392

Koerselman, W. (1989) *Hydrology and Nutrient Budgets of Fens in an Agricultural Landscape* (Utrecht) PhD Thesis Utrecht University

Koning, E.; Tjallingii, S.P. (1991) *Ecologie van de stad, een verkenning* (Den Haag) Platform Stadsecologie

Kooijman, A.M. (1993) *Changes in the Bryophyte Layer of Rich Fens as Controlled by Acidification and Eutrophication* (Utrecht) PhD Thesis University of Utrecht

Koole, S.L..; Berg, A.E. van den (2005) *Lost in the wilderness: Terror management, action control, and evaluations of nature* (j) Journal of Personality and Social Psychology 88(6): 1014-1028

Koomen, A.; Maas, G. (2004) *Het stuifzandlandschap als natuurverschijnsel* (j) Landschap 21(3): 159-169

Koppen, K. van (2002) *Echte natuur; een sociaaltheoretisch onderzoek naar natuurwaardering en natuurbescherming in de moderne samenleving* (Wageningen) Landbouwuniversiteit Wageningen

Korten, H. (2005) *Case study Crailo Sand Quarry Nature Bridge* **In:** Bohemen, H.D. van [ed.] *Ecological Engineering, bridging between ecology and civil engineering* (Boxtel) Aeneas Technical Publishers; 250-251

Koster, F. (1939) *Natuurleven in en om Amsterdam* (Amsterdam) Scheltema & Holkema's Boekhandel en Uitgevers

Koster, A. (2000) *Wilde bijen in het stedelijk groen* (j) De Levende Natuur 101: 213-215

Kowarik, I (1990) *Some responses of flora and vegetation to urbanization in Central Europe* **In:** Sukopp, H.; Hejný, S. [eds.] *Urban ecology* (The Hague) SPB Academic Publishing; 45-74

Krebs, C.J. (1994) *Ecology - The Experimental Analysis of Distribution and Abundance* (New York) Harper Collings College Publishers

Kromme Rijn Project (1974) *Het Kromme-Rijn landschap, een ecologische visie* (Amsterdam) Stichting Natuur en Milieu; Natuur en Milieu Reeks

Kuhn, M. (2003) *Greenbelt and Green Heart: separating and integrating landscapes in European city regions* (j) Landscape and Urban Planning 64 (1-2): 19-27

Kundzewic, Z.W. (2002) *Ecohydrology - seeking consensus on interpretation of the notion* (j) Hydrological Sciences Journal 47: 799-804

Kuttler, W. (1998) *Stadtklima* **In:** Sukopp, H.; Witting, R. [eds.] *Stadtökologie. Ein Fachbuch für Studium und Praxis.* (Stuttgart) G.Fisher

Kwak, R. G. M.; Berg, A. van den (2004) *Nieuwe Broedvogeldistricten van Nederland* (Wageningen) Alterra report 1006

L&V (1975) *Nota betreffende de relatie landbouw en natuur en landschapsbehoud* (Den Haag) Ministerie van Landbouw & Visserij; Tweede Kamer, zitting 1974-1975, 13285. nrs. 1-2.

Lack, D. (1954) *The natural regulation of animal numbers* (Oxford) Oxford University Press

Lack, D. (1966) *Population studies of birds* (Oxford) Oxford University Press

Lambert, A.M. (1971,1985) *The making of the Dutch Landscape - A historical geography of the Netherlands* (London) Seminar Press

Lamers, L.P.M.; Roozendaal, S.M.E. van; Roelofs, J.G.M. (1998) *Acidification of freshwater wetlands: combined effects of non-airborne sulfur pollution and desiccation* (j) Water Air Soil Pollution 105: 95-106

Lammers, G.W.; Zadelhoff, F.J. van (1996) *The Dutch national ecological network* **In:** Nowicki, P.; Bennett, G.; Middleton, D.; Rientjes, S; Wolters, R. [eds.] *Perspectives on ecological networks.* (Tilburg) European Centre for Nature Conservation

Lammers, W. (2003) *Kerncijfers voor de IBO studie Vogel- en Habitatrichtlijn* (Bilthoven) DLO, RIVM

References

Lammers, G.W.; Hinsberg, A. van; Loonen, W.; Reijnen, M.J.S.M.; Sanders, M.E. (2005) *Optimalisatie Ecologische hoofdstructuur; ruimte, milieu en watercondities voor duurzaam behoud van biodiversiteit* (Bilthoven) Milieu- en Natuurplanbureau

Landschap, Redactie (1988) *Vijf jaar 'Landschap'* (j) Landschap 5 (4): 231-234

Langevelde, F. van ; Claassen, F.; Schotman, A. (2002) *Two strategies for conservation planning in human-dominated landscapes* (j) Landscape and Urban Planning 58: 281-295

Larcher, F. [ed.] (2006) *Paessaggi agrari e forestali - La pianificazione, il progetto e la gestione dei territori rurali - Atti del Convegno* (Torino) Università degli Studi di Torino

Latour, J.B.; Reiling, R. (1993) *A multiple stress model for vegetation (MOVE): a tool for scenario studies and standard setting* (j) Sci Total Env. Suppl: 1513-1526

Laurie, M. (1976) *An introduction to landscape architecture* (New York) American Elsevier

Lazarus, R.S.; Folkman, S. (1984) *Stress, Appraisal, and Coping* (New York.) Springer

Le Roy, L.G. (1973) *Natuur uitschakelen - Natuur inschakelen* (Deventer) Ankh Hermes BV

Le Roy, L.G. (2002) *Natuur, Cultuur, Fusie. Nature, Culture, Fusion* In: Le Roy, Louis G.; Vollaard, Piet; Baukema, Esther; Mcintyre, Philippe Velez [eds.] (Rotterdam) Nai Uitgevers

Leenders, K.A.H.W. (1995) *Naar de climax van het gesloten landschap* In: Eerenbeemt, H.F.J.M. van den [ed.] *Geschiedenis van Noord-Brabant 1; traditie en modernisering 1796-1890.* (Amsterdam/Meppel) Boom; 142-151

Leeuw, J.J. de (1997) *Demanding divers Ecological energetics of food exploitation by diving ducks* (Lelystad) PhD Thesis University of Groningen; Rijkswaterstaat Directoraat IJsselmeergebied, Van Zee tot Land 61

Leeuwen, C.G. van; Doing Kraft, H. (1959) *Landschap en beplanting in Nederland - Richtlijnen voor de soortenkeuze bij beplantingen op vegetatiekundige grondslag* (Wageningen) Veenman

Leeuwen, C.G. van (1964) *The open- and closed theory as a possible contribution to cybernetics* (Leersum) Rijksinstituut voor Natuurbeheer

Leeuwen, C.G. van (1965) *Over grenzen en grensmilieu's* In: *Jaarboek 1964 Kon. Nederlandse Vereniging*; 53-54

Leeuwen, C.G. van (1965) *Het verband tussen natuurlijke en anthropogene landschapsvormen, bezien vanuit de betrekkingen in grensmilieus* (j) Gorteria 2: 93-105

Leeuwen, C.G. van (1966) *A relation theoretical approach to pattern and process in vegetation* (j) Wentia 15: 25-46

Leeuwen, C.G. (1971) *Ekologie* (Delft) Technische Universiteit Delft, faculteit Bouwkunde

Leeuwen, C.G. van (1975) *Ekologie - Ekologie 0 - Ekologie II* (Delft) THD - Bouwkunde

Leeuwen, B.H. van (1995) *Doelen voor algemene natuurkwaliteit Nodig en mogelijk* (j) Landschap 12(2): 55-58

Leeuwen, R. van (1998) *Het maken en de vorm. De NEN als vormgevingsvraagstuk* In: Feddes, F.; Herngreen, R.; Jansen, Sj.; Leeuwen, R. van; Sijmons, D. [ed.] *Oorden van onthouding - Nieuwe natuur in verstedelijkend Nederland* (Rotterdam) NAI; 158-166

Legget, Robert. F. (1973) *Cities and Geology* (New York) McGraw-Hill

LEI (2006) *LEI cost analysis report 2006 ['LEI rapportage inzake kostenanalyse 2006']* Landbouw Economisch Instituut

Lemaire, A.J.J.; Beringen, R.; Groen, C.L.G. (1997) *Verspreiding van doelsoorten [vaatplanten] in relatie tot de Ecologische Hoofdstructuur - Samenvatting FLORON-rapport nr 3* (j) Gorteria 23: 109-114

Lems, P.; Valkman, R. (2003) *Waarden van water, theoretisch kader* (Leeuwarden) NIDO

Lems, P.; Valkman, R. (2003) *Waarden van water, handreiking voor de praktijk* (Leeuwarden) NIDO

Lendering, Jona (2005) *Polderdenken De wortels van de Nederlandse overlegcultuur* (Amsterdam) Athenaeum

Leonard, M.L.; Horn, A.G. (2005) *Ambient noise and the design of begging signals* (j) Proceedings of the Royal Society of London B 272[1563]: 651-656

Levins, R. (1969) *Some demographic and genetic consequences of environmental heterogeneity for biological control* (j) Bulletin of the Entomology Society of America 71: 237-240

Liebrand, C.I.J.M.; Scherpenisse-Gutter, M.C.; Verbeek, P.J.M.; Peeters, G.M.T.; Goeij, A.A.M. de (2001) *Evaluatie van natuurvriendelijk maaibeheer langs rijkswegen in Limburg 1993-2001* (Nijmegen) Herhalingsonderzoek in opdracht van RWS-Directie Limburg, Bureau Natuurbalans-Limes Divergens en EurECO Ecologisch onderzoek en advies

Lier, H. N. van (1975) *Enkele aspecten van de projectstudie Midden-Brabant* (j) Mededelingen van de WLO 2(2): 16-22

Liere, L. van; Gulati, R.D. (1992) *Restoration and recovery of shallow eutrophic lake ecosystems in the Netherlands: epilogue* (j) Hydrobiologia 233: 283-287

Ligtelijn, V. (1999) *Aldo van Eyck; Werken* (Bussum) Thoth

Linden, P.J.H. van (1997) *A wall of tree-stumps as a fauna corridor* In: Canters, K.J.; Piepers, A.A.G.; Hendriks-Heerma, D. [eds.] *Habitat fragmentation and Infrastructure* (Delft) Proceedings International Conference on habitat fragmentation, infrastructure and the role of ecological engineering 17-21 September 1995. DWW-Delft; 409-418

Lindström, Å. (1995) *Stopover ecology of migrating birds: some unsolved questions* (j) Israel J. Zool. 41: 407-416

Litgens, G.; Herik, K. van den; Winden, A. van; Braakhekke, W. (2006) *Natuurlijke klimaatbuffers; Adaptatie aan klimaatverandering; Wetlands als waarborg* (Nijmegen) Bureau Stroming bv; Vereniging Natuurmonumenten; Vogelbescherming Nederland; Staatsbosbeheer; ARK Natuurontwikkeling; Waddenvereniging; http://www.vogelbescherming.nl/documents/pdf-files/hier_klimaatbuffers.pdf

Litjens, B.J.E. (1991) *Evaluatie wildviaducten A50* (Arnhem) LNV-NMF

LNV (1990) *Natuurbeleidsplan; Regeringsbeslissing* (Den Haag) Ministerie van Landbouw, Natuurbeheer en Visserij

LNV (1993) *Structuurschema Groene Ruimte - deel 3: Kabinetsstandpunt* (Den Haag) Ministerie van Landbouw Natuurbeheer & Visserij

LNV (1995) *Visie stadslandschappen - Discussienota* (Den Haag) Ministerie van Landbouw, Natuurbeheer en Visserij

LNV (2000) *Natuur voor Mensen, Mensen voor Natuur; Nota natuur, bos en landschap in de 21e eeuw* (Den Haag) Ministerie van Landbouw, Natuurbeheer en Visserij

LNV (2000) *Memorandum of Reply Birds Directive [Nota van antwoord Vogelrichtlijn - deel 1 - Algemeen]* (Den Haag) Ministry of Agriculture, Nature and Food Quality

LNV (2002) *Structuurschema Groene Ruimte 2 Samen werken aan groen Nederland* (Den Haag) Ministerie van Landbouw, Natuurbeheer en Visserij.

LNV (2004) *Agenda voor een Vitaal Platteland Meerjarenprogramma Vitaal Platteland 2004* (Den Haag) Ministerie van Landbouw, Natuur en Voedselkwaliteit

LNV (2004) *Sites document ['Gebiedendocument' Overzicht van habitattypen en soorten waarvoor gebieden zijn aangemeld en begrenzing van gebieden]* (Den Haag) Ministry of Agriculture, Nature and Food Quality

LNV (2004) *List document ['Lijstdocument', overzicht van gebiedsselectie voor de Habitatrichtlijn]* (Den Haag) Ministry of Agriculture, Nature and Food Quality

LNV (2004) *Reaction document ['Reactiedocument Aanmelding Habitatvogelrichtlijngebieden']* (Den Haag) Ministry of Agriculture, Nature and Food Quality

LNV (2004) *Justification document ['Verantwoordingsdocument' Selectiemethodiek voor aangemelde Habitatrichtlijnvogelgebieden]* (Den Haag) Ministry of Agriculture, Nature and Food Quality

LNV (2005) *Algemene Handreiking Natuurbeschermingswet 1998* (Den Haag) Ministerie van Landbouw, Natuur en Voedselkwaliteit

LNV (2005) *Handreiking Bestemmingsplan en Natuurwetgeving* (Den Haag) Ministerie van Landbouw, Natuur en Voedselkwaliteit

LNV (2005) *Handreiking Beheerplannen Natura 2000-gebieden* (Den Haag) Ministerie van Landbouw, Natuur en Voedselkwaliteit

LNV (2005) *Natura 2000 framework memorandum ['Natura 2000 contourennotitie']* Ministry of Agriculture, Nature and Food Quality

LNV (2006) *Natura 2000 profielendocument* (Den Haag) Ministerie van Landbouw, Natuur en Voedselkwaliteit

LNV (2006) *Reader's Guide to Natura 2000 sites documents ['Leeswijzer Natura 2000 gebiedendocumenten']* (Den Haag) Ministry of Agriculture, Nature and Food Quality

LNV (2006) *Natura 2000 targets document - Summary ['Natura 2000 doelendocument - samenvatting']* (Den Haag) Ministry of Agriculture, Nature and Food Quality

LNV (2006) *Natura 2000 targets document ['Natura 2000 doelendocument']* Ministry of Agriculture, Nature and Food Quality

LNV (2006) *Natura 2000 site documents ['Natura 2000 gebiedendocumenten']* (Den Haag) Ministry of Agriculture, Nature and Food Quality

LNV (2006) *Agenda for a living Countryside - Multi-year Programme 2007-2013 [Agenda voor een Vitaal Platteland]* (Den Haag) Ministry of Agriculture, Nature and Food Quality

LNV; V&W; VROM (2006) *Verhouding tussen de Kaderrichtlijn Water en de Vogel- en Habitatrichtlijnen - werkdocument* (Den Haag) Ministeries van Landbouw, Natuur en Voedselkwaliteit; Verkeer en Waterstaat; Volkshuisvesting, Ruimtelijke Ordening en Milieubeheer

LNV-EC (2005) *The National Ecological Network [NEN] of the Netherlands, update November 2005* (Ede) Ministerie van Landbouw, Natuur en Voedselkwaliteit

Logemann, D.; Schoorl, E.F. (1988) *Verbindingswegen voor plant en dier* (Utrecht) Stichting Natuur en Milieu

Londo, G. (1988) *Nederlandse phreatophyten* (Wageningen) Pudoc

Londo, G. (1997) *Natuurontwikkeling* (Leiden) Backhuys Publishers

Lörzing, H. (1992) *Van Bosplan tot Floriade - Nederlandse park- en landschapsontwerpen in de twintigste eeuw* (Rotterdam) 010

Louwe Kooijmans, L.P. (1997) *Paleo-ecologie van het rivierengebied; het prehistorische landschap als referentie voor natuurontwikkeling?* (j) Landschap 14(2): 147-158

Lovejoy, D. [ed.] (1973) *Land use and landscape planning* (Aylesbury) Leonard Hill Books

Lovelock, J. (1995) *The ages of Gaia, a biography of our living earth* (Oxford) Oxford University Press

Lucassen, E.C.H.E.T.; Smolders, A.J.P.; Crommenacker, J. van de; Roelofs., J.G.M. (2004) *Effects of stagnating sulphate rich groundwater on the mobility of phosphate in freshwater wetlands; a field experiment* (j) Arch. Hydrobiol. 160: 117-131

Lucht, F. van der; Verkleij, H. (2001) *Gezondheid in de grote steden: achterstanden en kansen* (Bilthoven) RIVM

Luck, M.; Wu, J. (2002) *A gradient analysis of urban landscape pattern: a case study from the Phoenix metropolitan region, Arizona, USA* (j) Landscape Ecology 17: 327-339

Lynch, K. (1974) *Site planning* (Cambridge) MIT

Maarel, E. van der (1974) *Iets over het Globaal Ecologisch Model voor de ruimtelijke ontwikkeling van Nederland* (j) Mededelingen van de WLO 1(4): 6-13

Maarel, E. van der (1976) *On the establishment of plant community boundaries* (j) Berichte der Deutsche Botanische Geselschaft 89: 415-443

Maarel, E. van der; Dauvelier, P.L. (1978) *Naar een Globaal Ecologisch Model voor de ruimtelijke ontwikkeling van Nederland* (Den Haag) Ministerie van Volkshuisvesting en Ruimtelijke Ordening

Maas, I.A.M.; Jansen, J. [eds.] (2000) *Psychische [on]gezondheid: determinanten en de effecten van preventieve interventies* (Bilthoven.) RIVM

Mabelis, A. (1990) *Natuurwaarden in cultuurlandschappen* (j) Landschap 7(4): 253-268

MacArthur, R.H.; Levins, R. (1964) *Competition, habitat selection and character displacement in a patchy environment* (j) Proc. Nat. Ac. Sci. 51: 1207-1210

MacArthur, R.H.; Pianka, E.R. (1966) *On optimal use of a patchy environment* (j) Amer. Nat. 100: 603-609

MacArthur, R.H.; Wilson, E.O. (1967) *The theory of island biogeography* (Princeton) Princeton University Press

Machado, J.R.; Ahern, J.; Saraiva, G. G.; Silva, E.A. da; Rocha, J.; Ferreira, J.C.; P.M.Sousa; Roquette, R. (1997) *Greenways Network for the Metropolitan are of Lisbon* **In:** Machado, J.R.; Ahern, J. [eds.] *Environmental Challenges in an Expanding Urban World and the Role of Emerging Information Technology* (Lisbon) CNIG; 281-289

Mackenbach, J.P. (2004) *Hoe belangrijk zijn innovaties in de gezondheidszorg geweest voor verbeteringen van de volksgezondheid?* **In:** Mackenbach, J.P. et al. [eds.] *Volksgezondheid gedetermineerd.* (Zutphen) Walburg Pers

Mader, H.J. (1981) *Der konflikt Strassen-Tierwelt aus ökologischer Sicht* (Bonn-Bad Godes-berg) Bundesanstalt für Naturschutz und Landschaftsforschung; Schriftenreihe für Landschaftpflege und Naturschutz 22.

Maier, S.F.; Watkins, L.R.; Fleshner, M. (1994) *The interface between behavior, brain, and immunity* (j) American Psychologist 49: 1004-1017

Maiman, R.; Decamps, H. (1989) *Ecology and Management of Aquatic-Terrestrial Ecotones* (Paris) Unesco; Man and Biosphere Series

Mak, Geert (1998) *Het ontsnapte land* (Amsterdam) Atlas

Mander, Ü.; Külvik, M.; Jongman, R.H.G. (2003) *Scaling in territorial ecological networks* (j) Landschap 20 (2): 113-127

Mann, C.C. (2006) *1491; de ontdekking van precolumbiaans Amerika* (Amsterdam) Nieuw Amsterdam / Manteau

Manzano, P.; Malo, J.E. (2006) *Extreme long distance seed dispersal through seed* (j) Front Ecol Environ. 4(5): 244-248; www.frontiersinecology.org

Margules, C.R.; Pressey, R.L.; Williams, P.H. (2002) *Representing biodiversity: data and procedures for identifying priority areas for conservation* (j) Journal of Biosciences 27: 309-326

Markesteijn, C.M.; Huele, R.; Bekker, G.J. (2006) *Exploration of the implications of the MJPO* (Delft) RWS/DWW; Report project 6400.2923

Mars, H. de (1996) *Chemical and Physical Dynamics of Fen Hydro-Ecology* (Utrecht) PhD Thesis University of Utrecht.

Mars, H. de; Garritsen, A.C. (1997) *Interrelationship between water quality and groundwater flow dynamics in a small wetland system along a sandy hill ridge* (j) Hydrological Processes 11: 335-351

Mars, H. de; Wassen, M.J. (1999) *Redox potentials in relation to water levels in different mire types in the Netherlands and Poland* (j) Plant Ecology 140: 41-51

Mason, C.F. (2000) *Thrushes now largely restricted to the built environment in eastern England* (j) Diversity and Distributions 6: 189-194

Maurer, B.A. (1996) *Energetics of avian foraging* In: Carey, C. [ed.] *Avian energetics and nutritional ecology* (New York) Chapman & Hall; 250-279

McDonnell, M.J.; Pickett, S.T.A.; Groffman, P.; Bohlen, P.; Pouyat, R.V.; Zipperer, W.C.; Parmelee, R.W.; Carreiro, M.M.; Medley, K. (1997) *Ecosystem processes along an urban-to-rural gradient* (j) Urban Ecosystems 1: 21-36

McGarical, K.; Marks, B.J. (1995) *Spatial Pattern Analysis Program for Quantifying Landscape Structure* (Portland, Oregon) Pacific Northwest Research Station, USDA-Forest Service

McHarg, I.L. (1967) *An ecological method for landscape architecture* (j) Landscape Architec-ture 57105-107

McHarg, I.L. (1971) *Design with nature* (Garden City) Doubleday & Co

McHarg, I.L. (1972) *An ecological method for landscape architecture* In: Shepard [ed.]

McHarg, I. (1982) *Ecological planning: The planner as catalyst* In: Sternlieb, Burchell & [ed.]; 13-15

McIntyre, N.E.; Knowles-Yanez, K.; Hope, D. (2000) *Urban ecology as an interdisciplinary field: differences in the use of "urban" between the social and natural sciences* (j) Urban Ecosystems 4: 5-24

McKeown, Th.F. (1976) *The role of medicine: dream, mirage or nemesis* (London) Nuffield Provincial Hospitals Trust

Meeus, J. (1988) *Lege decors en snelle clips; landbouw en landschap in het verenigd Europa na 1992* (j) Landschap 5(4): 286-294

Meijden, R. van der (1996) *Heukels' flora van Nederland* (Groningen) Wolters-Noordhoff

Meijden, R. van der (2005) *Heukels' Flora van Nederland* (Groningen) Wolters-Noordhoff

Melchers, M.; Timmermans, G. (1991) *Haring in het IJ* (Amsterdam) Stadsuitgeverij Amster-dam

Melosi, M.V. (2003) *The historical dimension of urban ecology: frameworks and concepts* In: Berkowitz, A.R.; Nilon, Ch. H.; Hollweg, K.S. [eds.] *Understanding urban ecosys-tems.* (New York) Springer; 187-200

Mels, T. (1999) *Wild landscapes; the cultural nature of Swedish National Parks* (Lund) Lund UP

637

Meltzer, J.; Westhoff, V. (1942) *Inleiding tot de plantensociologie* ('s Graveland) G.W. Breug-
hel
Mensvoort, K. van; Schwarz, M.; Gerritzen, M. (2005) *Next nature* (Amsterdam) BIS Publishers
Meshinev, T.; Apostolova, I.; Georgiev, V.; Dimitrov, V.; Petrova, A.; Veen, P.H. *Grasslands of
Bulgaria* (Sofia) Institute of Botany, Bulgarian Academy of Sciences in co-operation
with Royal Dutch Society for Nature Conservation
Metz, T.; Pflug, M. (1997) *Atlas van Nederland in 2005 - De nieuwe kaart* (Rotterdam) NAI
http://www.nieuwekaart.nl/
Meuleman, A.F.M.; Pedroli, G.B.M.; Runhaar, J.; Schot, P.P.; B., Vlaanderen; Wassen, M.J.;
[eds.] (1992.) *Hydro-ecologische voorspellingsmethoden voor beheer en beleid* (j)
Landschap Special Issue(2)
Meyer, V.J.; Hoekstra, M.J.; Inia, R.; Moerman, M.; Voort, S. van der (2000) *Atlas van de
Hollandse waterstad* (Delft) Technische Universiteit Delft, Faculteit Bouwkunde
Meyer, V.J. (2002) *De kern van de stedebouw - Deel 3, Het ontwerp van de openbare ruimte
Hoofdstuk Kunstwerken [in voorbereiding]* (Delft) Faculteit Bouwkunde; 5
Meyer, H.; Burg, L. van den (2005) *In dienst van de stad: 25 jaar werk van de stedebouwkun-
dige diensten van Amsterdam, Den Haag en Rotterdam* (Amsterdam) SUN
Millennium Ecosystem Assessment (2005) *Ecosystems & Human Well-being: Synthesis*
(Washington DC) Island Press
Miller, G.R.; Watson, A.; Jenkins, D. (1970) *Responses of Red Grouse populations to experi-
mental improvement of their food* In: Watson, A. [ed.] *Animal populations in relation
to their food resources* (Oxford) Blackwell; 323-334
MNP (2003) *Natuurbalans 2003* (Bilthoven) Milieu- en Natuurplanbureau, DLO, RIVM
MNP (2004) *Risico's in bedijkte termen; een thematische evaluatie van het Nederlandse
veiligheidsbeleid tegen overstromen* (Bilthoven) Milieu- en Natuurplanbureau - RIVM
rapportnummer 500799002 milieubalans@rivm.nl;
http://www.rivm.nl/bibliotheek/rapporten/500799002.html
MNP (2005) *Milieubalans 2005* (Bilthoven) Milieu- en Natuurplanbureau, DLO, RIVM
MNP (2005) *Natuurbalans 2005* (Den Haag) SDU Uitgevers
MNP (2006) *Plantenexoten in steden* (Bilthoven) Milieu- en Natuurplanbureau; Milieu- en
Natuurcompendium http://www.mnp.nl/mnc/i-nl-1193.html
MNP (2006) *Stedelijk gebied in Nederland, 1970-2000* (Bilthoven) Milieu- en Natuurplanbu-
reau; Milieu- en natuurcompendium: www.mnp.nl/mnc/i-nl-0063-print.html
MNP (2006) *Landscape typology* (Bilthoven) Milieu en Natuurplanbureau; Milieu- en natuur-
compendium: http://www.mnp.nl/mnc/i-en-1005.html
MNP (2006) *Identifiability of landscape elements and patterns* (Bilthoven) Milieu- en natuur-
planbureau; Milieu- en natuurcompendium; http://www.mnp.nl/mnc/i-en-1039.html
MNP (2006) *The international significance of landscapes* (Bilthoven) Milieu- en Natuurplanbu-
reau; Milieu- en natuurcompendium; www.mnp.nl/mnc/i-en-1034.html
MNP (2007) *De Vogel- en Habitatrichtlijnen in Nederland Een ex-ante evaluatie* (Bilthoven)
Milieu- en Natuurplanbureau
Molenaar, J.G. de (2002) *Wegverlichting en fauna, een quick scan van het proefproject alter-
natieve verlichting* Geciteerd in informatieblad Natuurmonumenten, februari 2002.
Motloch, J.L. (2001) *Introduction to Landscape Design* (New York) John Wiley & Sons
Mumford, L. (1961) *The City in History* (New York) Routledge
Myers, N.; Mittermeier, R.A.; Mittermeier, C.G.; Da Fonesca, G.A.B.; Kent, J. (2000) *Biodiver-
sity hotspots for conservation priorities* (j) Nature 40: 853-858
Nagy, K.A.; Siegfried, W.R.; Wilson, R.P. (1984) *Energy utilization by free-ranging Jackass
Penguins Spheniscus demersus* (j) Ecology 65: 1648-1655
Nagy, K.A. (1987) *Field metabolic rate and food requirement scaling in mammals and birds* (j)
Ecol. Monogr. 57: 111-128
Nagy, K.A.; Obst, B.S. (1991) *Body size effects of field energy requirements of birds: what
determines their field metabolic rates?* In: Bell, B.; et al. [eds.] *Acta XX Congressus
Internationalis Ornithologici* (Wellington New Zealand) New Zealand Ornithological
Congress Trust Board; 793-799

Natuhara, Y.; Imai, C. (1999) *Prediction of species richness of breeding birds by landscape-level factors of urban woods in Osaka Prefecture, Japan* (j) Biodiversity and Conservation 8: 239-253

Nes, R. van; Zijpp, N. J. van der (2000) *Scale-factor 3 for hierarchical road networks: a natural phenomenon?* (Delft) Trail Research School Delft University of Technology

Newton, I. (1980) *The role of food in limiting bird numbers* (j) Ardea 68: 11-30

Niemelä, J. (1999) *Is there a need for a theory of urban ecology?* (j) Urban Ecosystems 3: 57-65

Nienhuis, P.H.; Leuven, R.S.E.W.; Ragas, A.M.J. [eds.] (1998) *New concepts for sustainable management of river basins* (Leiden) Backhuys Publ.

Nieukerken, E.J. van; Loon, A.J. van (1995) *Biodiversiteit in Nederland* (Utrecht) NNM/ KNNV

Nieuwenhuijze, L. van; Toorn, M.W.M. van den; Vrijlandt, P. (1986) *Advies Landschapsbouw Groesbeek - Uitwerking van een methode voor de aanpak van landschapsstructuur-plannen* (Wageningen) Rijksinstituut voor onderzoek in de Bos- en Landschaps-bouw 'De Dorschkamp'

Nieuwenhuis, J.W.; Siffels, J.W.; Barendregt, A. (1991) *Hydro-ecological research for water management in the province of Noord-Holland, the Netherlands* In: Nachtnebel, H.P.; Kovar, K. [eds.] *Hydrological Basis of Ecologically Sound Management of Soil and Groundwater* (Wallingford) IAHS; 202; 269-278

Nieuwenhuizen, W.; Apeldoorn, R.C. van (1994) *Het gebruik van faunapassages door zoog-dieren bij Rijksweg 1 ter hoogte van Oldenzaal* (Den Haag) RWS/DWW, ABN-DLO; DWW-Versnipperingsreeks dl. 20

Nilsson, J; Grennfelt, J. P. [eds.] (1988) *Critical loads for sulphur and nitrogen* (Copenhagen) Nordic counsel of Ministers, Miljorapport 15: 1-418; UNECE/Nordic Council work-shop report, Skokloster, Sweden

NIPO (2004) *De stand van het land: de natuur* (Amsterdam) Volkskrant 9 april 2004

Nolan, B. [ed.] (1999) *Landscape - 9 + 1 Young Dutch landscape architects Rotterdam* NAI

Nolet, B.A.; Butler, P.J.; Masman, D.J.; Woakes, A.J. (1992) *Estimation of daily energy ex-penditure from heart rate and doubly labeled water in exercising geese* (j) Physiol. Zool. 65: 1188-1216

Noort, R. van; Aarts, W. (2000) *Beleving van het Amsterdamse oppervlaktewater door woon-bootbewoners* (Leiden) SWOKA

Noss, R.F. (1987) *Corridors in real landscapes: a reply to Simberloff and Cox* (j) Conservation Biology 1: 159-164

Noss, R.F. (1992) *The wildlands project: land conservation strategy* (j) Wild Earth(Special Issue): 10-25

Oberweger, K.; Goller, F. (2001) *The metabolic costs of bird song production* (j) Journal of Experimental Biology 204: 3379-3388.

Odum, E.P. (1971) *Fundamentals of ecology* (Philadelphia/London/Toronto) W.B. Saunders Co.

Odum, W.E.; Hoover, J.K. (1988) *A comparison of vascular plant communities in tidal freshwa-ter and saltwater marshes* In: Hook, D.D.; McKee, W. H.; Smith, H. K.; Gregory, J.; Burrell, V. G.; DeVoe, M. R.; Sojka, R. E.; Gilbert, S.; Banks, R.; Stolzy, L. H.; Brooks, C.; Matthews, T. D.; Shear, T. H. [eds.] Proceedings *The ecology and man-agement of wetlands* (Charleston, South Carolina, 1986); p. 526-534

Odum, E.P. (1997) *Ecology: A Bridge between Science and Society* (Massachusetts) Sinauer

Oers, R. van (2000) *Dutch Town Planning Overseas during VOC and WIC Rul [1600-1800]* (Zutphen) Walburg Pers

Oers, J.A.M. van [ed.] (2002) *'Gezondheid op Koers?' Volksgezondheid Toekomst Verkenning 2002* (Bilthoven) RIVM

Ogunseitan, O.A. (2005) *Topophilia and the quality of life* (j) Environmental Health Perspec-tives 113: 143-148

Öhman, A.; Mineka, S. (2001) *Fears, phobias, and preparedness: Toward an evolved module of fear and learning* (j) Psychological Review 108: 483-522

Okruszko, H. (1995) *Influence of hydrological differentiation of fens and their transformation after dehydration and on possibilities for restoration* In: Wheeler, In: B.D.; Shaw, S.C.; Fojt, W.J.; Robertson, R.A. [eds.] *Restoration of temperate wetlands* (Chiches-ter) Wiley; 113-119

Olde Venterink, H.G.M.; Wassen, M.J. (1997) *A comparison of six models predicting vegeta-tion response to hydrological habitat change* (j) Ecological Modelling 101: 347-361
Olwig, K. (1996) *Recovering the substantial nature of landscape* (j) Annals of the Association of American Geographers 86: 630-653
Oosterhoff, W. (1991) *Ontwerp voor een buurtpark in Breda - Hoe ecologie ook mooi kan zijn* (j) Groen 47(5)
Oostrom, C.G.J. van (1977) *Het relatie-onderzoek in de projectstudie Midden-Brabant* (j) WLO-Mededelingen 4(4): 10-11
Opdam, P. (1978) *Biogeografie van eilanden en de betekenis ervan voor de landschapsecolo-gie* (j) WLO-Mededelingen 5(1): 6-12
Opdam, P.; Rossum, T.A.W. van; Coenen, T.G. (1986) *Ecologie van kleine landschapsele-menten* (Leersum) RIN
Opdam, P. (1987) *De metapopulatie: model van een populatie in een versnipperd landschap* (j) Landschap 4(4): 288-306
Opdam, P.; Hengeveld, R. (1990) *Effecten [van versnippering] op planten- en dierpopulaties* **In:** RMNO [ed.] *De versnippering van het Nederlandse landschap: onderzoekspro-grammering vanuit zes disciplinaire benaderingen* (Rijswijk) RMNO-rapport 45
Opdam, P. (1991) *Metapopulation theory and habitat fragmentation: a review of holarctic breeding bird studies* (j) Landscape Ecol. 4: 93-106
Opdam, P. (1992) *Van patroon naar proces en terug; de relatie tussen landschapsecologisch onderzoek en toepassing* **In:** WLO Bestuur [ed.] *De Landschapsecologie, verslag WLO-Lustrumcongres 20 november 1992 Utrecht.* (Den Haag) SPB Academic Pu-blishing; 11-18
Opdam, P. (1993) *Van patroon naar proces en terug De relatie tussen landschapsecologisch onderzoek en toepassing* (j) Landschap 10: 47-54
Opdam, P. (2002) *Natuurbeleid, Biodiversiteit en de EHS: doen we het wel goed?* (Bilthoven, Wageningen) Milieu- en Natuurplanbureau-RIVM/Alterra, Werkdocument 2002/04
Opdam, P.; Foppen, R.; Vos, C. (2002) *Bridging the gap between ecology and spatial planning in landscape ecology* (j) Landscape Ecology 16: 767-779
Opdam, P.; Wiens, J. (2002) *Fragmentation, habitat loss and landscape management* **In:** Norris, K.; Pain, D. [eds.] *Conserving bird biodiversity.* (Cambridge) Cambridge Uni-versity Press, UK, pp. 202-223.
Opdam, P.; Reijnen, R.; Vos, C. (2003) *Robuuste verbindingen, nieuwe wegen naar natuur-kwaliteit* (j) Landschap 20(1): 31-37
Opdam, P.; Verboom, J.; Pouwels, R. (2003) *Landscape cohesion: an index for the conserva-tion potential of landscapes for biodiversity* (j) Landscape Ecology 18: 113-126
Opdam, P.; Wascher, D. (2004) *Climate change meets habitat fragmentation: linking land-scape andbiogeographical scale level in research and conservation* (j) Biological Conservation 117(3): 285-297
Opdam, P. (2006) *De ecologische hoofdstructuur; proeve van ontwikkelingsplanologie* (j) Stedebouw & Ruimtelijke ordening 87(2): 38-42
Orme, C.D.L.; Davies, R.G.; Burgess, M.; Eigenbrood, F.; Pickup, N.; Olson, V.A.; Webster, A.J.; Ding, T.S.; Rasmussen, P.C.; Ridgely, R.S.; Stattersfield, A.J.; Bennett, P.M.; Blackburn, T.M.; Gaston, K.J.; Owens, I.P.F. (2005) *Global hotspots of species rich-ness are not congruent with endemism or threat* (j) Nature 436(7053): 1016-1019
Oswald, F.; Baccini, P. (2003) *Netzstadt Designing the Urban* (Basel / Boston / Berlin) Birhäuser
Ottburg, F.G.W.A.; Smit, G.F.J. (2000) *Het gebruik door dieren van fauna passages van Direc-tie Utrecht* (Culemburg) Bureau Waardenburg/RWS-Utrecht.
Oudshoorn, H.M.; Schultz, B.; Urk, A. van; Zijderveld, P. (1998) *Sustainable Development of Deltas (SDD'98)* Proceedings *International conference at the occasion of 200 year Directorate-General for Public Works and Water Management* (Amsterdam) Delft University Press
Oxman, T.E.; Freeman, D.H.; Manheimer, E.D. (1995) *Lack of social participation or religious strength and comfort as risk factors for death after cardiac surgery in the elderly* (j) Psychosomatic Medicine 57[1]: 5-15.
Padding, P.; Scholten, H.J. (1988) *Ontwikkelingen in de landbouw, een ruimtelijk perspectief voor natuurontwikkeling?* (j) Landschap 5: 201-212

640

Paglia, C. (1991) *Sexual personal - Art and decadence from Nefertiti to Emily Dickinson* (London) Penguin Books

Palmboom, F. (1990) *Landschap en verstedelijking tussen Den Haag en Rotterdam* (Rotterdam) Stadsontwikkeling Rotterdam

Palmboom, F.; Bout, J. van den (1996) *Ypenburg: stadsdeel, buitenplaats en landschap tegelijk* (j) Groen 11: 36-39

Palmer, C. (1997) *Environmental ethics* Proceedings *Contemporary Ethical Issues* (Santa Barbara) ABC-CLIO

Parsons, R. (1991) *The potential influences of environmental perception on human health* (j) Journal of Environmental Psychology 11: 1-23.

Passmore, J. (1974) *Man's responsibility for nature - ecological problems and western traditions* (New York) Charles Scribers's Sons

Patri, T.; Streatfield, D.C.; Ingmire, T.J. (1970) *Early warning system - the Santa Cruz mountains regional pilot study* (Berkeley) University of California, Dept. of Landscape architecture

Pauly, D. (1995) *Anecdotes and the shifting baseline syndrome of fisheries* (j) Trends in Ecology & Environment 10: 430

Pedroli, G.B.M. (1989) *The Nature of Landscape; a Contribution to Landscape Ecology and Ecohydrology with Examples from the Strijper Aa Landscape, Eastern Brabant, the Netherlands* (Amsterdam) PhD Thesis University of Amsterdam

Pedroli, G.B.M. (1992) *Ecohydrologie; the State of the Art* (j) Landschap 9(2): 73-82

Pedroli, G.B.M.; Postma, R.; Rademakers, J.G.M.; Kerkhofs, M.J.J. (1996) *Welke natuur hoort bij de rivier? Naar een natuurstreefbeeld afgeleid van karakteristieke fenomenen van het rivierlandschap* (j) Landschap 13: 97-113

Pedroli, B.; Pinto-Correira, T.; Cornish, P. (2006) *Lanscape - what's in it? Trends in European landscape science and priority themes for concerted research* (j) Landscape Ecology 21: 421-430

Pekalska, E. (2005) *Dissimilarity representations in pattern recognition Concepts, theory and applications* (Delft) Delft University of Technology

Pelk, M. (2007) *The Natura 2000 Network in the Netherlands; Setting targets and conservation objectives* In: Jong, T.M. de; Dekker, J.N.M.; Posthoorn, R. [eds.] *Landscape ecology in the Dutch context: nature, town and infrastructure* (Zeist) KNNV-uitgeverij

Peskens, J.J. (1995) *Duindoornroosje, een geheime tuin in de Randstad* (j) Landschap 12(2): 47-49

Petts, G.E. (1996) *Water allocation to protect river ecosystems* (j) Regulated Rivers-Research & Management 12: 353-367

Pfister, H.P.; Keller, V.; Reck, H.; Georgii, B. (1997) *Bio-ökologische Wirksamkeit vor Grünbrücken ueber Verkehrwege* (Bonn-Bad Godesberg, Duitsland); Forschung Strassenbau und Strassen verkehrtechnik. Hft. 756.

Phillips, A. (2003) *Turning ideas on their head, the new paradigm for protected areas* (j) George Wright Forum 20: 8-32

Pickett, S.T.A.; Burch, W.R.; Dalton, S.E.; Foresman, T.W.; Grove, J. Morgan; Rowntree, R. (1997) *A conceptual framework for the study of human ecosystems in urban areas* (j) Urban ecosystems 1: 185-199

Pickett, S.T.A.; Cadanasso, M.L.; Grove, J.M.; Nilon, C.H.; Pouyat, R.V.; Zipperer, W.C.; Constanza, R. (2001) *Urban ecological systems: linking terrestrial ecological, physical, and socioeconomic components of metropolitan areas* (j) Annual Review of Ecology and Systematics 32: 127-157

Pienkowski, M.W.; Bignal, E.M.; Galbraith, C.A.; McCracken, DI.; Stillman, R.A.; Boobyer, M.G.; Curtis, D.J. (1996) *A simplified classification of land-type zones to assist the integration of biodiversity objectives in land-use policies* (j) Biological conservation 75: 11-25

Piepers, A.A.G. [ed.] (2001) *Infrastructure and nature; fragmentation and defragmentation; Dutch State of the Art Report for COST activity 341* (Delft); DWW-ontsnipperingsreeks nr. 39A

Pimm, S.L.; Russel, G.J.; Gittleman, J.L.; Brooks, T.M. (1995) *The future of biodiversity* (j) Science 269: 347-350

Pimm, S.L.; Raven, P. (2000) *Extinction by numbers* (j) Nature 403: 843-845

Pinkhof, M. (1905) *Door de stadsrietlanden* (j) De Levende Natuur 10: 64

Plate, C.R.; Liefaard, R.; Duuren, L. van (1992) *Veranderingen in de Nederlandse flora op basis van de Standaardlijst 1990* (j) Gorteria 18: 21-29

Poschlod, P.; Bonn, S. (1998) *Changing dispersal perspectives in the Central European land-scape since the last ice age: an explanation for the actual decrease of plant species richness in different habitats?* (j) Acta Botanica Neerlandica 47: 27-44

Pötz, H.; Bleuze, P. (1999) *Zichtbaar, tastbaar en zinvol* (Rotterdam) NAi publishers

Pouwels, R.; Jochem, R.; Reijnen, M.J.S.M.; Hensen, S.R.; Grift, J.G.M. van der (2002) *LARCH voor ruimtelijke ecologische beoordelingen van landschappen* (Wageningen) Alterra

Pouyat, R.V.; McDonnell, M.J. (1991) *Heavy metal accumulation in forest soils along an urban to rural gradient in southern NY, USA* (j) Water, Air & Soil Pollution 57/58: 797-807.

Prendergast, J.R.; Quinn, R.M.; Lawton, J.H.; Eversham, B.C.; Gibson, D.W. (1993) *Rare species, the coincidence of diversity hotspots and conservation strategies* (j) Nature 365: 335-337

Priemus, H. (2003) *Systeeminnovatie ruimtelijke ontwikkeling* (Delft) TUD Intreerede

Priemus, H. (2005) *Naar een systeeminnovatie voor ruimtelijke ontwikkeling [Towards a system innovation for spatial development]* (j) Nova Terra 5(3, October): 9-13

Pryor, R.J. (1968) *Defining the rural-urban fringe* (j) Soc. Forces 47: 202-215

Pullian, R. (1988) *Sources-sinks, and population regulation* (j) American Naturalist 132: 652-661

Putten, W.H. van der; Vet, L.E.M.; Harvey, J.A.; Wäckers, F.L. (2001) *Linking above- and belowground multitrophic interactions of plants, herbivores, pathogens and their antagonists* (j) Trends in Ecology and Evolution 16: 547-554

Pyšek, P. (1989) *On the richness of Central European urban flora* (j) Preslia 61: 329-334

PZH (2005) *Stedenbaan Zuidvleugel* (Den Haag) Provincie Zuid-Holland

Raemakers, I.P.; Schaffers, A.P.; Sykora, K.V.; Heijerman, T. (2001) *The importance of plant communities in road verges as a habitat for insects* In: Proceedings Section Experimental and Applied Entomology NEV 12; 101-106

Reest, P.J. van der; Bekker, G.J. (1990) *Kleine zoogdieren in wegbermen* (j) Meded.Vereniging voor Zoogdierkunde en Zoogdierbescherming. 1

Reid, W.V. (1998) *Biodiversity hotspots* (j) Trends in Ecology & Evolution 13: 275-280

Reijnen, M.J.S.M. (1995) *Disturbance by car traffic as threat to breeding birds in the Netherlands* (Leiden) PhD Thesis Universiteit Leiden

Reijnen, M.J.S.M.; Veenbaas, G.; Foppen, R.P.B. (1995) *Predicting the effects of motorway traffic on breeding bird populations* (Wageningen) Road and Hydraulic Engineering Division and DLO-Institute for Forestry and Nature Research

Reijnen, M.J.S.M.; R.Foppen (2006) *Impact of road traffic on breeding bird populations* In: Davenport, J.; Davenport, J. L. [eds.] *The ecology of transportation: Managing mobility for the environment* (New York/Berlin) Springer Environmental Pollution 10

Reijnen, M.J.S.M; Hinsberg, A. van; Lammers, G.W.; Sanders, M.E.; Loonen, W. (2007) *Optimising the Dutch National Ecological Network; Spatial and environmental conditions for a sustainable conservation of biodiversity* In: Jong, T.M. de; Dekker, J.N.M.; Posthoorn, R. [eds.] *Landscape ecology in the Dutch context: nature, town and infrastructure* (Zeist) KNNV-uitgeverij

Reneman, D.; Visser, M. (1999) *Mensenwensen De wensen van Nederlanders ten aanzien van natuur en groen in de leefomgeving* (Hilversum, Wageningen, Den Haag) Intomart, SC-DLO, LNV Reeks Operatie Boomhut nr. 6.

Renes, H. (2001) *Ministerie van LNV faalt met historische cultuurlandschappen* (j) Geografie 10(4): 47-49

Renes, J. (2003) *Nieuwe natuur in een oud landschap; cultuurhistorie en natuurontwikkeling in het rivierengebied* (j) De Levende Natuur 104: 272-274

Renes, J. (2005) *Wildparken in Nederland; sporen van een oude vorm van faunabeheer* (j) Historisch-Geografisch Tijdschrift 23: 21-34

Renes, H. (2006) *Landschap en de EHS; EHS in het landschap* (j) Landschap 23(3): 109-120

Renes, J. (2007) *Landscape in the Dutch 'National Ecological Network'* In: Jong, T.M. de; Dekker, J.N.M.; Posthoorn, R. [eds.] *Landscape ecology in the Dutch context: nature, town and infrastructure* (Zeist) KNNV-uitgeverij

Rensen-Bronkhorst, R. (1993) *Atlas van de flora van Eindhoven 1980-1989* (Eindhoven) KNNV

Reumer, J.W.F. (2000) *Stadsecologie; de stedelijke omgeving als ecosysteem* (Rotterdam) Natuurmuseum; Stadsecologische reeks

Ricketts, T.H.; Daily, G.C.; Ehrlich, P.R. (2002) *Does butterfly diversity predict moth diversity? Testing a popular indicator taxon at local scales* (j) Biological Conservation 103: 361-370

Ricklefs, R.E. (1996) *Avian energetics, ecology and evolution* **In:** Carey, C. [ed.] *Avian energetics and nutritional ecology* (New York) Chapman & Hall; 1-30

Riemsdijk, M.J. van [ed.] (1999) *Dilemma's in de bedrijfskundige wetenschap* (Assen) Van Gorcum

Rientjes, S.; Roumelioti, K. (2003) *Support for ecological networks in European nature conservation, an indicative social map* (Tilburg) ECNC

RIGO, Research en Advies (2004) *Leefbaarheid van wijken* (Den Haag) Ministerie van VROM

Rijnland, Hoogheemraadschap van (2002) *De Waterparagraaf in bestemmingsplannen* (Leiden) Hoogheemraadschap van Rijnland; http://www.watertoets.net/

RIVM; IKC Natuurbeheer; IBN-DLO; SC-DLO (1997) *Natuurverkenning 97* (Alphen a/d Rijn) Samsom H.D. Tjeenk Willink

RIVM (1998) *Natuurbalans 1998* (Alphen a/d Rijn) Samsom H.D. Tjeenk Willink

RIVM (2000) *Nationale Milieuverkenning 5 2000-2030* (Bilthoven/Alphen aan den Rijn.) RIVM/Samsom H.D. Tjeenk Willink

RIVM (2001) *Nationale Atlas Volksgezondheid* CBS Statline; (Bilthoven) Rijksinstituut voor Volksgezondheid en Milieu

RIVM (2005) *Natuurbalans 2005* (Bilthoven) Milieu en Natuur Planbureau

Rodenacker, W. G. (1970) *Methodisches Konstruieren* (Berlin / Heidelberg / New York) Springer-Verlag

Rodrigues-Iturbe, I. (2000) *Ecohydrology: A hydrological perspective of climate-soil-vegetation dynamics* (j) Water Resources Research 36: 3-9

Roelofs, J.G.M. (1983) *Impact of acidification and eutrophication on macrophyte communities in soft waters in the Netherlands; I Field observations; II Experimental Studies* (j) Aquatic Botany 17: 139-155; 389-411

Roelofs, J.G.M. (1991) *Inlet of alkaline river water into peaty lowlands: effects on water quality and Stratiotes aloides L. stands* (j) Aquatic Botany 39: 267-293

Roelofs, J.G.M.; Bobbink, R.; Brouwer, E.; Graaf, M.C.C. de (1996) *Restoration ecology of aquatic and terrestrial vegetation on non-calcareous sandy soils in the Netherlands* (j) Acta Botanica Neerlandica 45(4): 517-541

Roggema, R.E.; Wiersinga, W.A.; Zonderland, H.G.F. (1994) *Ecologische sturing van de ruimtelijke planvorming - Denken en doen in de Bredase praktijk* (j) Groen 50(4): 12-17

Rooijen, M. van (1984) *De groene stad; een historische studie over de groenvoorziening in de Nederlandse stad* (Den Haag) Stichting Cultuurfonds

Ross Ashby, W. (1957, 1956) *An Introduction to cybernetics* (New York) Wiley

Ross Ashby, W. (1965) *Design for a brain The origin of adaptive behaviour* (London) Chapman & Hall and Science Paperbacks

RPD (1966) *Tweede Nota Ruimtelijke Ordening* (Den Haag) Staatsuitgeverij

Runhaar, J. (1989) *Toetsing ecotopensysteem: Relatie tussen de vochtindicatie van de vegetatie en grondwaterstanden* (j) Landschap 6(2): 129-146

RWS (1998) *Delta's of the World* **In:** Oudshoorn, H.M.; Schultz, B.; Urk, A. van; Zijderveld, P. [eds.] Proceedings *Sustainable Development of Deltas SDD '98. International conference at the occasion of 200 year Directorate-General for Public Works and Water Management* (Delft) Delft University Press

RWS (1998) *Summary and Conclusions [SDD '98]* **In:** Oudshoorn, H.M.; Schultz, B.; Urk, A. van; Zijderveld, P. [eds.] Proceedings *International conference at the occasion of 200 year Directorate-General for Public Works and Water Management* (Amsterdam) Delft University Press

RWS (1998) *Sustainable Development of Deltas [SDD '98]* **In:** Oudshoorn, H.M.; Schultz, B.; Urk, A. van; Zijderveld, P. [eds.] Proceedings *Inetrnational conference at the occa-*

sion of 200 year Directorate-General for Public Works and Water Management
(Amsterdam) Delft University Press

Ryan, R.M.; Deci, E.L. (2000) *The darker and brighter sides of human existence: basic psychological needs as a unifying concept* (j) Psychological Inquiry 11[4]: 319-338

Ryan, R.M.; Deci, E.L. (2001) *On happiness and human potentials: a review of research on hedonic and eudaimonic well-being* (j) Annual Review of Psychology 52: 141-66

S&V (1970) *Bodem en planologie* (j) Stedebouw en Volkshuisvesting 51(aug.): 60 p + bijl.

Salet, W.; Faludi, A. [eds.] (2000) *The revival of strategic spatial planning* (Amsterdam) KNAW

Sanderson, E.W.; Jaiteh, M.; Levy, M.A.; Redford, K.H.; Wannebo, A.V.; Woolmer, G. (2002) *The Human Footprint and the Last of the Wild* (j) BioScience 52(10): 891-904

Saris, F.J.A.; Wilders, G.Th. A. (1981) *Landschapsecologisch onderzoek, knelpunten en wensen* (j) Mededelingen van de WLO 8(3): 97-105

Sassen, S. (2001) *The Global City: New York, London, Tokyo* (Oxford) Princeton University Press

Schama, S. (1995) *Landscape and memory* (New York) Knopf

Schaminee, J.H.J.; Stortelder, A.H.F.; Westhoff, V. (1995) *De vegetatie van Nederland deel 1 Grondslagen, methoden en toepassingen* (Leiden) Opulus press

Scheffer, M. (1998) *Ecology of shallow lakes* (xxx) Chapman & Hall; Population and Community Biology Series 22

Schimmel, H.J.W.; Thalen, D.C.P.; Westhoff, V. (1985) *Chris van Leeuwen, bouwmeester van het natuurbeheer* (j) De levende natuur 86(3): 66-73

Schoener, T.W. (1968) *Sizes of feeding territories among birds* (j) Ecology 49: 123-141

Schone, M.B.; Coeterier, J.F.; Toorn, M.W.M. van den (1997) *Autosnelwegen in het landschap Beleving door weggebruikers* (Wageningen) DLO-Staring Centrum

Schot, P.P. (1991) *Solute Transport by Groundwater Flow to Wetland Ecosystems* (Utrecht) PhD Thesis University of Utrecht

Schot, P.P.; Molenaar, A. (1992) *Regional changes in groundwater flow patterns and effects on groundwater composition* (j) Journal of Hydrology 130: 151-170

Schot, P.P.; Wassen, M.J. (1993) *Calcium concentrations in wetland groundwater in relation to water sources and soil conditions in the recharge area* (j) Journal of Hydrology 141: 197-217

Schouten, M.A.; Verwey, P.A.; Barendregt, A. (2007) *Biodiversity and the Dutch National Ecological Network* In: Jong, T.M. de; Dekker, J.N.M.; Posthoorn, R. [eds.] *Landscape ecology in the Dutch context: nature, town and infrastructure* (Zeist) KNNV-uitgeverij

Schouten, M.A.; Barendregt, A.; Verweij, P.A. (in prep. a) *Representing biodiversity: Spatial analysis of multi-taxon species composition in the Netherlands*

Schouten, M.A.; Barendregt, A.; Verweij, P.A. (in prep. b) *Species occurrence databases: inevitably biased, but towards what and does it matter?*

Schouten, M.A.; Verweij, P.A.; Ruiter, P.C. de; Barendregt, A. (in prep. c) *Determinants of species richness distribution in the Netherlands across multiple taxonomic groups*

Schrijnen, J. (2007) *The task of regional design* In: Jong, T.M. de; Dekker, J.N.M.; Posthoorn, R. [eds.] *Landscape ecology in the Dutch context: nature, town and infrastructure* (Zeist) KNNV-uitgeverij

Schroevers, P.J. (1982) *Landschapstaal; een stelsel van basisbegrippen voor de landschapsecologie* (Wageningen) Pudoc; Reeks Landschapsstudies 2

Schultz, E. (1992) *Waterbeheersing van de Nederlandse droogmakerijen* (Lelystad) PhD thesis, University of Technology Delft; Ministry of Transport, Public Works and Water Management, Directorate Flevoland; Van Zee tot Land 58

Scott, J.M.; Csuti, B.; Noss, R.F.; Butterfield, B.; Groves, C.; Anderson, H.; Caicco, S.; D'Erchia, F.; Edwards, T.C.; Ulliman, J.; Wright, R.G. (1993) *Gap analysis: a geographical approach to protection of biological diversity* (j) Wildlife Monographs 123: 1-31

Scott Findlay, C.; Bourdages, J. (1999) *Response Time of Wetland Biodiversity to Road Construction on Adjacent Lands* (j) Conservation Biology 14(1): 86-94

SCW (1972) *Begroeiing wegbermen en erosiebestrrijding [Dutch Road Construction Research Association: Roadside vegetations and erosion control]* (j) Studiecentrum Wegenbouw Communication, SCW Amhem 29

SCW (1975:) *Het onderhoud van begroeiing op wegbermen en taluds [Maintenance of ve-
getation on roadsides and slopes]* (j) Studiecentrum Wegenbouw Communication,
SCW Arnhem 37
SCW (1976) *Wegbeplantingen [Roadside plantings]* (j) Studiecentrum Wegenbouw Communi-
cation, SCW Arnhem 41
Sebba, R. (1991) *The landscapes of childhood The reflection of childhood's environment in
adult memories and in children's attitudes* (j) Environment and Behavior 23(4): 395
Seiberth, H. (1981) *Stadtökologie - Naturschutz und Landschapspflege in der Großstadt* **In:**
Andritzky, M.; Spitzer., K. [eds.] *Grün in der stadt.* (Hamburg) Rowohlt; 154-190
Seiler, A.; Hellding, J.O. (2006) *Mortality in wildlife due to transportation* **In:** Davenport, J.;
Davenport, J. L. [eds.] *The ecology of transportation: Managing mobility for the envi-
ronment* (New York/Berlin) Springer Environmental Pollution 10; 165-190
Sepp, K.; Kaasik, A.; Tederko, Z. (2002) *Development of national ecological networks in the
Baltic countries in the framework of the Pan-European Ecological network*
(Gland/Cambridge) IUCN Central Europe
Shear McCann, K. (2000) *The diversity-stability debate* (j) Nature 405: 228-233
Sheppard, C. (1995) *The shifting baseline syndrome* (j) Marine Pollution Bulletin 30: 766-767
Sievert, Th. (2003) *Cities without Cities: an interpretation of the Zwischenstadt* (London)
Routledge
Sigler, J. [ed.] (1995) *S, M, L, XL - Small, Medium, Large, Extra-Large - Office for Metropolitan
Architecture - Rem Koolhaas and Bruce Mau* (Rotterdam) 010
Sijmons, D.F.; Herngreen, G.F.W.; Kwakernaak, C.; Nieuwenhuijze, L. van; Philippa, M.;
Zadelhoff, F.J. van (1990) *Multiplex - "een bouwsteen voor zonering van de landelij-
ke ruimte"* (Den Haag) RPD
Sijmons, D.F. (1992) *Het casco-concept - een benaderingswijze voor de landschapsplanning*
(Utrecht) Hoofdafdeling Landschapsontwikkeling, Dir. Bos- en landschapsbouw
Sijmons, D. (1998) *= LANDSCHAP* (Amsterdam) Architectura + Natura
Sijmons, D. (2003) *Proposals for the studio of the government architect*
Simonds, J.O. (1961) *Landscape architecture - The shaping of man's natural environment*
(New York) McGraw-Hill
Simonds, J.O. (1997) *Landscape Architecture - A manual of site planning and design* (New
York) McGraw-Hill
Sival, F.P.; Grootjans, A.P. (1996) *Dynamics of seasonal bicarbonate supply in a dune slack:
effects on organic matter, nitrogen pool and vegetation succession* (j) Vegetatio 126:
39-50
Sival, F.P. (1997) *Dune Soil Acidification Threatening Rare Plant Species* (Groningen) PhD
Thesis University of Groningen
Sloep, P.B. (1983) *Patronen in het denken over vegetaties Een kritische beschouwing over de
relatietheorie* (Groningen) Regenboog
Smidt, J. de (2006) *De doelgroep van de WLO* (j) Landschap 23(4): 181
Smidt, J. de (2007) *Introduction* **In:** Jong, T.M. de; Dekker, J.N.M.; Posthoorn, R. [eds.] *Land-
scape ecology in the Dutch context: nature, town and infrastructure* (Zeist) KNNV-
uitgeverij
Smit, G.F.J. (1996) *Het gebruik van faunapassages bij rijkswegen Overzicht en onderzoeks-
plan* (Delft) Dienst Weg- en Waterbouwkunde; DWW-ontsnipperingsreeks deel 29
Snep, R.P.H. (2003) *How peri-urban areas act as source for nature quality in cities* (j) Landschap 20(5): 32
Sobolev, N. A.; Shvarts, E. A.; Kreindlin, M.L.; Mokievsky, V.O.; Zubakin, V.A. (1995) *Russia's
protected areas: a survey and identification of development problems* (j) Biodiversity
and Conservation 4(9): 964-83
Sonneman, J.A.; Walsh, C.J.; Breen, P.F.; Sharpe, A.K. (2001) *Effects of urbanization on
streams of the Melbourne region, Victoria, Australia II Benthic diatom communities*
(j) Freshwater Biology 46(4): 553-565
SOVON; CBS (2006) *Trends van vogels in het Nederlandse Natura 2000 netwerk* (Beek-
Ubbergen) SOVON Vogelonderzoek Nederland
Staatsbosbeheer (1985) *Landschapsarchitectuur* (Utrecht) Tentoonstelling van Staatsbosbeheer - Landschapsbouw

Staay, A. van der (1998) *Vignetten voor een landschap - Vignettes for a landscape* In: Harsema et al. [ed.]; 12-27

Standaard; Elmar (1999) *Zo heet dat: Beeldwoordenboek* (Antwerpen/Rijswijk) Standaard Multimedia/Elmar multimedia

Star, S.L.; Griesemer, J.R. (1989) *Institutional ecology, 'translations' and boundary objects: amateurs and professionals in Berkeley's Museum of vertebrate zoology, 1907-1939* (j) Social Studies of Science 19: 387-420

Steenhuis, M.; Hooimeijer, F.L. (2003) *Herinneringen aan twintig bewogen jaren* (j) Blauwe Kamer 2003(1): 52-57

Steiner, F.R. (1991) *The living landscape - An ecological approach to landscape planning* (New York) McGraw-Hill

Stephens, D.W.; Krebs., J.R. (1987) *Foraging theory* (Princeton, New Jersey) Princeton University Press

Sternberg, E.M. (2000) *The Balance Within: the science connecting health and emotions* (New York) Freeman

Stevens, S. (1997) *The legacy of Yellowstone* In: Stevens, S. [ed.] *Conservation through cultural survival; indigenous people and protected areas.* (Washington/Covelo) Island Press; 13-32

Stevers, R.A.M.; Runhaar, J.; Udo de Haes, H.A.; Groen, C.L.G. (1987) *Het CMLecotopensysteem, een landelijke ecosysteemtypologie toegespitst op de vegetatie* (j) Landschap 4(2): 135-150

Stolker-Nanninga, L.; Verschuuren, S. (1989) *A park city between court and port*

Stortelder, A.H.F.; Platjes, E.; Giesen, Th.G. (1995) *Ecologisch beheer van beplantingen langs rijkswegen in Drenthe* (Wageningen) IBN, report 150

STOWA (2002) *WaterNood Instrumentarium* (Lelystad) STOWA

Streefkerk, N. (1980) *Ecologie, planvorming en landschapsarchitectuur* (j) WLO-med. 7(4)

Stringer, L.A.; McAvoy, L.H. (1992) *The need for something different: spirituality and the wilderness adventure* (j) The Journal of Experimental Education 15[1]: 13-21

Stuyfzand, P.J. (1996) *Hydrochemistry and Hydrology of the Coastal Dune Area of the Western Netherlands* (Amsterdam) PhD Thesis Vrije Universiteit Amsterdam

Suhonen, J.; Jokimaki, J. (1988) *A biogeographical comparison of the breeding bird species assemblages in twenty Finnish urban parks* (j) Ornis Fennica 65: 76-83

Sukopp, H. (1990) *Stadtökologie; Das Beispiel Berlin* (Berlin) G.Fisher

Sukopp, H. (1990) *Urban ecology and its application in Europe* In: Sukopp, H.; Hejný, S. [eds.] *Urban ecology.* (The Hague) SPB Academic Publishing; 1-22

Sukopp, H.; Hejný, S. (1990) *Urban ecology* (The Hague) SPB Academic Publishing

Sukopp, H.; Wittig, R. (1993) *Stadtökologie* (Stuttgart) G. Fisher

Sukopp, H.; Numata, M.; Huber, A. [eds.] (1995) *Urban ecology as the basis of urban planning* (The Hague) SPB Academic Publishing

Sukopp, H. (1998) *Urban ecology - scientific and practical aspects* In: Breuste, J.; Feldmann, H.; Uhlmann, O. [eds.] *Urban ecology.* (Berlijn) Springer; 3-16

Sukopp, H. (2002) *On the early history of urban ecology in Europe* (j) Preslia Praha 74: 373-393

Sukopp, H. (2004) *Human-caused impact on preserved vegetation* (j) Landscape and Urban Planning 68: 347-355

Suter, W.; Eerden, M.R. Van (1992) *Simultaneous mass starvation of wintering diving ducks in Switzerland and the Netherlands: A wrong decision in the right strategy?* (j) Ardea 80: 229-242

Sutherland, W.J. (1996) *From individual behaviour to population ecology* (Oxford) Oxford University Press

Sykora, K.V.; Nijs, L.J. de; Pelsma, T.A.H.M. (1993) *Plantengemeenschappen van Nederlandse wegbermen* (Utrecht); KNNV-Natuurhist.Bibl.nr. 59

Sykora, K.V. (1998) *Wegen naar meer verscheidenheid* (Wageningen) Landbouwuniversiteit Wageningen, Inaugurale rede

Talbot, J.F.; Kaplan, S. (1986) *Perspectives on wilderness: reexamining the value of extended wilderness experiences* (j) Journal of Environmental Psychology 6: 177-188

Tamis, W.L.M.; Zelfde, M. van't; Meijden, R. van der; Groen, C.L.G.; Udo de Haes, H.A.
(2005) *Ecological interpretation of changes in the dutch flora in the 20th century* (j)
Biological Conservation 125: 211-224

Tardiff, B.; DesGranges, J.L. (1998) *Correspondance between bird and plant hotspots of the
St Lawrence river and influence of scale on their location* (j) Biological Conservation
84: 53-63

Taylor, S.E., R.L. Repetti; Seeman, T. (1997) *Health Psychology: what is an unhealthy environment and how does it get under the skin?* (j) Annual Review of Psychology 48:
411-447.

Teisman, G. (1995) *Complexe besluitvorming, een pluricentrisch perspectief op besluitvorming
over ruimtelijke investeringen* (Den Haag) Vuga

Teisman, G. (2006) *Stedelijke Netwerken, ruimtelijke ontwikkeling door het verbinden van
bestuurslagen [Urban Networks, spatial development through linking layers of
government]* (The Hague) Nirov

Tenner, W.A.; Belfroid, A.C.; Hattum, A.G.M. van; Aiking, H. (1997) *Ecologische aspecten bij
het bodemsaneringsbeleid in Amsterdam* (Amsterdam) IVM

Thayer, R.L. (1994) *Gray World, Green Heart, Technology, Nature and the Sustainable Landscape* (New York) Willey and Sons

Thayer, R.L. (1998) *Landscape as an Ecologically Revealing Landscape* (j) Landscape Journal Special Issue 1998: 118-130

Thijsse, J.P. (1899) *De Bijstad* (j) De Levende Natuur 4: 137-143

Thomas, K. (1983) *Man and the natural world - A history of modern sensibility* (New York)
Pantheon Books

Thomas jr., W.L. (1956 / 1974) *Man's role in changing the face of the earth, Vol 1* (Chicago)
University of Chicago Press

Thomas jr., W.L. (1956 / 1974) *Man's role in changing the face of the earth, Vol 2* (Chicago)
University of Chicago Press

Tilman, D. (2000) *Causes, consequences and ethics of biodiversity* (j) Nature 405: 208-211

Tjallingii, S. P.; Veer, A. A. de [eds.] (1982) *Perspectives in Landscape Ecology* (Wageningen)
Pudoc

Tjallingii, S.P. (1991) *Water en wijkontwerp, stadsecologische stappen in de planvorming* (j)
Landschap 8(1): 15-32

Tjallingii, S.P. (1995) *Ecopolis, strategies for ecologically sound urban development* (Leiden)
Backhuys Publishers

Tjallingii, S.P. (1996) *Ecological Conditions: strategies and structures in environmental planning* (Delft) PhD Thesis Technische Universiteit Delft Faculteit Bouwkunde

Tjallingii, S. (1998) *Stedelijk waterbeheer - een ecologisch perspectief* (Wageningen)

Tjallingii, S.P. (2000) *Ecology on the edge: Landscape and ecology between town and country*
(j) Landscape and Urban Planning 48: 103-119

Tjallingii, S. P. (2003) *Green and Red: Enemies or Allies? The Utrecht experience with green
structure planning* (j) Built Environment 29: 107-116

Tjallingii, S. P. (2005) *Carrying Structures: Urban Development Guided by Water and Traffic
Networks* In: Hulsbergen, E.D.; Klaasen, L.T.; Kriens, L. [eds.] *Shifting Sense, looking back to the future in spatial planning.* (Amsterdam) Techne Press; 355-368

Tjallingii, S. (2007) *From green belt to green structure; green areas and urban development
strategies in European cities* In: Jong, T.M. de; Dekker, J.N.M.; Posthoorn, R. [eds.]
Landscape ecology in the Dutch context: nature, town and infrastructure (Zeist)
KNNV-uitgeverij

Tjallingii, S.; Jonkhof, J. (2007) *Infranature; Infrastructure works, landscape and nature: experiences and debates in the Netherlands* In: Jong, T.M. de; Dekker, J.N.M.; Posthoorn, R. [eds.] *Landscape ecology in the Dutch context: nature, town and infrastructure* (Zeist) KNNV-uitgeverij

Tonneijck, A.E.G.; Blom-Zandstra, M. (2002) *Landschapselementen ter verbetering van de
luchtkwaliteit rond de Ruit van Rotterdam [Landscape elements for improvement of
the air quality around the area "de Ruit van Rotterdam"]* (Wageningen) Plant Research International

Toorn, M. van den (2005) *An integrated approach of sustainability in landscape architecture - The creation of new qualities in historic landscapes* (j) IASME TRANSACTIONS 2(6): 1028-1037

Toorn, M. van den (2006) *Landscape design at the regional level in Holland - Forestry and landscape planning at different levels of intervention* In: Larcher, F. [ed.]

Toorn, M. van den (2007) *Ecology and landscape architecture; Nature and culture in the Dutch landscape* In: Jong, T.M. de; Dekker, J.N.M.; Posthoorn, R. [eds.] *Landscape ecology in the Dutch context: nature, town and infrastructure* (Zeist) KNNV-uitgeverij

Toth, J. (1962) *A theoretical analysis of groundwater flow in small drainage basins* (j) J. Geophys. Res. 68: 4795-4812

Trautmann, A.; Vivier, E. (2001) *Agrin: a bridge between the nervous and immune systems* (j) Science 292: 1667-1668

Trepl, L. (1995) *Towards a theory of urban biocoenoses* In: Sukopp, H.; Numata, M.; Huber, A. [eds.] *Urban ecology as the basis of urban planning.* (The Hague) SPB Academic Publishing; 3-22

Trocme, M.; Cahill, S.; Vries, J.G. de; Farrall, H.; Folkeson, L.G.; Hichks, C.; Peymen, J. [eds.] (2003) *COST 341 - Habitat Fragmentation due to Transportation Infrastructure: The European Review* (Luxemburg) Office for official publications of the European Communities

Tromp, H.; Reitsma, J.M. (2001) *Flora- en vegetatie-inventarisatie in wegbermen van Rijkswaterstaat: Meetjaren 2000-2001* (Culemburg) Bureau Waardenburg i.o.v. RWS/DWW

Turner, T. (1992) *Open Space Planning in London, from standards per 1000 to green strategy* (j) Town Planning Review 63 (4): 365-386

Turner, T. (1998) *Landscape Planning and Environmental Impact Design* (London) UCL Press

Turner, R.K.; Bergh, J.C.J.M. van den; Söderquist, T.; Barendregt, A.; Straaten, J. van der; Maltby, E.; Ierland, E.C. van (2000) *Ecological-economic analysis of wetlands: scientific integration for management and policy* (j) Ecological Economics 35: 7-23

Turnhout, C.A.M. van; Foppen, R.P.B.; Leuven, R.S.E.W.; Siepel, H.; Esselink, H. (2007) *Scale-dependent homogenization: Changes in breeding bird diversity in the Netherlands over a 25-year period* (j) Biological Conservation 134: 505-516

UN Population Division (1999) *The World at Six Billion* (New York) United Nations Department of Economic and Social Affairs

UN Population Division (2003) *World Urbanization Prospects: The 2003 Revision* (New York) United Nations Department of Economic and Social Affairs

UNFPA (1996) *State of World Population 1996: Changing Places: Population Development and the Urban Future* (New York) United Nations Population Fund

UNFPA (2004) *State of World Population 2004: The Cairo Consensus at Ten: Population, Reproductive Health and the Global Effort to End Poverty* (New York) United Nations Population Fund

V&W (1990) *Tweede Structuurschema Verkeer en Vervoer* (Den Haag) Ministerie van Verkeer en Waterstaat

V&W (1998) *Vierde Nota waterhuishouding Regeringsbeslissing* (Den Haag) Ministerie van Verkeer & Waterstaat / Ando bv

V&W (2000) *Anders omgaan met water Waterbeleid in de 21e eeuw* (Den Haag) Ministerie van Verkeer en Waterstaat

V&W (2001) *Nationaal Verkeer en Vervoer Plan* (Den Haag) Ministerie van Verkeer en Waterstaat

Valentini, R.; Mateucci, G.; Dolman, A.J.; et al. (2000) *Respiration as the main determinant of carbon balance in European forests* (j) Nature 404: 861-865

Vane-Wright, R.I.; Humphries, C.I.; Williams, P.H. (1991) *What to protect? Systematics and the agony of choice* (j) Biological conservation 55: 235-254

Veenbaas, G.; Brandjes, G.J. (1999) *The use of fauna-passages along waterways under motorways* In: Dover, J.W.; Bunce, R.G.H. [eds.] Proceedings *Key Concepts in Landscape Ecology* (Preston, England) IALE UK; p. 315-320

Veer, M.A.C.; Reitsma, J.M.; Meijer, A.J.M. (1998) *Ontwikkelingen in bermvegetaties rijksweg A58, Onderzoeksresultaten 1993 t/m 1996 en een vergelijking met 1988* (Culemburg) Bureau Waardenburg en Rijkswaterstaat Zeeland

Veer, K. van der (?) *Nederland/Waterland [map]* (Amsterdam) Eurobook productions

648

Veling, K. (1997) *Verspreidingsonderzoek doelsoorten EHS, overkoepelende rapportage* (Wageningen) Vereniging Onderzoek Verspreiding Flora en Fauna

Ven, G.P. van de [ed.] (2004) *Man-made lowlands - History of water management and land reclamation in the Netherlands* (Utrecht) Matrijs

Vera, F. (1997) *Metaforen voor de wildernis* ('s-Gravenhage) Ministerie van Landbouw, Natuurbeheer en Visserij

Vera, F. (1998) *Ackeren tussen de Middeleeuwen en het Derde Millennium; natuurontwikkeling als reis door de tijd* In: Feddes, F.; Herngreen, R.; Jansen, S.; Leeuwen, R. van; Sijmons, D. [eds.] *Oorden van onthouding; nieuwe natuur in verstedelijkend Nederland.* (Rotterdam) NAi; 100-105

Verberk, W.C.E.P.; Duinen, G.A. van; Peeters, T.M.J.; Esselink, H. (2001) *Importance of variation in water-types for water beetle fauna [Coleoptera] in Korenburgerveen, a bog remnant in the Netherlands* (j) Proceedings Section Experimental and Applied Entomology NEV 12: 121-128

Verberk, W.C.E.P.; Brock, A. M.T.; Duinen, G.A. van; Es, M. van; Kuper, J.T.; Peeters, T.M.J.; Smits, M.J.A.; Timan, L.; Esselink, H. (2002) *Seasonal and spatial patterns in macroinvertebrate assemblage in a heterogeneous landscape* (j) Proceedings Section Experimental and Applied Entomology NEV 13: 35-43

Verboom, J.; Schotman, A.; Opdam, P.; Metz, J.A.J. (1991) *European Nuthatch metapopulations in a fragmented agricultural landscape* (j) Oikos 61: 149-156

Verboom, J.; Foppen, R.; Chardon, P.; Opdam, P.; Luttikhuizen, P. (2001) *Introducing the key patch approach for habitat networks with persistent populations: An example for marshland birds* (j) Biological Conservation 100: 89-101

Vereijken, P.H.; Jansen, M.J.W.; Akkermans, W.M.L.; Staritsky, I.G. (2005) *Kan het natuurbeleid ruimtelijk-economisch efficiënter?* (Wageningen) Plant Research International

Vereniging tot Behoud van Natuurmonumenten (1956) *Vijftig jaar natuurbescherming in Nederland - Gedenkboek uitgegeven ter gelegenheid van het Gouden jubileum van de Vereeniging tot Behoud van Natuurmonumenten in Nederland* (Amsterdam) Vereniging tot Behoud van Natuurmonumenten

Verhagen, R.; Diggelen, R. van; Bakker, J.P. (2003) *Natuurontwikkeling op minerale gronden - veranderingen in de vegetatie en abiotische omstandigheden gedurende de eerste tien jaar na ontgronden* (Groningen) Rijksuniversiteit Groningen, Lab. voor Plantenecologie

Verhoeven, J.T.A.; Keuter, A.; Logtestijn, R. van ; Kerkhoven, M.B. van ; Wassen, M. (1996) *Control of local nutrient dynamics in mires by regional and climatic factors: a comparison of Dutch and Polish sites* (j) Journal of Ecology 84: 647-656

Vermeulen, H.J.W. (1995) *Road-side verges: habitat and corridor for carabid beetles of poor sandy and open areas* (Wageningen) PhD Thesis Landbouwuniversiteit

Videler, J.J.; Nolet, B.A. (1990) *Costs of swimming measured at optimum speed: scale effects, differences between swimming styles, taxonomic groups and submerged and surface swimming* (j) Comp. Biochem. Physiol. 97A: 91-99

Vink, A.P.A. (1980) *Landschapsecologie en landgebruik* (Utrecht) Bohn, Scheltema & Holkema

Vista (1995) *Uit de klei getrokken* (Amsterdam) VISTA

Vista (1996) *Uit de klei getrokken* (Amsterdam) Report and poster commissioned by province North-Holland

Vista (1998) *Landschapsplan Zanderij Crailo* (Amsterdam) Commissioned by Goois Natuur Reservaat

Vista (2005a) *Inrichtingsplan Drijflanen-Wandelbos; part of Groene Mal Tilburg* (Tilburg) Concept report commissioned by Gemeente Tilburg, Gebiedsteam West.

Vista (2005b) *Update Econet Slotervaart* (Amsterdam) Concept report, commissioned by Stadsdeel Amsterdam Slotervaart

Vista (2005c) *Bloemendalerpolder, uniek woonlandschap aan de Vecht* (Amsterdam) Report, commissioned by AM Wonen bv and Ymere

Vista (2005d) *Toekomst Vliegbasis Soesterberg. Concept report* (Amsterdam) Hart van de Heuvelrug/Habiforum

Vista; DE LIJN (2005e) *Wij zijn Twente. Bestuurlijk manifest* (Enschede) Commissioned by Regio Twente

References

Vliegen, M.; Leeuwen, N. van (2005) *Bevolkingsconcentraties: van kleine kernen tot grote agglomeraties* (j) Bevolkingstrends 53(1): 4-21
Vogel, G.; Angermann, H.; Steen, J.C. van der (1970) *Sesam atlas bij de biologie* (Baarn) Bosch en Keuning
Vos, C.C.; Verboom, J.; Opdam, P.F.M.; Braak, C.J.F. ter (2001) *Towards ecologically scaled landscape indices* (j) American Naturalist 157: 24-51
Vreke, J.; Donders, J.L.; Langers, F.; Salverda, I.E.; Veeneklaas, F.R. (2006) *Potenties van groen! De invloed van groen in en om de stad op overgewicht bij kinderen en op het binden van huishoudens met midden- en hoge inkomens aan de stad [Potentials of greenery! The influence of green in and around the city on overweight and obesity and on the commitment of households with middle- and high incomes to the city]* (Wageningen) Alterra
Vries, J. de (1971) *Etymologisch woordenboek* (Utrecht) Het Spectrum
Vries, J. de; Woude, A. van der (1997) *The First Modern Economy: Success, Failure, and Perseverance of the Dutch Economy, 1500-1815* (Cambridge) Cambridge University Press
Vries, M.L. de (1996) *Nederland Waterland, Een nieuw leven voor gedempte grachten, vaarten, havens en beken* (Den Haag) Rijksdienst voor de Monumentenzorg
Vries, S. de; Bulens, D. (2001) *Rapportage project "Explicitering 300000 ha", fasen 1 en 2* (Wageningen) Alterra
Vries, S. de; Verheij, R. A.; Groenewegen, P. P.; Spreeuwenberg, P. (2003) *Natural environments -healthy environments? An exploratory analysis of the relationship between green space and health* (j) Environment and Planning A 35: 1717-1731
Vries, S. de; Hoogerwerf, M.; Regt, W. de (2004) *Analyses ten behoeve van een Groene Recreatiebalans voor Amsterdam* (Wageningen) Alterra-rapport 988
Vries, S.I. de; Bakker, I.; Mechelen, W. van; Hopman-Rock, M. (2007) *Determinants of activity-friendly neighborhoods for children: Results from the SPACE Study* (j) American Journal of Health Promotion 21(4): 312-316
Vrijlandt, P. (1976) *De ecologische betekenis van het landschap* In: Kerkstra et al. [ed.]; 28-37
Vrijlandt, P.; Kerkstra, K. (1976) *Mergelland - Landschap en mergelwinning - Deel A samenvattend rapport, Deel B achtergronden & Deel C kaartmateriaal* (Wageningen) Landbouwhogeschool, Afd. Landschapsarchitectuur
Vrijlandt, P.; Teer, M.; Toorn, M.W.M. van den; Zeeuw, P.H. de (1980) *Elektriciteitswerken in het landschap - probleemverkenning en conceptvorming* (Wageningen) Landbouwhogeschool, Afd. Landschapsarchitectuur
VRO (1977) *Derde Nota Ruimtelijke Ordening Deel 3: Nota landelijke gebieden Deel 2a: beleidsvoornemen* (Den Haag) Staatsuitgeverij
VRO (1977) *Derde Nota Ruimtelijke Ordening Deel 2: Verstedelijkingsnota Deel 2a: beleidsvoornemen* (Den Haag) Staatsuitgeverij
VROM (1988) *Vierde nota over de ruimtelijke ordening Deel d: Regeringsbeslissing* (Den Haag) SDU uitgeverij
VROM (1989) *Nationaal milieubeleidsplan 'kiezen of verliezen'* (Den Haag) VROM / SDU uitgeverij
VROM (1990) *Nationaal milieubeleidsplan plus* (Den Haag) VROM / SDU uitgeverij
VROM (1992) *Vierde nota over de ruimtelijke ordening Extra* (Den Haag) SDU
VROM (1998) *Nationaal Milieubeleidsplan 3* (Den Haag) Ministerie van Volkshuisvesting, Ruimtelijke Ordening en Milieubeheer
VROM (2001) *Een wereld en een wil: werken aan duurzaamheid Nationaal milieubeleidsplan 4* ('s-Gravenhage) Ministerie van Volkshuisvesting, Ruimtelijke Ordening en Milieu
VROM (2001) *Ruimte maken, ruimte delen Vijfde Nota over de Ruimtelijke Ordening* (Den Haag) Ministerie van Volkshuisvesting, Ruimtelijke Ordening en Milieubeheer
VROM (2004) *Nota Ruimte: Ruimte voor ontwikkeling PKB deel 3* ('s-Gravenhage) Ministerie van Volkshuisvesting, Ruimtelijke Ordening en Milieubeheer
VROM (2005) *Nota Ruimte: Ruimte voor ontwikkeling PKB deel 4* ('s-Gravenhage) Ministerie van Volkshuisvesting, Ruimtelijke Ordening en Milieubeheer
VROM (2006) *Kenmerken van stedelijke milieus in 2000/02*; www.vrom.nl/docs/bijlage5068.pdf

650

Vroom, M.J. (1982) *Syllabus KB-college landschapsarchitectuur - Deel 2* (Wageningen) Land-bouwhogeschool, Afd. Landschapsarchitectuur

Vroom, M.J. (1983) *Over de schoonheid van de natuur vergeleken met die van een polder-landschap* **In:** Boekhorst et al. [ed.]; 83-89

Vroom, M.J. [ed.] (1992) *Outdoor space - Environments designed by Dutch Landscape Archi-tects since 1945* (Amsterdam)

Vroom, M. (1993) *Vormgeven aan natuurontwikkeling* Proceedings *4e LOCUS - seminar* (Rotterdam)

Vroom, M.J. (2006) *Lexicon of garden and landscape architecture* (Basel) Birkhäuser

Vulink, J.Th.; Eerden, M.R. van (1998) *Hydrological conditions and herbivory as key operators for ecosystem development in Dutch artificial wetlands* **In:** Vries, M.F. Wallis De; Bakker, J.P.; Wieren, S.E. Van [eds.] *Grazing and Conservation Management.* (Dordrecht) Kluwer Academic Publ.; 217-252

Vulink, J.Th. (2001) *Hungry Herds. Management of temperate lowland wetlands by grazing* (Lelystad) PhD Thesis Groningen University; Rijkswaterstaat, Van Zee tot Land 66

Wagenaar, C. (1993) *The critical city: Lewis Mumford's view on Amsterdam* (j) Kunstlicht 14(3/4): 9-12

Ward Thompson, C. (2002) *Urban open space in the 21th century* (j) Landscape and Urban Planning 60: 59-72

Wardenaar, K. J.; Veen, P.J. (2007) *Landscape architecture and urban ecology; The use of Urban-ecological knowledge in landscape architecture* **In:** Jong, T.M. de; Dekker, J.N.M.; Posthoorn, R. [eds.] *Landscape ecology in the Dutch context: nature, town and infrastructure* (Zeist) KNNV-uitgeverij

Wardle, D.A. (2002) *Communities and ecosystems: linking the aboveground and belowground components* (Princeton) Princeton University Press

Warren-Rhodes, K.; Koenig, A. (2001) *Escalating Trends in the Urban Metabolism of Hong Kong: 1971-1997* (j) AMBIO 30[7]: 429-438

Wassen, M.J.; Barendregt, A.; Bootsma, M.C.; Schot, P.P. (1989) *Groundwater chemistry and vegetation of gradients from rich fen to poor fen in the Naardermeer [the Nether-lands]* (j) Vegetatio 79: 117-132

Wassen, M.J. (1990) *Water Flow as a Major Landscape Ecological Factor in Fen Develop-ment* (Utrecht) PhD Thesis University of Utrecht

Wassen, M.J.; Barendregt, A.; Palczynski, A.; de, Smidt J.T.; Mars, H. de (1990) *The relation-ship between fen vegetation gradients, groundwater flow and flooding in an undrained valley mire at Biebrza, Poland* (j) Journal of Ecology 78: 1106-1122

Wassen, M.J.; Barendregt, A.; Schot, P.P.; Beltman, B. (1990) *Dependency of local mesotro-phic fens on a regional groundwater flow system in a poldered river plain in the Netherlands* (j) Landscape Ecology 5(1): 21-38

Wassen, M.J. (1991) *Hydro-ecologie wordt volwassen* (j) Biovisie 71(10): 84

Wassen, M.J.; Barendregt, A. (1992) *Topographic position and water chemistry of fens in the Vecht river plain, the Netherlands* (j) Journal of Vegetation Science 3: 447-456

Wassen, M.J.; Diggelen, R. van; Verhoeven, J.T.A.; Wolejko, L. (1996) *A comparison of fens in natural and artificial landscapes* (j) Vegetatio 126: 5-26

Wassen, M.J.; Grootjans, A.P. (1996) *Ecohydrology: an interdisciplinary approach to wetland management and restoration* (j) Vegetatio 126: 1-4

Wassen, M.J.; Grootjans, A.P.; van, Wirdum G.; Wheeler, B.D.;[eds.] (1996) *Consequences of changes in the water cycle for ground water and surface water fed ecosystems: eco-hydrological approaches* (j) Vegetatio Special Issue 126

Wassen, M.J.; Witte, F.; Geelen, L.; Haan, M. de; Vos, C.C.; [eds.] (2001) *Herstel van duinval-leien* (j) Landschap Special Issue 18(3)

Wassen, M.J.; Coops, H; Klijn, F; Smits, A.J.M; Dirkx, G.H.P.; [eds] (2002) *Referentiebeelden voor Nat Nederland* (j) Landschap Special Issue 19(1)

Wassen, M.J.; Verhoeven, J.T.A. (2003) *Up-scaling, interpolation and extrapolation of biogeo-chemical and ecological processes* (j) Landschap 20(2): 63-78

Wassen, M. (2007) *Ecohydrology; Contribution to science and practice* **In:** Jong, T.M. de; Dekker, J.N.M.; Posthoorn, R. [eds.] *Landscape ecology in the Dutch context: na-ture, town and infrastructure* (Zeist) KNNV-uitgeverij

Watering, C.F. van de; Verkaar, J. J. (1987) *VersnipperingVerslag van een workshop gehou-den op 19 mei 1987 te Utrecht* (Utrecht) Dienst Weg- en Waterbouwkunde in sa-menwerking met het Centrum voor Milieukunde en de Stichting Natuur en Milieu.

Waterman, R.E. (1991) *Naar een integraal kustbeleid via bouwen met de natuur* (Delft) Wa-terman

Watson, A.; Moss, R. (1979) *Population cycles in the Tetraonidae* (j) Ornis Fennica 56: 87-109

Watson, A.; Moss, R. (1980) *Advances in our understanding of the population dynamics of Red Grouse from a recent fluctuation in numbers* (j) Ardea 68: 103-111

Weddle, A.E. (1973) *Applied analysis and evaluation techniques* In: Lovejoy, D. [ed.]; 51-83

Wee, B. van (2006) *Dure infrastructuur [Expensive infrastructure]* (j) Stedenbouw & Ruimtelij-ke Ordening 87(2): 376-37

Weeda, E. (1990) *Rode lijst Nederlandse hogere planten* (Leiden) Rijksherbarium

Weijden, W. van der; Middelkoop, N. (1998) *Heeft Nederland veel of weinig natuurgebied? Een internationale vergelijking* (j) Landschap 15: 225-230

Weiner, J. (1992) *Physiological limits to sustainable energy budgets in birds and mammals: ecological implications* (j) Trends Ecol. Evol. 7: 384-388

Werkcommissie Westen des Lands (1958) *De ontwikkeling van het westen des lands* (Den Haag) Rijksdienst voor het Nationale Plan

Werkgroep Theorie (1986) *Methoden en begrippen in de Landschapsecologie* (j) Landschap 3(3): 172-182

Werquin, A.; Duhem, B.,et al. (2005) *Green structure and urban planning - Final Report of COST action CII* (Luxemburg) Office for Official Publications of the European Com-munities

Westhoff, V.; Maarel, E. van der (1973) *The Braun-Blanquet Approach* In: Whittaker, R.H. [ed.] *Handbook of Vegetation Science* (The Hague) Junk; 617-726

Westhoff, V.; Held, A.J. den (1975) *Plantengemeenschappen van Nederland* (Zutphen) Thie-me en Cie

Westhoff, V. (1983) *De houding van de mens jegens plantengroei en landschap* In: Boekhorst, J.K.M., et al. [ed.]; 89-103

Wetenschappelijke Raad voor het Regeringsbeleid (1992) *Grond voor keuzen: vier perspec-tieven voor de landelijke gebieden in de Europese Gemeenschap* ('s-Gravenhage) WRR / Sdu

Wijffels, H. (2005) *Natuurbescherming in de 21e eeuw: oriëntaties en keuzen* (Nijmegen) UCM-DO

Williams, P.H.; Gaston, K.J. (1994) *Measuring more of biodiversity: can higher-taxon richness predict wholesale species richness?* (j) Biological Conservation 67: 211-217

Williams, P.; Faith, D.; Manne, L.; Sechrest, W.; Preston, C. (2005) *Complementarity analysis: Mapping the performance of surrogates for biodiversity* (j) Biological conservation 128: 253-246

Windt, H.J. van der (1995) *En dan: wat is natuur nog in dit land? Natuurbescherming in Neder-land 1880-1990* (Amsterdam/Meppel) Boom

Windt, H. van der; Wams, T.; Wiersinga, W. (2003) *De rots in het weiland, Biodiversiteit, ecologie en natuurontwikkeling in de stad* (j) Landschap. 20(3): 165-171

Winsum-Westra, M. van; Boer, T.A. de (2004) *[On]veilig in bos en natuur? Een verkenning van subjectieve en objectieve aspecten van sociale en fysieke veiligheid in bos- en natuurgebieden [Unsafe in the woods and nature? An exploration of subjective and objective aspects of social and physical safety in forest- and natural environmetns]* (Wageningen) Alterra

Wirdum, G. van (1979) *Dynamic aspects of trophic gradients in a mire complex* In: Anonymus [ed.] *Proceedings and Information 25* (the Hague) TNO-Committee on Hydrological Research; 62-82

Wirdum, G. van (1991) *Vegetation and hydrology of floating rich-fens* (Amsterdam) PhD The-sis University of Amsterdam

Wit , S. de (2003) *Typology of the lowlands* (Delft) University Press

Witmer, M. (1989) *Integral water management at regional level* (Utrecht) PhD Thesis Univer-sity of Utrecht

Witte, J.P.M.; Klijn, F.; Claessen, F.A.M.; C.L.G., Groen; Meijden, R. van der (1992) *A model to predict and assess the impacts of hydrological changes on terrestrial ecosystems, and its use on a climate scenario* (j) Wetland Ecology and Management 2: 69-83

Wittgenstein, L. (1953) *Philosophical investigations* (Oxford) Blackwell

Wittig, R.; Diesing, D.; Gödde, M. (1985) *Urbanophob - Urbanoneutral - Urbanophil Das Verhalten der Arten gegenüber dem Lebensraum Stadt* (j) Flora 177: 265-282

Wittig, R. (1991) *Ökologie der Großstadtflora* (Stuttgart/Jena) G. Fisher

WLO Bestuur [ed.] (1992) *De Landschapsecologie, verslag WLO-Lustrumcongres 20 november 1992 Utrecht* (Den Haag) SPB Academic Publishing

WNF (1993) *Levende Rivieren* (Zeist) Wereld Natuur Fonds

Woestenburg, M.; Kuypers, V.; Timmermans, W. (2006) *Over de brug, hoe Zanderij Crailo een natuurbrug kreeg [About the bridge, how the Crailo Sand Quarry got a nature bridge]* (Hilversum, Wageningen) Goois Natuurreservaat, Blauwdruk Publishers

Wolschke - Bulmahn, J. (2004) *The native plant enthusiasm - Ecological Panacea or Xenophobia?* **In:** ECLAS [ed.] Proceedings *A critical light on landscape architecture* (Ås, Norway) ECLAS Dept. of Landscape architecture and Spatial Planning

Wolters-Noordhof [ed.] (2001) *De Grote Bosatlas 2002/2003 Tweeënvijfstigste editie + CD-Rom* (Groningen) WN Atlas Productions

Wood, D. (1989) *Unnatural illusions: Some words about visual resource management* (j) Landscape Journal 7(2): 192-205

Wooley, J.B.; Owen, R.B. (1978) *Energy costs of activity and daily energy expenditure in the Black Duck* (j) J. Wildl. Manage 42: 739-745

World Health Organization (2000) *Air Quality Guidelines for Europe [Second Edition]* (Copenhagen) WHO Regional Office for Europe

World Health Organization (2001) *The World Health Report 2001 - Mental Health: New Understanding, New Hope* (Geneva) WHO

World Resources Institute (1996) *World Resources 1996-97: The urban environment* (Washington) World Resources Institute, United Nations Environment Programme, United Nations Development Programme, and the World Bank.

Woud, A. van der (1987) *Het lege land, de ruimtelijke orde van Nederland 1798-1848* (Amsterdam) Olympus

WRR (1998) *Ruimtelijke ontwikkelingspolitiek* (Den Haag) Sdu; Wetenschappelijke Raad voor het Regeringsbeleid

Wymenga, E.; Jalving, R.; Stege, E. ter (1996) *Vegetatie en weidevogels in relatienotagebieden in Nederland: een tussentijdse analyse van de natuurwetenschappelijke resultaten van beheersovereenkomsten in Nederlandse relatienotagebieden* (Veenwouden) Altenburg & Wymenga ecologisch onderzoek bv

Wynhoff, I. (1998) *Lessons from the reintroduction of Maculinea teleius and M nausithous in the Netherlands* (j) Journal of Insect Conservation 2: 47-57

Wynne-Edwards, V.C. (1962) *Animal dispersion in relation to Social Behaviour* (Edinburgh, London) Oliver & Boyd

Wynne-Edwards, V.C. (1970) *Feedback from food resources to population regulation* **In:** Watson, A. [ed.] *Animal populations in relation to their food resources* (Oxford) Blackwell; 413-427

Zadelhoff, F.J. van (1978) *Het Onderzoeksproject "Nederlandse Duinvalleien"* (j) WLO-Mededelingen 5(2): 24-30

Zadelhoff, F.J. van; Zande, A.N. van der (1991) *Natuurbeleidsplan en onderzoek - een toelichting* (j) Landschap 8(1): 59-72

Zalewski, M. (2000) *Ecohydrology - the scientific background to use ecosystem properties as management tools toward sustainability of water resources* (j) Ecological Engineering 16: 1-8

Zalewski, M.; Robarts, R. (2003) *Ecohydrology - new paradigm for integrated water resources management* (j) SIL news 40: 1-5

Zandbelt, D.; Berg, R. van den (2003) *Schiphol Werkweek Deltanet* (Delft) Vereniging Deltametropool

Zande, A.N. van der (2004) *Tussen Onland en Verwonderland Victor Westhoff Lezing 2004* Universitair Centrum voor Milieuwetenschappen en Duurzame Ontwikkeling (Nijmegen) University Press

653

Zande, A.N. van der (2006) *Landschap vol betekenissen Over het omgaan met historie in de ruimtelijke inrichting Oratie op 20 april 2006* (Wageningen) WUR

Zande, A. N. van der (2007) *Heritage interests as a context of landscape ecology; is the WLO really developing into a community of Landscape Scientists?* **In:** Jong, T.M. de; Dekker, J.N.M.; Posthoorn, R. [eds.] *Landscape ecology in the Dutch context: nature, town and infrastructure* (Zeist) KNNV-uitgeverij

Zandvoort ordening & advies (1994) *Sustainable Twente (NL) - Growth and green combined: towards an integrated approach of townplanning and landscape planning* s.l.

Zapparoli, M. (1997) *Urban development and insect biodiversity of the Rome area, Italy* (j) Landscape and Urban Planning 38: 77-86

Zijlstra, M.; Loonen, M.J.J.E.; Eerden, M.R. van; Dubbeldam, W. (1991) *The Oostvaarder-splassen as a key-moulting site for Greylag Geese Anser anser in western Europe* (j) Wildfowl 42: 45-52

Zoest, J. van (2001) *Stadsecologie: de stad en de mens* **In:** Halm, H. van; Timmermans, G.; Koningen, H.; Bouman, R.; Melchers, M.; Kazus, J. [eds.] *De wilde stad, 100 jaar natuur van Amsterdam, een eeuw Koninklijke Nederlandse Natuurhistorische Vereniging, afdeling Amsterdam 1901-2001* (Utrecht) KNNV; 140-144

Zoest, J. van (2006) *Leven in de stad - Betekenis en toepassing van natuur in de stedelijke omgeving* (Utrecht) KNNV uitgeverij

Zoest, J. van (2007) *Driving forces in urban ecology* **In:** Jong, T.M. de; Dekker, J.N.M.; Posthoorn, R. [eds.] *Landscape ecology in the Dutch context: nature, town and infrastructure* (Zeist) KNNV-uitgeverij

Zonderwijk, P. (1978) *Het behoud van flora en fauna in het cultuurlandschap* **In:** *Wetenschap in dienst van het natuurbehoud [...]* Uitgave 50 jarig bestaan Natuurwetenschappe-lijke Commissie van de Natuurbeschermingsraad

Zonderwijk, P. (1979) *De bonte berm; de rijke flora en fauna langs onze wegen* (Ede) Zomer en Keuning

Zonderwijk (1991) *Zeldzame planten in wegbermen* (Ede) CROW

Zonneveld, J.I.S. (1974) *Tussen bergen en zee De wordingsgeschiedenis der lage landen* (Utrecht) Oosthoek

Zonneveld, I.S. (1982) *Groei en bloei van de landschapsecologie* (j) WLO-Mededelingen 9 (2): 82-90

Zonneveld, I.S. (1989) *Theorieën en concepten: een tussentijdse balans* (j) Landschap 6(1): 65-77

Zonneveld, I.S. (1995) *Land Ecology; an Introduction to Landscape Ecology as a Base for Land Evaluation, Land Management and Conservation* (Amsterdam) SPB Academic Publishing

Zonneveld, I. S. (2000) *Count your blessings? Twenty-five years of landscape ecology* **In:** Klijn, J. A.; Vos, W. [eds.] *From Landscape ecology to landscape science* (Dordrecht) Kluwer; 31-41

Zuiderwijk, A. (1989) *Reptielen in wegbermen, een analyse van 106 locaties* (Amsterdam) Universiteit van Amsterdam, in opdracht van DWW

Zwarts, L.; Drent, R.H. (1981) *Prey depletion and the regulation of predator density: Oyster-catchers [Haematopus ostralegus] feeding on Mussels [Mytilus edulis]* **In:** Jones, N.V.; Wolff, W.J. [eds.] *Feeding and survival strategies of estuarine organisms* (New York) Plenum Press; 193-216

Zwarts, L.; Wanink, J.H. (1993) *How the food supply harvestable by waders in the Wadden Sea depends on the variation in energy density, body weight, biomass, burying depth and behaviour of tidal-flat invertebrates* (j) Neth. J. Sea Res. 31: 441-476

Zwarts, L. (1996) *Waders and their estuarine food supplies* (Lelystad) PhD Thesis University of Groningen; Rijkswaterstaat, Van Zee tot Land 60

KEY WORDS

Italics means 'from figure'.
Plural means 'more general than singular'

661

663

666

683

686

687